SCIENCE

A CLOSER LOOK

Macmillan
McGraw-Hill

Partnerships

The American Museum of Natural History in New York City is one of the world's preeminent scientific, educational, and cultural institutions, with a global mission to explore and interpret human cultures and the natural world through scientific research, education, and exhibitions. Each year the Museum welcomes around four million visitors, including 500,000 schoolchildren in organized field trips. It provides professional development activities for thousands of teachers; hundreds of public programs that serve audiences ranging from preschoolers to seniors; and an array of learning and teaching resources for use in homes, schools, and community-based settings. Visit www.amnh.org for online resources.

As the education arm of the **National Science Foundation**–funded Joint Oceanographic Institutions, JOI Learning brings the excitement of discovery and the scientific process to your classroom through expedition-based science, math, reading, and social studies activities. JOI Learning and Macmillan/McGraw-Hill have formed a partnership to give teachers and students access to JOI's resources through www.macmillanmh.com.

The National Science Digital Library (NSDL) is funded by the **National Science Foundation** as an online library of resources for science, technology, engineering, and mathematics education. NSDL has partnered with Macmillan/McGraw-Hill to provide teaching and learning resources that will help develop teachers' content knowledge in the topic and skill areas addressed at each grade level.

Students with print disabilities may be eligible to obtain an accessible, audio version of the pupil edition of this textbook. Please call Recording for the Blind & Dyslexic at 1-800-221-4792 for complete information.

B

Program Authors

Dr. Jay K. Hackett

Professor Emeritus of Earth Sciences

University of Northern Colorado
Greeley, CO

Dr. Richard H. Moyer

Professor of Science Education and Natural Sciences

University of Michigan–Dearborn
Dearborn, MI

Dr. JoAnne Vasquez

Elementary Science Education Consultant

NSTA Past President

Member, National Science Board and NASA Education Board

Mulugheta Teferi, M.A.

Principal, Gateway Middle School

Center of Math, Science, and
   Technology
St. Louis Public Schools
St. Louis, MO

Dinah Zike, M.Ed.

Dinah Might Adventures LP
San Antonio, TX

Kathryn LeRoy, M.S.

**Executive Director
Division of Mathematics and
Science Education**

Miami-Dade County Public
   Schools
Miami, FL

Dr. Dorothy J. T. Terman

Science Curriculum Development Consultant

**Former K–12 Science and
Mathematics Coordinator**

Irvine Unified School District
Irvine, CA

Dr. Gerald F. Wheeler

**Executive Director
National Science Teachers
Association**

Bank Street College of Education
New York, NY

Contributors and Reviewers

Lori Gilchrist
Sharon Elementary
Suwanee, GA

Connie Grubbs
Varner Elementary
Powder Springs, GA

Tasha Hamil
Cumming Elementary
Cumming, GA

Nancy Hayes
Educational Consultant
Lemont, IL

Carol Johnson
Jane D. Hull Elementary
Chandler, AZ

Jerry D. Kelley, Ed.S.
Chestatee Elementary
Forsyth, GA

Andrew C. Kemp
Jefferson County Public Schools
Louisville, KY

Heather W. Kemp
Middletown Elementary
Louisville, KY

Tricia Reda Kerr
Science Specialist, EXCEL Program
The Ohio State University
Columbus, OH

Barbara Kingston
Blessed Sacrament School
Jackson Heights, NY

Gina Koger
Carroll County Public School
Westminster, MD

Bonnie Kohler
L'Anse Creuse Public Schools
Harrison Township, MI

Heather LeBlanc
Chestatee Elementary
Gainesville, GA

Larry Lebofsky
Senior Research Scientist
Lunar and Planetary Laboratory
University of Arizona
Tucson, AZ

Richard MacDonald
Science Curriculum Leader
Hampton City Schools
Hampton, VA

Brenda S. Martin
Coal Mountain Elementary
Cumming, GA

Rebecca Martin
Westridge Elementary
Frankfort, KY

Corinne Masters
Natoma Elementary
Natoma, KS

Tiah E. McKinney
Albert Einstein Fellow
National Science Foundation
Arlington, VA

Sharon Meyer
Barnesville Elementary
Barnesville, OH

Janiece Mistich
Tchefuncte Middle School
Mandeville, LA

Anthony Molock
Cascade Elementary
Atlanta, GA

Sandy Morris
Department of Learning Services
Wichita, KS

Terri Oatis-Wilson
Peyton Forest Elementary
Atlanta, GA

Brenda A. Oulsnam
Clayton County Schools (retired)
Jonesboro, GA

Jim Peters
Science Resource Teacher
Carroll County Board of Education
Westminster, MD

Sharon Pinion
Sawnee Elementary
Cumming, GA

Amy Quick
Joseph M. Carkenord Elementary
Chesterfield, MI

Stacey Race
Sharon Elementary
Suwanee, GA

Gloria R. Ramsey
Mathematics/Science Specialist
Memphis City Schools
Memphis, TN

Anna Reitz
Forsyth County Schools
Cumming, GA

Steve A. Rich
Science Coordinator
Georgia Youth Science & Technology
   Center
Carrollton, GA

Maureen Riordan
Fairway Elementary
Wildwood, MO

Richard Ruiz
Jane D. Hull Elementary
Chandler, AZ

Ruth M. Ruud
Millcreek Township School District
Erie, PA

Sarah Rybarczyk
Joseph M. Carkenord Elementary
Chesterfield, MI

Laura W. Schaefer
Coordinator, School Partnerships
Missouri Botanical Garden
St. Louis, MO

Rhonda Segraves
Settles Bridge Elementary
Suwanee, GA

Ursula M. Sexton
Senior Research Associate/Educational
   Consultant
WestEd
San Ramon, CA

Rita Jane Shelton
Louisa Middle School
Louisa, KY

Matt Silberglitt
Science Assessment Specialist
Minnesota Department of Education
Roseville, MN

William L. Siletti
Packer Collegiate Institute
Brooklyn, NY

Georgia Ann Smith
Sunflower Elementary
Lenexa, KS

Victoria L. Thom
Baker Elementary
Acworth, GA

Shannon Tribble
Daves Creek Elementary
Cumming, GA

Shirley Whorley
Science Coordinator, K–12
Roanoke City Public Schools
Roanoke, VA

Laura Wilkowski
Science Consultant
Midland, MI

Dr. Sharon Wynstra
Science Coordinator
Rockford Public Schools
Rockford, IL

Brad Yohe
Science Supervisor
Carroll County Public Schools
Westminster, MD

Table of Contents

Overview

Print and Technology Resources .... Tviii
Inquiry-Based Activities .... Txii
Standards-Based Content .... Txiii
The Learning Cycle .... Txiv
Differentiated Instruction .... Txvi
A Manageable Organization .... Txviii
Assessment .... Txvix
Program Overview .... Txx
Components Chart .... Txxii

Teaching with the Student Edition

Student Edition Table of Contents v
Be a Scientist 1
Science Safety 14

Life Science

UNIT A Living Things

CHAPTER 1 A Look at Living Things

Chapter Planner 18A
Activity Planner 18B
Lesson 1 Planner 20A
Lesson 2 Planner 30A
Lesson 3 Planner 42A
Lesson 4 Planner 52A
Chapter Review 64

CHAPTER 2 Living Things Grow and Change

Chapter Planner 66A
Activity Planner 66B
Lesson 1 Planner 68A
Lesson 2 Planner 80A
Lesson 3 Planner 90A
Chapter Review 98

UNIT B Ecosystems

CHAPTER 3 Living Things in Ecosystems

Chapter Planner 104A
Activity Planner 104B
Lesson 1 Planner 106A
Lesson 2 Planner 118A
Lesson 3 Planner 132A
Chapter Review 146

CHAPTER 4 Changes in Ecosystems

Chapter Planner 148A
Activity Planner 148B
Lesson 1 Planner 150A
Lesson 2 Planner 160A
Lesson 3 Planner 172A
Chapter Review 182

Earth Science

UNIT C Earth and Its Resources

CHAPTER 5 Earth Changes

Chapter Planner 188A
Activity Planner 188B
Lesson 1 Planner 190A
Lesson 2 Planner 202A
Lesson 3 Planner 212A
Chapter Review 222

CHAPTER 6 Using Earth's Resources

Chapter Planner 224A
Activity Planner 224B
Lesson 1 Planner 226A
Lesson 2 Planner 238A
Lesson 3 Planner 248A
Lesson 4 Planner 258A
Chapter Review 270

UNIT D Weather and Space

CHAPTER 7 Changes in Weather

Chapter Planner 276A
Activity Planner 276B
Lesson 1 Planner 278A
Lesson 2 Planner 288A
Lesson 3 Planner 302A
Chapter Review 312

CHAPTER 8 Planets, Moons, and Stars

Chapter Planner 314A
Activity Planner 314B
Lesson 1 Planner 316A
Lesson 2 Planner 326A
Lesson 3 Planner 336A
Lesson 4 Planner 346A
Chapter Review 354

Physical Science

Unit E Matter

CHAPTER 9 Observing Matter

Chapter Planner 360A
Activity Planner 360B
Lesson 1 Planner 362A
Lesson 2 Planner 372A
Lesson 3 Planner 382A
Chapter Review 392

CHAPTER 10 Changes in Matter

Chapter Planner 394A
Activity Planner 394B
Lesson 1 Planner 396A
Lesson 2 Planner 406A
Lesson 3 Planner 416A
Chapter Review 424

Unit F Forces and Energy

CHAPTER 11 Forces and Motion

Chapter Planner 430A
Activity Planner 430B
Lesson 1 Planner 432A
Lesson 2 Planner 442A
Lesson 3 Planner 452A
Lesson 4 Planner 462A
Chapter Review 474

CHAPTER 12 Forms of Energy

Chapter Planner 476A
Activity Planner 476B
Lesson 1 Planner 478A
Lesson 2 Planner 488A
Lesson 3 Planner 498A
Lesson 4 Planner 510A
Chapter Review 520

Student Edition Reference

Reference Table of Contents R1
Science Handbook R2
Health Handbook R14
Foldables R27
Glossary R29
Index R45

Teacher Resources

Teacher Resources TR1
Graphic Organizers. TR3
Teacher Aids TR18
  United States Map. TR18
  World Map TR19
  Graph Paper TR20
  Calendar TR21
  Rulers. TR22
  Periodic Table of Elements TR23
Activity Rubrics TR24
Writing Rubrics TR26
Materials TR34
Bibliography TR37
Science Yellow Pages TR40
Scope and Sequence TR64
Correlations. TR80
  Correlation to National Standards TR80

SCIENCE
A CLOSER LOOK

A wealth of resources that brings science to life

Student and Teacher Editions ▼

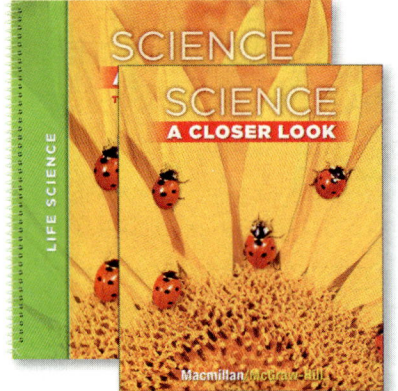

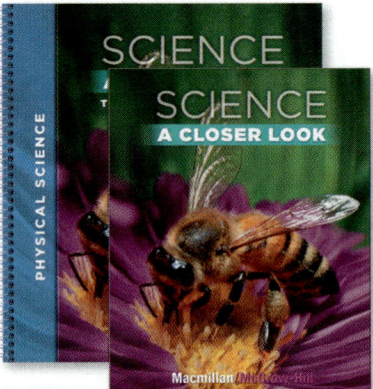

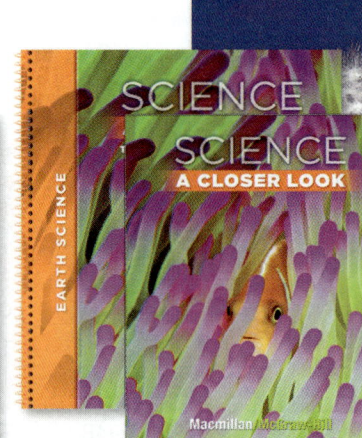

Grade 1

Grade 2

Grade 3

Kindergarten

Pre-K also available

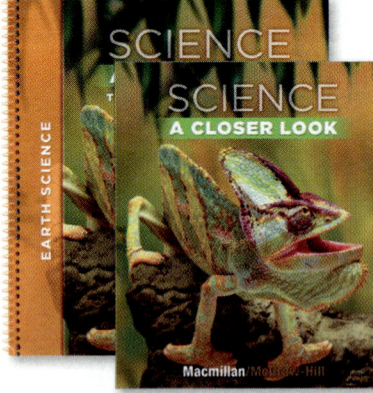

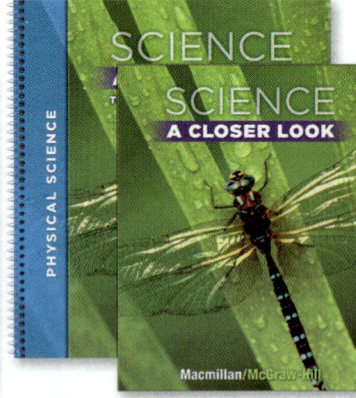

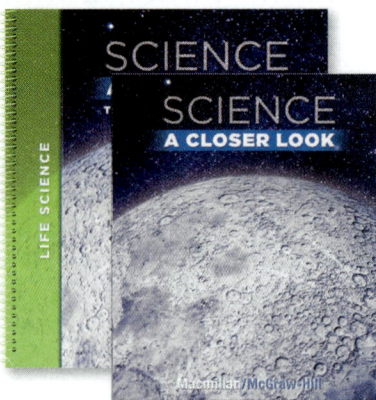

Grade 4

Grade 5

Grade 6

Activity Resources ▶

Materials Kits support Science Activities.

Grab 'n Go Activity Bags make preparing for activities quick and easy.

Activity Flipcharts provide flexibility for your Science Center.

The program for today's teacher delivers...

activities & content
Standards-based instruction with activities that work.

access for all
Support for all learners.

ease of use
"Fast Track" to fit a busy schedule.

Instructional Resources ▼

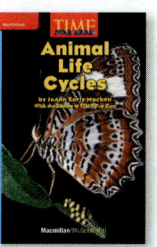

Reading Resources make science accessible and fun to read.

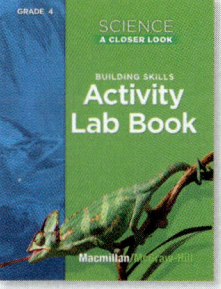

Key Resources support instruction, build skills, and promote comprehension.

Supporting Resources provide for a wide variety of student needs.

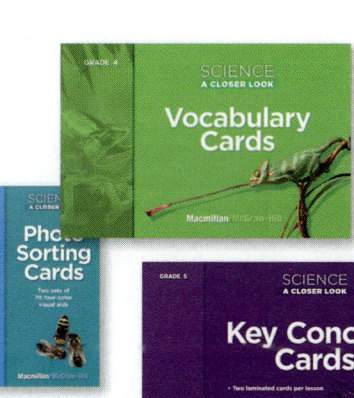

also with **full technology support**

Technology
for the Student

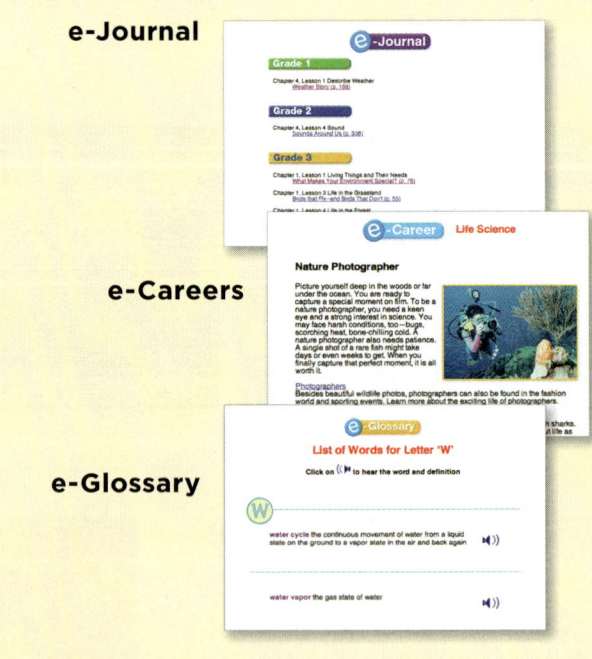

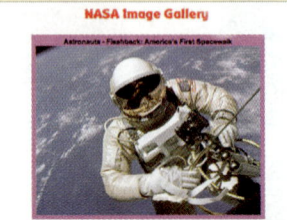

visit us at www.macmillanmh.com

Provides supportive activities and enriching content for the entire program

Practice and Activities ▼

Science in Motion ▶
Animated key concepts with assessment

Operation: Science Quest CD-ROMs ▶
Interactive learning simulations and problem solving modules

Student Navigator CD-ROM ▶
Student Edition on CD

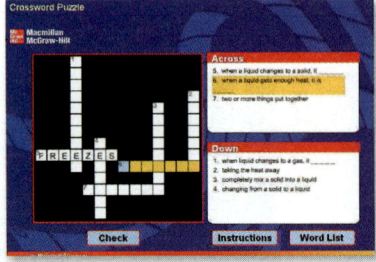

◀ **PuzzleMaker CD-ROMs**
Creates puzzles to reinforce science vocabulary

◀ **Science Songs Audio CDs**
Reinforce science content with familiar tunes

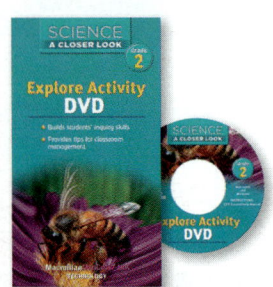

◀ **Science Activity DVDs**
Video modeling of the explore activities

e-Review
Narrated, animated version, with interactive quizzes

e-Journal

e-Careers

e-Glossary

NASA Image Gallery

Technology for the Teacher

Planning and Instruction ▼

Instructional Navigator CD-ROMs ▶
Electronic Teacher's Edition with lesson planning and printable resources

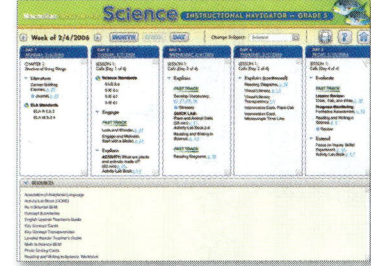

Professional Development DVDs ▶

Classroom Presentation Toolkit CD-ROMs ▶
PowerPoint presentations for every lesson

Progress Reporter ▶
Student assessments via online testing

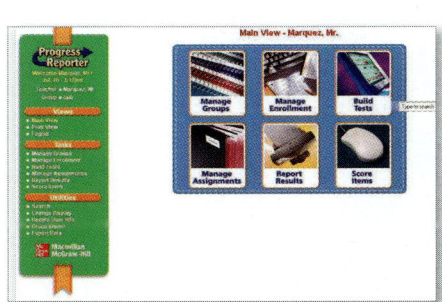

visit us at www.macmillanmh.com

Offers a rich collection of professional development resources, leveled book selections and links to topic-related sites

Teacher Resources

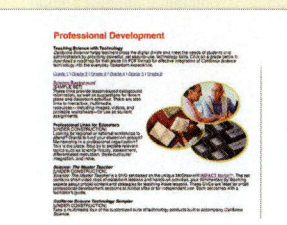

National Science Digital Library

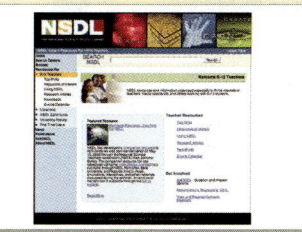

Other Partnerships

Bibliography

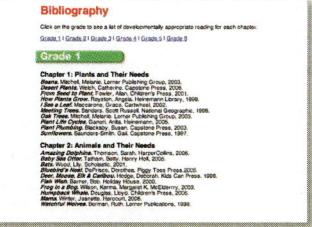

Inquiry-based Activities

- **Provide a variety of inquiry experiences**
- **Support structured, guided, and open inquiry**
- **Foster conceptual understanding**

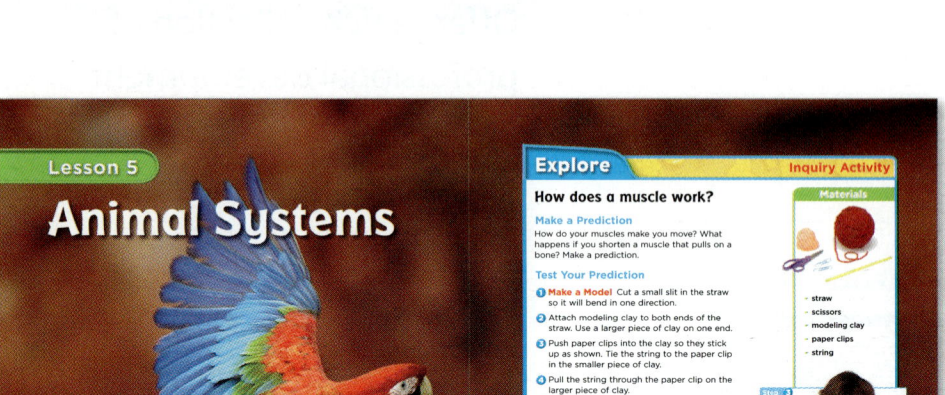

Explore Activities begin every lesson

Inquiry Investigations extend science learning

Skill Builder Activities develop inquiry skills

full technology support

- Science Activity DVDs

Standards-based Content

- **Develops the big ideas of science**
- **Provides depth of understanding**
- **Supports reading skills**

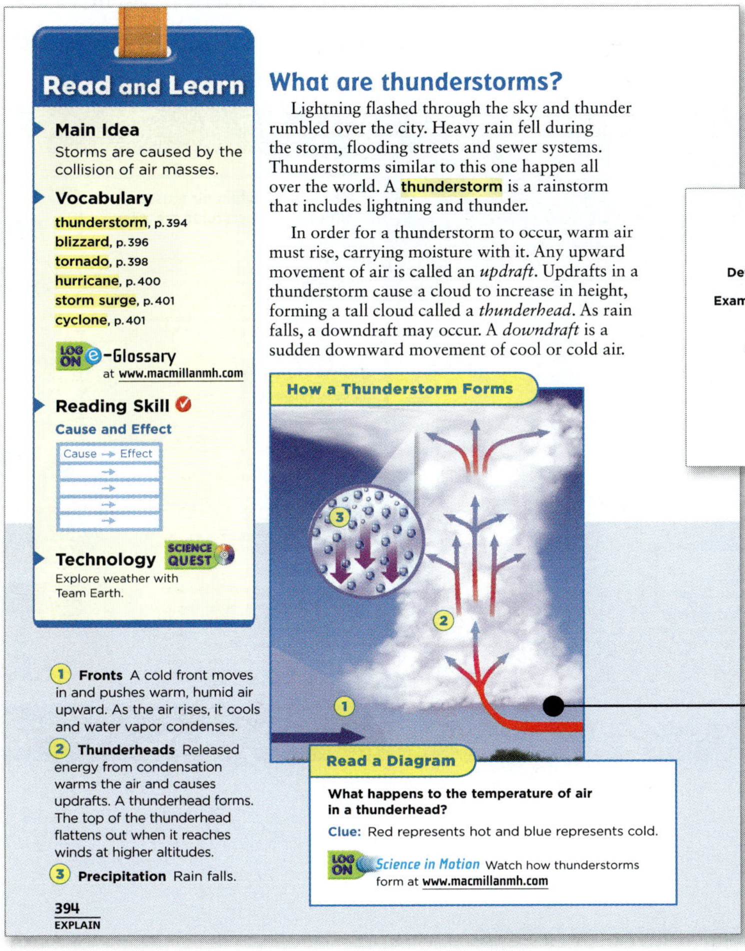

Read and Learn

▶ **Main Idea**
Storms are caused by the collision of air masses.

▶ **Vocabulary**
thunderstorm, p. 394
blizzard, p. 396
tornado, p. 398
hurricane, p. 400
storm surge, p. 401
cyclone, p. 401

LOG ON **e-Glossary**
at www.macmillanmh.com

▶ **Reading Skill** ✓
Cause and Effect

Cause	→	Effect
	→	
	→	
	→	

▶ **Technology** **SCIENCE QUEST**
Explore weather with Team Earth.

What are thunderstorms?

Lightning flashed through the sky and thunder rumbled over the city. Heavy rain fell during the storm, flooding streets and sewer systems. Thunderstorms similar to this one happen all over the world. A **thunderstorm** is a rainstorm that includes lightning and thunder.

In order for a thunderstorm to occur, warm air must rise, carrying moisture with it. Any upward movement of air is called an *updraft*. Updrafts in a thunderstorm cause a cloud to increase in height, forming a tall cloud called a *thunderhead*. As rain falls, a downdraft may occur. A *downdraft* is a sudden downward movement of cool or cold air.

How a Thunderstorm Forms

(1) **Fronts** A cold front moves in and pushes warm, humid air upward. As the air rises, it cools and water vapor condenses.

(2) **Thunderheads** Released energy from condensation warms the air and causes updrafts. A thunderhead forms. The top of the thunderhead flattens out when it reaches winds at higher altitudes.

(3) **Precipitation** Rain falls.

394
EXPLAIN

Read a Diagram

What happens to the temperature of air in a thunderhead?

Clue: Red represents hot and blue represents cold.

LOG ON **Science in Motion** Watch how thunderstorms form at www.macmillanmh.com

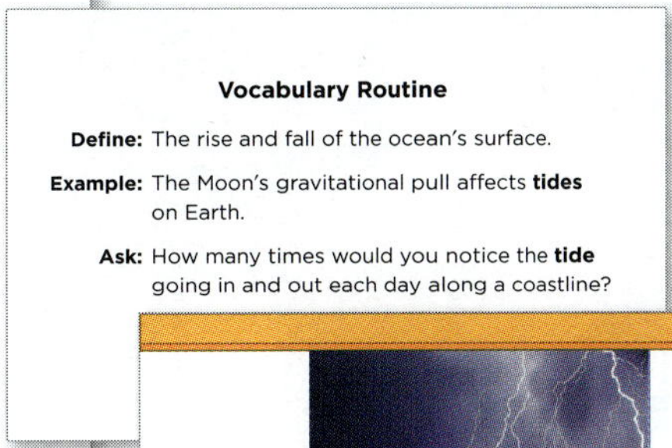

Vocabulary Routine

Define: The rise and fall of the ocean's surface.

Example: The Moon's gravitational pull affects **tides** on Earth.

Ask: How many times would you notice the **tide** going in and out each day along a coastline?

thunderstorm
(thun′dər stôrm′)

© Macmillan/McGraw-Hill
Grade 5

Research-based learning model stimulates student interest

Highlighted vocabulary supported by lesson graphics

Lesson visuals build conceptual understanding

---

full technology support

 Science in Motion Watch how thunderstorms form at www.macmillanmh.com

 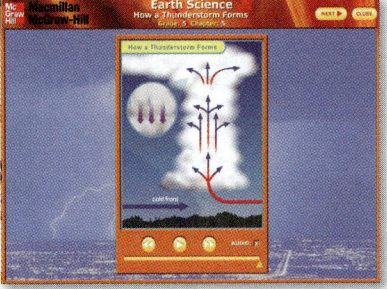 

◀ **Science in Motion** animated key concepts with assessment

plus...

- **Operation: Science Quest CD-ROM**
- **Student Navigator**

The Learning Cycle

Engage

Stimulates student interest and prepares them for the lesson.

1

Explore

Provides a hands-on experience around which the lesson concept is developed.

2

Lesson 3

Severe Storms
Tucson, AZ

Look and Wonder

On any given day, more than 40,000 thunderstorms are rumbling somewhere on Earth. What causes these spectacular storms?

Explore — Inquiry Activity

What happens when air masses of different temperatures meet?

Form a Hypothesis
What happens to an air mass when it meets another air mass of the same temperature or of a cooler temperature? Write your answer as a hypothesis in the form "If an air mass meets another air mass of the same or of a cooler temperature, then . . ." Like air, water flows and carries heat. Using water as a model for air can help you test your hypothesis.

Test Your Hypothesis
1. ⚠ **Be Careful.** Using scissors, cut the cardboard so it fits tightly in the clear box. Wrap the cardboard in aluminum foil.
2. Pour 4 cups of cold water into one container and 4 cups of warm water into the other one. Place a few drops of blue food coloring into the cold water and red into the warm water.
3. Hold the cardboard tightly against the bottom of the box. Pour the cold water on one side and the warm water on the other.
4. **Observe** Watch the box from the side as you remove the cardboard.
5. Now repeat the same test with warm water in both containers and food coloring in only one.

Draw Conclusions
6. What are the variables in this experiment?
7. **Infer** Which test looked like it formed storms? Why?

Explore More
Will a greater difference in temperature between the warm and cold water increase the observable effects? Form a hypothesis and test it.

Materials
- scissors
- cardboard
- clear plastic box
- aluminum foil
- cold water
- 2 containers
- warm water
- food coloring

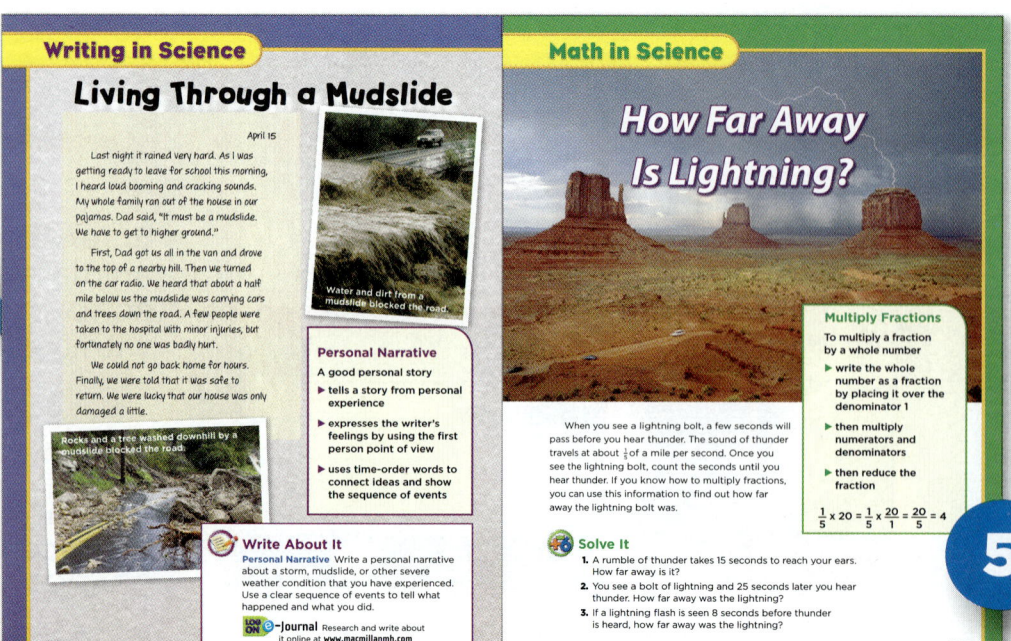

Writing in Science

Living Through a Mudslide

April 15

Last night it rained very hard. As I was getting ready to leave for school this morning, I heard loud booming and cracking sounds. My whole family ran out of the house in our pajamas. Dad said, "It must be a mudslide. We have to get to higher ground."

First, Dad got us all in the van and drove to the top of a nearby hill. Then we turned on the car radio. We heard that about a half mile below us the mudslide was carrying cars and trees down the road. A few people were taken to the hospital with minor injuries, but fortunately no one was badly hurt.

We could not go back home for hours. Finally, we were told that it was safe to return. We were lucky that our house was only damaged a little.

Water and dirt from a mudslide blocked the road.

Rocks and a tree washed downhill by a mudslide blocked the road.

Personal Narrative
A good personal story
- tells a story from personal experience
- expresses the writer's feelings by using the first person point of view
- uses time-order words to connect ideas and show the sequence of events

Write About It
Personal Narrative Write a personal narrative about a storm, mudslide, or other severe weather condition that you have experienced. Use a clear sequence of events to tell what happened and what you did.
🔵 **e-Journal** Research and write about it online at www.macmillanmh.com

Math in Science

How Far Away Is Lightning?

When you see a lightning bolt, a few seconds will pass before you hear thunder. The sound of thunder travels at about $\frac{1}{5}$ of a mile per second. Once you see the lightning bolt, count the seconds until you hear thunder. If you know how to multiply fractions, you can use this information to find out how far away the lightning bolt was.

Multiply Fractions
To multiply a fraction by a whole number
- write the whole number as a fraction by placing it over the denominator 1
- then multiply numerators and denominators
- then reduce the fraction

$$\frac{1}{5} \times 20 = \frac{1}{5} \times \frac{20}{1} = \frac{20}{5} = 4$$

🔵 **Solve It**
1. A rumble of thunder takes 15 seconds to reach your ears. How far away is it?
2. You see a bolt of lightning and 25 seconds later you hear thunder. How far away was the lightning?
3. If a lightning flash is seen 8 seconds before thunder is heard, how far away was the lightning?

Extend

Links the development of science big ideas to other curriculum areas.

5

Explain

Introduces vocabulary and makes the science content understandable through words and visuals.

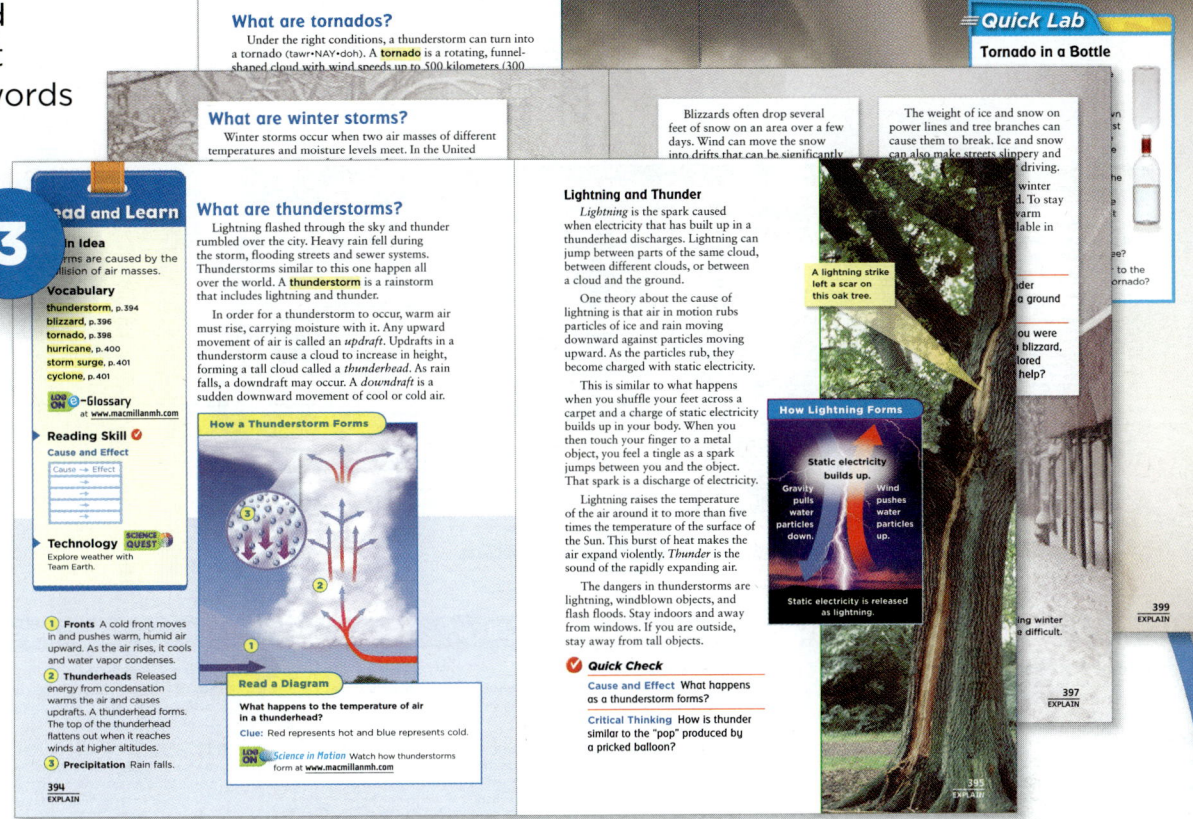

Evaluate

Assesses student understanding and provides opportunities for re-teaching.

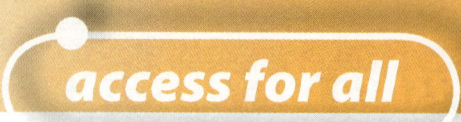

Differentiated Instruction

- **Provides access to tested science concepts**
- **Includes diverse instructional tools to reach all students**
- **Offers strategies for English Language Learners**

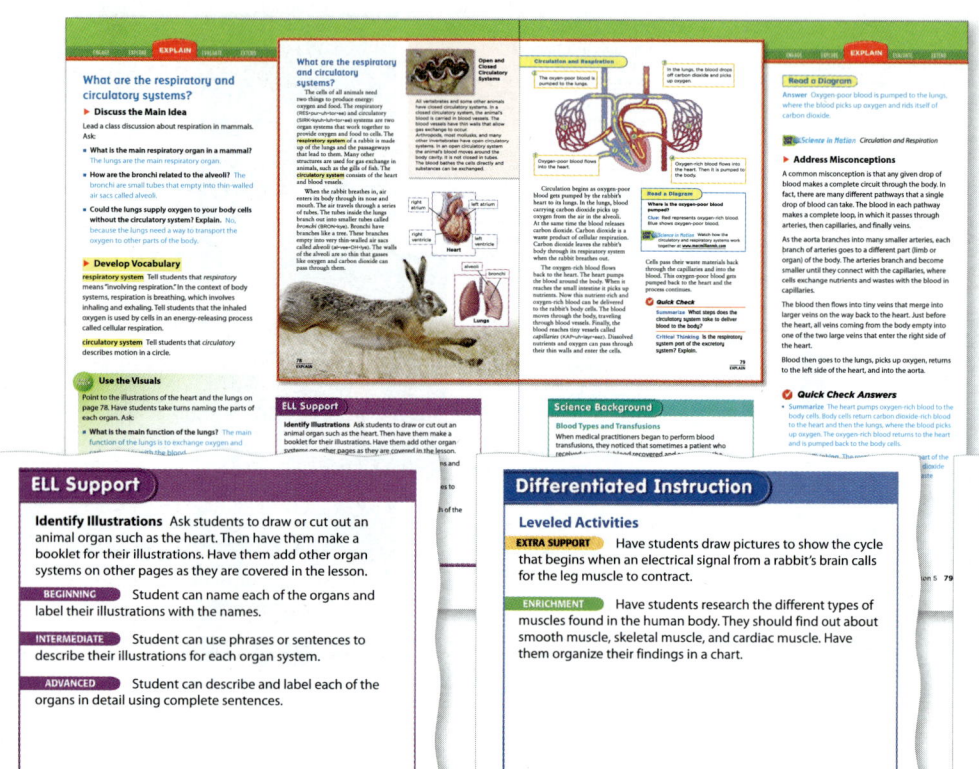

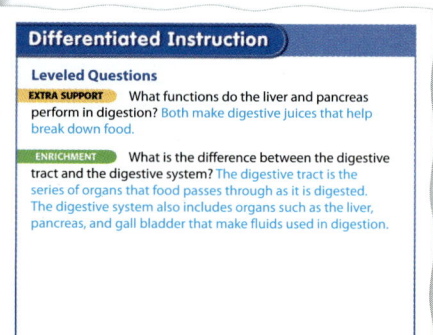

Teacher's Edition provides complete support for the teacher

ELL Support

Identify Illustrations Ask students to draw or cut out an animal organ such as the heart. Then have them make a booklet for their illustrations. Have them add other organ systems on other pages as they are covered in the lesson.

BEGINNING Student can name each of the organs and label their illustrations with the names.

INTERMEDIATE Student can use phrases or sentences to describe their illustrations for each organ system.

ADVANCED Student can describe and label each of the organs in detail using complete sentences.

Differentiated Instruction

Leveled Activities

EXTRA SUPPORT Have students draw pictures to show the cycle that begins when an electrical signal from a rabbit's brain calls for the leg muscle to contract.

ENRICHMENT Have students research the different types of muscles found in the human body. They should find out about smooth muscle, skeletal muscle, and cardiac muscle. Have them organize their findings in a chart.

Differentiated Instruction

Leveled Questions

EXTRA SUPPORT What functions do the liver and pancreas perform in digestion? Both make digestive juices that help break down food.

ENRICHMENT What is the difference between the digestive tract and the digestive system? The digestive tract is the series of organs that food passes through as it is digested. The digestive system also includes organs such as the liver, pancreas, and gall bladder that make fluids used in digestion.

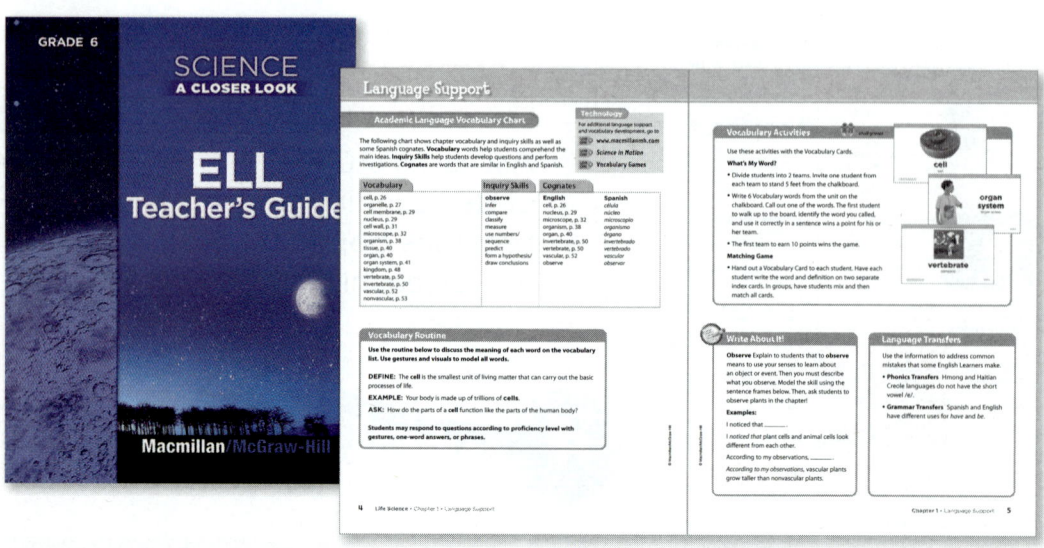

ELL Teacher's Guide supports language acquisition

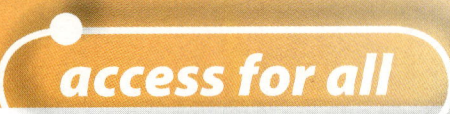

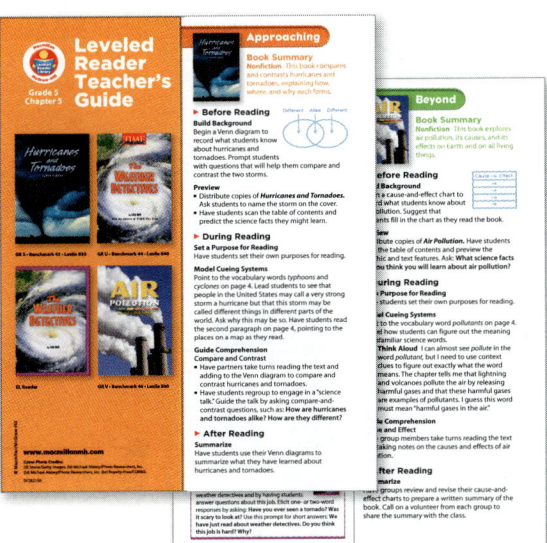

Leveled Readers
deliver multi-level science content in trade book format

Leveled Reader Teacher's Guides
provide additional reading strategies for using the Leveled Readers

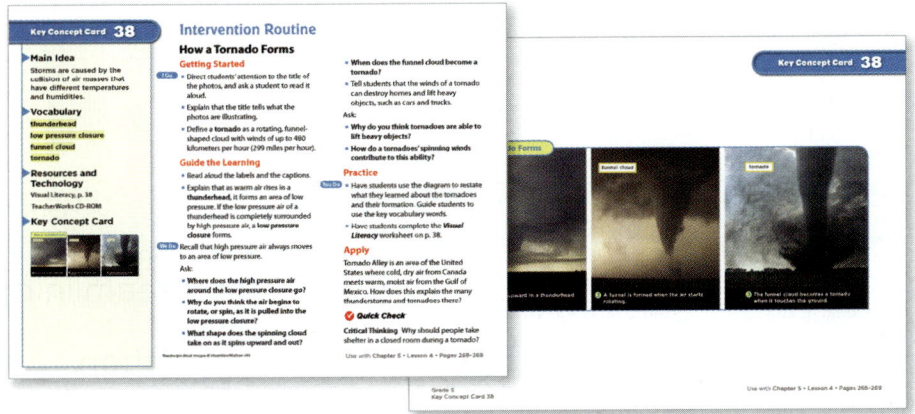

Key Concept Cards
allow for small group or one-on-one instruction for essential concepts

full technology support

 e-Review Summaries and quizzes online at www.macmillanmh.com

◄ **e-Review**
contains animated summaries and assessment

plus...

• **Leveled Readers Audio Selections**

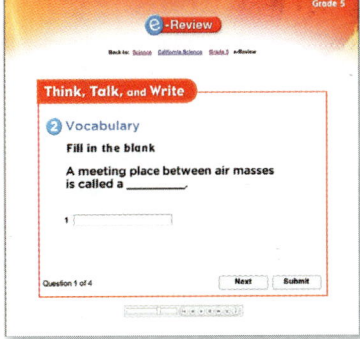

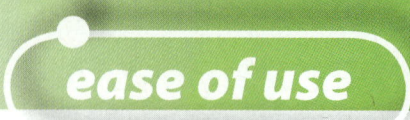
ease of use

A Manageable Organization

- Makes planning easy
- Delivers lesson resources quickly and effectively
- Maximizes the use of instructional time

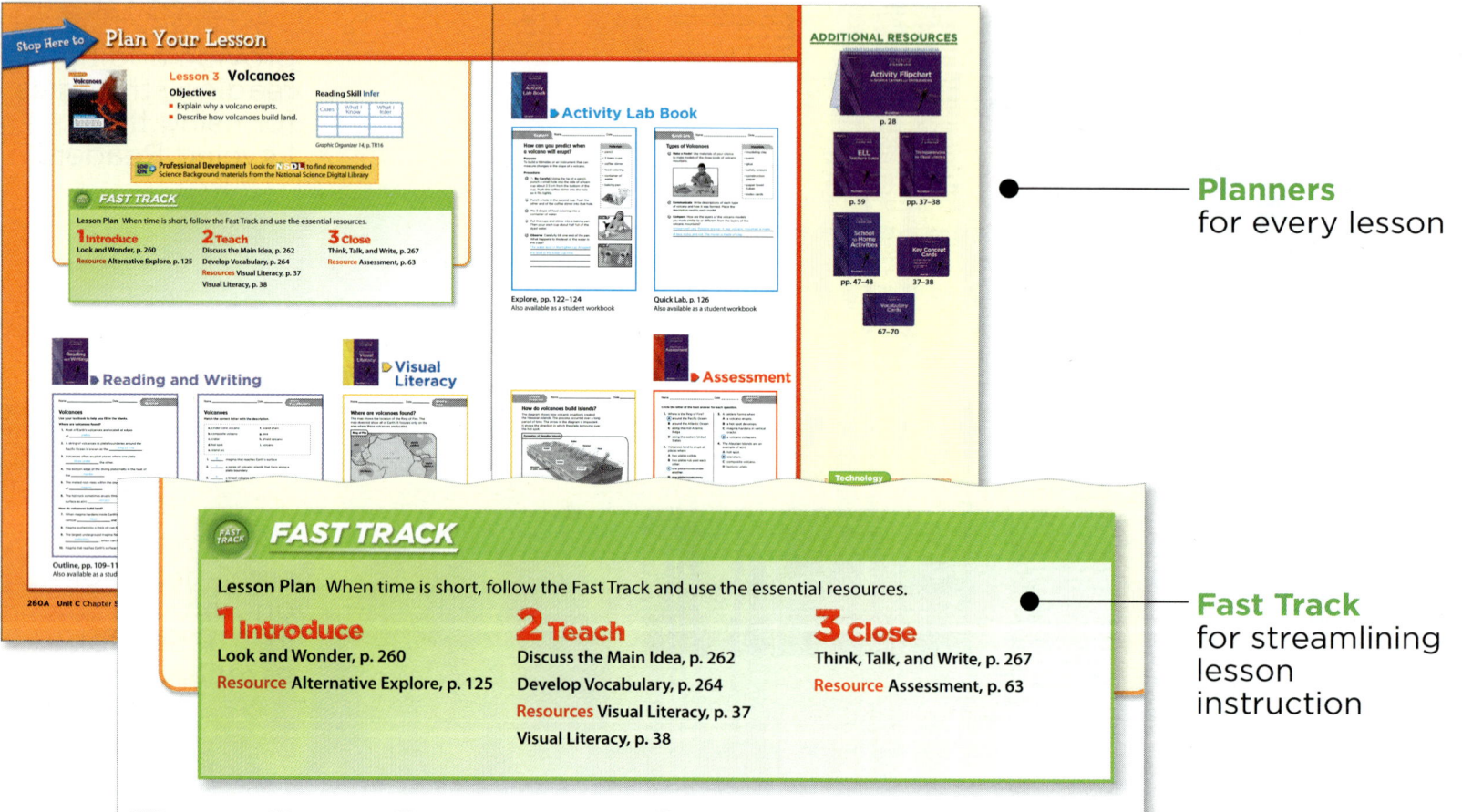

Planners
for every lesson

FAST TRACK

Lesson Plan When time is short, follow the Fast Track and use the essential resources.

1 Introduce
Look and Wonder, p. 260
Resource Alternative Explore, p. 125

2 Teach
Discuss the Main Idea, p. 262
Develop Vocabulary, p. 264
Resources Visual Literacy, p. 37
Visual Literacy, p. 38

3 Close
Think, Talk, and Write, p. 267
Resource Assessment, p. 63

Fast Track
for streamlining lesson instruction

full technology support

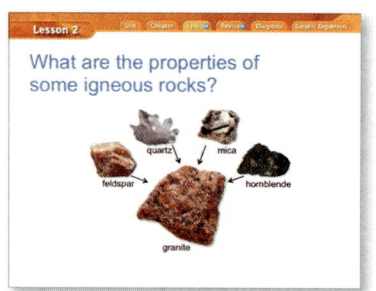

◄ **Classroom Presentation Toolkit CD-ROM**
PowerPoint delivery system for each chapter

plus...

- **Instructional Navigator CD-ROM**

Assessment

- **Includes a variety of assessment options**
- **Contains tools to assess student understanding**
- **Helps inform instruction and tracks student progress**

1 Introduce

▶ **Assess Prior Knowledge**

Have students discuss what they know about severe storms. Ask students what severe storms they have actually experienced, and what storms they may have seen on television or at the movies. List student responses on the board. Possible answers: thunderstorms, severe winter snowstorms, tornados, hurricanes

- **What makes some storms severe?** Possible answers: high winds, heavy rains or snow, lightning, flooding

Entry level Assessments to help determine student readiness

Formative Assessment

Approaching Have students state what instruments scientists use to gather weather data.

On Level Have students explain why meteorologists need data from the upper levels of the atmosphere.

Challenge Have students describe the weather conditions that are required for the formation of a hurricane.

Key Concept Cards For student intervention, see the prescribed routines on **Key Concept Cards 55–56.**

Think, Talk, and Write

1. **Main Idea** What causes storms?

2. **Vocabulary** Tornados and hurricanes are examples of a(n) _____.

3. **Cause and Effect** What causes a hurricane to form?

   Cause → Effect
   →
   →
   →
   →

4. **Critical Thinking** What is the reason why most thunderstorms do not become cyclones?

5. **Test Prep** What is a storm surge?
   - **A** a circular pattern of winds
   - **B** a bulge of water in the ocean
   - **C** a winter storm with freezing rain
   - **D** a large region of cold air

6. **Test Prep** Which of the following is a storm with a low-pressure center?
   - **A** thunderstorm
   - **B** ice storm
   - **C** tornado
   - **D** blizzard

Formative Assessments to check for understanding during the lesson

Summative Assessments to determine extent of student learning

Chapter Review with test prep

Chapter Tests

GRADE 5

SCIENCE
A CLOSER LOOK

BUILDING SKILLS
Assessment

Macmillan/McGraw-Hill

Lesson Review

Visual Summary

Thunderstorms and winter storms occur when two air masses of different temperatures and moisture levels meet.

Cyclones, such as hurricanes and tornados, are storms with a low-pressure center and circular winds.

Meteorologists use different kinds of instruments to gather data about weather variables.

Think, Talk, and Write

1. **Main Idea** What...
2. **Vocabulary** Tor... are examples of...
3. **Cause and Effect** hurricane to form...

4. **Critical Thinking** reason why most do not become c...

5. **Test Prep** What is a storm surge?
   - A a circular pattern of winds
   - B a bulge of water in the ocean
   - C a winter storm with freezing rain
   - D a large region of cold air

6. **Test Prep** Which of the following is a storm with a low-pressure center?
   - A thunderstorm
   - B ice storm
   - C tornado
   - D blizzard

Make a FOLDABLES Study Guide

Make a Trifold Book. Use the titles shown. Discuss the topic on the inside of each fold.

Writing Link

Fictional Narrative
Write about what it would be like to work as a meteorologist. Discuss the daily tasks that you would do.

Social Studies Link

Tornado Frequency
Research and write a report about Tornado Alley. Include information about how people that live in Tornado Alley prepare for tornados.

-Review Summaries and quizzes online at www.macmillanmh.com

403
EVALUATE

★ *full technology support*

- **Test Generator CD-ROM**
- **Progress Reporter CD-ROM**

	Pre-K	Kindergarten	Grade 1	Grade 2
Life Science	**Unit A** **Be a Scientist** **Unit B** **Plants**	**Unit A** **Plants** • Parts of Plants • What Plants Need and How They Grow • Leaves and Flowers and How We Use Plants	**Unit A** **Plants** Chapter 1 **Plants Are Living Things** Chapter 2 **Plants Grow and Change**	**Unit A** **Plants and Animals** Chapter 1 **Plants** Chapter 2 **Animals**
	Unit C **Animals**	**Unit B** **Animals** • Animals Are Everywhere • Animal Needs • How Animals Grow and Change	**Unit B** **Animals and Their Homes** Chapter 3 **All About Animals** Chapter 4 **Places to Live**	**Unit B** **Habitats** Chapter 3 **Looking at Habitats** Chapter 4 **Kinds of Habitats**
Earth Science	**Unit D** **Our Earth**	**Unit C** **Our Earth, Our Home** • Soil and Rocks • Land and Water • Resources and Recycling	**Unit C** **Our Earth** Chapter 5 **Looking at Earth** Chapter 6 **Caring for Earth**	**Unit C** **Our Earth** Chapter 5 **Land and Water** Chapter 6 **Earth's Resources**
	Unit E **Sky and Weather**	**Unit D** **Weather and Sky** • Look at Weather • Seasons • Sun, Moon, Stars	**Unit D** **Weather and Sky** Chapter 7 **Weather and Seasons** Chapter 8 **The Sky**	**Unit D** **Weather and Sky** Chapter 7 **Observing Weather** Chapter 8 **Earth and Space**
Physical Science	**Unit F** **Matter and Motion**	**Unit E** **Exploring Matter** • Paper and Cloth • Wood, Metal, and Clay • Water	**Unit E** **Matter** Chapter 9 **Matter Everywhere** Chapter 10 **Changes in Matter**	**Unit E** **Matter** Chapter 9 **Looking at Matter** Chapter 10 **Changes in Matter**
		Unit F **Moving Right Along** • Wheels and Motion • Gravity and Sounds • Magnets	**Unit F** **Motion and Energy** Chapter 11 **On the Move** Chapter 12 **Energy Everywhere**	**Unit F** **Motion and Energy** Chapter 11 **How Things Move** Chapter 12 **Using Energy**

Grade 3 | Grade 4 | Grade 5 | Grade 6

Life Science

Grade 3	Grade 4	Grade 5	Grade 6
Unit A **Living Things**	**Unit A** **Living Things**	**Unit A** **Diversity of Life**	**Unit A** **Diversity of Life**
Chapter 1 **A Look at Living Things**	Chapter 1 **Kingdoms of Life**	Chapter 1 **Cells and Kingdoms**	Chapter 1 **Classifying Living Things**
Chapter 2 **Living Things Grow and Change**	Chapter 2 **The Animal Kingdom**	Chapter 2 **Parents and Offspring**	Chapter 2 **Cells**
Unit B **Ecosystems**	**Unit B** **Ecosystems**	**Unit B** **Ecosystems**	**Unit B** **Patterns of Life**
Chapter 3 **Living Things in Ecosystems**	Chapter 3 **Exploring Ecosystems**	Chapter 3 **Interactions in Ecosystems**	Chapter 3 **Genetics**
Chapter 4 **Changes in Ecosystems**	Chapter 4 **Surviving in Ecosystems**	Chapter 4 **Ecosystems and Biomes**	Chapter 4 **Ecosystems**

Earth Science

Grade 3	Grade 4	Grade 5	Grade 6
Unit C **Earth and Its Resources**	**Unit C** **Earth and Its Resources**	**Unit C** **Earth and Its Resources**	**Unit C** **Earth and Its Resources**
Chapter 5 **Earth Changes**	Chapter 5 **Shaping Earth**	Chapter 5 **Our Dynamic Earth**	Chapter 5 **Changes over Time**
Chapter 6 **Using Earth's Resources**	Chapter 6 **Saving Earth's Resources**	Chapter 6 **Protecting Earth's Resources**	Chapter 6 **Conserving Our Resources**
Unit D **Weather and Space**	**Unit D** **Weather and Space**	**Unit D** **Weather and Space**	**Unit D** **Weather and Space**
Chapter 7 **Changes in Weather**	Chapter 7 **Weather and Climate**	Chapter 7 **Weather Patterns**	Chapter 7 **Weather and Climate**
Chapter 8 **Planets, Moons, and Stars**	Chapter 8 **The Solar System and Beyond**	Chapter 8 **The Universe**	Chapter 8 **Astronomy**

Physical Science

Grade 3	Grade 4	Grade 5	Grade 6
Unit E **Matter**	**Unit E** **Matter**	**Unit E** **Matter**	**Unit E** **Matter**
Chapter 9 **Observing Matter**	Chapter 9 **Properties of Matter**	Chapter 9 **Comparing Kinds of Matter**	Chapter 9 **Classifying Matter**
Chapter 10 **Changes in Matter**	Chapter 10 **Matter and Its Changes**	Chapter 10 **Physical and Chemical Changes**	Chapter 10 **Chemistry**
Unit F **Forces and Energy**	**Unit F** **Forces and Energy**	**Unit F** **Forces and Energy**	**Unit F** **Forces and Energy**
Chapter 11 **Forces and Motion**	Chapter 11 **Forces**	Chapter 11 **Using Forces**	Chapter 11 **Exploring Forces**
Chapter 12 **Forms of Energy**	Chapter 12 **Energy**	Chapter 12 **Using Energy**	Chapter 12 **Exploring Energy**

SCIENCE
A CLOSER LOOK

Components	K	1	2	3	4	5	6
Student Edition*		•	•	•	•	•	•
Teacher's Edition*	•	•	•	•	•	•	•
Student Edition Unit Big Books		•	•				
Reading and Writing		•	•	•	•	•	•
Math		•	•	•	•	•	•
Activity Lab Book		•	•	•	•	•	•
Visual Literacy		•	•	•	•	•	•
Assessment		•	•	•	•	•	•
Transparencies for Visual Literacy		•	•	•	•	•	•
ELL Teacher's Guide		•	•	•	•	•	•
Vocabulary Cards	•	•	•	•	•	•	•
Key Concept Cards		•	•	•	•	•	•
Leveled Readers	•	•	•	•	•	•	•
Leveled Reader Teacher's Guide		•	•	•	•	•	•
Kindergarten Flipbook	•						
Literature Big Books	•	•	•				
A to Z Activity Book	•						
Science Projects in a Pocket		•	•				
Science Resource Book	•						
Floor Puzzles	•						
Science on the Go	•	•					
Photo Sorting Cards	•	•	•				
Equipment Kits	•	•	•	•	•	•	•
Activity Flipchart		•	•	•	•	•	•
Science Fair Handbook		•	•	•	•	•	•
Online Student Edition		•	•	•	•	•	•
Student Edition on CD-ROM		•	•	•	•	•	•
Online Teacher's Edition	•	•	•	•	•	•	•
Instructional Navigator CD-ROM	•	•	•	•	•	•	•
Progress Reporter CD-ROM		•	•	•	•	•	•
Presentation Toolkit CD-ROM		•	•	•	•	•	•
Science Activity DVD		•	•	•	•	•	•
Science Songs Audio CD	•	•	•				
PuzzleMaker CD-ROM		•	•	•	•	•	•
Operation: Science Quest CD-ROM		•	•	•	•	•	•
Professional Development for Science, The Master Teacher Series	•	•	•	•	•	•	•
Teacher's Desk Reference	•	•	•	•	•	•	•
Companion Web site	•	•	•	•	•	•	•

Pre-K Components: Teacher's Edition, Flipbook, Big Science Readers, Photo Cards, Science Song Posters, and Science Songs CD
* also available in Spanish

Be a Scientist

AMERICAN
MUSEUM OF
NATURAL
HISTORY

The Scientific Method 2

What Do Scientists Do? 4

Form a Hypothesis 5

How Do Scientists Test a Hypothesis? 6

Test Your Hypothesis. 7

Analyze the Data 9

How Do Scientists Analyze Data? 9

How Do Scientists Draw Conclusions? 10

Draw Conclusions 11

Focus on Skills 12

Life Science

UNIT A Living Things

Unit Literature Monarch Butterfly 16

CHAPTER 1

A Look at Living Things **18**

Lesson 1 Living Things and Their Needs 20

 Reading in Science 28

Lesson 2 Plants and Their Parts 30

Inquiry Investigation 40

Lesson 3 Animals and Their Parts 42

Inquiry Skill Builder 50

Lesson 4 Classifying Animals 52

Writing in Science 62

Math in Science 63

Chapter 1 Review 64

CHAPTER 2

Living Things Grow and Change **66**

Lesson 1 Plant Life Cycles 68

Inquiry Skill Builder 78

Lesson 2 Animal Life Cycles 80

Writing in Science 88

Math in Science 89

Lesson 3 From Parents to Young 90

 Reading in Science 96

Chapter 2 Review 98

Careers in Science 100

UNIT B Ecosystems

Unit Literature Once Upon a Woodpecker 102

CHAPTER 3

Living Things in Ecosystems 104

Lesson 1 Food Chains and Food Webs 106

Inquiry Skill Builder 116

Lesson 2 Types of Ecosystems 118

Reading in Science 130

Lesson 3 Adaptations 132

Inquiry Investigation 144

Chapter 3 Review 146

CHAPTER 4

Changes in Ecosystems 148

Lesson 1 Living Things Change Their Environments . . 150

Inquiry Skill Builder 158

Lesson 2 Changes Affect Living Things 160

Writing in Science 170

Math in Science 171

Lesson 3 Living Things of the Past 172

Reading in Science 180

Chapter 4 Review 182

Careers in Science 184

Earth Science

UNIT C Earth and Its Resources

Unit Literature One Cool Adventure 186

CHAPTER 5

Earth Changes **188**

Lesson 1 Earth's Features 190

Inquiry Skill Builder 200

Lesson 2 Sudden Changes to Earth 202

Reading in Science 210

Lesson 3 Weathering and Erosion 212

Writing in Science 220

Math in Science 221

Chapter 5 Review 222

CHAPTER 6

Using Earth's Resources **224**

Lesson 1 Minerals and Rocks 226

Writing in Science 236

Math in Science 237

Lesson 2 Soil 238

Inquiry Skill Builder 246

Lesson 3 Fossils and Fuels 248

Reading in Science 256

Lesson 4 Air and Water Resources 258

Inquiry Investigation 268

Chapter 6 Review 270

Careers in Science 272

viii

UNIT D Weather and Space

Unit Literature What a Difference Day Length Makes .. 274

CHAPTER 7

Changes in Weather. 276

Lesson 1   Weather. 278

Inquiry Skill Builder 286

Lesson 2   The Water Cycle 288

Reading in Science 300

Lesson 3   Climate and Seasons 302

Writing in Science 310

Math in Science 311

Chapter 7 Review 312

CHAPTER 8

Planets, Moons, and Stars 314

Lesson 1   The Sun and Earth 316

Writing in Science 324

Math in Science 325

Lesson 2   The Moon and Earth 326

Inquiry Investigation 334

Lesson 3   The Planets 336

Inquiry Skill Builder 344

Lesson 4   The Stars. 346

Reading in Science 352

Chapter 8 Review 354

Careers in Science 356

Physical Science

UNIT E Matter

Unit Literature The Good Ship Popsicle Stick 358

CHAPTER 9

Observing Matter 360

Lesson 1 Properties of Matter 362

Reading in Science 370

Lesson 2 Measuring Matter 372

Inquiry Skill Builder 380

Lesson 3 Solids, Liquids, and Gases 382

Writing in Science 390

Math in Science 391

Chapter 9 Review 392

CHAPTER 10

Changes in Matter 394

Lesson 1 Changes of State 396

Inquiry Skill Builder 404

Lesson 2 Physical Changes 406

Reading in Science 414

Lesson 3 Chemical Changes 416

Inquiry Investigation 422

Chapter 10 Review 424

Careers in Science 426

UNIT F Forces and Energy

Unit Literature Jump Rope 428

CHAPTER 11

Forces and Motion **430**

Lesson 1 Position and Motion 432

Reading in Science 440

Lesson 2 Forces 442

Inquiry Investigation 450

Lesson 3 Work and Energy 452

Inquiry Skill Builder 460

Lesson 4 Using Simple Machines 462

Writing in Science 472

Math in Science 473

Chapter 11 Review 474

CHAPTER 12

Forms of Energy **476**

Lesson 1 Heat 478

Inquiry Skill Builder 486

Lesson 2 Sound 488

Inquiry Investigation 496

Lesson 3 Light 498

Reading in Science 508

Lesson 4 Electricity 510

Writing in Science 518

Math in Science 519

Chapter 12 Review 520

Careers in Science 522

Activities and Investigations

Life Science

CHAPTER 1

Explore Activities

How do living and nonliving
things differ? 21
How are plants alike? 31
How do an animal's structures
help it meet its needs? 43
How can you classify animals? 53

Quick Labs

Observe Cells 26
Observe Stems. 35
Observe Animal Structures 47
Model a Backbone. 55

Inquiry Skills and Investigations

What do plants need to survive?. . . 40
Classify 50

CHAPTER 2

Explore Activities

What does a seed need
to grow? 69
How does a caterpillar grow
and change? 81
Which traits are passed on from
parents to their young? 91

Quick Labs

Fruits and Seeds 73
A Bird's Life Cycle 85
Inherited Traits 93

Inquiry Skills and Investigations

Form a Hypothesis 78

CHAPTER 3

Explore Activities

What kind of food do owls need? . 107
Can ocean animals live and grow
in fresh water? 119
Does fat help animals survive in
cold environments? 133

Quick Labs

Observe Decomposers. 114
Water Temperatures 127
Storing Water. 137

Inquiry Skills and Investigations

Communicate 116
How does camouflage help some
animals stay safe? 144

CHAPTER 4

Explore Activities

How can worms change their
environment? 151
How can a flood affect plants?. . . . 161
How do fossils tell us about
the past? 173

Quick Labs

Model Pollution 155
A Changing Ecosystem 167
Fossil Mystery. 177

Inquiry Skills and Investigations

Use Numbers 158

snake

Sun

hawk

Activities and Investigations

Earth Science

CHAPTER 5

Explore Activities

Does land or water cover more
of Earth's surface? 191
How does sudden movement
change the land? 203
How can rocks change in
moving water? 213

Quick Labs

Your State's Features 195
A Model Volcano 207
Materials Settle 217

Inquiry Skills and Investigations

Make a Model 200

CHAPTER 6

Explore Activities

How do a mineral's color and
mark compare? 227
What makes up soil? 239
How do some fossils form? 249
How is Earth's water made clean? 259

Quick Labs

Classify Rocks 231
Classify Soils 243
Model Imprints 251
Record Water Use 265

Inquiry Skills and Investigations

Use Variables 246
What things pollute the air? 268

CHAPTER 7

Explore Activities

How can you tell air is
around you? 279
How do raindrops form? 289
How do temperature and
precipitation patterns compare? . . 303

Quick Labs

Make a Windsock 283
Cloud in a Jar 293
Compare Climates 307

Inquiry Skills and Investigations

Interpret Data 286

CHAPTER 8

Explore Activities

How do shadows change
throughout the day? 317
How does the Moon's shape
seem to change? 327
How do the planets move
through space? 337
Why do we see stars at night? . . . 347

Quick Labs

Model Earth's Rotation 319
Make a Flip Book 331
Sizing Up Planets 339
Make a Constellation 349

Inquiry Skills and Investigations

Why does the Moon's shape
appear to change? 334
Observe 344

Activities and Investigations

Physical Science

CHAPTER 9

Explore Activities

How do you describe objects? 363
How can you measure length? 373
How are solids different
from liquids? 383

Quick Labs

Classify Matter 367
Measure Mass and Volume....... 377
Compare Solids, Liquids,
and Gases 387

Inquiry Skills and Investigations

Measure 380

CHAPTER 10

Explore Activities

What happens when ice
is heated? 397
How can you change matter? 407
How can matter change?........ 417

Quick Labs

Condense Water Vapor 401
Separating Mixtures 412
A Chemical Change............. 419

Inquiry Skills and Investigations

Predict 404
How can physical and chemical
changes affect matter? 422

CHAPTER 11

Explore Activities

How can you describe an
object's position?................ 433
How can pushes affect the way
objects move? 443
What is work?.................. 453
How can a simple machine help
you lift objects? 463

Quick Labs

Measure Speed.................. 438
Observe Gravity 447
Using Energy 457
Inclined Planes................. 469

Inquiry Skills and Investigations

How does distance affect the pull
of a magnet on metal objects?.... 450
Infer 460

CHAPTER 12

Explore Activities

What happens to air when it
is heated? 479
How can you make sounds? 489
How does light move? 499
What makes a bulb light? 511

Quick Labs

Heating Water and Soil 481
Changing Sounds 493
Mixing Colors 505
Conductors and Insulators 516

Inquiry Skills and Investigations

Experiment 486
How does sound move through
different types of matter? 496

Be a Scientist

Chameleons can change
color to communicate.

Giant Madagascan chameleon

Be a Scientist

The Scientific Method

Objectives
- Identify the steps in the scientific method.
- Learn how scientists form and test a hypothesis.

1 Introduce

▶ **Assess Prior Knowledge**

Discuss with students whether they've seen a movie or a television nature program about animal life on an island. Ask:

- **What animals were shown and described?** Answers will vary.

- **Can you describe some of the special things about these animals?** Answers will vary.

Look and Wonder

Invite students to share their responses to the Look and Wonder statement and questions:

- **What would it be like to live on a tropical island? What kinds of things might you see there?**

- Write ideas on the board and note any misconceptions that students may have. Address these misconceptions as you teach *Be a Scientist.*

RESOURCES and TECHNOLOGY
▶ **Activity Lab Book,** pp. 1–3
▶ **Activity Flipchart,** p. 1

Be a Scientist

The Scientific Method

Look and Wonder

Madagascar is a tropical island off the coast of Africa. It is home to plants and animals found nowhere else on Earth! What would it be like to live on a tropical island? What kinds of things might you see there?

2
ENGAGE

ELL Support

Preview the Lesson Use the pictures on pages 2–11 to preview the selection with students. Point to people, animals, and items in the pictures, as you say their names. Have students repeat after you. For each spread, use complete sentences to describe what the scientists are doing. Encourage students to ask questions. Review the pictures again. Have students take turns describing what is happening in each picture.

BEGINNING   Have students use one or two words to describe the picture.

INTERMEDIATE   Have students use phrases or short sentences to describe the picture.

ADVANCED   Have students use complete sentences to describe the picture.

Explore

What do you know about animals that live in Madagascar?

▶ How do you look for animals in their natural habitat?
▶ What kinds of animals would you see in a forest?
▶ What does an animal need to live in a forest?
▶ How do scientist find answers to these questions?

Chris Raxworthy

Paule Razafimahatratra

Meet two scientists who are curious about the natural world and everything that lives in it. Chris Raxworthy and Paule Razafimahatratra study animals that live in Madagascar. They work at the American Museum of Natural History in New York City and at the University of Antananarivo in Madagascar.

3
EXPLORE

Explore

Purpose This selection helps students understand the steps scientists take as they use the scientific method to learn more about animals and their habitat.

Madagascar is an island off the coast of Africa. Here, in the tropical forests live a diverse array of plants and animals. Because the island is isolated, many of the plants and animals found here are found nowhere else. Scientists exploring the island often discover new species. Chris Raxworthy and Paule Razafimahatratra (Paul Raaz-ah Fee-ma Ha-tra-tra) are scientists who are researching one species, the giant Madagascan chameleon. In this selection students will follow Chris and Paule as they use the steps in the scientific method to determine in what kind of habitat the giant chameleon can be found.

Inquiry Discuss with students how scientists learn about animals. Ask:

■ **How do you look for animals in their natural habitat?** Possible answers: Learn where the animals live in their habitat (under the ground, in a tree, and so on), go to that place and quietly look for them, use binoculars to find/see them.

■ **What kinds of animals would you see in a forest?** Accept all reasonable answers. Possible answers include: chipmunks, squirrels, snakes, birds, raccoons, groundhogs, rabbits, and so on.

■ **What does an animal need to live in a forest?** Possible answers: food, water, and shelter. Students might also mention a climate that matches the animal's lifestyle.

■ **How do scientists find answers to these questions?** Possible answers: They go to places where the animals live and observe them. They take notes about what the animal eats, where it lives, and how it behaves.

Alternative Explore

What do you know about studying animals?

• **Materials:** encyclopedias, the Internet, if available, drawing supplies

• Encourage students to explore animals further. As an alternative or supplement to the Explore questions, have them use the Alternative Explore worksheet.

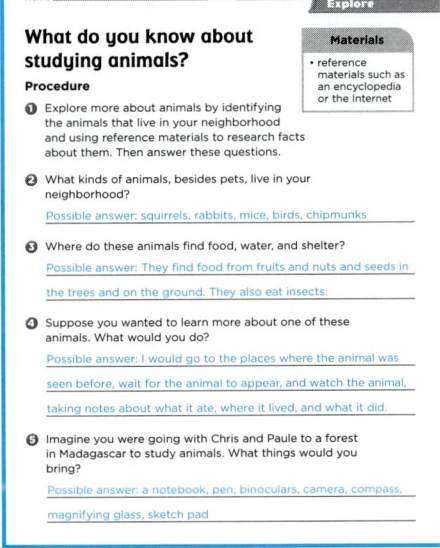

Name _____ Date _____

Alternative Explore

What do you know about studying animals?

Materials
• reference materials such as an encyclopedia or the Internet

Procedure

❶ Explore more about animals by identifying the animals that live in your neighborhood and using reference materials to research facts about them. Then answer these questions.

❷ What kinds of animals, besides pets, live in your neighborhood?
Possible answer: squirrels, rabbits, mice, birds, chipmunks

❸ Where do these animals find food, water, and shelter?
Possible answer: They find food from fruits and nuts and seeds in the trees and on the ground. They also eat insects.

❹ Suppose you wanted to learn more about one of these animals. What would you do?
Possible answer: I would go to the places where the animal was seen before, wait for the animal to appear, and watch the animal, taking notes about what it ate, where it lived, and what it did.

❺ Imagine you were going with Chris and Paule to a forest in Madagascar to study animals. What things would you bring?
Possible answer: a notebook, pen, binoculars, camera, compass, magnifying glass, sketch pad

Activity Lab Book, p. 3

What do scientists do?

▶ Discuss the Main Idea

Point out to students that we all use methods every day to complete various tasks. Call on volunteers to describe the step-by-step method they use to put on their shoes and socks. Explain that scientists, too, use a method when they do their research. Direct students to the chart that outlines the scientific method. Call on a volunteer to read the steps aloud. Discuss how scientists are using the scientific method to find out more about animals and where they live. Ask:

- **What are Chris and Paule doing in Madagascar?** They want to find out about the animals that live there.

- **What will Chris and Paule use to learn about animals?** They will use the scientific method.

- **How does the scientific method help them?** Possible answers: It helps them explain how things happen in the natural world. It helps them think about questions and how to answer them.

▶ Develop Vocabulary

hypothesis Tell students that a hypothesis is a possible answer scientists have for why something is the way it is or for why something happens the way it does. A hypothesis is the basis for exploring the question further.

chameleon *Word Origin* The word *chameleon* comes from the Greek word *khamaileon. Khamai* means "on the ground," and *leon* means "lion." *Khamai* refers to the place where one would expect to find a lizard.

What do scientists do?

Chris and Paule want to find out about the many amazing animals that live in Madagascar. Much of the island has never been explored by scientists. New plants and animals are discovered all the time.

The scientific method is a process that scientists use to investigate the world around them. It helps them answer questions about the natural world.

Right now, Chris and Paule are studying a lizard called a giant Madagascan chameleon. Chris has observed these chameleons in dry forests. He wants to know where else in Madagascar the chameleons live.

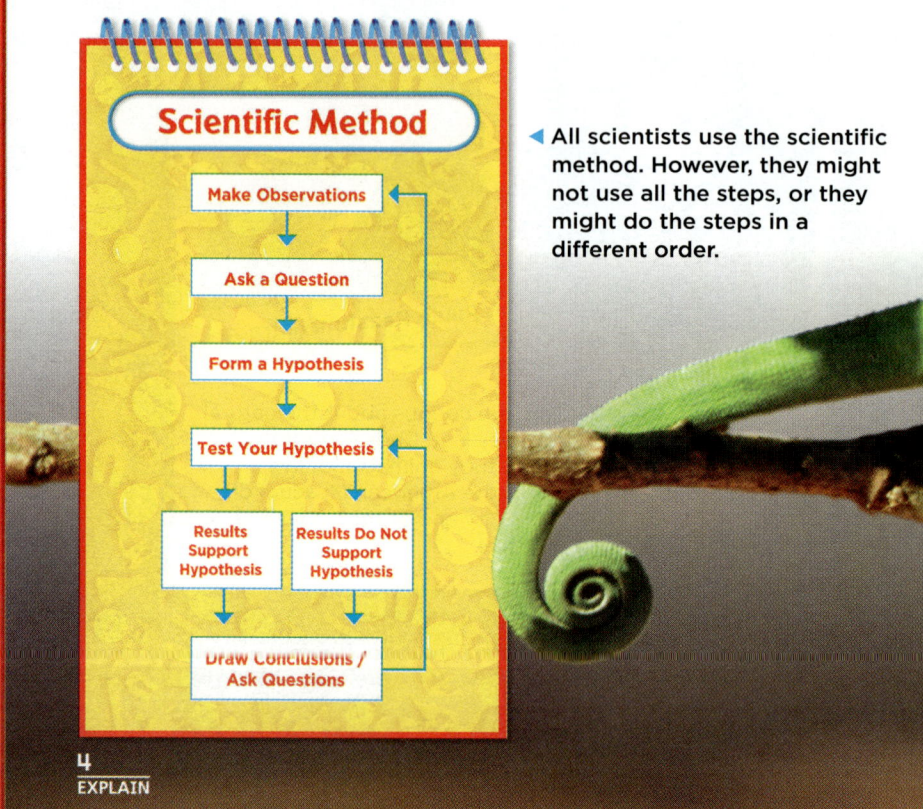

Scientific Method

Make Observations

Ask a Question

Form a Hypothesis

Test Your Hypothesis

Results Support Hypothesis   Results Do Not Support Hypothesis

Draw Conclusions / Ask Questions

◀ All scientists use the scientific method. However, they might not use all the steps, or they might do the steps in a different order.

4
EXPLAIN

Science Background

Scientific Method The scientific method begins with observations and prior knowledge. Scientists develop a question related to their observations and use what they already know to form a hypothesis. Then they make a plan to test their hypothesis. They collect data by making observations, conducting experiments, and/or making and using a model. They organize their data, analyze it, and use it to test their hypothesis. The hypothesis is either supported and then subjected to further testing, or rejected and replaced with a new hypothesis. Scientists document each step so that other scientists can evaluate, replicate, and use the results for their own research. The scientific method is typically iterative.

Chris knows that variables such as temperature and rainfall affect where animals live. A variable is something that can change.

Chris uses this information to form a hypothesis. A hypothesis is a statement that can be tested to answer a question.

Here is Chris's hypothesis. If a place has temperatures between 10 and 40 degrees Celsius and between 50 and 150 centimeters of rainfall every year, then giant Madagascan chameleons could live there.

Form a Hypothesis

1. Ask lots of "why" questions.

2. Look for connections between important variables.

3. Suggest possible explanations for those connections.

▶ Make sure the explanations can be tested.

giant Madagascan chameleon ▲

▶ **Discuss the Main Idea**

Have students identify the initial step scientists take when using the scientific method. Call on a volunteer to read the steps in *Form a Hypothesis* aloud. Ask:

■ **Can you tell me what a hypothesis is?** Possible answers: a prediction, an answer to a question

■ **What does Chris know about the Madagascan chameleon?** He knows it lives in dry forests.

■ **What does Chris want to find out?** He wants to know where else in Madagascar this chameleon species lives.

■ **What are two variables that affect where an animal lives?** temperature and rainfall

■ **What hypothesis did Chris formulate to answer his question?** If a place has a certain temperature and amount of rainfall, then giant Madagascan chameleons could live there.

▶ **Explore the Main Idea**

ACTIVITY   Have students work in groups. Write the following on the board: *Why do plants grow towards the light? Why do flowers have scents?* Ask each group to choose one question. Have groups discuss their question and share what they know. Have them formulate a hypothesis to answer their question and decide how they might test it. Call on each group to present their findings to the class.

Differentiated Instruction

Leveled Activities

EXTRA SUPPORT   Have students research the characteristics of a dry tropical forest. Have them make a drawing of a typical dry forest habitat. Ask volunteers to use their drawings to describe the dry tropical forest to the rest of the class.

ENRICHMENT   Have students research the characteristics of a tropical rain forest. Have them make a drawing of a typical rain-forest habitat. Ask volunteers to use their drawings to describe the tropical rain forest to the rest of the class. Call on students to identify how the dry forest and rain forest are alike and different.

Science Background

Madagascar Madagascar is known for unusual plants and animals found nowhere else on Earth. By studying these species, scientists hope to learn more about how life evolved on the island and to contribute to the task of finding and saving these species before they become extinct. Some of the island's forests are disappearing largely due to agriculture practices and to a growing human population that is gradually taking over wildlife areas. By collecting data on Madagascar's species, scientists can compare healthier regions to those in danger and help focus efforts on conserving Madagascar's biodiversity.

Be a Scientist   **5**

How do scientists test a hypothesis?

▶ Discuss the Main Idea

Call on volunteers to review the steps of the scientific method that Chris and Paule have followed so far. Explain that once scientists have a hypothesis, they can begin to think of a plan to collect data that will help them either accept or reject the hypothesis. Call on a volunteer to read the steps in *Test Your Hypothesis* aloud. Ask:

- **How does the map Chris makes help him find the chameleons?** Possible answers: The map helps Chris find places that have the same rainfall and temperature as the places where he already found the chameleons.

- **What strategy will Chris and Paule use to collect the data?** Possible answer: They will go out into the field to look for chameleons.

- **What are some other ways scientists collect data?** Possible answers: perform an experiment, make a computer model

▶ Develop Vocabulary

Madagascan/Malagasy Write *Madagascan* and *Malagasy* on the board. Point out that *Madagascan* and *Malagasy* are synonyms, words that have the same meaning. Both words are adjectives, words used to describe a person, place or thing; and nouns, a person, place or thing. Elicit that as nouns these words describe a person from Madagascar. When the words are used as an adjective, it means "of or pertaining to Madagascar." Call on students to use the words as nouns and as adjectives in sentences they compose.

How do scientists test a hypothesis?

The giant Madagascan chameleon is about as long as a banana. It is hard to find in the dense forest, though, because it hides. People in Madagascar say you can never find a chameleon when you are looking for one!

Where should Chris and Paule look for chameleons? In order to find out, they study their data about temperature and rainfall. Data is information. They put this data into a computer and make a map. The computer colors yellow all the areas that are likely to have chameleons. Those areas have similar temperatures and rainfall to places where chameleons have been found before. Chris predicts that if they go to those areas, they will find giant Madagascan chameleons.

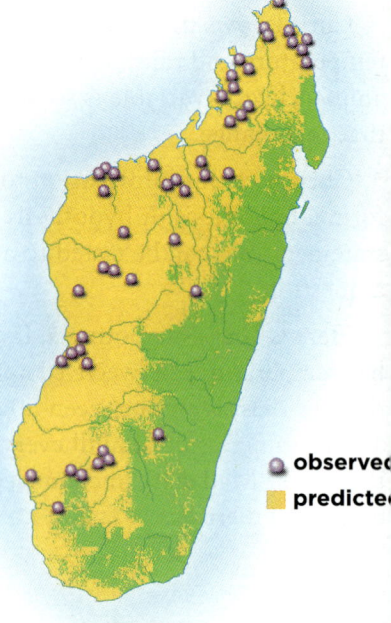

○ observed
▣ predicted

▲ The purple dots on this map show where giant Madagascan chameleons have been seen before. The yellow areas show where Chris and Paule think the chameleons live.

◀ Chris uses his headlamp to find chameleons at night when they sleep.

6
EXPLAIN

Science Background

Field biologists are scientists who go to different locations to learn more about how organisms are related to and depend on one another. They typically sample several areas and use the data they collect to identify how many species (both plant and animal) live in the ecosystem and estimate the population or distribution range of each species. Field biologists take notes to record the important details about each species they find. Sometimes they use satellite data to understand how an ecosystem can change over time. This information can then be combined with field observations to allow scientists to investigate the impact of environmental change on the distribution of species.

Chris and Paule choose new places to look for chameleons. They choose places that are in the yellow areas on the map. They collect data in these places to test their hypothesis. They use procedures that other scientists can repeat. That way other scientists can check Chris's and Paule's results.

"We wear headlamps and search at night, when the chameleons are sleeping and are easier to find," Chris explains. "We look up in the branches for pale-colored comma shapes." Every time they find a chameleon, Chris and Paule make careful notes and take photographs. They record the exact date, time, and place in their field journals.

Test Your Hypothesis

1. Think about the different kinds of data that could be used to test the hypothesis.

2. Choose the best method to collect this data.

   - **perform an experiment** (in the lab)
   - **observe the natural world** (in the field)
   - **make and use a model** (on a computer)

3. Then plan a procedure and gather data.

   ▶ Make sure that the procedure can be repeated.

Chris and Paule record data when they find giant Madagascan chameleons and other lizards.

7
EXPLAIN

▶ Discuss the Main Idea

Discuss with students some of the problems scientists might have collecting data in a dense forest. Ask:

- **What problems might Chris and Paule have looking for a green chameleon in a dense, green forest?** Possible answer: The chameleons hide well and are the same color as the plants, so they can be difficult to find.

- **How do Chris and Paule solve this problem?** They collect their data at night when the chameleons are sleeping and are easier to find.

- **What procedures do Chris and Paule follow when they find a chameleon?** They take photographs. They record the exact date, time, and place in a journal.

Science Background

Chameleons have features that distinguish them from other lizards. Their mittenlike grasping feet are well adapted to tree climbing. Their tongues can be twice the length of their bodies. They can shoot them out quickly to precisely capture an insect. Each of the chameleon's eyes can operate independently: It can keep one eye on a predator and use the other to follow an insect buzzing nearby. Chameleons are also known for their ability to change color. Chameleons do not only change color to camouflage themselves relative to their surroundings. Their skin also changes in response to temperature, light, and mood. Communication with other chameleons is also an important reason behind these color changes.

▶ Discuss the Main Idea

Call on volunteers to review the steps of the scientific method that Chris and Paule have followed so far and what they have done at each step. Point out that the next step is to analyze the data that they collected. Call on a volunteer to read the steps in *Analyze the Data.* Ask:

- **What data do Chris and Paule have?** They have the numbers of chameleons found and where they were found.

- **How can Chris and Paule be sure that the chameleons they found are giant Madagascan chameleons?** Possible answers: They look at their photos and notes about the chameleons and make sure they have all the features of a giant Madagascan chameleon.

- **What did Chris and Paule do with their data?** They marked every place on the map that they found a giant Madagascan chameleon. Then they looked for patterns.

- **When Chris and Paule analyzed their data, what did they find?** They found chameleons in the places they predicted they would find them.

CHRIS'S FIELD JOURNAL

April 9th, 2006
Ambohibola Forest
15mm rain measured in rain gauge
Temperature range from 20-34°C
Heavy afternoon rain shower

This deciduous forest has large trees and cut tree stumps. The forest edge is burnt. Hunting and cattle grazing occur in the forest. It has many small streams that dry up in the winter.

○ observed
▢ predicted
🦎 new observations

Draw Direct students to the pictures on pp. 8–9. Have each student choose a chameleon to draw and color. Then have students display their drawings and describe them to the group. They can describe features such as size, color, and shape. Provide assistance when necessary by using questions, such as, *What color is the chameleon? How big is the chameleon?*

BEGINNING Have students describe their drawings using one- or two-word phrases.

INTERMEDIATE Have students describe their drawings using phrases or short sentences.

ADVANCED Have students describe their drawings using complete sentences.

giant Madagascan chameleon
(Furcifer oustaleti)
Found at 10:45 A.M. in grassland
with scattered trees. Laid 17
eggs 14x8mm.

Madagascan
day gecko
(Phelsuma
madagascariensis)
on a tree trunk
at 11:30 A.M., in a
small clump of
trees growing by
a small stream.

Analyze the Data

1. Organize the data as a chart, table, graph, diagram, map, or group of pictures.

2. Look for patterns in the data. These patterns can show how important variables in the hypothesis affect one another.

▶ Make sure to check the data by comparing it to data from other sources.

How do scientists analyze data?

Part of testing a hypothesis is looking for patterns in the data that has been collected. Chris and Paule study the information from all of the locations they visited. They mark the seven places on the map where they found a giant Madagascan chameleon. Then they look for patterns in their data.

They observe that the chameleons they found were in the yellow area on the map. They talk about the temperatures and rainfall in the places where they found the chameleons.

9
EXPLAIN

How do scientists analyze data?

▶ Develop Vocabulary

analyze Explain that when scientists analyze data they are examining it in great detail in order to understand it better or discover more about it. They ensure their data is accurate, compare and contrast different data sets, and look for patterns.

data Explain that data is information that is obtained from observations. The data is analyzed and used by scientists to draw conclusions.

Be a Scientist **9**

How do scientists draw conclusions?

▶ Discuss the Main Idea

Point out that the final step in the scientific method is to draw conclusions. Call on a volunteer to read the steps in *Draw Conclusions* aloud. Ask:

- **What did the data Chris and Paule analyzed show?** It showed that if a place has a certain temperature and amount of rainfall, then giant Madagascan chameleons could live there.

- **Did the data allow Chris and Paule to accept or reject Chris's hypothesis?** The analysis of the data they collected allowed Chris and Paule to accept the hypothesis.

- **What new questions resulted from Chris and Paule's research?** What other variables affect where giant Madagascan chameleons live? Could they search for other living things in the same way? Which places on the island are home to the greatest number of plants and animals?

- **Why is it important for Chris and Paule to share their results?** Possible answers: Other scientists can learn from their work. Biologists can make conservation plans for Madagascar. Chris and Paule can use the results to continue their own research.

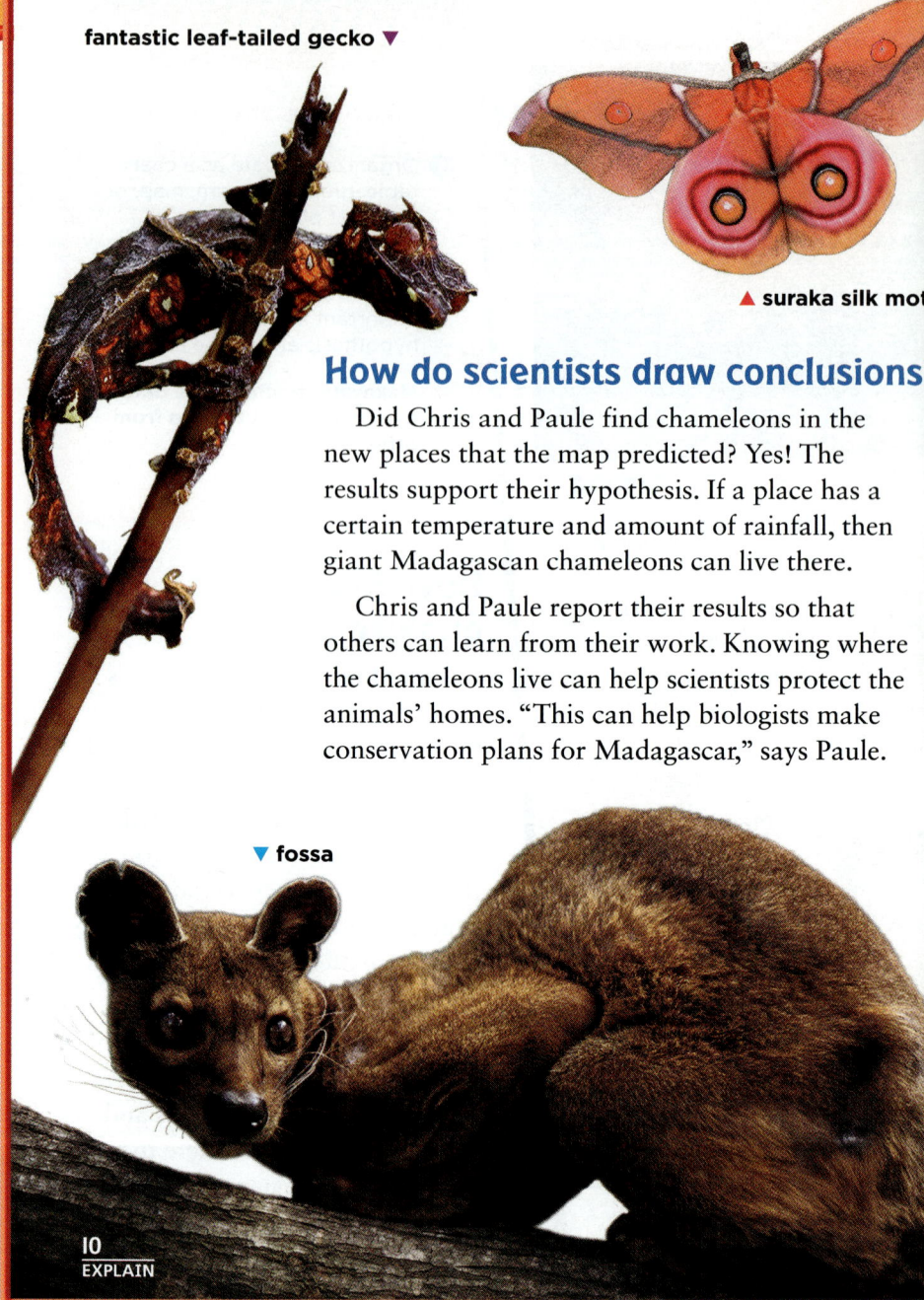

fantastic leaf-tailed gecko ▼

▲ suraka silk moth

How do scientists draw conclusions?

Did Chris and Paule find chameleons in the new places that the map predicted? Yes! The results support their hypothesis. If a place has a certain temperature and amount of rainfall, then giant Madagascan chameleons can live there.

Chris and Paule report their results so that others can learn from their work. Knowing where the chameleons live can help scientists protect the animals' homes. "This can help biologists make conservation plans for Madagascar," says Paule.

▼ fossa

10
EXPLAIN

Science Background

Conservation While extinction is a natural part of Earth's history, scientists have observed an increase in the number of species becoming extinct in recent years. One of the largest threats to endangered species is humans. As human population increases by several million each year, the demand on resources, such as land, limits the resources available to other species. Conservation efforts are helping to preserve biodiversity and the health of the planet. Scientists in conservation "hotspots" such as Madagascar provide recommendations to the government to find a balance between sustainable development (economic and social goals that do not exhaust all the natural resources) and the preservation of ecosystems.

Homework Activity

Using the Scientific Method

Have students outline how they would research this question using the scientific method: *Do squirrels have a preference for the kind of tree they live in?* When students have completed the assignment, call on volunteers to present their outlines to the rest of the class. Discuss with the class whether or not the scientific method was appropriately utilized in their plan and have students suggest what is missing or what needs to be clarified.

Malagasy tree frog ▲

Chris and Paule's results lead them to new questions. What other variables affect where giant Madagascan chameleons live? The animals shown on this page all live in Madagascar. Could scientists search for these living things in the same way? Which places on the island are home to the greatest number of plants and animals? New questions can lead to a new hypothesis and to learning new things. Learning more about the animals that live in Madagascar will help protect them.

Think, Talk, and Write

1. Why is the scientific method useful to scientists?

2. What other questions about animals can you think of? Choose one and form a hypothesis that can be tested.

▼ **red millipede**

Draw Conclusions

1. Decide if the data clearly support or do not support the hypothesis.

2. If the results are not clear, rethink how the hypothesis was tested and make a new plan.

3. Write down the results to share with others.

▶ Make sure to ask questions.

ring-tailed lemur ▶

EVALUATE

▶ **Develop Vocabulary**

biologist *word origin* The word *biologist* is a combination of the Greek word ***bios***, meaning "life," and the suffix ***-logia***, meaning "study of." A biologist is one who studies life.

conservation Point out to students that *conservation* means "the preservation, management, and care of natural and cultural resources." Call on students to create sentences using *conservation*.

3 Close

▶ **Think, Talk, and Write**

1. Answers should include the idea that the scientific method provides strong procedures that guide how we investigate and answer questions about natural phenomena.

2. Answers will vary. Accept reasonable answers. For example, if a squirrel buries an acorn, will it grow to be a tree? Hypothesis: I predict if the acorn is not buried, it will not grow into a tree.

Formative Assessment

Approaching Have students identify the question Chris and Paule wanted to answer.

On-Level Have students list the steps in the scientific method. Next to each step have them identify what Chris and Paule did at that step.

Challenge From what they have read, have students make inferences about the needs of other animals that live alongside the giant Madagascan chameleon in the dry tropical forest.

Science Background

Species of Madagascar Over eighty percent of the plants and animals on Madagascar are endemic, or unique, to the island. In fact, two-thirds of the world's chameleon species and all sixty-eight species of lemurs identified today are native to Madagascar. Because of their geographic isolation, species on this island evolved independently from other species found in Africa and Asia. The baobab tree, for example, is well adapted to the dry, arid climate of the island's western region, thanks to its ability to store a thousand gallons of water in its barrel-shaped trunk. Seven species of baobab are native to Madagascar, but only one species of this tree is found in Africa.

Focus on Skills

Objective
Understand and use inquiry skills.

Using Inquiry Skills

Explain that inquiry skills can help students and scientists organize and use the information they gather. These skills are also useful in other areas of study, such as history, mathematics, and health. Ask:

- **Which inquiry skill do you use now?** Possible answers: I observe when I look at science pictures and diagrams. I experiment to find the information I need to answer my questions about the observations I made.

Each Focus on Skills extension features a particular inquiry skill: Classify, p. 64; Form a Hypothesis, p. 78; Communicate, p. 116; Use Numbers, p. 158; Make a Model, p. 200; Use Variables, p. 246; Interpret Data, p. 286; Observe, p. 344; Measure, p. 380; Predict, p. 404; Infer, p. 460; and Experiment, p. 486.

▶ Learn It

As you read the skills to the class, ask:

- **Why is *Classify* an important inquiry skill?** When things are organized into groups or categories, they are easier to understand.

- **Why is *Make a Model* an important inquiry skill?** When I make a model, I can examine the structure of something to understand it better.

- **Which inquiry skill helps us to understand and analyze the information we learn?** I interpret data so I can understand the information I have gathered

▶ Try It

Assign one inquiry skill to each of 12 small groups. Ask each group to define the skill and prepare a class presentation about the uses for that skill in science and in other classroom studies. Encourage groups to use graphics or posters to explain the skills.

Focus on Skills

Scientists use many skills as they work through the scientific method. Skills help them gather information and answer questions they have about the world around us. Here are some skills they use.

sea star

Observe Use your senses to learn about an object or event.

Classify Place things with similar properties into groups.

Form a Hypothesis Make a statement that can be tested to answer a question.

Use Numbers Order, count, add, subtract, multiply, and divide to explain data.

Communicate Share information with others.

Make a Model Make something to represent an object or event.

lizard

goldfish

beetle

animal	what I observed

▲ **Observe** the animals on these pages. Then make a chart to **communicate** your observations.

12
EXTEND

ELL Support

Use Realia Write a skill on the board, read the word out loud, and have students repeat. Read the definition from pages 12–13 with students. Use gestures and images to demonstrate how the skill is used: *I observe that the apple is red.* Encourage students to come up with examples using real objects from the classroom. Repeat for each skill.

BEGINNING   Students can identify the skills being demonstrated by naming it or pointing to the word on the board.

INTERMEDIATE   Students can use a phrase or short sentence to demonstrate an inquiry skill.

ADVANCED   Students can demonstrate more than two inquiry skills, such as *observe* and *infer*.

hedgehog

parrot

dragonfly

Use Variables Identify things that can control or change the outcome of an experiment.

Interpret Data Use information that has been gathered to answer questions or solve a problem.

Measure Find the size, distance, time, volume, area, mass, weight, or temperature of an object or event.

Predict State possible results of an event or experiment.

Infer Form an idea from facts or observations.

Experiment Perform a test to support or disprove a hypothesis.

snail

Animal Young	
Animal	**Average Number of Young**
beetle	75
sea star	2,000,000
lizard	14
hedgehog	4
gazelle	1

▲ Use this chart to **infer** how an animal's size affects how many young it has at a time.

Inquiry Skill Builder

In each chapter of this book, you will find an Inquiry Skill Builder. These features will help you build the skills you need to become a great scientist.

gazelle

13
EXTEND

▶ Apply It

Inquiry skills are also used throughout the *Explore* features within each lesson and in the *Be a Scientist* extensions.

Have students take turns telling the class how they have used inquiry skills in science. Ask:

- **How have you used the *Communicate* skill?** I communicate with others when I explain something.

- **How have you used the *Measure* skill?** I have to measure temperature and length.

- **How have you used the *Predict* skill?** I predict when I make a guess based on what I think will happen.

- **How have you used the *Infer* skill?** I infer when I am able to find information, even though it is not specifically given.

- **How has the *Use Numbers* skill helped you?** I use numbers when I calculate differences and similarities in the data I am gathering.

- **How has the *Use Variables* skill helped you?** I can control and change what I do in an experiment when I use variables.

- **How have you used the *Form a Hypothesis* skill?** I form a hypothesis when I give myself a possible answer to a question I want to test.

Integrate Math

Measure the Distance

Ask students to measure the distance from their homes to school, using any unit of measurement they choose.

Suggest that students use numbers; for instance, feet, meters, city blocks, or buildings, to calculate the distance. Have students take turns giving the distance and the unit of measurement they used.

Science Safety

Safety Tips

Objective Identify the reasons safety procedures are important.

▶ Talk About It

Encourage students to share their experiences with rules and to discuss why rules are made. Ask:

- **What kinds of rules do you have at home?**

Write students' responses on chart paper. Ask:

- **Why do people make rules?** Students should respond that rules are created to keep them safe.

▶ Learn About It

Have a volunteer read the first sentence on page 14. Ask students to list other safety symbols they know, such as stop signs. Invite them to look through their books and find the **Be Careful!** notation. Ask:

- **Why do you need to be careful when doing this activity?**

Discuss the types of science activities students may do in class, and encourage them to propose safety procedures. Have a volunteer read the remainder of page 14. For each safety tip, ask students to explain the rational behind the rule.

▶ Try It

Divide the class into five groups and assign one safety tip from page 14 to each group. Have each group create a poster to explain and illustrate their safety tip, and encourage them to present their posters to the class.

RESOURCES and TECHNOLOGY

▶ **Activity Lab Book**, pp. v–vi

Safety Tips

In the Classroom

- Read all of the directions. Make sure you understand them. When you see "⚠ **Be Careful**," follow the safety rules.

- Listen to your teacher for special safety directions. If you do not understand something, ask for help.

- Wash your hands with soap and water before an activity.

- Be careful around a hot plate. Know when it is on and when it is off. Remember that the plate stays hot for a few minutes after it is turned off.

- Wear a safety apron if you work with anything messy or anything that might spill.

- Clean up a spill right away, or ask your teacher for help.

- Dispose of things the way your teacher tells you to.

- Tell your teacher if something breaks. If glass breaks, do not clean it up yourself.

- Wear safety goggles when your teacher tells you to wear them. Wear them when working with anything that can fly into your eyes or when working with liquids.

- Keep your hair and clothes away from open flames. Tie back long hair, and roll up long sleeves.

- Keep your hands dry around electrical equipment.

- Do not eat or drink anything during an experiment.

- Put equipment back the way your teacher tells you to.

- Clean up your work area after an activity, and wash your hands with soap and water.

In the Field

- Go with a trusted adult—such as your teacher, or a parent or guardian.

- Do not touch animals or plants without an adult's approval. The animal might bite. The plant might be poison ivy or another dangerous plant.

> **Responsibility**
> Treat living things, the environment, and one another with respect.

14

Integrate Writing

Introduction to the Science Kit

Distribute Science Kit items to small groups of students. Choose items that are likely to be unfamiliar, such as goggles, funnels, hand lenses, or droppers. Have students discuss what each item is and how it may be used by scientists.

Assemble students, show each item and ask:

- **How might you use this item during a science activity?**

For items that are unfamiliar to students, name the item and explain how it is used. Ask them to choose one item, draw it, label it, and write a sentence to describe how it is used.

National Science Education Standards

The following Fundamental Concepts and Principles for Grades K–4 Content Standard B are covered in this unit:

- Objects have many observable properties. Those properties can be measured using tools.

- Objects are made of one or more materials. Objects can be described by the properties of the materials from which they are made, and those properties can be used to separate or sort a group of objects or materials.

- Materials can exist in different states—solid, liquid, and gas. Some common materials, such as water, can be changed from one state to another by heating or cooling.

CHAPTER 9

Observing Matter

Main Idea Matter can be described by physical properties and measured with tools that record standard units. Matter can be in the form of a solid, a liquid, or a gas.

Lesson 1 Properties of Matter
Matter is anything that has volume and mass. You can use properties to describe and identify matter.

Lesson 2 Measuring Matter
Matter can be measured using tools that record standard units.

Lesson 3 Solids, Liquids, and Gases
Solids, liquids, and gases are three forms of matter.

CHAPTER 10

Changes in Matter

Main Idea Matter can undergo physical changes, such as changes in state, and chemical changes. Mixtures and solutions result from physical changes; different kinds of matter result from chemical changes.

Lesson 1 Changes of State
Adding or removing heat can cause matter to change state.

Lesson 2 Physical Changes
Matter looks different after a physical change, but it is still the same kind of matter. You can mix matter together to form mixtures and solutions.

Lesson 3 Chemical Changes
Chemical changes cause different kinds of matter to form.

CHAPTER 9
Observing Matter

 CD-ROM

Instructional Navigator CD-ROM
Interactive Lesson Planner, Teacher's Edition, Worksheets, and Links to Online Resources

Presentation Toolkit CD-ROM

Test Generator

PuzzleMaker CD-ROM

 DVD

Science: Master Teacher DVD Set

Science Activity DVD

 **www.macmillanmh.com**

 Science in Motion Particles of a Solid, Liquid, and Gas

Online Teacher's Edition

Leveled Reader Database

Translated Concept Summaries

Progress Reporter Assessments

Professional Development

e-Glossary

e-Journal

e-Review

NSDL National Science Digital Library

CHAPTER 10
Changes in Matter

 CD-ROM

Instructional Navigator CD-ROM
Interactive Lesson Planner, Teacher's Edition, Worksheets, and Links to Online Resources

Presentation Toolkit CD-ROM

Test Generator

PuzzleMaker CD-ROM

 DVD

Science: Master Teacher DVD Set

Science Activity DVD

 **SCIENCE QUEST** *Physical and Chemical Changes*

 **www.macmillanmh.com**

 Science in Motion Heating Water

Online Teacher's Edition

Leveled Reader Database

Translated Concept Summaries

Progress Reporter Assessments

Professional Development

e-Glossary

e-Journal

e-Review

NSDL National Science Digital Library

◀ Science Activity DVD

CHAPTER 9
Observing Matter

 CD-ROM

Student Navigator CD-ROM

PuzzleMaker CD-ROM

 DVD

Science Activity DVD

 **LOG ON** www.macmillanmh.com

 LOG ON *Science in Motion* Particles of a Solid, Liquid, and Gas

 LOG ON

Online Student Edition

Online Vocabulary Games

Translated Concept Summaries

 LOG ON

e-Careers

e-Glossary

e-Journal

e-Review

CHAPTER 10
Changes in Matter

 CD-ROM

Student Navigator CD-ROM

PuzzleMaker CD-ROM

 DVD

Science Activity DVD

 **SCIENCE QUEST** *Physical and Chemical Changes*

 LOG ON www.macmillanmh.com

 LOG ON *Science in Motion* Heating Water

 LOG ON

Online Student Edition

Online Vocabulary Games

Translated Concept Summaries

 LOG ON

e-Careers

e-Glossary

e-Journal

e-Review

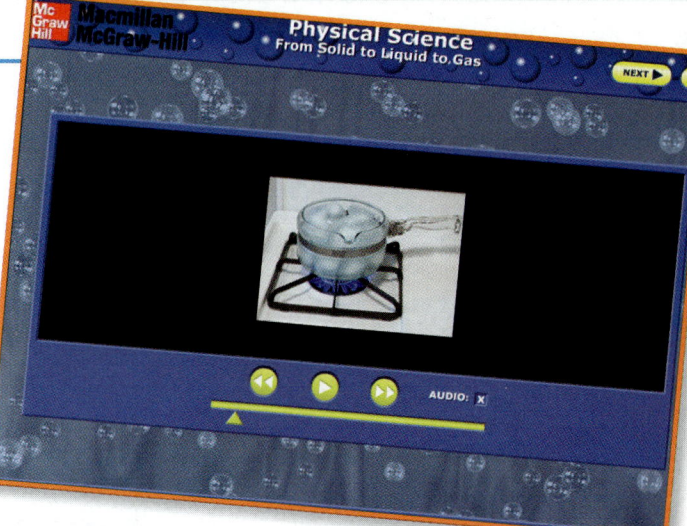

Science in Motion *Heating Water* ▶

Literature
Magazine Article

Objective
■ Understand the importance of imagination to science.

Robert McDonald and the crew on *Mjollnir*

358

The Good Ship Popsicle Stick

Genre: Magazine Article A magazine article that gives factual information about a specific topic is nonfiction. Ask:

■ **What are the clues that an article is true, or nonfiction?** Possible answer: The article tells about real people and something they did.

Before Reading

Direct students' attention to the photograph of the ship on page 358. Ask:

■ **What kind of ship do you think this is?** Accept all reasonable responses.

■ **What do you think this ship is made of?** Students should recognize that the ship is made of small pieces of wood. They might possibly recognize the popsicle sticks.

During Reading

Have students look for details about why and how the ship was built. Ask:

■ **Why did Captain Rob build the ship?** to show children that they can do anything

■ **How did the ship get its name?** It was named for the hammer of Thor, the Viking god of thunder.

RESOURCES and TECHNOLOGY

▶ **Reading and Writing,** p. 167

LOG ON e-Journal Online research and writing

ELL Support

Use Sentence Strips Give students an opportunity to reread the article. Have volunteers read portions of the article aloud, and guide less proficient students in reading some sentences or paragraphs aloud. Then ask students to complete the following sentence strips: *Captain Robert floated his ship on a river in the city of ____.* Amsterdam *The ship he built was made of ____ and ____.* popsicle sticks. glue *The name of the ship comes from a ____story.* Viking

BEGINNING Students can point to or name the materials used to make the ship and the name of the ship.

INTERMEDIATE Students can write short sentences and phrases to describe the ship.

ADVANCED Students can use complete sentences to explain why Captain Bob built the ship.

Writing Rubric p. TR26

The Good Ship Popsicle Stick

from *Time for Kids*

September 2, 2005

On August 16, former Hollywood stuntman Robert McDonald performed a record-breaking stunt. He floated a ship made of wooden ice cream sticks on a river that flows through the city of Amsterdam, in the Netherlands. And the boat didn't sink!

McDonald built the 50-foot-long replica of a Viking ship with 15 million ice cream sticks and more than two tons of glue. The 13-ton cruiser is named *Mjollnir* (MIL•ner), after the hammer of Thor, the Viking god of thunder.

Captain Rob is the president of the Sea Heart Ship Foundation. The group's goal is to spread fun to kids in hospitals around the world. "I have a dream to show children they can do anything," he says. "If they can dream it, they can do it."

 ### Write About It

Response to Literature This article is about a ship made from ice cream sticks. What words are used to describe the ship? Choose an object around you. Then use words to tell about it.

 e-Journal Write about it online at www.macmillanmh.com

359

After Reading

Explain to students that this article is about just one person's dream to do something unusual. Have students spend a few minutes of quiet time to reflect on what kinds of dreams they have and what they would like to do in the future. Ask:

- **If you could do anything you want to do, what would it be?** Accept all reasonable responses.

- **How would you go about making your dream come true?** Accept all reasonable responses.

Write About It

Students should recognize that the 50-foot-long ship is made up of 15 million wooden ice-cream sticks held together with more than two tons of glue.

If students have trouble describing objects around them, ask:

- **What is the size and shape of the object?**

- **What is the object made of?**

- **What color is the object?**

More to Read

States of Matter: A Question and Answer Book, by Fiona Bayrock (Fact Finders, 2006)

Differentiated Instruction

Leveled Activities

EXTRA SUPPORT Encourage students to identify three different objects in the classroom and then make a list of all the materials that make up each object. Tell students they can work together in pairs to engage all students in the class. Have students compare their lists with other members of the class, especially if they chose the same objects.

ENRICHMENT Have students use encyclopedias and the Internet, if available, to research other myths, such as those of the Vikings, Native Americans, Romans, and Greeks. Encourage students to write a brief report on the myth they most enjoyed reading about. Also ask them to make a model or a drawing of some aspect of the myth to share with the class.

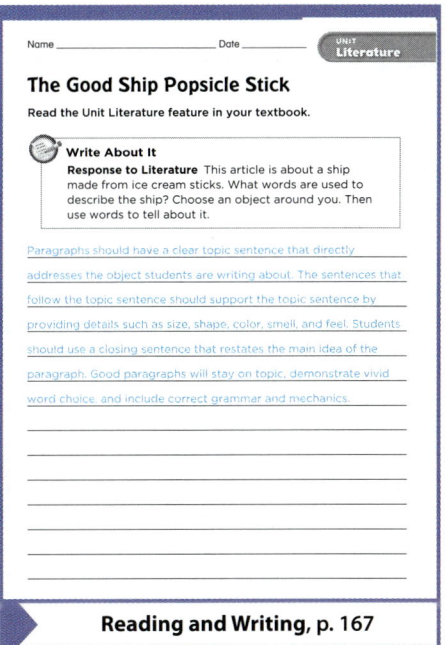

Reading and Writing, p. 167

Presentation Toolkit CD-ROM Lesson Presentations

Instructional Navigator CD-ROM Interactive Lesson Planner, Teacher's Edition, Worksheets, and Online Resources

Lesson	OBJECTIVES AND READING SKILLS	VOCABULARY	RESOURCES AND TECHNOLOGY
1 Properties of Matter PAGES 362–369	▪ Define matter as anything that has mass and takes up space. ▪ Describe properties of matter and understand that properties can be used to identify matter.	matter volume mass property element	▶ Reading and Writing, pp. 169–172 ▷ Visual Literacy, pp. 55–56 ▶ Activity Lab Book, pp. 152–155 ▶ Transparencies, pp. 55–56
PACING: 2 days FAST TRACK: 1 day	**Reading Skill** Main Idea and Details _Graphic Organizer 1_		
2 Measuring Matter PAGES 372–379	▪ Measure matter using tools that record standard units. ▪ Compare and contrast weight and mass.	metric system pan balance gravity weight	▶ Reading and Writing, pp. 175–178 ▷ Visual Literacy, pp. 57–58 ▶ Activity Lab Book, pp. 156–159 ▶ Transparencies, pp. 57–58
PACING: 2 days FAST TRACK: 1 day	**Reading Skill** Summarize _Graphic Organizer 5_		
3 Solids, Liquids, and Gases PAGES 382–389	▪ Define the three common states of matter: solid, liquid, and gas. ▪ Explain the properties of solids, liquids, and gases.	states of matter solid liquid gas	▶ Reading and Writing, pp. 179–182 ▶ Math, pp. 15–16 ▷ Visual Literacy, pp. 59–60 ▶ Activity Lab Book, pp. 163–166 ▶ Transparencies, pp. 59–60 ▶ Science in Motion, _Particles of a Solid, Liquid, and Gas_
PACING: 2 days FAST TRACK: 1 day	**Reading Skill** Classify _Graphic Organizer 11_		

| **Chapter 9 Review** PAGES 392–393 | ▪ Summarize chapter concepts. | **Resources** ▶ Assessment, pp. 108–111, 115–118 ▶ Reading and Writing, pp. 185–186 | **Technology** Progress Reporter Assessments e-Review |

PACING Assumes a day is a 25- to 35-minute session

LOG ON www.macmillanmh.com for more planning resources and www.macmillanmh.com/nsdl/ for science resources from **NSDL**

Activity Planner

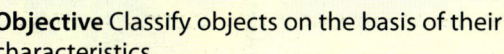

Science Activity DVD Explore Activity demos

Materials included in the Activity Kit are listed in *italics*.

EXPLORE Activities

Explore *p. 363* | PACING: 20 minutes

Objective Observe the properties of common objects.

Skills observe, communicate, infer, experiment

Materials classroom objects, *hand lens*

⭐ **PLAN AHEAD** Model describing an object using various senses without naming the object.

Explore *p. 373* | PACING: 20 minutes

Objective Compare different units of measurement of length.

Skills measure, communicate, interpret data, infer

Materials paper and pencil

⭐ **PLAN AHEAD** Clear an area in the classroom of desks so that students have room to work.

Explore *p. 383* | PACING: 20 minutes

Objective Develop operational definitions of solids and liquids.

Skills observe, experiment, classify

Materials *block,* beaker, plastic spoon, water, salt, hand soap, safety goggles, clay

⭐ **PLAN AHEAD** Provide waste containers for used water, salt, and hand soap.

QUICK LAB Activities

Quick Lab *p. 367* | PACING: 15 minutes

Objective Classify objects on the basis of their characteristics.

Skills communicate, classify, interpret data

Materials ten objects

Quick Lab *p. 377* | PACING: 15 minutes

Objective Verify predictions of mass and volume with measurements.

Skills predict, measure, interpret data

Materials *pan balance* with masses, *measuring cup* and water, *toy car,* golf ball, *marble*

⭐ **PLAN AHEAD** Acquire toy cars made of metal for this activity.

Quick Lab *p. 387* | PACING: 15 minutes

Objective Distinguish the properties of a solid, a liquid, and a gas.

Skills observe, communicate

Materials three sealable plastic bags, water, rock

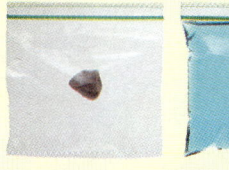

⭐ **PLAN AHEAD** Remind students to hold bags filled with water over a container to avoid spills.

FOR MORE ACTIVITIES

Activity Flipchart for Science Centers and Workstations

Focus on Skills	**Everyday Science**
Teach the inquiry skill: Measure, p. 380	Water Planet, p. 64

See **Activity Lab Book Teacher's Guide** for more support.

For a comprehensive list of consumable and non-consumable materials, see the back of the unit tab.

English Language Learner Support

Academic Language

English language learners need help in building their understanding of the academic language used in daily instruction and science activities. The following strategies will help to increase students' language proficiency and comprehension of content and instruction words.

Strategies to Reinforce Academic Language

- **Use Context** Academic language should be explained in the context of the task. Use gestures, expressions, and visuals to support meaning.

- **Use Visuals** Use charts, transparencies, and graphic organizers to explain key labels to help students understand classroom language.

- **Model** Use academic language as you demonstrate the task to help students understand instruction.

Academic Language Vocabulary Chart

The following chart shows chapter vocabulary and inquiry skills as well as some Spanish cognates. **Vocabulary** words help students comprehend the main ideas. **Inquiry Skills** help students develop questions and perform investigations. **Cognates** are words that are similar in English and Spanish.

Vocabulary	Inquiry Skills	Cognates	
		English	**Spanish**
matter, p. 364	observe, p. 363	observe, p. 363	*observar*
property, p. 365	communicate, p. 363	communicate, p. 363	*comunicar*
volume, p. 365	infer, p. 363		
mass, p. 365	experiment, p. 363	infer, p. 363	*inferir*
element, p. 368	measure, p. 373	property, p. 365	*propiedad*
metric system, p. 374	interpret data, p. 373	volume, p. 365	*volumen*
pan balance, p. 376	classify, p. 383	mass, p. 365	*masa*
gravity, p. 378		element, p. 368	*elemento*
weight, p. 378		interpret data, p. 373	*interpretar datos*
states of matter, p. 384		metric system, p.374	*sistema métrico*
solid, p. 384		gravity, p. 378	*gravedad*
liquid, p. 386		classify, p. 383	*clasificar*
gas, p. 387		solid, p. 384	*sólido*
		liquid, p. 386	*líquido*
		gas, p. 387	*gas*

Vocabulary Routine

Use the routine below to discuss the meaning of each word on the vocabulary chart. Use gestures and visuals to model all words.

Define **Volume** describes how much space an object takes up.

Example A big marker has more **volume** than a small crayon.

Ask What else has more **volume** than a crayon?

Students may respond to questions according to proficiency level with gestures, one-word answers, or phrases.

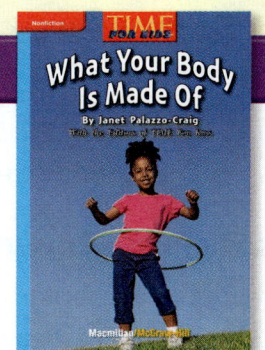

ELL Leveled Reader

What Your Body Is Made Of
by Janet Palazzo-Craig

Summary Go on a journey to discover the human body.

Reading Skill
Draw Conclusions

Vocabulary Activities

Help students describe the properties of objects.

BEGINNING Display a variety of objects having different properties. Direct students' attention to the chart on page 365. Draw a similar chart on the board, and guide students to complete it for each object. Ask questions such as: *What color is the _____? Does the _____ feel smooth or rough?*

INTERMEDIATE Direct students' attention to the chart on page 365. Draw a similar chart on the board. Display an object, and have students call out words that describe the color, shape, feel, and taste. List the words on the board. Encourage students to say a sentence summarizing the properties of the object. Repeat the activity with other objects.

ADVANCED Have students form groups. Provide small and big rocks. Students compare the mass, volume, and properties of the two types of rocks. They report their findings to another group: *The big rocks have more volume than the small ones.*

Language Transfers

Grammar Transfer
Phrasal verbs are not used in Korean and Spanish. Native speakers of these languages might confuse related phrasal verbs.
The matter takes out space.

Phonics Transfer
Vietnamese, Hmong, and Haitian Creole do not have the /yü/ sound as in *volume.*

CHAPTER 9

Observing Matter

 THE BIG IDEA What are some ways you can describe matter?

Chapter Preview Read the summaries at the end of each lesson. Predict what the lessons will be about.

▶ **Assess Prior Knowledge**

Before reading the chapter, create a **KWL** chart with students. Read the Big Idea question and then ask:

- How is matter described by its properties?
- What properties of matter can be measured?
- How do three states of matter differ in volume and shape?

Observing Matter

What We **K**now	What We **W**ant to Know	What We **L**earned
Matter makes up objects.	What is matter?	
You can use a ruler to measure things.	What properties of matter can be measured?	
Water is a liquid.		

Answers shown represent sample student responses.

Follow the **Instructional Plan** at right after assessing students' prior knowledge of chapter content.

RESOURCES and TECHNOLOGY

- **School to Home Activities**, pp. 73–78
- **Reading and Writing**, pp. 168–186
- **Assessment**, pp. 109–120
- **Presentation Toolkit CD-ROM**
- **PuzzleMaker CD-ROM**
- **www.macmillanmh.com**
- **e-Journal**

CHAPTER 9

Observing Matter

Lesson 1
Properties of Matter 362

Lesson 2
Measuring Matter . . 372

Lesson 3
Solids, Liquids, and Gases 382

The Big Idea
What are some ways you can describe matter?

360   Castle Geyser in Yellowstone National Park

Differentiated Instruction

Instructional Plan

Chapter Concept All matter has mass and volume.

EXTRA SUPPORT Students who need to know the basic properties of matter should cover all of **Lesson 1**, pp. 362–371, before continuing with the rest of the chapter.

ON LEVEL Students who know basic properties of matter can just cover the topic of elements in **Lesson 1**, pp. 368–369, and then go to **Lesson 2**, pp. 372–381, to explore mass, and contrast mass and weight.

ENRICHMENT Students who are ready to go further go to **Lesson 3**, pp. 382–391, to contrast the forms of matter from the understanding that matter is made up of tiny particles.

Key Vocabulary

matter
anything that takes up space and has mass (p. 364)

property
a characteristic of something (p. 365)

state of matter
a form of matter, such as solid, liquid, or gas (p. 384)

solid
matter that has a definite volume and shape (p. 384)

liquid
matter that has a definite volume but not a definite shape (p. 386)

gas
matter that does not have a definite volume or a definite shape (p. 387)

More Vocabulary

volume, p. 365

mass, p. 365

element, p. 368

metric system, p. 374

pan balance, p. 376

gravity, p. 378

weight, p. 378

361

Vocabulary Preview

■ Have a volunteer read the **Key Vocabulary** words aloud to the class. Ask students to find one or two words in the chapter by using the given page references. Add these words and their definitions to a class "Word Wall."

■ Encourage students to use the illustrated glossary in the student edition's reference section. Guide students to explore the **e-Glossary,** which offers audio pronunciations, definitions, and sentences using the vocabulary words

Science Leveled Readers

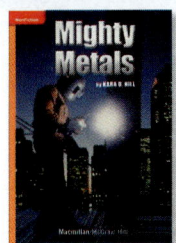

APPROACHING

Mighty Metals Examine matter in the form of metals such as gold and iron.
ISBN: 978-0-02-285877-3

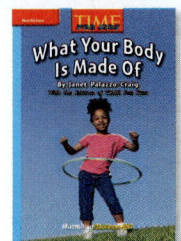

ON LEVEL

What Your Body Is Made Of Go on a journey to discover the human body.
ISBN: 978-0-02-285883-4

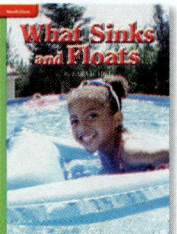

BEYOND

What Sinks and Floats Understand matter and its density.
ISBN: 978-0-02-285885-8

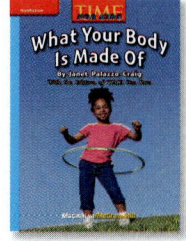

ELL

What Your Body Is Made Of Uses sheltered language of On-Level Reader.
ISBN: 978-0-02-283437-1

See teaching strategies in the Leveled Reader Teacher's Guide. To order, call 1-800-442-9685.

 Leveled Reader Database Online Readers, searchable by topic, reading level, and keywords

Plan Your Lesson

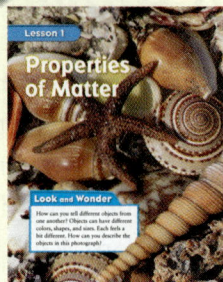

Lesson 1 Properties of Matter

Objective

- Define matter as anything that has mass and takes up space.
- Describe properties of matter and understand that properties can be used to identify matter.

Reading Skill Main Idea and Details

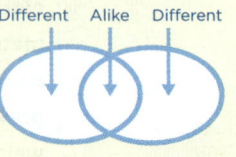

Different   Alike   Different

Graphic Organizer 1, p. TR 3

LOG ON **Professional Development** Look for **NSDL** to find recommended Science Background materials from the National Science Digital Library

FAST TRACK

Lesson Plan When time is short, follow the Fast Track and use the essential resources.

1 Introduce
Look and Wonder, p. 362

Resource Alternative Explore, p. 154

2 Teach
Discuss the Main Idea, p. 364
Discuss the Main Idea, p. 366

Resource Visual Literacy, p. 55

3 Close
Think, Talk, and Write, p. 369

Resource Assessment, p. 125

▶ Reading and Writing

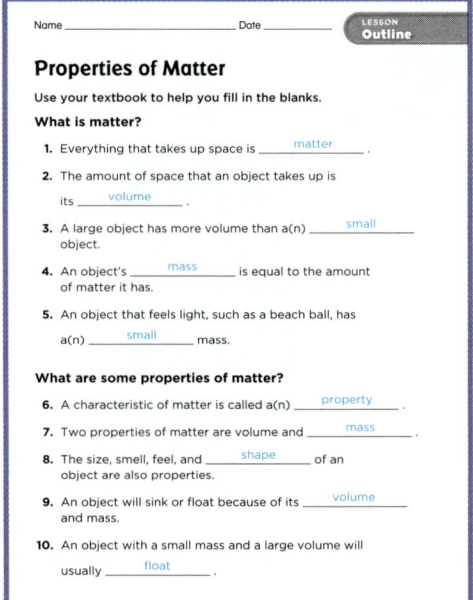

Outline, pp. 169–170
Also available as a student workbook

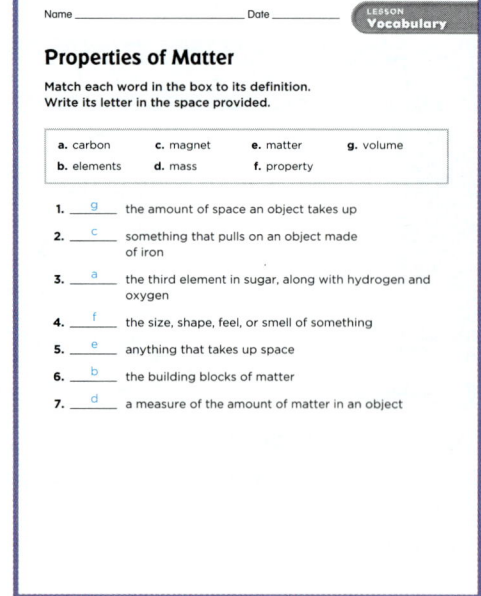

Vocabulary, p. 171
Also available as a student workbook

 **Visual Literacy**

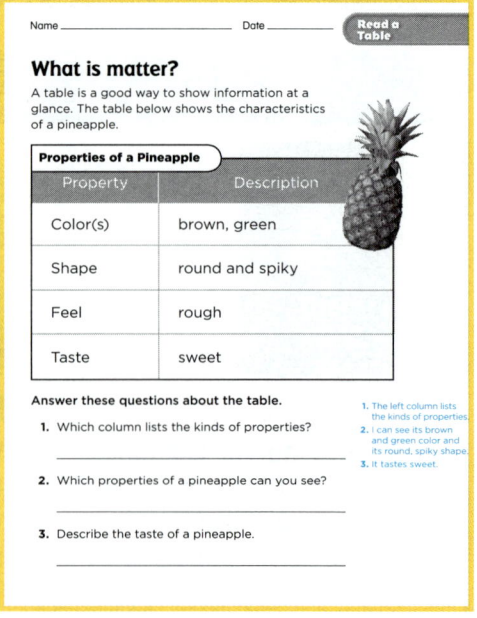

Read a Table, p. 55
Also available as a transparency

▶ Activity Lab Book

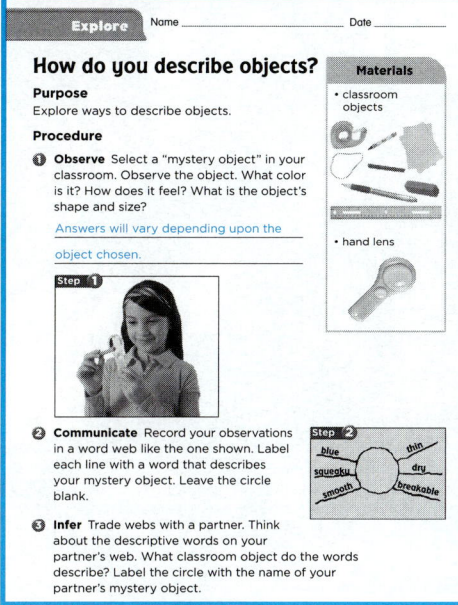

| Explore | Name _____ Date _____ |

How do you describe objects?

Purpose
Explore ways to describe objects.

Procedure

1. **Observe** Select a "mystery object" in your classroom. Observe the object. What color is it? How does it feel? What is the object's shape and size?

Answers will vary depending upon the object chosen.

Materials
• classroom objects

• hand lens

Step 1

2. **Communicate** Record your observations in a word web like the one shown. Label each line with a word that describes your mystery object. Leave the circle blank.

Step 2

3. **Infer** Trade webs with a partner. Think about the descriptive words on your partner's web. What classroom object do the words describe? Label the circle with the name of your partner's mystery object.

Explore, pp. 152–153
Also available as a student workbook

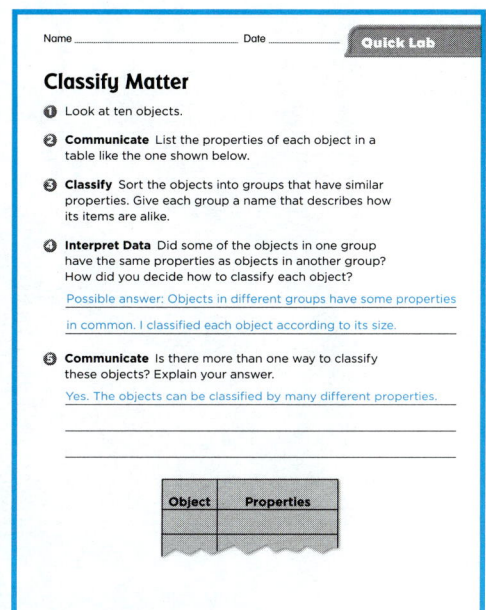

| Name _____ Date _____ | Quick Lab |

Classify Matter

1. Look at ten objects.

2. **Communicate** List the properties of each object in a table like the one shown below.

3. **Classify** Sort the objects into groups that have similar properties. Give each group a name that describes how its items are alike.

4. **Interpret Data** Did some of the objects in one group have the same properties as objects in another group? How did you decide how to classify each object?

Possible answer: Objects in different groups have some properties in common. I classified each object according to its size.

5. **Communicate** Is there more than one way to classify these objects? Explain your answer.

Yes. The objects can be classified by many different properties.

Object	Properties

Quick Lab, p. 155
Also available as a student workbook

▶ Assessment

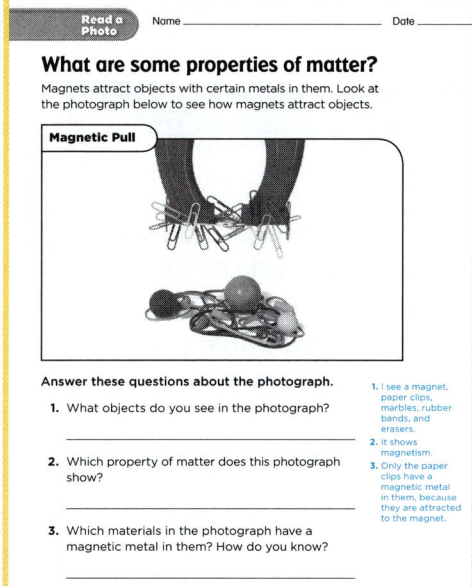

| Read a Photo | Name _____ Date _____ |

What are some properties of matter?

Magnets attract objects with certain metals in them. Look at the photograph below to see how magnets attract objects.

Magnetic Pull

Answer these questions about the photograph.

1. What objects do you see in the photograph?

2. Which property of matter does this photograph show?

3. Which materials in the photograph have a magnetic metal in them? How do you know?

1. I see a magnet, paper clips, marbles, rubber bands, and erasers.
2. It shows magnetism.
3. Only the paper clips have a magnetic metal in them, because they are attracted to the magnet.

Read a Photo, p. 56
Also available as a transparency

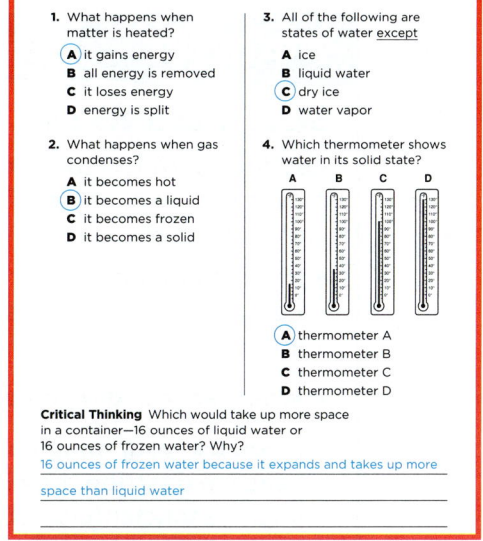

| Name _____ Date _____ | Lesson 1 Test |

Circle the letter of the best answer for each question.

1. What happens when matter is heated?
   - (A) it gains energy
   - B all energy is removed
   - C it loses energy
   - D energy is split

2. What happens when gas condenses?
   - A it becomes hot
   - (B) it becomes a liquid
   - C it becomes frozen
   - D it becomes a solid

3. All of the following are states of water <u>except</u>
   - A ice
   - B liquid water
   - (C) dry ice
   - D water vapor

4. Which thermometer shows water in its solid state?

   A   B   C   D

   - (A) thermometer A
   - B thermometer B
   - C thermometer C
   - D thermometer D

Critical Thinking Which would take up more space in a container—16 ounces of liquid water or 16 ounces of frozen water? Why?

16 ounces of frozen water because it expands and takes up more space than liquid water

FAST TRACK **Lesson Test, p. 125**
Also available as a student workbook

ADDITIONAL RESOURCES

Activity Flipchart for Science Centers and Workstations

p. 41

ELL Teacher's Guide

p. 103

Transparencies for Visual Literacy

pp. 55–56

School to Home Activities

pp. 73–74

Key Concept Cards

55–56

Vocabulary Cards

133–137

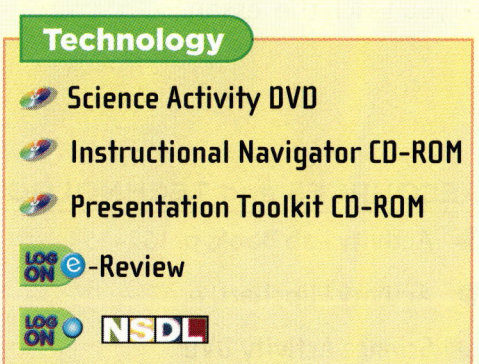

Technology

- 💿 **Science Activity DVD**
- 💿 **Instructional Navigator CD-ROM**
- 💿 **Presentation Toolkit CD-ROM**
- 🔵 **e-Review**
- 🔵 **NSDL**

Lesson 1 Properties of Matter

Objectives
- Define matter as anything that has mass and takes up space.
- Describe properties of matter and understand that properties can be used to identify matter.

1 Introduce

▶ **Assess Prior Knowledge**

Have students name their five senses. Then have them describe what they can find out about something by using their senses. Ask:

- **Can you use all your senses to describe every object? Give an example to support your answer.** Possible answer: No; the senses of touch, smell, taste, and sight can be used to describe an orange, but hearing would not describe it.

- **How do you describe a pencil if your eyes are closed?** Possible answer: You can use touch to describe the shape and size of the pencil as well as how smooth it is.

FAST TRACK Look and Wonder

Invite students to share their responses to the Look and Wonder statement and question:

- **How can you describe the objects in this photograph?** Possible answer: They can be described by their shapes, their colors, their sizes, their textures.

Write ideas on the board and note any misconceptions that students may have. Address these misconceptions as you teach the lesson.

RESOURCES and TECHNOLOGY
▶ Activity Lab Book, p. 152–154
▶ Activity Flipchart, p. 41
💿 Science Activity DVD

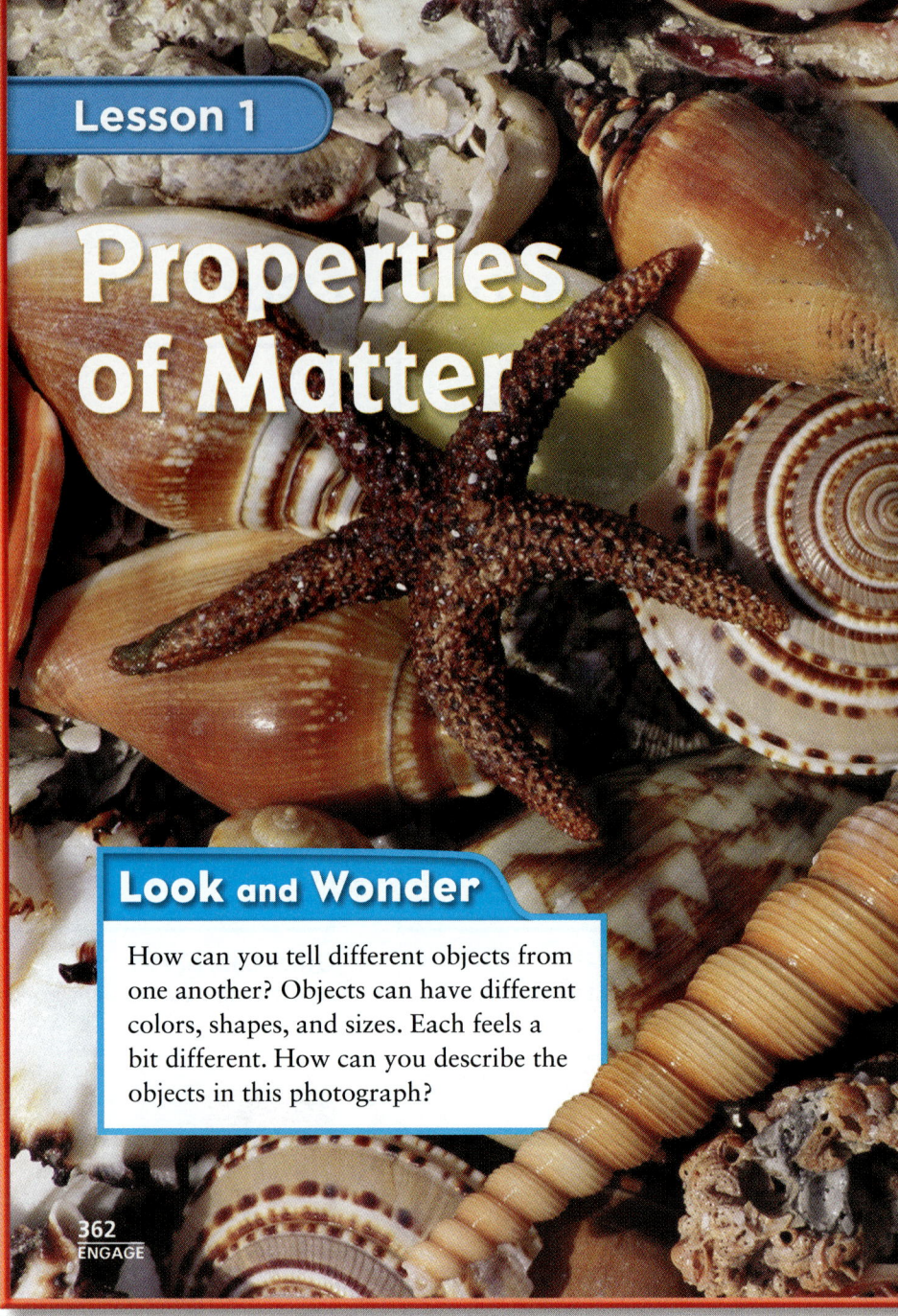

Lesson 1

Properties of Matter

Look and Wonder

How can you tell different objects from one another? Objects can have different colors, shapes, and sizes. Each feels a bit different. How can you describe the objects in this photograph?

362
ENGAGE

Warm Up

Start with a Visual

Cut out colorful pictures from magazines. Have each student draw a picture of objects from one of the magazine pictures and examine them. Ask:

- **If you chose one object in your picture, how would you describe it?** Possible answer: I would describe its color and shape.

- **What sense(s) did you use to make your description?** sight

- **If you use your imagination, can you think of another sense to describe the picture?** Possible answer: I could imagine touching the object and sensing that it was smooth or rough.

Explore   Inquiry Activity

How do you describe objects?

Purpose
Explore ways to describe objects.

Procedure

1 **Observe** Select a "mystery object" in your classroom. Observe the object. What color is it? How does it feel? What is the object's shape and size?

2 **Communicate** Record your observations in a word web like the one shown. Label each line with a word that describes your mystery object. Leave the circle blank.

3 **Infer** Trade webs with a partner. Think about the descriptive words on your partner's web. What classroom object do the words describe? Label the circle with the name of your partner's mystery object.

Draw Conclusions

4 Were you able to guess your partner's mystery object? Was your classmate able to guess your mystery object?

5 What helped you most in figuring out your partner's object?

Explore More

Experiment How might your web be different if you were blindfolded and could only touch the mystery object? Try it to find out.

Materials

classroom objects

hand lens

Step 1

Step 2

blue   thin
squeaky   dry
smooth   breakable

363
EXPLORE

Explore   👥 pairs   🕐 20 minutes

Plan Ahead Prepare a plan for students who are challenged using one or more senses. Provide a way for them to describe the object they choose.

Purpose This activity will help students describe observable properties of an object. They will then infer the identity of another object from its listed properties.

Structured Inquiry

1 **Observe** Tell students to avoid staring at the object they choose while describing it. The descriptions should be the only clues about the identity of the object.

2 **Communicate** Allow students to vary the number of lines used within a prescribed range.

3 **Infer** Have students continue to infer objects until they guess which one was chosen.

4 Students will eventually be able to infer which object was chosen.

5 Possible answers: its color, its size

Guided Inquiry   Explore More

Experiment Instead of color and other descriptions of visual appearance, properties related to texture should be listed. For example, instead of saying the object is shiny, the description will say it feels smooth. Some aspects of the descriptions, such as size and shape, will be the same.

Open Inquiry

Have students relate how their descriptions would change if the object were in a box and they could neither see it nor feel it. Ask:

How could you find out about the object?

Alternative Explore

What are the properties of an object?

Have one student secretly choose an object. Have another student ask questions about it. Questions should relate to observable properties, and the student should be able to answer them with "yes" or "no." Allow the student who is guessing to ask up to 10 questions. Whenever the student thinks he or she knows what the object is, he or she should guess. Have students take turns choosing objects and asking questions.

Alternative Explore   Name _____ Date _____

What are the properties of an object?

Purpose
In this activity, you will describe the properties of an object so that someone else can infer what it is.

Procedure

1 Work with a partner. You and your partner will take turns secretly choosing an object in the classroom.

2 Choose an object and have your partner ask up to 5 questions about its properties. Each question must only have a "yes" or "no" answer.

3 **Record Data** Use the table below to help you.

Guess From Properties		
Property	Question	Yes or No

4 **Communicate** What object did your partner choose?
Answers will vary depending upon the object chosen.

5 **Interpret Data** How did asking questions about the properties of the object help you to identify it?
The answers to the questions helped describe the object.

Activity Lab Book, p. 154

2 Teach

Read and Learn

Main Idea Have students observe the pictures in the lesson. Ask them what they think they will learn in the lesson based on their observations.

Vocabulary Have students read aloud the sentences containing the vocabulary words. Show the students a glass that is half-full of water. Ask students to use each vocabulary word correctly in a sentence related to the glass and its contents.

Reading Skill Main Idea and Details
Graphic Organizer 1 Have students fill in a Main Idea and Details graphic organizer as they read each two pages of the lesson. They can use the Quick Check questions to find the main idea and details.

What is matter?

FAST TRACK **Discuss the Main Idea**

Have student volunteers find the topic sentence in each paragraph on pages 364–365 and change that statement into a question. Have other students answer each question. Discuss the answers with the class.

▶ **Address Misconceptions**

A common misconception is that matter with greater volume always has greater mass. Mass is a property of a material, and materials vary greatly in the mass they have in a certain volume of material.

Have students use a balance to find the mass of 100 mL of various materials.

RESOURCES and TECHNOLOGY
- Reading and Writing, pp. 169–171
- Visual Literacy, p. 55
- PuzzleMaker CD-ROM
- Presentation Toolkit CD-ROM
- e-Glossary

Read and Learn

What is matter?

▶ **Main Idea**
Matter is anything that has volume and mass. You can use properties to describe and identify matter.

▶ **Vocabulary**
matter, p. 364
volume, p. 365
mass, p. 365
property, p. 365
element, p. 368

LOG ON **e-Glossary**
at www.macmillanmh.com

▶ **Reading Skill** ✓
Main Idea and Details

Look around. Do you see things with different colors, sizes, and shapes? Things differ in the way they look, feel, sound, and smell. All the things around you are alike in one way, however. All are made of matter (MAT•uhr).

Matter is anything that takes up space. You are matter. This book is matter. Even the air you breathe is matter. All of these things take up space.

What can you see, hear, and touch at the beach?

364
EXPLAIN

Science Background

What Are Physical Properties? The properties listed in the text are all physical properties, which means that they can be observed without changing the identity of the material. Color, shape, and size are physical properties. Even if physical properties change, the change does not affect the composition of the material. Matter also has chemical properties, which are those that are observed by changing the identity of the material. The ability of a material to burn or to rust are examples of chemical properties.

See **Science Yellow Pages** in the Teacher Resources section for background information.

LOG ON **Professional Development** For more Science Background and resources from **NSDL** visit **www.macmillanmh.com/nsdl/**

Volume

Volume (VOL•yewm) describes how much space an object takes up. It tells how big or small an object is. This beach ball takes up more space than this bowling ball. The beach ball has more volume.

Mass

All objects have mass. **Mass** is a measure of the amount of matter in an object. An object with a large mass feels heavy. An object with a small mass feels light. This bowling ball feels heavier than this beach ball. This is because the bowling ball contains more matter. The bowling ball has more mass.

Volume and mass are properties (PROP•uhr•teez) of matter. A **property** is a characteristic of something. The way an object looks, tastes, smells, sounds, and feels are other properties that you can observe.

▲ This beach ball has more volume but less mass than this bowling ball.

Properties of a Pineapple

Property	Description
Color(s)	brown, green
Shape	round and spiky
Feel	rough
Taste	sweet

Read a Table

How does a pineapple taste?

Clue: Headings help you find information.

 **Quick Check**

Main Idea and Details What are two properties of all types of matter?

Critical Thinking Why is sound not matter?

365
EXPLAIN

Differentiated Instruction

Leveled Activities

EXTRA SUPPORT   Ask students to cut pictures of objects out of magazines or catalogs and place them facedown. Have them choose two pictures and predict which of the two objects would have more mass. Then have them predict which object would have more volume.

ENRICHMENT   Have students write a paragraph, with examples, explaining that, for the same material, an increase in mass results in an increase in volume. For example, a pond and a puddle both contain water. The pond contains a greater volume of water, so it contains a greater mass of water.

ENGAGE   EXPLORE   **EXPLAIN**   EVALUATE   EXTEND

▶ **Develop Vocabulary**

matter *Scientific vs. Common Use* Remind students of a common use of *matter* as a verb: being of importance, as in saying that something "matters." Relate this to *matter* in the lesson.

volume *Scientific vs. Common Use* Tell students that *volume* can refer to a book, especially one belonging to a set, such as an encyclopedia. Point out that the larger the volume of a book, the more space it takes up.

mass *Word Origin* The word *mass* comes from the Latin word **massa,** which means "kneaded dough, lump." A lump is a relatively large mass of something.

property *Scientific vs. Common Use* Point out that a common use of *property* means "something owned." Ask students to list things they consider to be their property, such as a jacket or a bicycle.

▶**Use the Visuals**

Ask:

■ **What might you buy in a store that has more mass in a small bag than something else has in a large bag?** Possible answer: a small bag of canned foods and a large bag of cereal

Read a Table

Answer sweet

✔ **Quick Check Answers**

• **Main Idea and Details** Mass and volume are properties of matter.

• **Critical Thinking** It has no mass or volume.

What are some properties of matter?

FAST TRACK ## Discuss the Main Idea

Have students read the definition of *property* on page 365 and give several examples. Tell students that some properties of matter involve how the matter behaves. Have them observe what happens when you bring a magnet near a piece of floatable wood and a metal paper clip. Then have them observe what happens when you drop these objects into a glass of water. Ask:

- **What is one property the paper clip has that wood does not?** Possible answer: It is attracted to a magnet.

- **What is one property that wood has that the paper clip does not?** Possible answer: Wood can float in water.

▶ Use the Visuals

Refer students to the visual on p. 366. Ask:

- **Why does a metal anchor sink and a life preserver float?** Possible answer: The life preserver has less mass per volume compared to the metal anchor.

▶ Address Misconceptions

Students may think that all metals are attracted to magnets. Metals and alloys such as stainless steel, tin, and gold are not attracted. Metals such as aluminum and copper are only weakly attracted. Steel, iron, cobalt, and nickel are attracted.

> **FACT** Only some metals are attracted to magnets.

RESOURCES and TECHNOLOGY

▶ **Activity Lab Book,** p. 155

💿 **Presentation Toolkit CD-ROM**

What are some properties of matter?

The world is full of many kinds of matter. We use properties to tell them apart. An object may be hot or cold. It could feel smooth or rough, wet or dry. Here are some properties that help us describe and identify matter.

Sinking and Floating

Some matter sinks in water. Some matter floats. For example, a rock sinks in water and an apple floats. Metal objects usually sink, while wooden objects often float. Objects sink or float because of their mass and volume. Objects with a lot of mass and little volume tend to sink. Objects with little mass and a lot of volume tend to float.

A life preserver floats on water. ▼

An anchor sinks in water.

366
EXPLAIN

Differentiated Instruction

Leveled Questions

EXTRA SUPPORT   Will a feather sink or float in water? Why? float; it has a relatively small mass per volume

ENRICHMENT   Imagine a coin and a piece of paper in the Sun. Which object might be hot if you touch it? Why? The coin might be hot because it is metal, which conducts heat from the Sun.

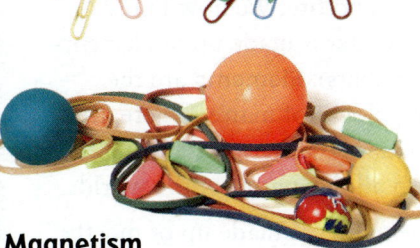

Magnetism

Magnets have a special property. Magnets pull on, or *attract*, certain metals, such as iron. They do not attract wood, plastic, or water. Put a magnet near an object made of iron. What happens? The magnet pulls on the object, and then the object "sticks" to the magnet.

Conducting Heat

Some matter *conducts* heat. This means that some kinds of matter let heat move through them easily. For example, heat moves easily through metals such as iron and copper. This is why these materials make good cooking pots. Heat moves from the stove through the metal pot. The pot gets warm. Wood does not heat up quickly. This is why wood makes good pot handles and cooking spoons.

FACT Only some metals are attracted to magnets.

≡Quick Lab

Classify Matter

① Look at ten objects.

② **Communicate** List the properties of each object in a table like the one shown.

③ **Classify** Sort the objects into groups that have similar properties. Give each group a name that describes how its items are alike.

④ **Interpret Data** Did some of the objects in one group have the same properties as objects in another group? How did you decide how to classify each object?

⑤ **Communicate** Is there more than one way to classify these objects? Explain your answer.

Object	Properties

✓ Quick Check

Main Idea and Details Name three properties of matter.

Critical Thinking What properties of plastic make it useful as a bowl but not as a cooking pan?

≡Quick Lab
small groups   15 minutes

Objective Observe properties of objects, and use these properties to classify the objects.

Materials ten objects

① Have objects available for students to use, but allow them to use items of their own choosing, also.

② Advise students to allow adequate room in their tables for listing several properties.

③ Names will vary depending on the properties used for classification.

④ Some objects do share properties with objects in other groups.

⑤ Yes; objects have many properties, and not all of them are used in this classification.

▶ Explore the Main Idea

ACTIVITY In the classroom, have students use magnets to find out which objects, or parts of objects, are attracted to magnets. Use students' results to make a class list of these objects. Emphasize that the objects attracted to the magnet contain the element iron. Explain to students that they should not try their magnets near electronic devices such as computers or cell phones.

✓ Quick Check Answers

- **Main Idea and Details** Possible answer: magnetism, volume, mass

- **Critical Thinking** Plastic does not dissolve and is hard enough to make into a shape. Plastic will melt when heated to a certain temperature.

ELL Support

Classify Write the following words on the board as headings: *Plastic*, *Metal*, and *Wood*. Have students repeat the words after you. Discuss the meanings of the words. Then have students work in small groups to name classroom objects in the three different categories. Write the names of the objects on the board under each heading: *Plastic*, *Metal*, or *Wood*.

BEGINNING Students can point to or name an object or picture of an object in one of the three categories.

INTERMEDIATE Students can use phrases and short sentences to describe plastic, metal, or wooden objects.

ADVANCED Students can describe plastic, metal, or wooden objects using complete sentences.

Lesson 1   **367**

What is matter made of?

▶ Discuss the Main Idea

As a demonstration, cut a piece of aluminum foil into smaller and smaller pieces. Have a student volunteer look at the pieces using a hand lens and report on what he or she sees. Ask:

- Is the foil still aluminum after it is cut? How do you know? Yes; the properties are still the same.

▶ Develop Vocabulary

element *Scientific vs. Common Use* Point out that a common use of the word *element* means "a part of." Ask students to relate this use to how the word is used in the lesson. An element is a part of other materials.

▶ Use the Visuals

Refer students to the examples of different elements in the visual on page 368. Point out that neon is the gas inside the tube. The tube itself is made of some other material. Ask:

- What property is shown by all the elements except neon? Possible answer: They are solid.

- How do you think the properties of these elements affect how they are used? Possible answer: All these elements except neon are hard and might be used to build things.

✔ Quick Check Answers

- **Main Idea and Details** Possible answer: because all materials on Earth are made up of elements

- **Critical Thinking** An iron nail is made mostly of one element, while water is made of two elements. An iron nail and water have different properties.

Elements

A few elements are shown here.

iron

silver

gold

aluminum

carbon

neon

What is matter made of?

People once thought that all matter was made up of combinations of water, air, earth, and fire. We now know that all matter is made up of elements (EL•uh•muhnts). **Elements** are the building blocks of matter. There are more than 100 different elements. They make up all the matter in the world.

Some matter is made up of mostly one element. An iron nail contains mostly the element iron. Aluminum (uh•LEW•muh•nuhm) foil contains mostly the element aluminum.

Most matter on Earth is made up of more than one element. Water is made up of the elements hydrogen and oxygen. Sugar is made up of hydrogen, oxygen, and a third element called carbon. Elements join in different ways and in different amounts to form everything in our world.

✔ Quick Check

Main Idea and Details Why are elements called the building blocks of matter?

Critical Thinking How is an iron nail different from water?

Homework Activity

List Physical Properties

Have students choose two objects in their homes and list all the physical properties they can for the objects. Have them draw a Venn diagram that shows the properties the objects have in common and the properties that vary between the objects.

Lesson Review

Visual Summary

Matter is anything that has **volume** and **mass**.

Matter can be described and identified by its **properties**.

All matter is made up of **elements**.

Make a FOLDABLES Study Guide

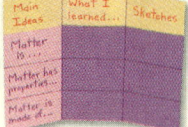

Make a Trifold Book. Use it to summarize what you learned about matter and its properties.

Think, Talk, and Write

1 **Main Idea** What are some properties of matter?

2 **Vocabulary** What is matter?

3 **Main Idea and Details** Choose two objects. List all the properties you can to describe each one.

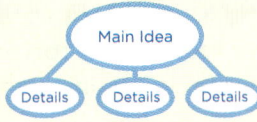

4 **Critical Thinking** What property of glass makes it a good material for windows?

5 **Test Prep** What are the building blocks of matter?
A liquids
B elements
C wood
D water

Writing Link

Writing That Describes
Suppose you brought your favorite toy to school and lost it. Write a notice to hang on your classroom bulletin board. What properties of the toy will you describe?

Math Link

Make a Table
Collect five small objects. Predict which ones will sink and which ones will float. Then put the objects in a tank of water. Record your results in a table.

 e-Review Summaries and quizzes online at www.macmillanmh.com

369
EVALUATE

Formative Assessment

Approaching Give students a photograph and a property. Have them name objects in the photograph that have a property they choose.

On-Level Have students prepare a table that lists properties discussed in the lesson in one column and that defines the property, in their own words, in a second column.

Challenge Have students work in pairs. Have one student name a property. Have the other student name two samples of matter that have that property.

Key Concept Cards For student intervention, see the prescribed routine on **Key Concept Cards 55 and 56.**

---

3 Close

Lesson Review

▶ **Visual Summary**

Have students save their lesson Study Guides. They will use them at the end of the chapter for review.

▶ **Make a FOLDABLES Study Guide**

See pp. R27–R28 for instructions.

FAST TRACK **Think, Talk, and Write**

1 **Main Idea** Possible answers: color, shape, mass, volume

2 **Vocabulary** anything that has mass and takes up space

3 **Main Idea and Details**

Sample answer:

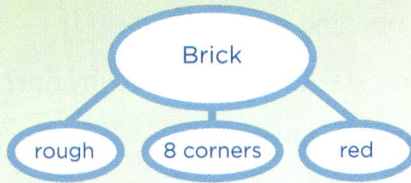

4 **Critical Thinking** Possible answer: Glass is clear.

5 **Test Prep** B

Writing Link Students' descriptions should contain enough properties to clearly identify the object and prevent confusion about the lost object and other objects in the school.

Math Link Students' tables should have three columns, one each for the objects, predictions, and results, and a row for the head and each object. The graph should contain one bar for the number of objects that float and one bar for those that sink.

RESOURCES and TECHNOLOGY

▶ **Reading and Writing,** p. 172

▶ **School to Home Activities,** pp. 73–74

▶ **e-Review** Narrated Summary and Quiz

▶ **Progress Reporter** Assessments

Reading in Science

Objective
- Identify main ideas and their supporting details.

Meet Neil deGrasse Tyson

Genre: Nonfiction

- **Who is this article about?** Neil deGrasse Tyson
- **What job does he have?** He's an astrophysicist, a scientist who studies how the universe works.

Before Reading

Skim the first paragraph with students. Ask:

- **What does an astrophysicist study?** An astrophysicist studies the universe and how it works.

Direct students to the opening question Ask:

- **Do you think you are "star dust"? Why or why not?**

Write students' responses on the board.

Meet
Neil deGrasse TYSON

Did you know that you are "star dust"? Neil deGrasse Tyson can tell you what that means. He is a scientist at the American Museum of Natural History in New York. He studies how the universe works.

Your body is full of hydrogen, carbon, and many other elements. All these elements were first formed in stars a long time ago. How did these elements make their way from the stars to your body?

Most elements form inside the dense and fiery centers of stars. Hydrogen combines to form all of the other elements in these conditions. Throughout their lives, stars scatter elements into space. Over millions of years, these elements combine to form new stars, planets, or even living things like you!

Neil is an astrophysicist. An astrophysicist is a scientist who studies how the universe works. ▶

370
EXTEND

ELL Support

Sequence of Events Draw a sequence chart on the board. Ask: **What elements are in your body?** hydrogen, carbon, etc. **Find them on the page. Where do they come from?** stars Draw stars in the first box. Elicit what should be in the other two boxes. Help students describe the sequence using *first, next,* and *last.* Teach students how to use context clues to understand terms such as *make their way.*

BEGINNING   Students can use gestures and short phrases to answer questions about the chart.

INTERMEDIATE   Students can use short sentences and phrases to explain how elements in our bodies come from stars.

ADVANCED   Students can use their own words to tell why we are "star dust" and then explain the sequence.

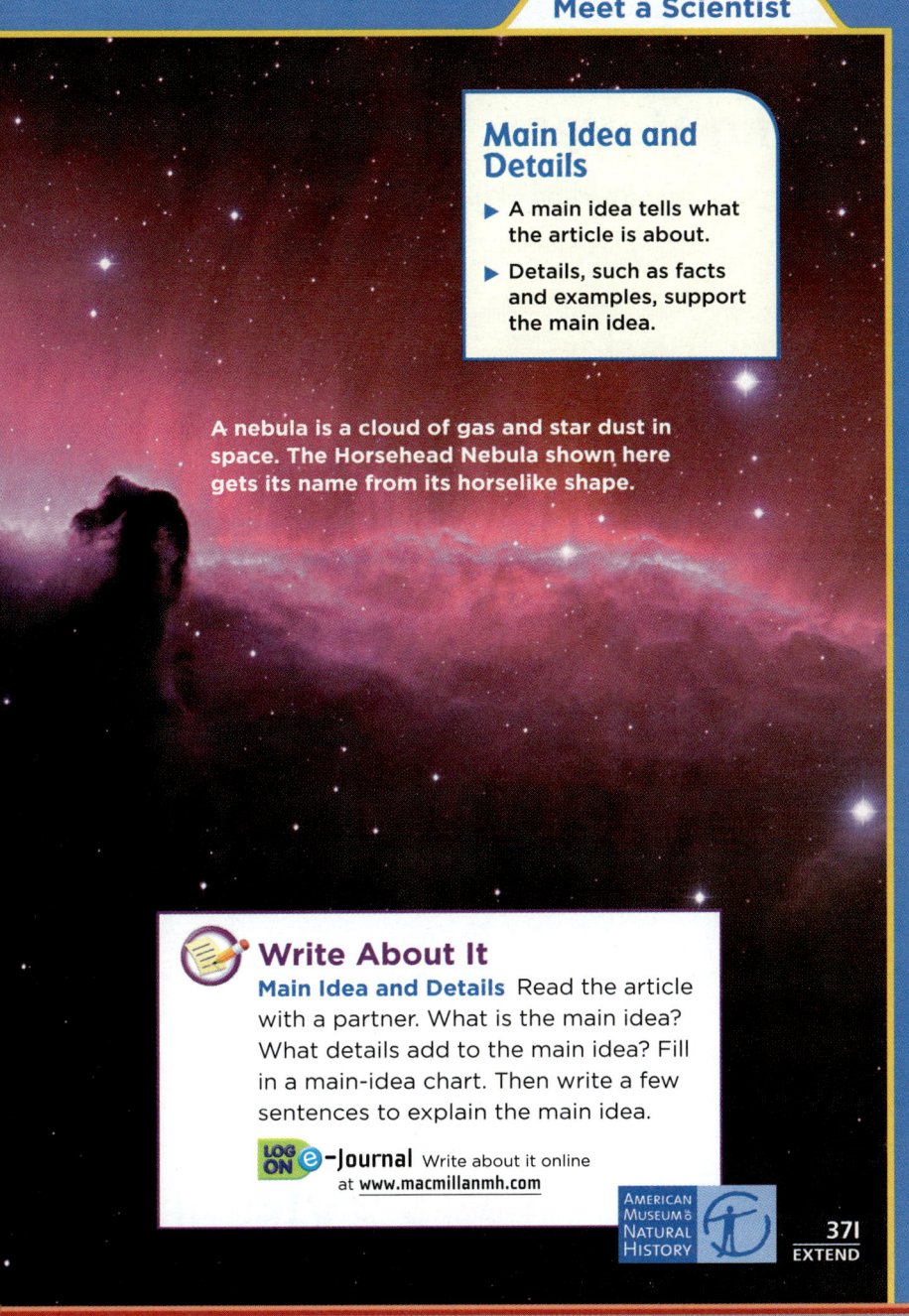

Main Idea and Details

▶ A main idea tells what the article is about.

▶ Details, such as facts and examples, support the main idea.

A nebula is a cloud of gas and star dust in space. The Horsehead Nebula shown here gets its name from its horselike shape.

Write About It

Main Idea and Details Read the article with a partner. What is the main idea? What details add to the main idea? Fill in a main-idea chart. Then write a few sentences to explain the main idea.

LOG ON **e-Journal** Write about it online at **www.macmillanmh.com**

AMERICAN MUSEUM OF NATURAL HISTORY

371 EXTEND

During Reading

Read the article together. Encourage students to look for details that tell about the elements we are made of. Ask questions that relate to the elements.

■ **Where do elements form?** in the center of stars

■ **What elements are found in your body?** Answers will vary but should include hydrogen and carbon.

Have students think about the main idea of each paragraph as they read. Ask:

■ **What are some important things that we learned by reading this paragraph?**

■ **Can you say that you are star dust now?** Yes, because the elements in my body formed in stars.

After Reading

Display Graphic Organizer 1. Explain to students that knowing what the most important information, or main idea, is will help them better understand what they read. Have students look at the second paragraph. Ask:

■ **What's the main idea of this paragraph?** Write, *Elements were first formed in stars.* Then have students look for details that support this idea. Have students use words and phrases from the story to complete the Graphic Organizer. Then have students each write a question that can be answered from the article.

Write About It

Main Idea: Our bodies are full of elements first formed in stars. Details: Elements of hydrogen, carbon, and calcium, among others, were first formed in the stars a long time ago. Over millions of years, elements combine to form new stars, planets, or living things.

Integrate Reading

Word Web

Write the word *elements* on the board. Ask students which elements they read about in this article. Use their responses to build a branch of a word web. Then have students look through the article to find other words and phrases that relate to the word *elements*. Ask volunteers to write them on the web. When you're finished with the word web, use it to have a discussion about what students have learned about elements by reading the article.

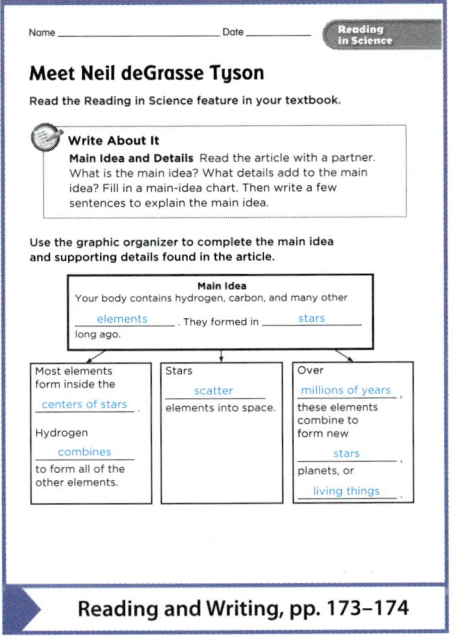

Reading and Writing, pp. 173–174

P-TR26 Writing Rubric

Plan Your Lesson

Lesson 2 Measuring Matter

Objective

- Measure matter using tools that record standard units.
- Compare and contrast weight and mass.

Reading Skill Summarize

Summary

Graphic Organizer 5, p. TR7

 LOG ON **Professional Development** Look for **NSDL** to find recommended Science Background materials from the National Science Digital Library

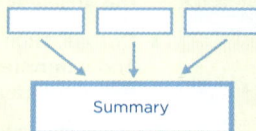

FAST TRACK

Lesson Plan When time is short, follow the Fast Track and use the essential resources.

1 Introduce

Look and Wonder, p. 372

Resource Alternative Explore, p. 158

2 Teach

Discuss the Main Idea, p. 374

Discuss the Main Idea, p. 376

Discuss the Main Idea, p. 378

Resource Visual Literacy, p. 57

3 Close

Think, Talk, and Write, p. 379

Resource Assessment, p. 113

▶ Reading and Writing

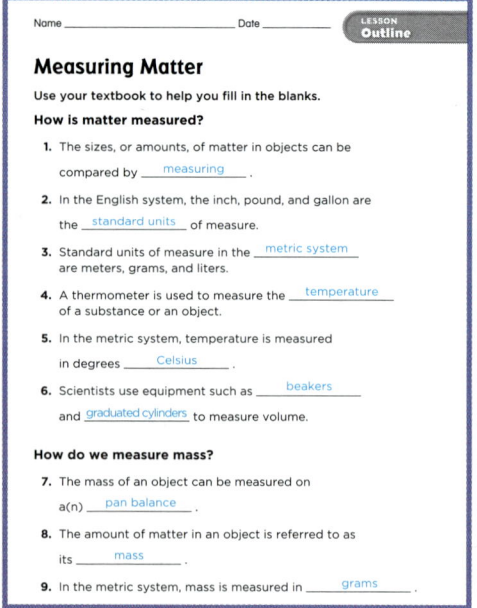

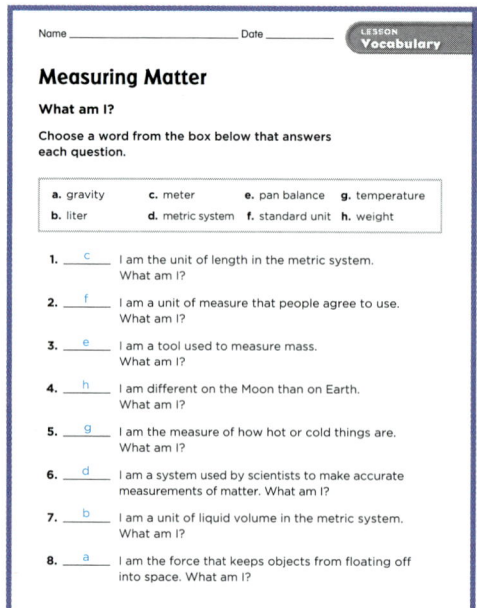

Outline, pp. 175–176
Also available as a student workbook

Vocabulary, p. 177
Also available as a student workbook

▶ Visual Literacy

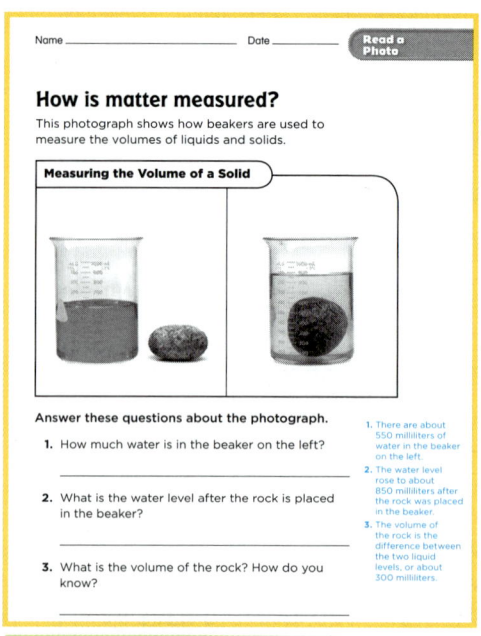

FAST TRACK **Read a Photo, p. 57**
Also available as a transparency

▶ Activity Lab Book

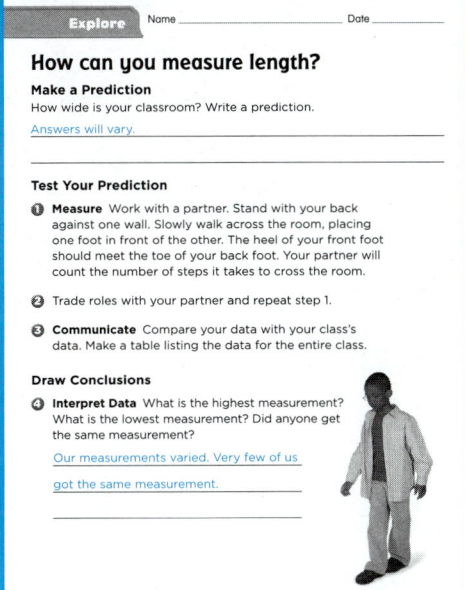

Explore   Name _____ Date _____

How can you measure length?

Make a Prediction

How wide is your classroom? Write a prediction.

Answers will vary.

Test Your Prediction

❶ **Measure** Work with a partner. Stand with your back against one wall. Slowly walk across the room, placing one foot in front of the other. The heel of your front foot should meet the toe of your back foot. Your partner will count the number of steps it takes to cross the room.

❷ Trade roles with your partner and repeat step 1.

❸ **Communicate** Compare your data with your class's data. Make a table listing the data for the entire class.

Draw Conclusions

❹ **Interpret Data** What is the highest measurement? What is the lowest measurement? Did anyone get the same measurement?

Our measurements varied. Very few of us

got the same measurement.

Explore, pp. 156–157
Also available as a student workbook

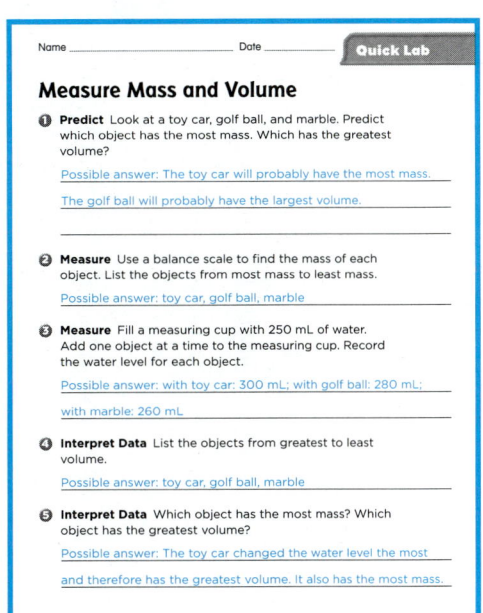

Name _____ Date _____   **Quick Lab**

Measure Mass and Volume

❶ **Predict** Look at a toy car, golf ball, and marble. Predict which object has the most mass. Which has the greatest volume?

Possible answer: The toy car will probably have the most mass.

The golf ball will probably have the largest volume.

❷ **Measure** Use a balance scale to find the mass of each object. List the objects from most mass to least mass.

Possible answer: toy car, golf ball, marble

❸ **Measure** Fill a measuring cup with 250 mL of water. Add one object at a time to the measuring cup. Record the water level for each object.

Possible answer: with toy car: 300 mL; with golf ball: 280 mL;

with marble: 260 mL

❹ **Interpret Data** List the objects from greatest to least volume.

Possible answer: toy car, golf ball, marble

❺ **Interpret Data** Which object has the most mass? Which object has the greatest volume?

Possible answer: The toy car changed the water level the most

and therefore has the greatest volume. It also has the most mass.

Quick Lab, p. 159
Also available as a student workbook

▶ Assessment

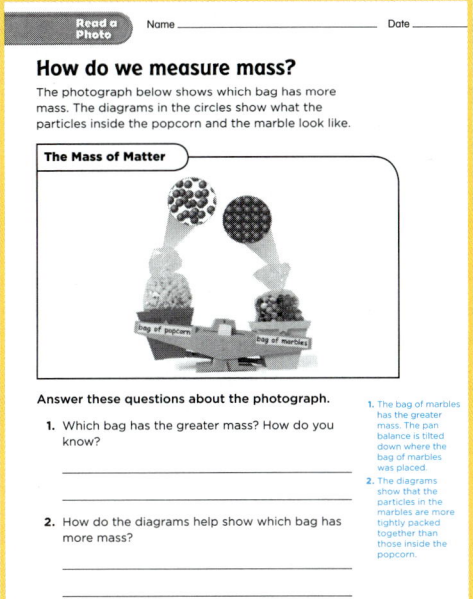

Read a Photo   Name _____ Date _____

How do we measure mass?

The photograph below shows which bag has more mass. The diagrams in the circles show what the particles inside the popcorn and the marble look like.

The Mass of Matter

Answer these questions about the photograph.

1. Which bag has the greater mass? How do you know?

_____

_____

2. How do the diagrams help show which bag has more mass?

_____

_____

1. The bag of marbles has the greater mass. The pan balance is tilted down where the bag of marbles was placed.

2. The diagrams show that the particles in the marbles are more tightly packed together than those inside the popcorn.

Read a Photo, p. 58
Also available as a transparency

Name _____ Date _____   **Lesson 2 Test**

Circle the letter of the best answer for each question.

1. What unit of measure do people agree to use?

   A metric system
   B Fahrenheit
   C Celsius
   Ⓓ standard unit

2. In the metric system, what are units of length based on?

   Ⓐ meter
   B liter
   C gram
   D inch

3. Which instrument will best measure mass?

   A graduated cylinder
   B thermometer
   Ⓒ pan balance
   D tape measure

4. In the metric system, what unit is used to measure a liquid's volume?

   A meter
   Ⓑ liter
   C gram
   D kilo

Critical Thinking

Which of the three units of measure—meter, centimeter, or millimeter—would be best for measuring the football field? Why?

The meter would be best for measuring the football field because

it is the longest unit and would provide the most accurate

measurement.

FAST TRACK   **Lesson Test, p. 113**
Also available as a student workbook

ADDITIONAL RESOURCES

p. 42

p.104

pp. 57–58

pp. 75–76

57–58

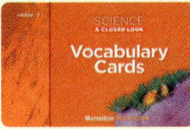

138–141

Technology

🔹 **Science Activity DVD**

🔹 **Instructional Navigator CD-ROM**

🔹 **Presentation Toolkit CD-ROM**

LOG ON **e-Review**

LOG ON **NSDL**

Lesson 2 Measuring Matter

Objectives
- Measure matter using tools that record standard units.
- Compare and contrast weight and mass.

1 Introduce

▶ Assess Prior Knowledge

Point out to students that certain properties of matter can be measured. Ask:

- **What properties can you measure?**
  Possible answers: height, weight

- **What tools do you use to measure properties?**
  Possible answers: ruler, scale

Look and Wonder

Invite students to share their responses to the Look and Wonder statement and question:

- **Why is it important to know how to measure matter?** Possible answers: Measured amounts can be compared; you might need to know how much of something you have.

Write ideas on the board and note any misconceptions that students may have. Address these misconceptions as you teach the lesson.

RESOURCES and TECHNOLOGY
- **Activity Lab Book**, pp. 156–158
- **Activity Flipchart**, p. 42
- **Science Activity DVD**

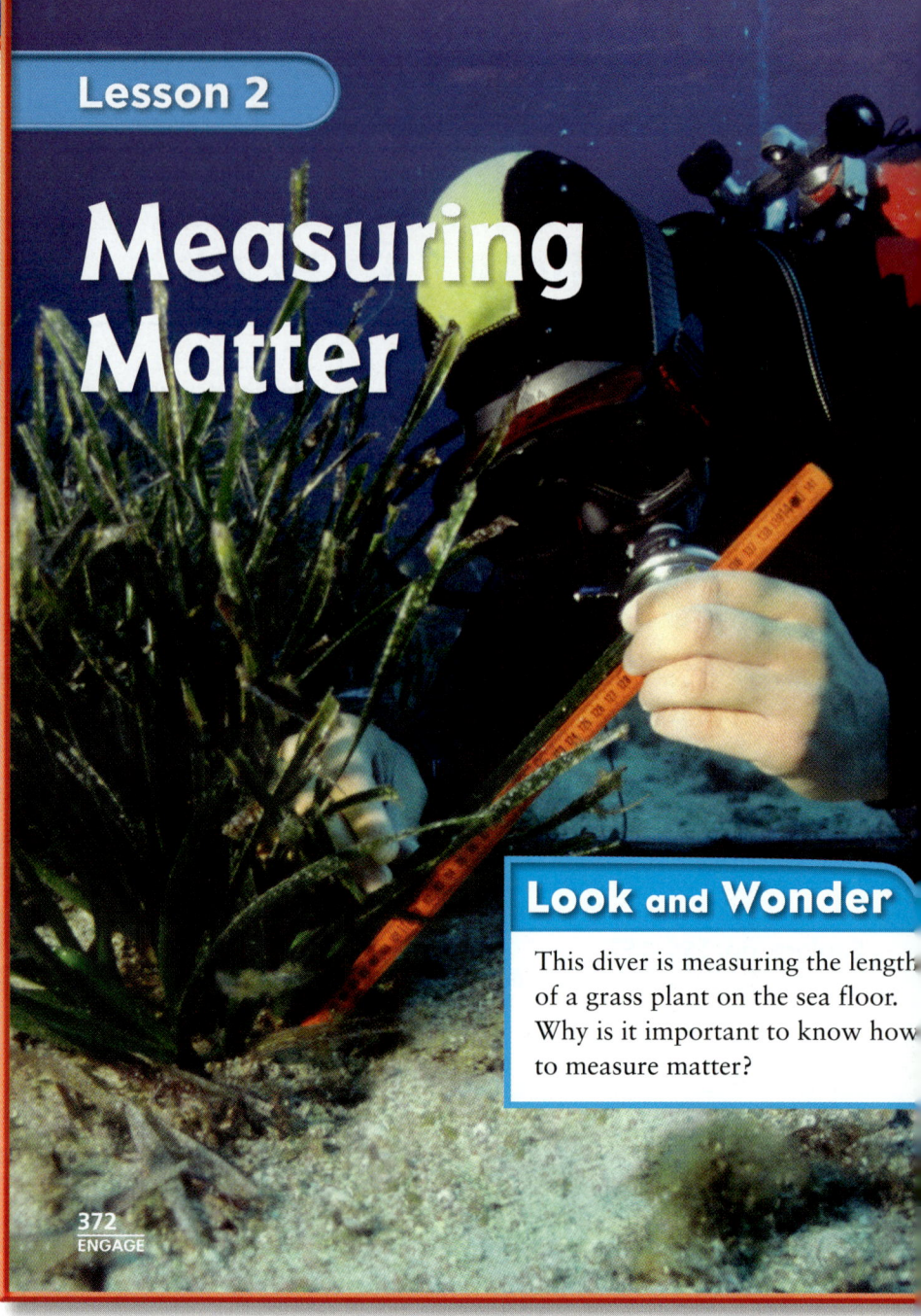

Lesson 2

Measuring Matter

Look and Wonder

This diver is measuring the length of a grass plant on the sea floor. Why is it important to know how to measure matter?

372
ENGAGE

Warm Up

Start with a Demonstration

Show students a road map and a piece of paper. Point out that the property of length can be measured, no matter how large or small the measurement. Use a metric ruler to measure the length of the paper. Emphasize that the measurement includes both a number and a unit. Using the road map, show students that sometimes a model and a scale must be used because the property is too large or too small to easily measure directly.

Explore

Inquiry Activity

How can you measure length?

Make a Prediction

How wide is your classroom? Write a prediction.

Test Your Prediction

1. **Measure** Work with a partner. Stand with your back against one wall. Slowly walk across the room, placing one foot in front of the other. The heel of your front foot should meet the toe of your back foot. Your partner will count the number of steps it takes to cross the room.

2. Trade roles with your partner and repeat step 1.

3. **Communicate** Compare your data with the class's data. Make a table listing the data for the entire class.

Draw Conclusions

4. **Interpret Data** What is the highest measurement? What is the lowest measurement? Did anyone get the same measurement?

5. **Infer** Why were there different measurements? Why is it useful to use measuring tools, such as a ruler?

Explore More

Measure Scientists use the metric system to measure matter. Predict how wide your classroom is in meters and centimeters. Then use a metric ruler to measure the width of your classroom. How do your measurements compare with your predictions?

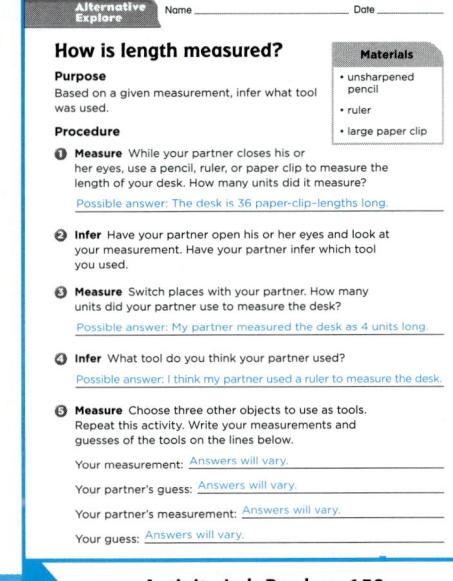

373
EXPLORE

Alternative Explore

How is length measured?

Materials unsharpened pencil, ruler, large paper clip, student desk

While one student closes his or her eyes, have another student measure the length of a desk. From the measurement, have the first student guess which tool was used. Have students change roles and repeat, using tools of their own choosing.

Alternative Explore Name _____ Date _____

How is length measured?

Purpose
Based on a given measurement, infer what tool was used.

Procedure

1. **Measure** While your partner closes his or her eyes, use a pencil, ruler, or paper clip to measure the length of your desk. How many units did it measure?
   Possible answer: The desk is 36 paper-clip-lengths long.

2. **Infer** Have your partner open his or her eyes and look at your measurement. Have your partner infer which tool you used.

3. **Measure** Switch places with your partner. How many units did your partner use to measure the desk?
   Possible answer: My partner measured the desk as 4 units long.

4. **Infer** What tool do you think your partner used?
   Possible answer: I think my partner used a ruler to measure the desk.

5. **Measure** Choose three other objects to use as tools. Repeat this activity. Write your measurements and guesses of the tools on the lines below.

   Your measurement: Answers will vary.

   Your partner's guess: Answers will vary.

   Your partner's measurement: Answers will vary.

   Your guess: Answers will vary.

Materials
• unsharpened pencil
• ruler
• large paper clip

Activity Lab Book, p. 158

Explore

pairs · 20 minutes

Plan Ahead Move any classroom furniture that might interfere with performing the activity. Acquire metric rulers or metersticks for each pair of students.

Purpose This activity will help students understand units of measurement and why these units must be standardized. They will also predict measurements and compare and contrast their predictions with actual measurements. Students will be introduced to and use the metric system.

Structured Inquiry

Make a Prediction Predictions should include a number and a unit of length.

1. **Measure** Advise students to be sure their feet form a straight line and are not angled compared to each other and that they walk in a straight line. Based on your students' abilities, decide how they should record fractional numbers of feet or if they should use only the nearest whole number.

4. **Interpret Data** Students should notice that different sizes of feet will yield different results for the same length measured.

5. **Infer** The lengths of students' feet differ. A standard of measurement, such as a ruler, must give the same measurement every time.

Guided Inquiry Explore More

Measure Assist students with using the metric ruler. Emphasize the importance of placing the ruler accurately as it is moved from place to place. Help students read the markings on the ruler.

Open Inquiry

Have students hypothesize whether their results would be easier to compare if they used unsharpened pencils instead of their feet to measure distance. Ask:

Can you design and carry out an experiment to test the "unsharpened pencil" hypothesis?

Lesson 2 **373**

2 Teach

Read and Learn

Main Idea Have students examine the pictures in the lesson and compare the measuring tools they see with those they have used. Ask them what they think they will learn from the lesson about making measurements.

Vocabulary Have students read aloud the vocabulary terms. Have them write sentences with two of the terms used correctly. The sentences should show how the terms are related.

Reading Skill **Summarize**
Graphic Organizer 5 Have students fill in a Summarize graphic organizer as they read each two pages of the lesson. They can use the Quick Check questions to determine the summaries.

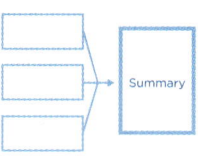

How is matter measured?

Discuss the Main Idea

Show students a basketball. Ask:

- **How could you measure the distance around the ball?** Use a tool that measures length, such as a tape measure.

- **How could you measure the volume of the ball?** Possible answer: Place the ball in a known volume of water and measure how much the water volume increases.

- **How could you measure the temperature of the ball?** Possible answer: Use a thermometer on the surface of the ball.

RESOURCES and TECHNOLOGY

- **Reading and Writing,** pp. 175–177
- **Visual Literacy,** p. 57
- **PuzzleMaker CD-ROM**
- **Presentation Toolkit CD-ROM**
-  **e-Glossary**

Read and Learn

▶ **Main Idea**
Matter can be measured using tools that record standard units.

▶ **Vocabulary**
metric system, p. 374
pan balance, p. 376
gravity, p. 378
weight, p. 378

 e-Glossary
at www.macmillanmh.com

▶ **Reading Skill** ✓
Summarize

```
[ ]
[ ]  →  Summary
[ ]
```

How is matter measured?

Many properties of matter can be observed or measured with tools. You can look closely at an object with a hand lens. You can measure its length and width with a ruler. You can use a thermometer to measure its temperature.

Measuring is a way to compare sizes or amounts. People use tools marked with standard units to measure matter. A *standard unit* is a unit of measurement that people agree to use, such as feet or miles. A common system of standard units is the **metric system** (MET•rik SIS•tuhm). Scientists use the metric system.

Length
You measure length to find out how long something is. You have probably used rulers to measure how tall you are. In the metric system, length is measured in units called meters.

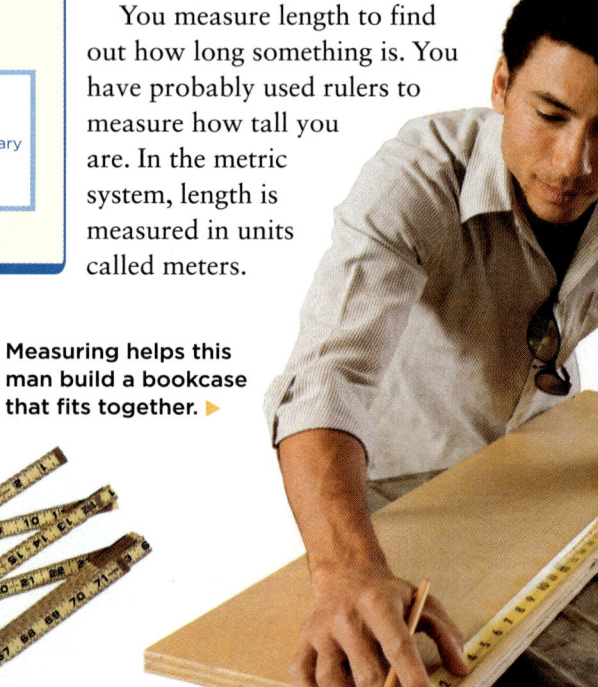

Measuring helps this man build a bookcase that fits together. ▶

Science Background

Standard Units For a unit to be used, it must be standard, or the same. For example, a tool that measures the length of two identical objects at two different locations must measure the same length. Therefore, units must be defined in terms of something that does not change. The meter was first defined in terms of the circumference of Earth. A meter is now defined as the distance traveled by light in a certain fraction of a second.

See **Science Yellow Pages** in the Teacher Resources section for background information.

Professional Development For more Science Background and resources from **NSDL** visit **www.macmillanmh.com/nsdl/**

Measuring the Volume of a Solid

Read a Photo

How can you measure the volume of this rock?
Clue: Look how the water level changes.

Volume

Volume describes how much space an object takes up. You probably have used measuring cups to measure the volume of liquids. You can also use beakers or graduated cylinders. In the metric system a liquid's volume is measured in units called liters.

You can measure the volume of a solid, too. First, measure some water. Then place a solid object completely under the water. Subtract the original water level from the new water level. The difference is the solid's volume.

▲ The volume of a liquid may be measured using a graduated cylinder, beaker, or measuring cup.

 Quick Check

Summarize What are three measurements you could make to describe matter?

Critical Thinking Why is it important to use standard units?

375
EXPLAIN

Differentiated Instruction

Leveled Activities

EXTRA SUPPORT   Have students use thermometers to measure the temperature of several different locations in the classroom.

ENRICHMENT   Have students design and conduct an experiment to show why the method shown on p. 375 for measuring the volume of a rock could not be used to measure a volume of salt. Experiments will show that the salt would dissolve.

▶ Develop Vocabulary

metric system   Tell students that another term sometimes used instead of *metric system* is SI, from the French *Systeme Internationale.* The units for length, area, and volume are the same in the two systems.

▶ Address Misconceptions

A common misconception is that units of length measure only how tall something is. Any measurement that shows how far it is from one point to another is a measurement of length. Have students name other terms used to indicate measurements of length.

FACT▶ **Measurements such as height, depth, circumference, radius, and width are measurements of length.**

Show students several objects of different shapes and ask them to list the properties of the object they could measure using a ruler.

▶ Use the Visuals

Refer students to the visual on measuring the volume of a rock on page 375. Ask:

■ **Why would this method be helpful if you needed to know the volume of something with an uneven shape?** Possible answer: because it can give an accurate volume, no matter the shape of the object

Read a Photo

Answer Measure the volume before and after the rock is placed in the water. The difference between these volumes is the volume of the rock.

✓ Quick Check Answers

- **Summarize** Possible answers: length, volume, temperature
- **Critical Thinking** so that scientists around the world can compare measurements

How do we measure mass?

FAST TRACK Discuss the Main Idea

Show students a pan balance and show them how to use it to measure the mass of several small objects. Ask:

- **When are the two sides of the balance at the same height?** when both sides of the balance contain the same mass

- **An object is placed on one balance pan. The other pan contains two 10-gram masses and one 5-gram mass when the pans are the same height. What is the mass of the object?** 25 grams

▶ Develop Vocabulary

pan balance *Word Origin* The word *balance* comes from the Latin *bis,* which means "twice," and *lanx,* which means "a dish or scale." Have students relate this word origin to a pan balance.

▶ Use the Visuals

Refer students to the visuals on pages 376–377. Ask:

- **How can you tell when the masses on the pans of a balance are equal?** When the pans are equal in height.

RESOURCES and TECHNOLOGY

▶ **Activity Lab Book,** p. 159

💿 **Presentation Toolkit CD-ROM**

How do we measure mass?

You can use a **pan balance** to measure mass. Remember that mass is a measure of the amount of matter in an object. To find an object's mass, you balance it with objects whose masses you know. First, place the object on one end of a pan balance. Then add the known masses to the other side until both sides are level. When the two sides are level, you know the mass of the object.

In the metric system, mass is measured in grams. A gram is close to the amount of mass in two small paper clips. A kilogram is the same as 1,000 grams.

Objects with the same volume do not always have the same mass. A marble is about the same size as a piece of popcorn. However the marble has a greater mass. How is that possible?

▲ Gram masses can be used to find the mass of an object.

This pan balance measures mass. ▼

376
EXPLAIN

Differentiated Instruction

Leveled Questions

EXTRA SUPPORT   **If you are told that a measurement is 25 grams, is this a measurement of length or of mass?** mass

ENRICHMENT   **What is one object that is smaller than a shoe box but has more mass?** Possible answer: a book

Matter is made up of tiny particles. In some objects the particles are close together. In other objects they are farther apart. The particles inside a marble are packed together more tightly than those inside a piece of popcorn. A marble has more particles than a piece of popcorn. It has more mass.

✔ Quick Check

Summarize How can you measure mass using a pan balance?

Critical Thinking How could you measure the mass of a liquid with a balance?

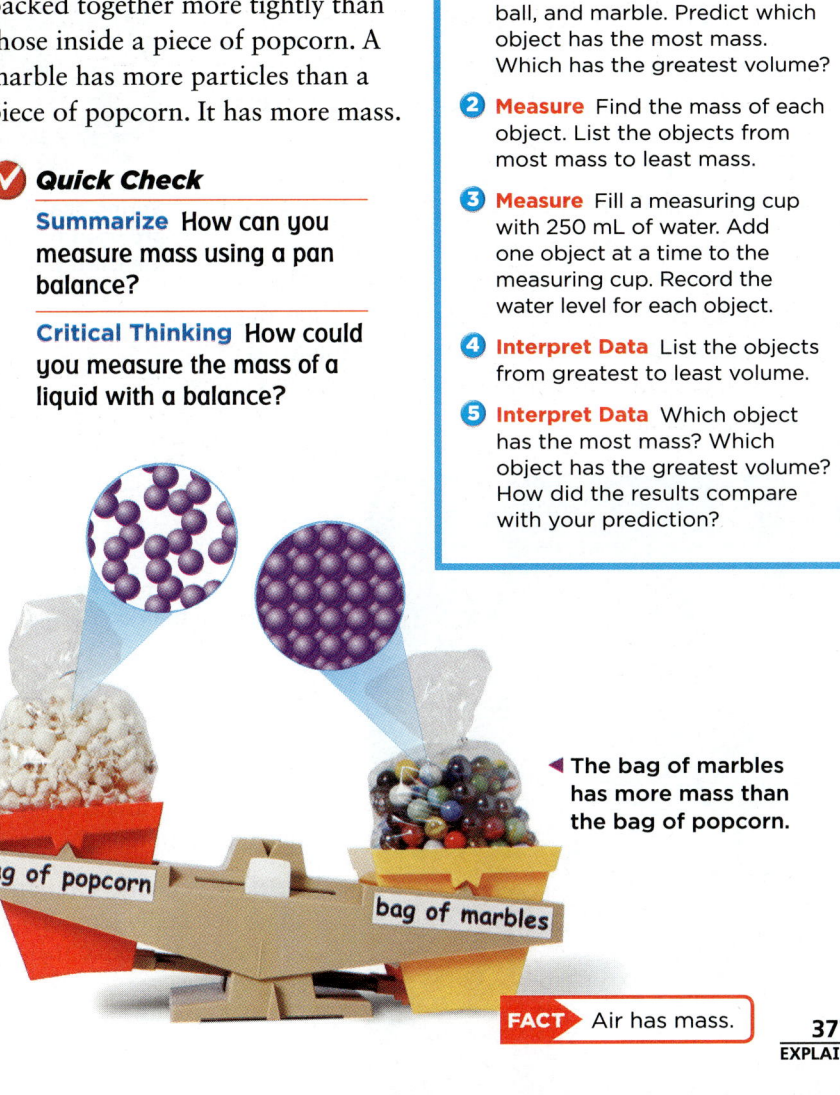

The bag of marbles has more mass than the bag of popcorn.

ag of popcorn

bag of marbles

FACT ▶ Air has mass.

377
EXPLAIN

≡ Quick Lab

Measure Mass and Volume

1 Predict Look at a toy car, golf ball, and marble. Predict which object has the most mass. Which has the greatest volume?

2 Measure Find the mass of each object. List the objects from most mass to least mass.

3 Measure Fill a measuring cup with 250 mL of water. Add one object at a time to the measuring cup. Record the water level for each object.

4 Interpret Data List the objects from greatest to least volume.

5 Interpret Data Which object has the most mass? Which object has the greatest volume? How did the results compare with your prediction?

≡ Quick Lab   small groups   15 minutes

Objective Predict and measure mass and volume, and compare the measurements with the predictions.
Materials pan balance with masses, measuring cup, water, metal toy car, golf ball, marble

1 The toy cars should be metal, not plastic. Encourage students to hold the items before predicting. Holding one item in each hand can help them make an accurate prediction of relative mass and volume.

2 Help students make sure the pans balance before they place any masses on them. Answers will depend on the masses of the objects used.

3 Most measuring cups require estimating volume. Show students how to estimate volume from the water level in the cup.

4 Perform a sample calculation of volume for students. Help them subtract the volume of water before adding an object from the volume of the water and the object. Answers depend on the volume of the objects used.

5 Assist students in comparing their results with their predictions.

✔ Quick Check Answers

- **Summarize** Place the object on one of the pans and known masses on the other pan until the pans balance in height.

- **Critical Thinking** Subtract the mass of an empty container from the mass of the container with the liquid.

ELL Support

Explain Write the word *measurements* on the board and have students repeat it after you. Explain what measurements are and why they are important. Review the terms *weight*, *mass*, and *volume*. Write them on the board as well and discuss them with students. Ask students if they can name tools for measuring *weight*, *mass*, and *volume*. Write their responses on the board. Elicit *scale*, *pan balance*, and *measuring cup*.

BEGINNING Students can point to or name different tools for measuring weight, mass, and volume.

INTERMEDIATE Students can use phrases and short sentences to describe how to use a scale, a pan balance, or a measuring cup.

ADVANCED Students can describe how to measure weight, mass, and volume in complete sentences.

How are mass and weight different?

▶ Discuss the Main Idea

Discuss with students that weight is the effect of gravity on the mass in an object. Tell them that although the planet Uranus is bigger than Earth, the force of gravity on Uranus is slightly less than it is on Earth. Ask:

- **How would your mass on Uranus compare to your mass on Earth?** It would be the same.

- **How would your weight on Uranus compare to your weight on Earth?** My weight would be greater on Earth than it would be on Uranus.

▶ Develop Vocabulary

gravity *Word Origin* Explain to students that the word *gravity* comes from the Latin word ***gravitatem,*** which means "weight or heaviness." Ask students to explain how weight relates to gravity. Weight is the effect of gravity on mass.

weight *Word Origin* The word *weight* comes from the Old English word ***gewiht.***

✓ *Quick Check Answers*

- **Summarize** Weight depends on the pull of gravity, while mass is the same no matter what the gravity is.

- **Critical Thinking** Yes; the pull of gravity on the Moon is less than the pull of gravity on Earth.

How are mass and weight different?

What happens when you leap into the air? Do you float away? No, you come back to the ground. This happens because of gravity (GRAV·i·tee). **Gravity** is a pulling force that holds you on Earth. Gravity keeps you and everything on Earth from floating into space.

You can measure how much Earth's gravity pulls on you. This measurement is your weight (WAYT). **Weight** is a measure of the pull of gravity on you. Weight can be measured with a spring scale. Weight is different from mass. If you visited the Moon, your mass would stay the same. The matter inside you would not change. However your weight would change. This is because the pull of the Moon's gravity is weaker than the pull of Earth's gravity. Your weight on the Moon would be less than your weight on Earth. Your mass would be the same.

Spring scales are used to measure weight.

◀ The pull of gravity is weaker on the Moon than on Earth.

✓ *Quick Check*

Summarize How is weight different from mass?

Critical Thinking Do you think you could jump higher on the Moon? Explain.

378
EXPLAIN

Homework Activity

Working Measurements

Have students interview family members and find out how they use measurements in their daily lives. Have them find out why measurements are important at home and at work. Have them report their findings in a visual display.

Lesson Review

Visual Summary

Properties of matter, such as **length** and **volume**, can be measured and observed with tools.

Mass can be measured with a pan balance.

Mass stays the same. The **weight** of an object depends on the force of gravity.

Make a FOLDABLES **Study Guide**

Make a Layered-Look Book. Use it to summarize what you learned about measuring.

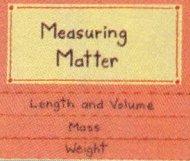
Measuring Matter
Length and Volume
Mass
Weight

Think, Talk, and Write

❶ **Main Idea** How is matter measured?

❷ **Vocabulary** What is gravity?

❸ **Summarize** Does a large object always have a lot of mass? Explain your answer.

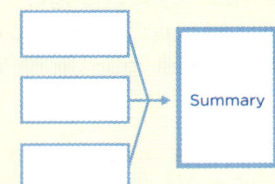
Summary

❹ **Critical Thinking** Why is it important to measure accurately?

❺ **Test Prep** What tool would you use to measure weight?
A thermometer
B hand lens
C spring scale
D ruler

  **Math Link**

Metric Measurements
What tool would you use to measure the length of a pencil in centimeters? Use this tool to measure the length of four objects. List them in order from shortest to longest.

 Social Studies Link

Do Research
About 5,000 years ago people began using standard weights. Some early weights were the shekel (SHEK•uhl) and the mina (MYE•nuh). Find out more about early systems of measurement.

 -Review Summaries and quizzes online at www.macmillanmh.com

379
EVALUATE

Formative Assessment

Approaching Have students make a table that shows the property measured, what tool is used to measure it, and what units are used for the measurement.

On-Level Give students an example of a food-product label. Have them list the nutrition information given in grams or milligrams, and tell what type of measurement this represents.

Challenge Have students select several product labels and list measurements that are on the package. If two measurements, such as mass and volume, are present, list both of them. Have them make a bar graph showing the types of measurements on the packages.

Key Concept Cards For student intervention, see the prescribed routine on **Key Concept Cards 57 and 58.**

---

3 Close

Lesson Review

▶ **Visual Summary**

Have students save their lesson Study Guides. They will use them at the end of the chapter for review.

▶ **Make a** FOLDABLES **Study Guide**

See pp. R27–R28 for instructions.

FAST TRACK **Think, Talk, and Write**

❶ **Main Idea** Matter is measured using tools marked with standard units.

❷ **Vocabulary** Gravity is a pulling force that holds you on Earth.

❸ **Summarize**

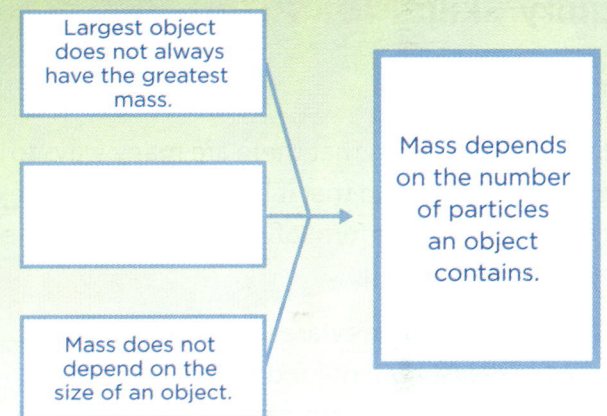
Largest object does not always have the greatest mass.

Mass does not depend on the size of an object.

Mass depends on the number of particles an object contains.

❹ **Critical Thinking** Possible answer: It is important to measure accurately when cutting materials to build objects so that the pieces fit together.

❺ **Test Prep** C

 Math Link A metric ruler; students' lists will vary depending on the items chosen to measure.

Social Studies Link Student research will probably reflect that early measurements were based on common objects that were not standardized.

RESOURCES and TECHNOLOGY

▶ **Reading and Writing, p. 178**

▶ **School to Home Activities, pp. 75–76**

-Review Narrated Summary and Quiz

Progress Reporter Assessments

Focus on Skills

Objective

- Measure and compare the mass of water as a solid and as a liquid.

Materials measuring cup, ice cubes, pan balance, plastic wrap, tape measure, thermometer

Plan Ahead Gather enough materials for each small group. Set aside time for measuring mass every 30 minutes. Store the ice in coolers until it is needed.

EXTEND This activity will teach students to measure the mass of water as a solid and then as a liquid. Students will then compare the masses of water in two states.

Inquiry Skill: Measure

▶ Learn It

- Explain to students that there are many ways to measure things and that it is important that they choose the correct form of measurement for the designated purpose.

- Students should be aware that there are many units of measurement for different tools; for example, measurements can be taken in inches or centimeters.

▶ Try It

1. Emphasize that in an experiment only one thing should change so that it can be measured correctly. In this activity, the relationship between the state of matter and its mass is measured. It is important that the amount of water remains constant in the cup.

2. Students can determine whether the mass has changed by comparing the current mass to the original mass they recorded.

4. In answering the question, students should find that water from the melted ice cubes should have the same mass as the original ice cubes, as long as water vapor does not escape from the cup.

Focus on Skills

Inquiry Skill: Measure

You have learned that matter is anything that takes up space and has mass. Water is matter that is important to life on Earth. It is found on Earth as solid ice and liquid water. It is even found in the air. What happens to water's mass as it changes from a chunk of solid ice to liquid water? Scientists **measure** things to answer questions like this.

measuring cup

▶ Learn It

When you **measure**, you find such things as the mass, volume, length, or temperature of an object. You can also measure distances and time. Scientists use many tools to measure things. Some of these tools are shown on this page. Scientists use measurements to describe and compare objects or events.

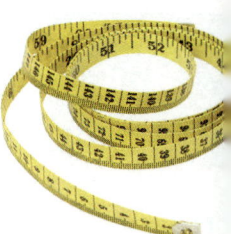

tape measure

pan balance

thermometer

Integrate Math

Comparing Masses

Ask students to find two objects that are about the same size. Before students measure the objects, have them make a prediction. Ask:

- **Which object has the greater mass?**

Have students use a balance to find the mass of each object. Then have students write a comparison statement for the two objects, such as, "The mass of the pen is greater than the mass of the pencil" or "pen's mass > pencil's mass." Have them make two more comparisons so that they measure at least six objects.

Skill Builder

Try It

You know that scientists **measure** things to answer questions. You can measure, too. Answer this question. Do ice cubes have the same mass after they melt?

To start, place several ice cubes in a cup. Then cover the cup with plastic wrap so the water stays inside the cup.

Measure mass by placing the cup on one end of a pan balance. Add masses to the other side of the pan balance until both sides are level. Record the mass on a chart.

Time	Mass

Measure the mass every $\frac{1}{2}$ hour until the ice is completely melted.

Now use your measurements to answer the question. Do ice cubes have the same mass after they melt?

Apply It

Now **measure** to answer this question: Does ice cream have the same mass after it melts? How do you know?

**381
EXTEND**

▶ Apply It

Have students set up an experiment and measure the mass of the ice cream as a solid and then as a liquid. Students can conclude that the mass of the solid is the same as the mass of the same matter in liquid form. **Be Careful!** Remind students to never eat anything in a lab environment. Food used in experiments can be contaminated by measuring tools, chemicals, and other hazardous materials.

Have students use their results to extend the concept introduced in the experiments. Ask: **What happens to the mass of a liquid when the liquid becomes a gas?** Students should infer that the mass stays the same. Point out that the mass also stays the same when the matter freezes or condenses. In general, mass stays the same during any change of state.

Ask students whether they can use their results to predict that other properties stay the same during a change of state. Tell students that many properties do change when a substance changes state. For example, when ice melts, its shape changes.

Focus on Skills   Name _____   Date _____

Measure

You have learned that matter is anything that takes up space and has mass. Water is matter that is important to life on Earth. It is found on Earth as solid ice and liquid water. It is even found in the air. What happens to water's mass as it changes from a chunk of solid ice to liquid water? Scientists measure things to answer questions like this.

Learn It

When you measure, you find such things as the mass, volume, length, or temperature of an object. You can also measure distances and time. Scientists use many tools to measure things. Some of these tools are shown on this page. Scientists use measurements to describe and compare objects or events.

Activity Lab Book, pp. 160–162

RESOURCES and TECHNOLOGY

▶ **Activity Lab Book,** pp. 160–162

▶ **Activity Flipchart,** p. 43

🔵 **Instructional Navigator CD-ROM**

Stop Here to ▶ Plan Your Lesson

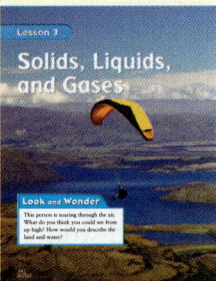

Lesson 3 Solids, Liquids, and Gases

Objective

- Define the three common states of matter: solid, liquid, and gas.
- Explain the properties of solids, liquids, and gases.

Reading Skill Classify

Graphic Organizer 11, p. TR13

 **Professional Development** Look for **NSDL** to find recommended Science Background materials from the National Science Digital Library

FAST TRACK

Lesson Plan When time is short, follow the Fast Track and use the essential resources.

1 Introduce
Look and Wonder, p. 382

Resource Alternative Explore, p. 165

2 Teach
Discuss the Main Idea, p. 384

Discuss the Main Idea, p. 386

Discuss the Main Idea, p. 388

Resource Visual Literacy, p. 59

3 Close
Think, Talk, and Write, p. 389

Resource Assessment, p. 114

▶ Reading and Writing

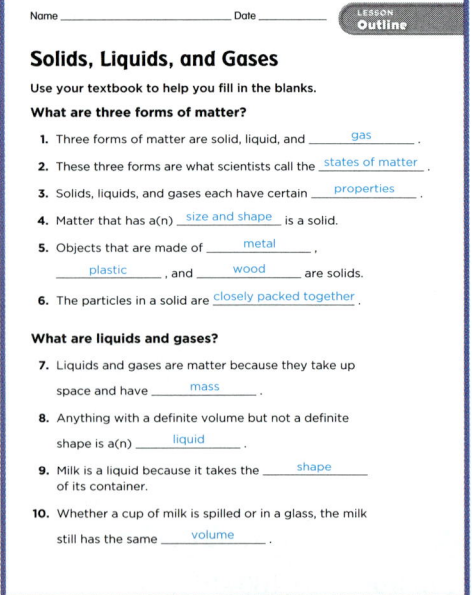

Outline, pp. 179–180
Also available as a student workbook

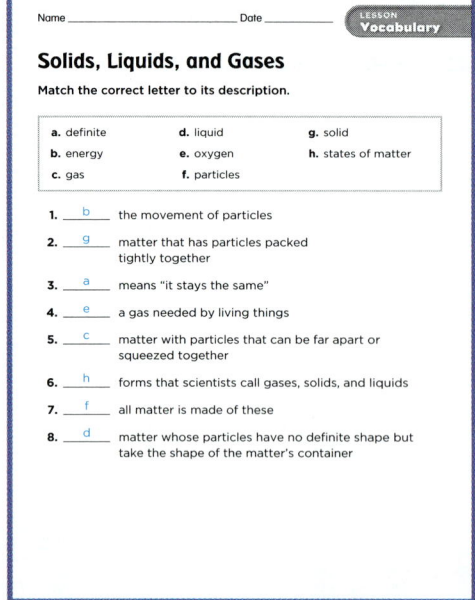

Vocabulary, p. 181
Also available as a student workbook

 ▶ **Visual Literacy**

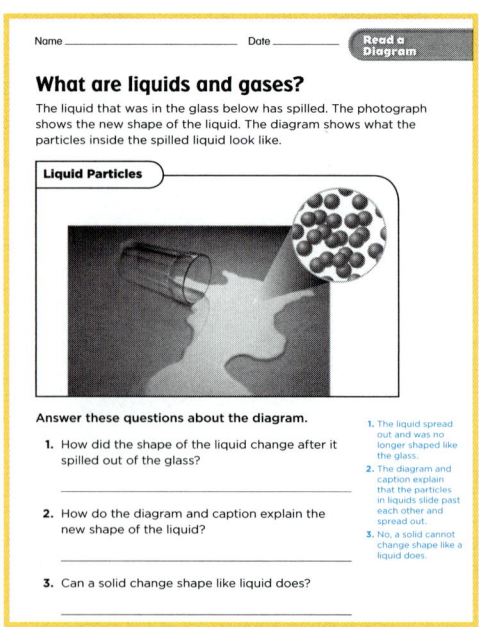

Read a Diagram, p. 59
Also available as a transparency

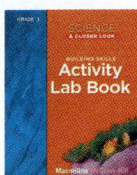

▶ Activity Lab Book

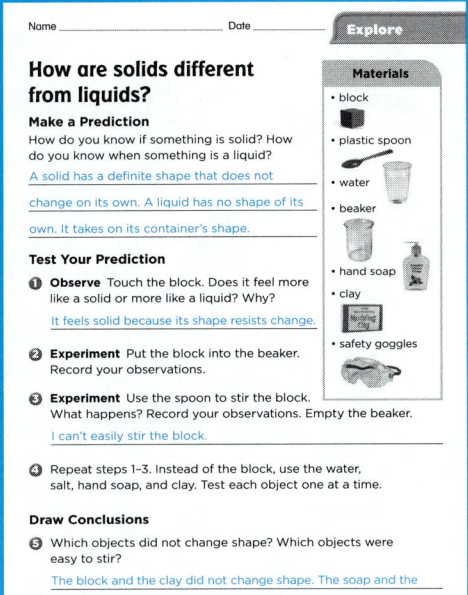

Explore, pp. 163–164
Also available as a student workbook

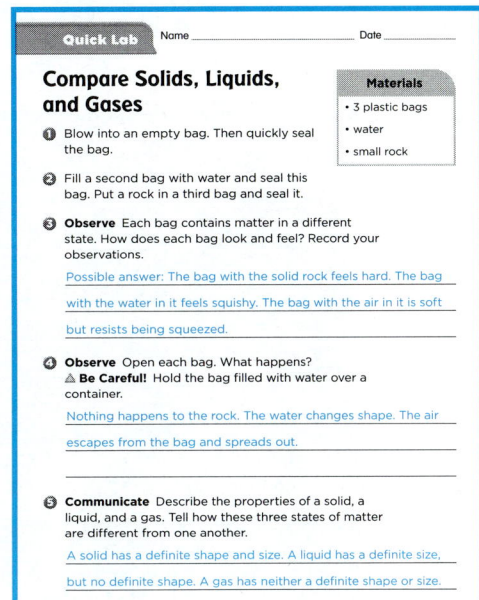

Quick Lab, p. 166
Also available as a student workbook

▶ Assessment

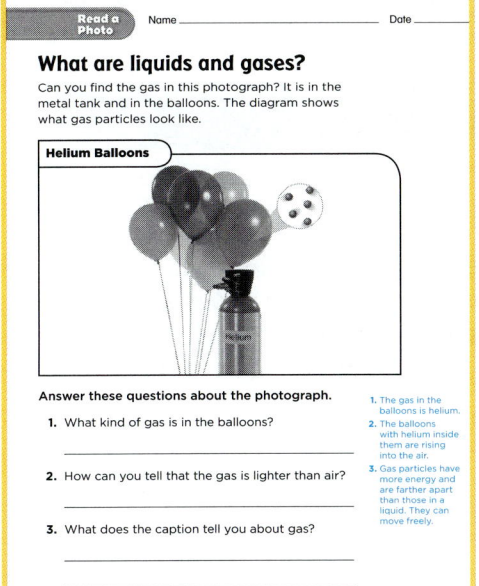

Read a Photo, p. 60
Also available as a transparency

Lesson Test, p. 114
Also available as a student workbook

ADDITIONAL RESOURCES

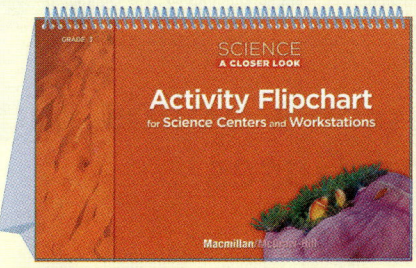

p. 44

p. 105

pp. 59–60

pp. 77–78

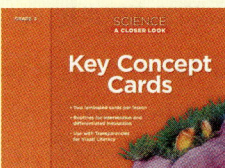

59–60

142–145

Technology

🔴 **Science Activity DVD**

🔴 **Instructional Navigator CD-ROM**

🔴 **Presentation Toolkit CD-ROM**

🔵 **Science in Motion** *Particles of a Solid, Liquid, and Gas*

🔵 **e-Review**

🔵 **NSDL**

Lesson 3 Solids, Liquids, and Gases

Objectives
- Define the three common states of matter: solid, liquid, and gas.
- Explain the properties of solids, liquids, and gases.

1 Introduce

▶ Assess Prior Knowledge

Explain to students that they are going to learn about solids, liquids, and gases. Ask:

- **What does it tell you about an airplane to say that it is solid?** Possible answers: It does not change shape easily; it is hard.

- **How do you know water is a liquid?** Possible answer: It flows from place to place.

- **Why do you fill a balloon with a gas like air instead of a solid?** Possible answer: The gas takes the same shape as the balloon.

Have students save their answers and revise them at the end of the lesson.

FAST TRACK   Look and Wonder

Invite students to share their responses to the Look and Wonder statement and questions:

- **What do you think you could see from up high?** Possible answer: a larger amount of land than can be seen from on the ground

- **How would you describe the land and water?** Possible answer: The shape of the land surface changes, but the water surfaces are flat.

Note any misconceptions that students may have. Address these misconceptions as you teach the lesson.

RESOURCES and TECHNOLOGY
▶ **Activity Lab Book,** pp. 163–165
▶ **Activity Flipchart,** p. 44
▶ **Science Activity DVD**

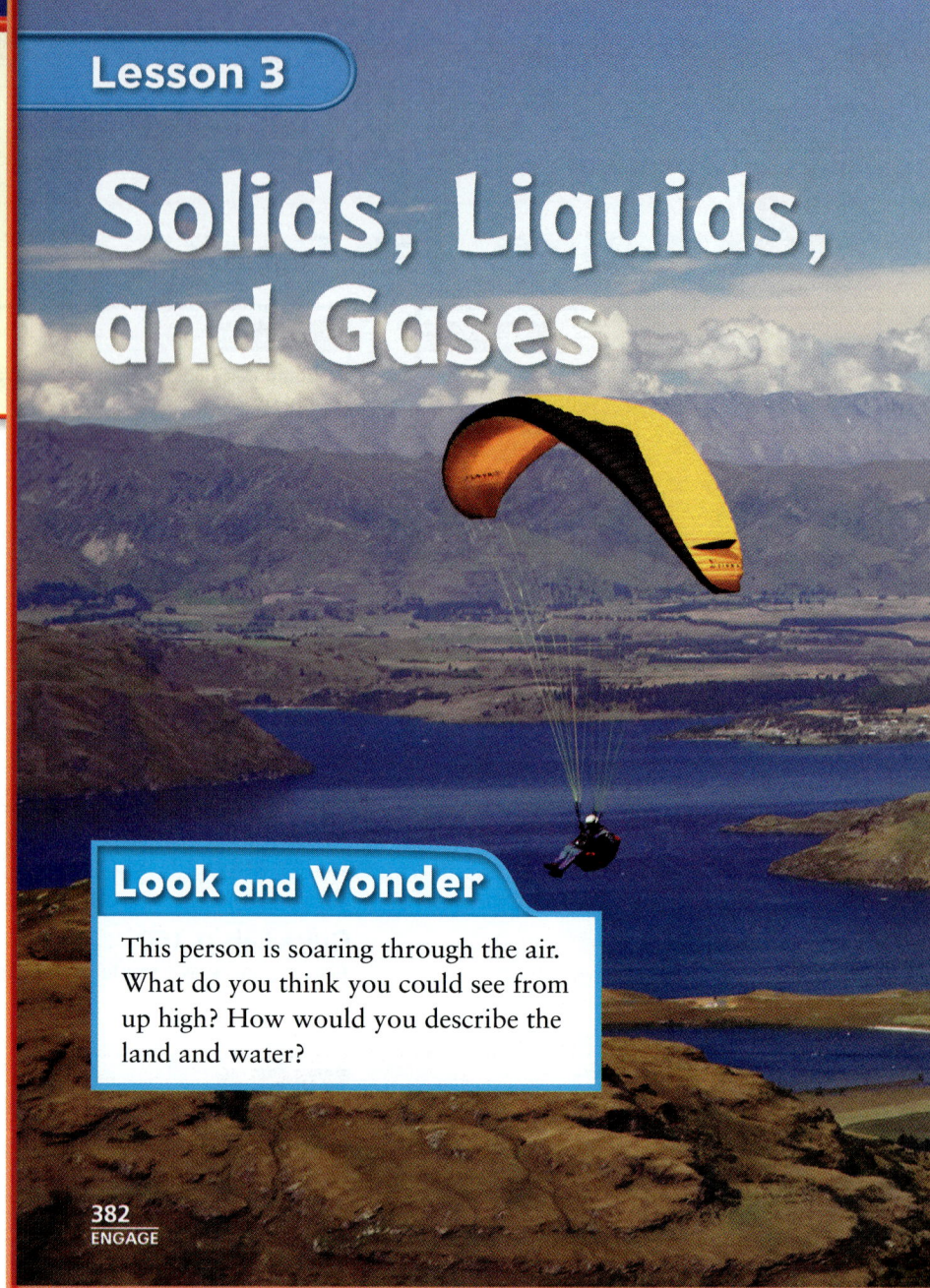

Lesson 3

Solids, Liquids, and Gases

Look and Wonder

This person is soaring through the air. What do you think you could see from up high? How would you describe the land and water?

382
ENGAGE

Warm Up

Start with a Book

In small groups, have students read *Solids, Liquids, and Gases* by Ginger Garrett (Children's Press, ISBN 0-516-24663-1). Have students make three columns on a piece of paper and label them *Solid, Liquid,* and *Gas.* As they read the book, have students list properties and examples of these different states of matter.

Explore

Explore
Inquiry Activity

How are solids different from liquids?

Make a Prediction
How do you know if something is solid? How do you know when something is a liquid?

Test Your Prediction

1. **Observe** Touch the block. Does it feel more like a solid or more like a liquid? Why?

2. **Experiment** Put the block into the beaker. Record your observations.

3. **Experiment** Use the spoon to stir the block. What happens? Record your observations. Empty the beaker.

4. Repeat steps 1–3. Instead of the block, use the water, salt, hand soap, and clay. Test each object one at a time.

Draw Conclusions

5. Which objects did not change shape? Which objects were easy to stir?

6. **Classify** Which objects are solids? Which are liquids?

7. Explain how solids are different from liquids.

Explore More

Experiment What would happen if you put each object in the freezer? What would happen if you put each object in a warm place? Form a hypothesis and test it.

Materials

block   beaker

plastic spoon   water

salt   hand soap

safety goggles   clay

Step 2

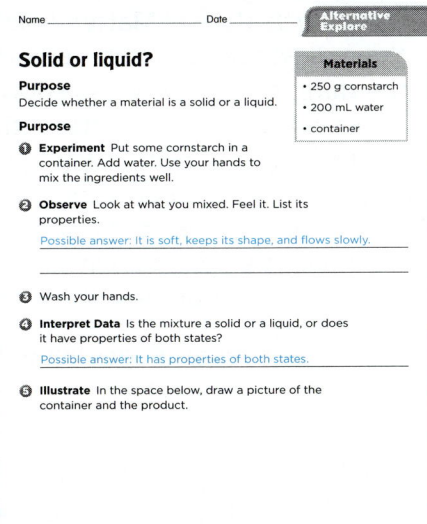
Step 4

383
EXPLORE

Alternative Explore

Solid or liquid?

Materials 250 g cornstarch, 200 mL water, container

Have students thoroughly mix the water and the cornstarch in the container. Have them examine the product and decide whether they think it is a solid or a liquid or if it has properties of both states. Have them list what these properties are.

Name _____ **Date** _____

Alternative Explore

Solid or liquid?

Purpose
Decide whether a material is a solid or a liquid.

Purpose

1. **Experiment** Put some cornstarch in a container. Add water. Use your hands to mix the ingredients well.

2. **Observe** Look at what you mixed. Feel it. List its properties.
   Possible answer: It is soft, keeps its shape, and flows slowly.

3. Wash your hands.

4. **Interpret Data** Is the mixture a solid or a liquid, or does it have properties of both states?
   Possible answer: It has properties of both states.

5. **Illustrate** In the space below, draw a picture of the container and the product.

Materials
- 250 g cornstarch
- 200 mL water
- container

Activity Lab Book, p. 165

Explore

pairs   20 minutes

Plan Ahead Have waste containers available for used water, salt, and hand soap. Have containers available for students to return their blocks and clay for later use. Make sure any student who is visually challenged is paired with a sighted student who can explain her or his observations.

Purpose This activity will help students investigate the properties of solids and liquids.

Structured Inquiry

Make a Prediction Possible predictions: Something is solid if it keeps the same shape. The shape of a liquid changes to fit the shape of its container.

1. **Observe** The block feels solid; it is hard and smooth.

2. **Experiment** The block keeps its shape.

3. **Experiment** Still, nothing happens to the block.

4. Have students test the materials in the following order: salt, clay, water, hand soap. The beaker should be dry when testing the salt and clay.

5. The water, salt, and hand soap changed their shape. They were also easiest to stir.

6. **Classify** liquids: water, hand soap; solids: salt, clay

7. Solids keep their shape, while liquids take the shape of their container (beaker). Liquids are easier to stir. Salt is a solid made up of tiny pieces. It takes the shape of its container and is easily stirred.

Guided Inquiry Explore More

Experiment If you put a cup of water in the freezer, the liquid water would change to a solid. If you put the cup in a warm place, the water would seem to disappear. The liquid water is changing into a gas.

Open Inquiry

Have students extend the activity to include any changes in volume of the items tested. Ask: **Do solids or liquids change volume when their containers are changed?**

2 Teach

Read and Learn

Main Idea Have students examine the pictures in the lesson and list five things in the pictures that they think are solids and liquids. Have them also list anything they infer to be a gas. Ask them what they think they will learn from the lesson about these states of matter.

Vocabulary Have students read aloud the vocabulary terms and their definitions. Have them restate the sentences in their own words.

Reading Skill Classify
Graphic Organizer 11 Have students fill in the Classify graphic organizer as they read each two pages of the lesson. They can use the Quick Check questions to determine each of the classifications.

What are three forms of matter?

FAST TRACK **Discuss the Main Idea**

Show students a solid, such as a student desk. Ask:

- **When you move the desk from place to place, what happens to its shape?** It stays the same.

- **What happens to the amount of space it takes up?** It stays the same.

- **If your desk is a solid, what can you say about the shape and volume of a solid?** The shape and volume stay the same.

RESOURCES and TECHNOLOGY

- **Reading and Writing, pp. 179–181**
- **PuzzleMaker CD-ROM**
- **Presentation Toolkit CD-ROM**
- **e-Glossary**

Read and Learn

▶ **Main Idea**
Solids, liquids, and gases are three forms of matter.

▶ **Vocabulary**
states of matter, p. 384
solid, p. 384
liquid, p. 386
gas, p. 387

LOG ON **e-Glossary**
at www.macmillanmh.com

▶ **Reading Skill** ✓
Classify

What are three forms of matter?

Matter comes in many forms. Look at the picture below. The canoe is a solid. The river is made up of water, a liquid. The air is made of gases. Solids, liquids, and gases make up three forms of matter. Scientists call these forms **states of matter**. Each of these states of matter has certain properties.

Solids

Most of the things you see around you are solids (SOL•idz). A **solid** is matter that takes up a definite amount of space and has its own shape. Definite means "it stays the same." This book is a solid. Pencils, desks, and pillows are solids, too. If you put a pencil into a jar or box, it stays the same. It has a definite size and shape.

How are these boys using three states of matter?

384
EXPLAIN

Science Background

The Fourth State of Matter Solid, liquid, and gas are the three most common states of matter on Earth, and they are the states students need to know about at this grade level. However, most of the matter in the universe is a fourth type of matter known as *plasma*. Plasma consists of charged particles, which are ionized atoms and electrons. The matter in stars is mostly plasma. Even outer space is not a complete vacuum; it contains thin plasma. On Earth, plasma is present in such things as neon signs and the instruments that produce lasers.

See **Science Yellow Pages** in the Teacher Resources section for background information.

LOG ON **Professional Development** For more Science Background and resources from **NSDL** visit **www.macmillanmh.com/nsdl/**

Remember that matter is made up of tiny particles. These particles are too small to see. In a solid these particles are packed closely together. They do not have a lot of room to move around. The particles stay in place so the solid keeps its shape.

The particles in this solid horseshoe cannot move much.

▲ Solids can be hard or soft. This goalie's helmet is hard, but his leg pads are soft.

▲ Even though you can change the shape of clay, it is still a solid.

 **Quick Check**

Classify What are three solids you use every day?

Critical Thinking A rubber band can change its shape when it is stretched. Do you think a rubber band is a solid or a liquid? Explain your answer.

385
EXPLAIN

▶ **Develop Vocabulary**

states of matter *Word Origin* This term comes from the Latin *status,* which means "matter of position or standing," and *materia,* which means "substance from which something is made." Have students relate how "matter of position or standing" relates to states of matter.

solid *Word Origin* The word *solid* comes from the Latin *solidus*, which means "firm, whole, entire." Have students relate this meaning to the fact that solids do not spread out or break up into parts.

▶ **Address Misconceptions**

A common misconception is that all solids are hard. Point out that pillows and marshmallows are both soft, but they are solids. A solid object tends to resist a change in shape, but solids can be deformed, sometimes permanently.

FACT Any sample of matter that has a definite volume and a definite shape is a solid.

Have students choose five solids. Have them sequence the solids from softest to hardest.

Quick Check Answers

- **Classify** Possible answers: cup, cereal, book
- **Critical Thinking** It is solid because its volume stays the same, even though it can be stretched.

Differentiated Instruction

Leveled Activities

EXTRA SUPPORT Divide the class into three groups. Have each group think of a way to model the movement of particles in different states of matter.

ENRICHMENT Have students develop a "What Am I?" game in which one student describes properties of a sample of matter, including describing the state it is in. Other students use this description to identify the material.

What are liquids and gases?

FAST TRACK ### Discuss the Main Idea

Remind students that shape and volume were used to define a solid. Ask:

- **When a liquid is moved from one container to another, what happens to its shape? What happens to its volume?** It changes to take the shape of its container; its volume stays the same.

- **What happens to the shape of a gas when it is moved from a tank to a balloon? What happens to its volume?** Its shape changes from the shape of the tank to the shape of the balloon and from a small volume in the tank to a larger volume in the balloon.

▶ Develop Vocabulary

liquid *Word Origin* The word *liquid* comes from the Latin word *liquidus,* which means "fluid, liquid, moist."

gas *Word Origin* The word *gas* comes from the Greek *khaos,* which refers to empty space. Have students relate these meanings to the way particles move in a gas.

▶ Use the Visuals

Refer students to the visuals on pages 386–387 that show how particles move in a liquid and in a gas. Have students explain how to determine the state of matter of something if you know how its particles move.

Read a Diagram

Answer The particles in a liquid are able to move past one another. They can spread out to fill a container.

 Science in Motion *Particles of a Solid, Liquid, and Gas*

RESOURCES and TECHNOLOGY

▷ **Visual Literacy,** p. 59

▶ **Activity Lab Book,** p. 166

💿 **Presentation Toolkit CD-ROM**

What are liquids and gases?

Liquids and gases are two other states of matter. Like solids, they take up space and have mass.

Liquids

A **liquid** is matter that has a definite volume but not a definite shape. A liquid takes the shape of its container. Water, shampoo, and milk are some liquids. When milk is inside a carton, it takes the shape of the carton. When you pour milk into a glass, it takes the shape of the glass. If you spill the milk, it will spread out over the floor. If you were able to mop up the milk and put it back into the carton, it would still be the same amount of milk. The volume of the milk stays the same. Only its shape changes.

▲ Liquids take the shape of their containers. Liquids also take up a definite amount of space inside their containers.

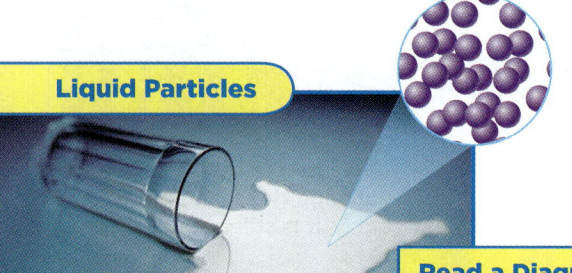
Liquid Particles

◀ The particles in a liquid are able to slide past one another. That is why liquids can change shape.

Read a Diagram

How would you describe the particles in a liquid?

Clue: Illustrations can help show things that are hard to see.

LOG ON *Science in Motion* Watch how the particles of a liquid move at **www.macmillanmh.com**

386
EXPLAIN

Classroom Equity

Encourage all students to get involved. To encourage more students to volunteer answers in this lesson, wait three to five seconds before calling on someone to answer a question. As an experiment, have students write their names on index cards. Then, take the deck, shuffle it, and turn up the cards one by one until you have called on each student.

Gases

You cannot always see gases, but they are all around you. A **gas** is matter that has no definite shape or volume. A gas takes the shape and volume of its container.

Think about balloons being blown up with a helium tank. Helium is a gas. When it is in the tank, it has a small volume. It has the shape of the tank. When the gas is used to fill balloons, it spreads out. It then has a much greater volume. It also changes shape. It takes the shape of the balloons.

▲ The particles in a gas have more energy than the particles in a liquid. In a gas, the particles of matter can move about freely.

Helium

≡ Quick Lab

Compare Solids, Liquids, and Gases

gas

liquid

solid

① Blow into an empty bag. Then quickly seal the bag.

② Fill a second bag with water and seal this bag. Put a rock in a third bag and seal it.

③ **Observe** Each bag contains matter in a different state. How does each bag look and feel? Record your observations.

④ **Observe** Open each bag. What happens?
⚠ **Be Careful.** Hold the bag filled with water over a container.

⑤ **Communicate** Describe the properties of a solid, a liquid, and a gas. Tell how these three states of matter are different from one another.

✔ Quick Check

Classify List three liquids you drink every day.

Critical Thinking Suppose a balloon filled with helium bursts. What would happen to the gas?

387
EXPLAIN

≡ Quick Lab

small groups 15 minutes

Objective Observe the properties of the states of matter under the same circumstances.

Materials 3 sealable plastic bags, water, rock

① Have students seal all but about one inch of the bag before blowing into it.

② **Be Careful!** Have paper towels available to clean up spills. Emphasize that spills on the floor must be cleaned up immediately.

③ Possible answers: The air and liquid change shape in the bags when squeezed; the rock does not change shape when its bag is squeezed.

④ Possible answers: The water and the air come out of the bags. The rock stays the same in the bag.

⑤ A solid has definite shape and volume. A liquid has definite volume and no definite shape. A gas has no definite shape or volume.

✔ Quick Check Answers

- **Classify** Possible answers: water, soda, milk

- **Critical Thinking** The helium gas particles would come out of the balloon and spread in all directions into the air.

How do you use all the states of matter?

▶ Discuss the Main Idea

Bring a bicycle into the classroom. Have students feel the solid parts of the bicycle and squeeze the tires to show that the air in them can be compressed into a smaller volume. Have them look at the oil on the chain, but do not allow them to touch it. Ask:

- **How do you know that the handlebars are solid?** They have definite shape and volume.

- **How do you know that the oil on the chain is a liquid?** It has a definite volume but changes shape.

- **You do not see what is in the tire. How do you know it is a gas?** It takes the shape of the tire and can change volume.

✔ Quick Check Answers

- **Classify** solid, liquid, and gas

- **Critical Thinking** Possible answers: You use solids to make a car, liquids to drink, and gases to breathe.

How do you use all the states of matter?

Solids, liquids, and gases are all around you. You use them in many ways. Many of the foods you eat are solids. Your body needs water, a liquid. You need oxygen, a gas from the air. Oxygen helps you get the energy you need from the food you eat.

You use the states of matter in other ways, too. You can find three states of matter on a bicycle, for example. Many parts of the bicycle are made of solids. The handlebars, seat, and the rubber of the tires are solids. The tires are filled with air, a gas. The oil on the bicycle chain is a liquid.

▲ Oil, a liquid, helps a bicycle chain move smoothly.

▲ You pump air into the tires to inflate them.

✔ Quick Check

Classify What are three states of matter found on a bicycle?

Critical Thinking How do you use the different states of matter?

The bicycle frame is solid. It has to be solid to keep the bicycle together. ▶

Homework Activity

Demonstrate States of Matter in Cars

Have students work with an adult to find out how solids, liquids, and gases are used in an automobile. They will probably find out that most of the automobile is made from solids; liquids include gasoline, oil, washer fluid, and liquid in the radiator; and gases include the air in the tires and exhaust gases. They might find out that gasoline is changed to a gas before it burns. Have students present their findings in an oral or a written report or in a visual display.

Lesson Review

Visual Summary

 A **solid** is matter that has a definite volume and shape.

 A **liquid** is matter that has a definite volume but no definite shape.

 A **gas** is matter that has no definite volume or shape.

Make a FOLDABLES Study Guide

Make a Layered-Look Book. Use it to summarize what you learned about states of matter.

Think, Talk, and Write

1 Main Idea What are the properties of a solid, liquid, and gas?

2 Vocabulary What is matter that has no definite shape or size?

3 Classify What kind of matter is this book? What kind of matter is water? What kind of matter is air?

4 Critical Thinking Compare solids, liquids, and gases. How are they alike? How do they differ?

5 Test Prep Matter that spreads out to fill its container is a

A gas.
B liquid.
C mass.
D solid.

 **Math Link**

Solve a Problem
A helium tank can inflate 126 large balloons. It can inflate three times as many small balloons. How many small balloons can the tank inflate?

 Art Link

Make a Poster
Draw diagrams that show the differences among solids, liquids, and gases. Write a brief explanation of each diagram.

 e-Review Summaries and quizzes online at www.macmillanmh.com

389
EVALUATE

Formative Assessment

Approaching Have students illustrate a use of a solid, a liquid, and a gas.

On-Level Ask students to explain why modeling clay is a solid, even though they can change its shape by pressing on it.

Challenge Have students use research materials to find out and report about how one state can change to another state and the role energy plays in these changes. Point out that energy added to matter makes its particles move faster, and when energy is released by matter, its particles slow down.

 **Key Concept Cards** For student intervention, see the prescribed routine on **Key Concept Cards 59 and 60.**

---

3 Close
Lesson Review

▶ **Visual Summary**

Have students save their lesson Study Guides. They will use them at the end of the chapter for review.

▶ **Make a FOLDABLES Study Guide**

See pp. R27–R28 for instructions.

FAST TRACK **Think, Talk, and Write**

1 Main Idea Solids have definite shape and volume. Liquids have definite volume but no definite shape. Gases have no definite shape or volume.

2 Vocabulary gas

3 Classify

Book = Solid	Water = Liquid
Air = Gas	

4 Critical Thinking Possible answer: A solid differs from a liquid and a gas because it keeps its shape and volume. One way that a gas and a liquid are alike is that they both change shape to fit the shape of their containers. A gas differs from a liquid because the volume of a gas changes to fit its container, while the volume of a liquid does not.

5 Test Prep A

Math Link 378 small balloons

Art Link Diagrams should include descriptions of shape and volume and how far apart particles are and how they move.

RESOURCES and TECHNOLOGY

▶ **Reading and Writing,** p. 182

▶ **School to Home Activities,** pp. 77–78

▶ **Science in Motion,** *Particles of a Solid, Liquid, and Gas*

▶ **e-Review** Narrated Summary and Quiz

▶ **Progress Reporter** Assessments

Writing in Science

Objective
- Write a description of an everyday object.

Describe Matter

Learn It

Discuss how "describing words" refer to an object's properties. To describe something, you must know its properties. A good description includes information about a combination of properties that apply only to that object.

Try It

- Have students read the first paragraph on this page and answer the questions about a pizza's properties. Write the describing words for the pizza on the board. Ask students to explain why these are describing words. Ask volunteers to group the details about the pizza in an order that makes sense.

Apply It

- Make illustrations of some of the pizza descriptions. Compare and contrast the pictures of the different pizzas. Then have students work in groups to write menu descriptions for one or two different kinds of pizzas. Have groups share with the class.

 Write About It

- Have students work individually. Suggest they refer to the "Descriptive Writing" box as needed to make sure they include all the parts of a good description.

- Invite student pairs to exchange papers and identify the describing words in each other's descriptions.

RESOURCES and TECHNOLOGY

▶ **Reading and Writing,** pp. 183–184

e-Journal Online research and writing

Writing Rubric P. TR26

Writing in Science

Describe Matter

You can describe matter in many ways. How would you describe a pizza to someone who has never seen one? How does it look? How does it smell? These are some of the pizza's observable properties. How big is the pizza? What is its mass? These are some of its measurable properties. Is it a solid or a liquid? This is its state of matter.

Descriptive Writing

A good description

▶ includes describing words to tell how something looks, sounds, feels, smells or tastes

▶ uses details to create a picture for the reader

▶ groups together details in an order that makes sense

 **Write About It**

Descriptive Writing Think of an object you use every day, such as your book bag. How would you describe it to someone who has never seen it before? Use the object's properties to write a description of the object.

e-Journal Write about it online at www.macmillanmh.com

Integrate Writing

Classifying Matter

- Have students choose a solid, such as salt, a rubber ball, or a baseball bat.

- Ask students to write a description of the solid. Tell them to think about which details are necessary for someone to classify the object as a solid.

- Post descriptions on the bulletin board.

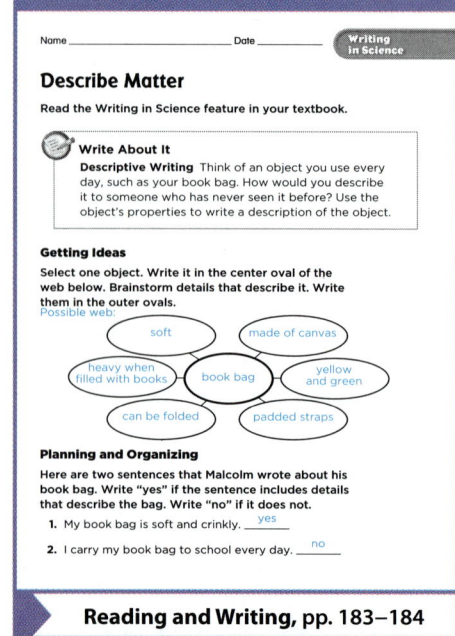

Name _____ Date _____ Writing in Science

Describe Matter

Read the Writing in Science feature in your textbook.

Write About It
Descriptive Writing Think of an object you use every day, such as your book bag. How would you describe it to someone who has never seen it before? Use the object's properties to write a description of the object.

Getting Ideas
Select one object. Write it in the center oval of the web below. Brainstorm details that describe it. Write them in the outer ovals.
Possible web:

- soft
- made of canvas
- heavy when filled with books
- book bag
- yellow and green
- can be folded
- padded straps

Planning and Organizing
Here are two sentences that Malcolm wrote about his book bag. Write "yes" if the sentence includes details that describe the bag. Write "no" if it does not.

1. My book bag is soft and crinkly. ___yes___
2. I carry my book bag to school every day. ___no___

▶ **Reading and Writing,** pp. 183–184

Measuring Perimeter

Solids come in many shapes and sizes. They can be round like a ball or square like a brick. They can be huge like a skyscraper or tiny like a grain of sand. You can measure the distance around a solid. The distance around a solid object is called the *perimeter*.

Find the Perimeter

▶ To find the perimeter of an object, add the lengths of all of its sides.

6 + 2 + 6 + 2 = 16
This rectangle's perimeter is 16.

 ### Solve It

Find the perimeter of the red square. Find the perimeter of the blue triangle. How can you find the perimeter of the entire house? Try it.

391
EXTEND

Math in Science

Objective

■ Demonstrate the method of finding the perimeter of a solid.

Measuring Perimeter

Learn It

Review shapes with students. Remind them that triangles have three sides, squares have four equal sides, and rectangles have two pairs of equal sides. Ask:

■ **How would I find the perimeter of one of these shapes?** add the lengths of all the sides

Try It

■ Draw a rectangle, slightly wider than it is tall, on the board. Then draw a diagonal line connecting two of the corners. Label the vertical lines *3*, the horizontal lines *4*, and the diagonal *5*.

■ **What is the perimeter of the rectangle?**
3 + 4 + 3 + 4 = 14

■ **What is the perimeter of the triangle?** 3 + 4 + 5 = 12

Apply It

■ Draw a rectangle, slightly taller than it is wide, on the board. Tell students this represents a plot of land. Add the footprint of a house, a path to the sidewalk, and a tree or two. Label the vertical lines *30* and the horizontal lines *24*. Tell students the piece of land measures 30 meters by 24 meters.

■ **Have students calculate the perimeter of the property.** 30 + 24 + 30 + 24 = 108 meters

Solve It

■ 36; 23; Total the length of the bottom and the two sides of the rectangle and the two sides of the triangle: 9 + 9 + 9 + 7 + 7 = 41

RESOURCES and TECHNOLOGY

▶ **Math in Science, pp. 15–16**

Integrate Math

Classroom Perimeters

• Have students use a cloth tape measure to find the length and width of various rectangular objects in the classroom, such as desktops or books. Tell them to write down the measurements.

• Have students calculate the perimeter of the objects they measured.

• Students can check their calculations by measuring the perimeter of the objects with the tape measure.

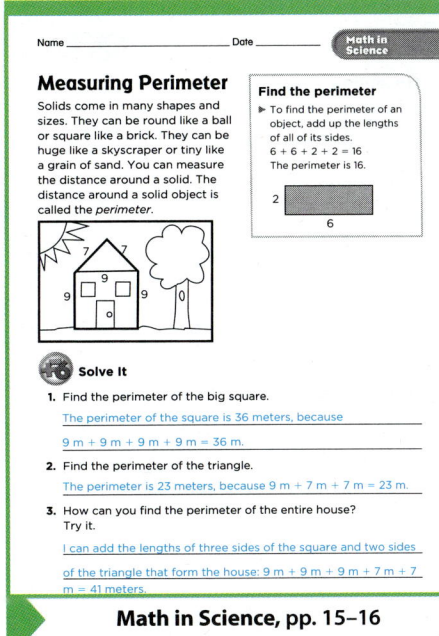

Measuring Perimeter

Solids come in many shapes and sizes. They can be round like a ball or square like a brick. They can be huge like a skyscraper or tiny like a grain of sand. You can measure the distance around a solid. The distance around a solid object is called the *perimeter*.

Find the perimeter
▶ To find the perimeter of an object, add up the lengths of all of its sides.
6 + 6 + 2 + 2 = 16
The perimeter is 16.

Solve It
1. Find the perimeter of the big square.
The perimeter of the square is 36 meters, because 9 m + 9 m + 9 m + 9 m = 36 m.
2. Find the perimeter of the triangle.
The perimeter is 23 meters, because 9 m + 7 m + 7 m = 23 m.
3. How can you find the perimeter of the entire house? Try it.
I can add the lengths of three sides of the square and two sides of the triangle that form the house: 9 m + 9 m + 9 m + 7 m + 7 m = 41 meters.

Math in Science, pp. 15–16

Visual Summary

Have students look at the pictures to review the main ideas of the chapter.

Vocabulary

1. gas
2. volume
3. metric system
4. liquid
5. gravity
6. solid
7. mass
8. elements
9. properties
10. matter

Make a FOLDABLES Study Guide

See pages R27–R28 in the back of the Teacher's Edition for more information on Foldables. Shown below is a back view of a sample study guide for this chapter review.

CHAPTER 9 Review

Visual Summary

Lesson 1 Matter is anything that has volume and mass. You can use properties to describe and identify matter.

Lesson 2 Matter can be measured using tools that record standard units.

Lesson 3 Solids, liquids, and gases are three forms of matter.

Make a FOLDABLES Study Guide

Glue your lesson study guides to a piece of paper as shown. Use your study guide to review what you have learned in this chapter.

Vocabulary

Fill each blank with the best term from the list.

elements, p. 368	**matter**, p. 364
gas, p. 387	**metric system**, p. 374
gravity, p. 378	**properties**, p. 365
liquid, p. 386	**solid**, p. 384
mass, p. 365	**volume**, p. 365

1. Matter with no definite shape or volume is a _____.

2. The amount of space an object takes up is its _____.

3. Scientists make measurements using the _____.

4. If matter has a definite volume, but not a definite shape, it is in a _____ state.

5. The pulling force that holds you on Earth is called _____.

6. Matter with a definite shape and volume is a _____.

7. The amount of matter in an object is its _____.

8. All matter is made up of _____.

9. Size and color are examples of _____.

10. Anything that has mass and volume is _____.

e-Review Summaries and quizzes online at **www.macmillanmh.com**

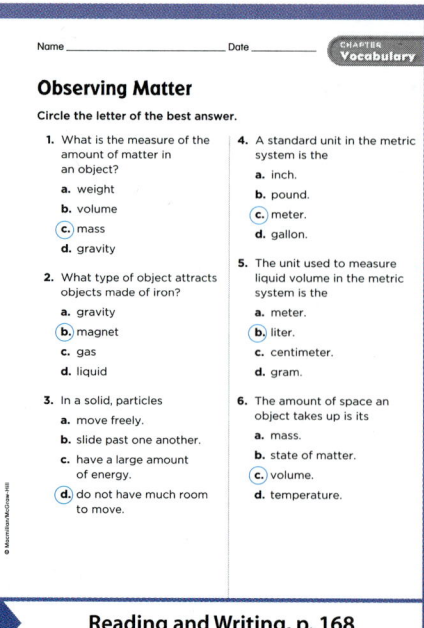

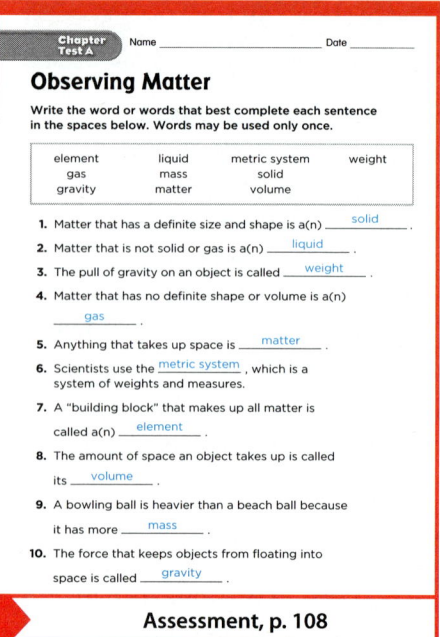

Skills and Concepts

Answer each of the following in complete sentences.

11. Summarize Name three properties of an object that you can measure using the metric system. What standard units would you use for each?

2. Descriptive Writing Write a brief description of a solid, a liquid, and a gas. Include a diagram with your description.

3. Measure What steps do you take to measure the mass of an object with a pan balance?

4. Critical Thinking Where can you find the three states of matter in a car?

5. What properties might the two objects shown below have in common? How do you think their properties might be different?

gold

aluminum

16. What are some ways you can describe matter?

What Is It Made Of?

▶ Make a book about some of the matter that surrounds you every day—the clothes you wear.

▶ Choose some of your favorite articles of clothing. Then describe their physical properties. What materials are they made of? What colors are they? What other properties do they have?

▶ Put a picture or drawing of each piece of clothing on a page in your book. Include a description of the properties of the clothing next to each item.

▶ Choose two pieces of clothing. How are they the same? How are they different? Use their properties to describe their similarities and differences.

Test Prep

1. In the metric system the volume of a liquid is measured in

A liters.

B inches.

C centimeters.

D meters.

393

Skills and Concepts

11. Summarize temperature (degrees Celsius), length (meters), volume (liters), mass (grams).

12. Descriptive Writing A solid has a definite volume and shape. A liquid has a definite volume. gas does not have a definite volume or shape.

13. Measure Place the object on one end of a pan balance. Then add known masses to the other side until both sides are even. When the two sides are even, add the amounts of the known masses. This sum is the mass of the object.

14. Critical Thinking Possible answer: Solids include the body of the car; liquids include gasoline, oil, and windshield-wiper fluid; gases include the air inside the inflated tires of the car.

15. Both objects are solids, and both are made up of mostly one element. Answers will vary, but students should consider properties such as hardness, size, shape and ability to float.

 16. Students should use information from the chapter to answer.

Performance Assessment

Scoring Rubric

4 points Student has (1) listed a variety of articles of clothing to investigate; (2) described the properties of those materials; (3) made a book that includes drawings and descriptions (4) compared and contrasted the properties of two articles of clothing.

3 points Student has correctly given three of the four possible responses.

2 points Student has correctly given two of the four possible responses.

1 point Student has correctly given one of the four possible responses.

Test Prep

1. A

Summative Assessment and Intervention

Assessment provides tests for Chapter 9.

Leveled Readers may be used to reteach lesson content in an alternative format. The leveled readers deliver chapter content in different readability levels. The back cover of each reader provides comprehension building activities specific to the book content. (see p. 361)

Key Concept Cards 55–60 contain prescribed routines for student intervention.

Presentation Toolkit CD-ROM Lesson Presentations

 Instructional Navigator CD-ROM Interactive Lesson Planner, Teacher's Edition, Worksheets, and Online Resources

Lesson	OBJECTIVES AND READING SKILLS	VOCABULARY	RESOURCES AND TECHNOLOGY
1 Changes of State PAGES **396–403** PACING: 2 days FAST TRACK: 1 day	■ Measure and record the temperature of water in different states. ■ Identify the effect of heating and cooling matter. **Reading Skill** Predict *What I Predict / What Happens* *Graphic Organizer 3*	melt boil evaporate condense freeze	▶ Reading and Writing, pp. 188–191 ▷ Visual Literacy, pp. 61–62 ▶ Activity Lab Book, pp. 167–170 Transparencies, pp. 61–62 **Operation: Science Quest,** *Physical and Chemical Changes* **Science in Motion,** *Heating Water*
2 Physical Changes PAGES **406–413** PACING: 2 days FAST TRACK: 1 day	■ Define physical changes as those that do not change the identity of a material. ■ Describe how to make and separate mixtures. **Reading Skill** Draw Conclusions *Text Clues / Conclusions* *Graphic Organizer 13*	physical change mixture solution	▶ Reading and Writing, pp. 192–195 ▷ Visual Literacy, pp. 63–64 ▶ Activity Lab Book, pp. 174–177 Transparencies, pp. 63–64
3 Chemical Changes PAGES **416–421** PACING: 2 days FAST TRACK: 1 day	■ Describe chemical changes. ■ Understand that chemical changes are part of our everyday life. **Reading Skill** Infer *Clues / What I Know / What I Infer* *Graphic Organizer 14*	chemical change	▶ Reading and Writing, pp. 198–201 ▷ Visual Literacy, pp. 65–66 ▶ Activity Lab Book, pp. 179–181 Transparencies, pp. 65–66

Chapter 10 Review PAGES **424–425**	■ Summarize chapter concepts.	**Resources** ▶ Assessment, pp. 121–124, 128–131 ▶ Reading and Writing, pp. 202–203	**Technology** Progress Reporter Assessments e-Review

PACING   Assumes a day is a 25- to 35-minute session

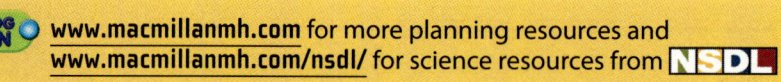

www.macmillanmh.com for more planning resources and
www.macmillanmh.com/nsdl/ for science resources from **NSDL**

 **Science Activity DVD** Explore Activity demos

Materials included in the Activity Kit are listed in *italics*.

EXPLORE Activities

Explore *p. 397* | PACING: 30 minutes

Objective Observe a change of state from a solid to a liquid.
Skills measure, communicate, infer, predict
Materials, *thermometer*, *plastic cup* of ice, *spoon*

⭐ **PLAN AHEAD** Provide waste containers for water if no sink is available.

Explore *p. 407* | PACING: 20 minutes

Objective Explore ways that matter changes without becoming a different substance.
Skills observe, experiment, infer
Materials paper, *clay*, ice cubes, scissors

⭐ **PLAN AHEAD** Caution students to handle scissors carefully.

Explore *p. 417* | PACING: 30 minutes

Objective Observe a chemical change that produces a gas.
Skills observe, measure, experiment, infer
Materials *vinegar*, *flour*, *baking soda*, 2 *balloons*, 2 *plastic bottles*, *goggles*, *funnel*, *measuring cup* and *spoons*

⭐ **PLAN AHEAD** Collect clear, half-liter bottles with narrow openings.

QUICK LAB Activities

Quick Lab *p. 401* | PACING: 15 minutes

Objective Demonstrate how water vapor condenses into a liquid.
Skills observe, infer
Materials *plastic cup*, ice cubes, water

⭐ **PLAN AHEAD** Increase humidity in dry climates by boiling water in the classroom.

Quick Lab *p. 412* | PACING: 15 minutes

Objective Demonstrate the separation of a mixture.
Skills experiment, observe
Materials *marbles*, *sand*, *metal paper clips*, *bowl*, colander, magnet

⭐ **PLAN AHEAD** Remind students that an experiment involves a series of steps that will be done.

Quick Lab *p. 419* | PACING: 15 minutes

Objective Observe an example of a chemical change.
Skills observe, experiment, infer
Materials dull pennies, *bowl*, *salt*, *vinegar*, spoon or stirrer, teaspoon

⭐ **PLAN AHEAD** Collect dull pennies for students to use.

FOR MORE ACTIVITIES

Activity Flipchart
for Science Centers and Workstations

Focus on Skills	Be a Scientist	Learning Lab	Everyday Science
Teach the inquiry skill: Predict, p. 46	Physical and chemical changes, p. 49		Matter, p. 65; Physical/Chemical Changes, p. 66

See **Activity Lab Book Teacher's Guide** for more support.
For a comprehensive list of consumable and non-consumable materials, see the back of the unit tab.

Technology

For additional language support and vocabulary development, go to **www.macmillanmh.com**

- **Science in Motion**
  Heating Water

- **Vocabulary Games**

Academic Language

English language learners need help in building their understanding of the academic language used in daily instruction and science activities. The following strategies will help to increase students' language proficiency and comprehension of content and instruction words.

Strategies to Reinforce Academic Language

- **Use Context** Academic language should be explained in the context of the task. Use gestures, expressions, and visuals to support meaning.

- **Use Visuals** Use charts, transparencies, and graphic organizers to explain key labels, to help students understand classroom language.

- **Model** Use academic language as you demonstrate the task, to help students understand instruction.

Academic Language Vocabulary Chart

The following chart shows chapter vocabulary and inquiry skills as well as some Spanish cognates. **Vocabulary** words help students comprehend the main ideas. **Inquiry Skills** help students develop questions and perform investigations. **Cognates** are words that are similar in English and Spanish.

Vocabulary	Inquiry Skills	Cognates	
		English	**Spanish**
melt, p. 398	measure, p. 397	communicate, p. 397	*comunicar*
boil, p. 399	communicate, p. 397	infer, p. 397	*inferir*
evaporate, p. 399	infer, p. 397	predict, p.397	*predecir*
condense, p. 400	predict, p. 397	evaporate, p. 399	*evaporar*
freeze, p. 401	observe, p. 407	condense, p. 400	*condensar*
physical change, p. 408	experiment, p. 407	observe, p. 407	*observar*
mixture, p. 410		solution, p. 411	*solución*
solution, p. 411			
chemical change, p. 418			

Vocabulary Routine

Use the routine below to discuss the meaning of each word on the vocabulary chart. Use gestures and visuals to model all words.

Define Evaporate means to change from a liquid into a gas slowly.

Example If you get wet on a hot day, the water on your body will evaporate and you will be dry again.

Ask Does water evaporate faster on a hot day or on a cool day?

Students may respond to questions according to proficiency level with gestures, one-word answers, or phrases.

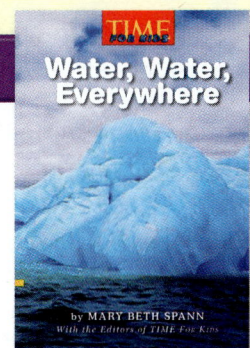

ELL Leveled Reader

Water, Water, Everywhere by Mary Beth Spann

Summary Study Earth's water cycle and water's changing states.

Reading Skill Compare and Contrast

Vocabulary Activities

Help students understand the different states of water.

BEGINNING Walk students through the photographs on pages 398–401. Write on the board the terms *melt*, *boil*, *evaporate*, *condense*, and *freeze*. Students point to or name the state of water for each of the following clues you give: *Dripping ice cube.* **(melt)** *Water bubbling.* **(boil)** *Clothes drying.* **(evaporate)** *Dewdrops on the grass.* **(condense)** *Ice cubes.* **(freeze)**

INTERMEDIATE Walk students through the photographs on pages 398–401. Write on the board the terms *melt, boil, evaporate, condense,* and *freeze*. Show a bottle of water. Ask: *How can this water change state?* Students give concrete examples, using the words on the board.

ADVANCED Assign each student one of the five water words: *melt, boil, evaporate, condense,* and *freeze*. The student draws a picture showing an example of the word, with a sentence describing what is going on. Group students with different words and have them compare their pictures. Have groups discuss how water can change.

Language Transfers

Grammar Transfer
In Hmong and Vietnamese, pronouns may be overused with nouns:
The water, it boils and evaporates.

Phonics Transfer
Hmong does not have the /oi/ sound as in *boil*.

Changes in Matter

THE BIG IDEA In what ways can matter change?

Chapter Preview Examine the titles in the chapter. Use these titles to predict what the lessons will be about.

▶ Assess Prior Knowledge

Before reading the chapter, create a **KWL** chart with students. Read the Big Idea question and then ask:

- How does matter change from one state to another?
- What type of change occurs when the identity of a material stays the same?
- What type of change happens when a match burns?

Changes In Matter

What We **K**now	What We **W**ant to Know	What We **L**earned
When a solid melts, it changes to a liquid.	What causes a solid to melt?	
A rock broken into pieces is still the same kind of rock.	What kind of change changes the size of something?	
Oxygen and iron form rust.		

Answers shown represent sample student responses.

Follow the **Instructional Plan** at right after assessing students' prior knowledge of chapter content.

RESOURCES and TECHNOLOGY

- **School to Home Activities**, pp. 81–86
- **Reading and Writing**, pp. 187–203
- **Assessment**, pp. 121–133
- **Presentation Toolkit CD-ROM**
- **PuzzleMaker CD-ROM**
-  **www.macmillanmh.com**
-  **e-Journal**

Changes in Matter

Lesson 1
Changes of State... 396
Lesson 2
Physical Changes.. 406
Lesson 3
Chemical Changes.. 416

The Big Idea
In what ways can matter change?

394

Differentiated Instruction

Instructional Plan

Chapter Concept Some changes in matter are reversible.

EXTRA SUPPORT Students who need to describe physical changes of matter should cover **Lesson 2**, pp, 406–415, in sequence after completing **Lesson 1**, pp. 396–405.

ON LEVEL Students who can describe physical changes can cover just mixtures and separating mixtures in **Lesson 2**, pp. 410–413, after covering change of state in **Lesson 1**, pp. 396–405.

ENRICHMENT For students who are ready to go further, **Lesson 3**, pp. 416–423, builds on chemical changes from Grade 2 by offering comparisons with physical changes.

Key Vocabulary

melt
to change from a solid to a liquid (p. 398)

boil
to change from a liquid to a gas (p. 399)

physical change
a change in the way matter looks (p. 408)

mixture
different kinds of matter mixed together (p. 410)

solution
when one or more kinds of matter are mixed evenly into another kind of matter; a solution can be a solid, a liquid, or a gas (p. 411)

chemical change
a change that causes different kinds of matter to form (p. 418)

More Vocabulary

evaporate, p. 399

condense, p. 400

freeze, p. 401

395

Vocabulary Preview

- Have a volunteer read the **Key Vocabulary** words aloud to the class. Ask students to find one or two words in the chapter by using the given page references. Add these words and their definitions to a class "Word Wall."

- Encourage students to use the illustrated glossary in the student edition's reference section. Guide students to explore the **e-Glossary,** which offers audio pronunciations, definitions, and sentences using the vocabulary words.

Science Leveled Readers

ALSO ON AUDIO CD
Leveled Reader Library

APPROACHING
Glassmaking Photos tell the history and step-by-step process of this ancient art
ISBN: 0-02-284668-9

ON LEVEL
Water, Water, Everywhere Study Earth's water cycle and water's changing states.
ISBN: 0-02-284669-7

BEYOND
Chocolate Here's a lesson about chocolate and the chocolate-making process.
ISBN: 0-02-284671-9

ELL
Water, Water, Everywhere Uses sheltered language of On-Level Reader.
ISBN: 0-02-283484-2

See teaching strategies in the Leveled Reader Teacher's Guide. To order, call 1-800-442-9685.

 Leveled Reader Database Online Readers, searchable by topic, reading level, and keywords

Plan Your Lesson

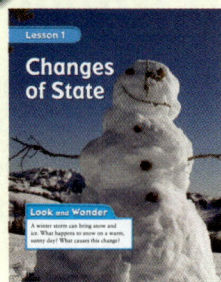

Lesson 1 Changes of State

Objective

- Measure and record the temperature of water in different states.
- Identify the effects of heating and cooling matter.

Reading Skill Predict

What I Predict	What Happens

Graphic Organizer 3, p. TR5

LOG ON **Professional Development** Look for **NSDL** to find recommended Science Background materials from the National Science Digital Library

FAST TRACK

Lesson Plan When time is short, follow the Fast Track and use the essential resources.

1 Introduce
Look and Wonder, p. 396

Resource Alternative Explore, p. 169

2 Teach
Discuss the Main Idea, p. 398
Develop Vocabulary, p. 400

Resource Visual Literacy, p. 61

3 Close
Think, Talk, and Write, p. 403

Resource Assessment, p. 125

Reading and Writing

LESSON Outline Name _____ Date _____

Changes of State

Use your textbook to help you fill in the blanks.

What happens when matter is heated?

1. When something melts, it changes from a(n) _____solid_____ to a(n) _____liquid_____.

2. Matter gains _____energy_____ when it is heated.

3. Particles in solids are held _____close together_____

4. Particles in liquids _____flow around_____ one another.

5. When something boils, it changes from a(n) _____liquid_____ to a(n) _____gas_____.

6. Heat causes particles in a liquid to move _____faster_____ and _____spread out_____

7. Liquids can slowly change into a gas, a process known as _____evaporation_____

8. Water in the form of a gas is called _____water vapor_____.

What happens when matter is cooled?

9. When a solid, a liquid, or a gas is cooled, it _____loses_____ energy.

10. When a gas cools to the right temperature, it will _____condense_____.

Outline, pp. 188–189
Also available as a student workbook

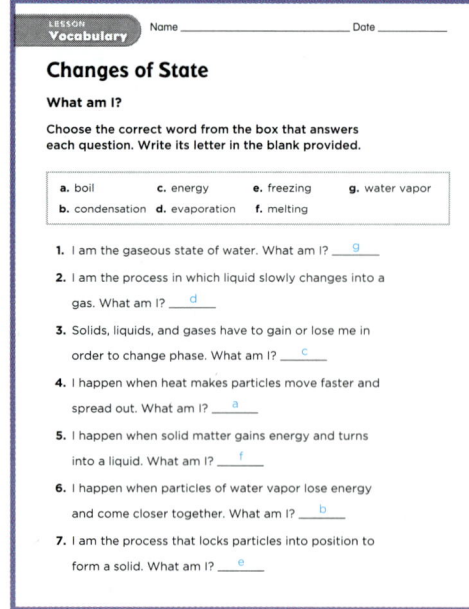

LESSON Vocabulary Name _____ Date _____

Changes of State

What am I?

Choose the correct word from the box that answers each question. Write its letter in the blank provided.

a. boil	c. energy	e. freezing	g. water vapor
b. condensation	d. evaporation	f. melting	

1. I am the gaseous state of water. What am I? __g__

2. I am the process in which liquid slowly changes into a gas. What am I? __d__

3. Solids, liquids, and gases have to gain or lose me in order to change phase. What am I? __c__

4. I happen when heat makes particles move faster and spread out. What am I? __a__

5. I happen when solid matter gains energy and turns into a liquid. What am I? __f__

6. I happen when particles of water vapor lose energy and come closer together. What am I? __b__

7. I am the process that locks particles into position to form a solid. What am I? __e__

Vocabulary, p. 190
Also available as a student workbook

Visual Literacy

Name _____ Date _____ **Read a Diagram**

What happens when matter is heated?
This diagram shows water in its three different states. Notice how the water looks different in each photograph.

Heating Water

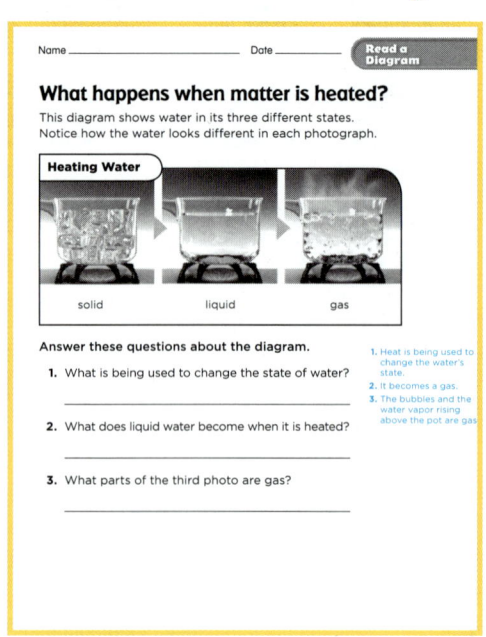

solid   liquid   gas

Answer these questions about the diagram.

1. What is being used to change the state of water?

2. What does liquid water become when it is heated?

3. What parts of the third photo are gas?

1. Heat is being used to change the water's state.
2. It becomes a gas
3. The bubbles and the water vapor rising above the pot are gas

FAST TRACK **Read a Diagram, p. 61**
Also available as a transparency

▶ Activity Lab Book

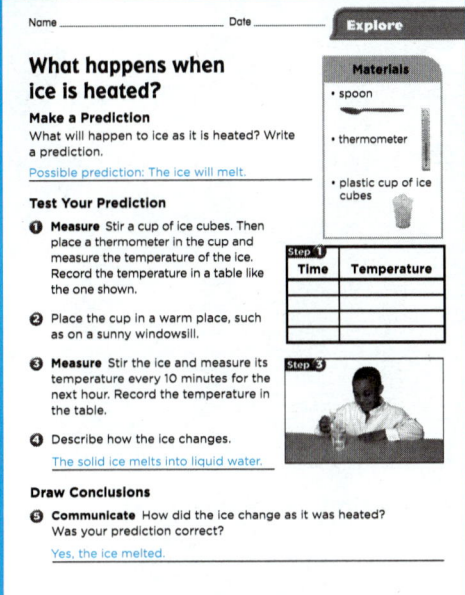

Explore, pp. 167–168
Also available as a student workbook

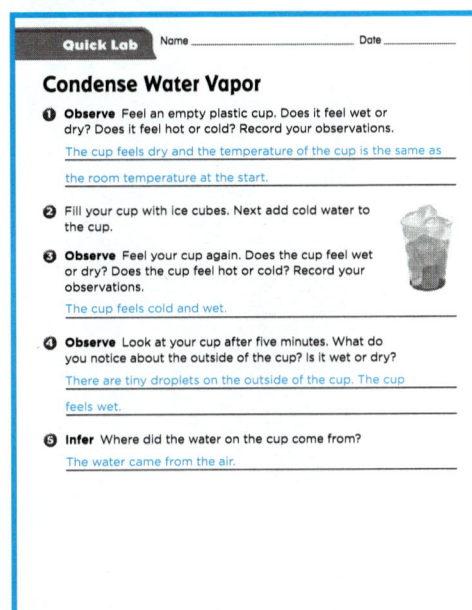

Quick Lab, p. 170
Also available as a student workbook

▶ Assessment

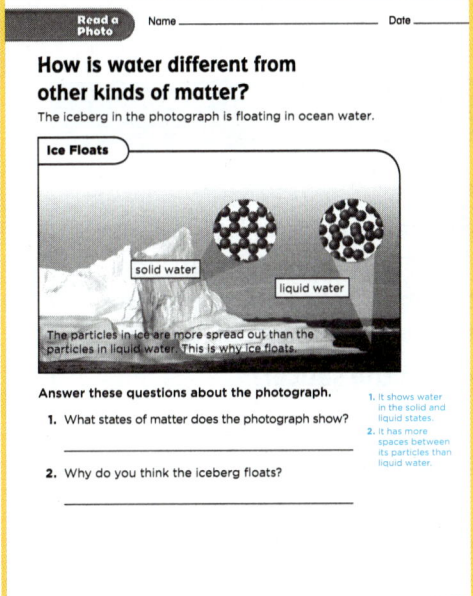

Read a Photo, p. 62
Also available as a transparency

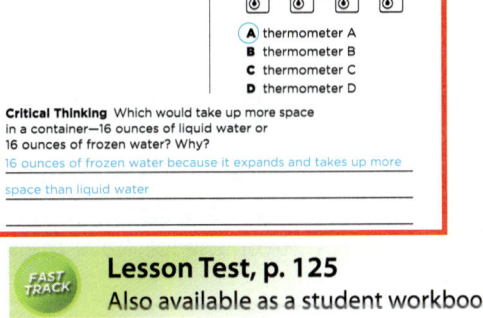

Lesson Test, p. 125
Also available as a student workbook

ADDITIONAL RESOURCES

p. 46

p. 115

pp. 57–58

pp. 75–76

57–58

146–150

Technology

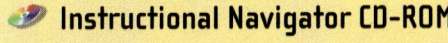

🌀 **Science Activity DVD**

🌀 **Instructional Navigator CD-ROM**

🌀 **Presentation Toolkit CD-ROM**

SCIENCE QUEST *Physical and Chemical Changes*

LOG ON 🔵 **Science in Motion** *Heating Water*

LOG ON ℮ **-Review**

LOG ON 🔵 **NSDL**

Lesson 1 Changes of State

Objectives
- Measure and record the temperature of water in different states.
- Identify the effects of heating and cooling matter.

1 Introduce

▶ **Assess Prior Knowledge**

Have students discuss familiar changes of state. Ask:

- **What happens to ice cream when it sits outside the freezer?** It changes from a solid to a liquid.

- **What happens to melted candle wax after the flame goes out?** It changes from a liquid to a solid.

Have students list other examples of changes of state.

Look and Wonder

Invite students to share their responses to the Look and Wonder question:

- **What happens to snow on a warm, sunny day? What causes this change?** Possible answer: The snow melts; it warms up as energy is added to it.

Write ideas on the board and note any misconceptions that students may have. Address these misconceptions as you teach the lesson.

RESOURCES and TECHNOLOGY
▶ **Activity Lab Book,** pp. 167–169
▶ **Activity Flipchart,** p. 45
💿 **Science Activity DVD**

Lesson 1

Changes of State

Look and Wonder

A winter storm can bring snow and ice. What happens to snow on a warm, sunny day? What causes this change?

396
ENGAGE

Warm Up

Start with a Demonstration

Add hot water to a beaker or clear plastic cup until it is about half full. Place a glass saucer on top of the beaker or cup, and place two ice cubes on the saucer. After two minutes, pick up the saucer and have students observe the bottom of it. Ask:

- **What do you see on the bottom of the saucer?** drops of water

- **Where do you think the water came from?** Possible answer: from water in the air

Tell students that they will learn about how materials change from one state to another. If students think the water came from the ice, have them revisit this response later in the lesson.

Explore

Inquiry Activity

What happens when ice is heated?

Make a Prediction

How does ice change as it is heated? Write a prediction.

Test Your Prediction

1. **Measure** Place a thermometer in a cup of ice. Measure the temperature of the ice. Record the temperature in a table like the one shown.

2. Place the cup in a warm place, such as on a sunny windowsill.

3. **Measure** Stir the ice and measure its temperature every 10 minutes for the next hour. Record the temperature in the table.

4. Describe how the ice changes.

Draw Conclusions

5. **Communicate** How did the ice change as it was heated? Was your prediction correct?

6. **Infer** What happened to the temperature of the water as the ice melted? At what temperature does ice melt?

Explore More

Predict What will happen to the water as it continues to sit in the warm place after the ice has melted? Test your prediction and find out.

Materials

thermometer

plastic cup of ice

spoon

Step 1	
Time	Temperature

Step 3

397
EXPLORE

Explore

small groups · 30 minutes

Plan Ahead Have coolers available to store ice until it is needed. Make sure enough thermometers are available. Have waste containers for melted ice if a sink is not available.

Purpose This activity helps students observe a change in state from solid to liquid water.

Structured Inquiry

Make a Prediction Possible prediction: The ice will melt as it is heated.

1. **Measure** **Be Careful!** Caution students to be careful when using the thermometers, as they can break easily.

5. **Communicate** Answers will vary depending on the predictions. The ice melted.

6. **Infer** The temperature stayed the same until all of the ice melted. All the energy was used to melt the ice, not lower the temperature.

Guided Inquiry Explore More

Predict Possible prediction: The temperature of the water will increase; some of the water will evaporate.

Open Inquiry

Ask students how the results might be different if they had used warm water instead of cold to make the ice. Have them think of their own question about how water temperature would affect the results. Ask students to make a plan and carry out an experiment to answer the question. Ask:

How does the starting temperature of the water affect the results of the experiment?

Alternative Explore

How does energy affect state?

Materials room-temperature water

Have each student place a drop of room-temperature water on his or her forearm, and spread it around. Have them observe how the water feels on their skin. Have them allow the water to evaporate and explain why this change of state occurs.

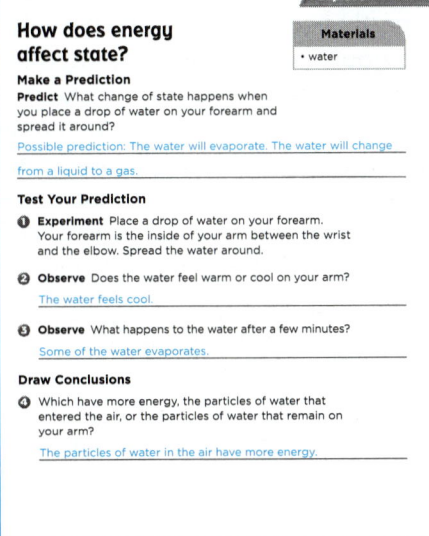

Name _____ Date _____

Alternative Explore

How does energy affect state?

Materials
• water

Make a Prediction
Predict What change of state happens when you place a drop of water on your forearm and spread it around?

Possible prediction: The water will evaporate. The water will change from a liquid to a gas.

Test Your Prediction
1. **Experiment** Place a drop of water on your forearm. Your forearm is the inside of your arm between the wrist and the elbow. Spread the water around.

2. **Observe** Does the water feel warm or cool on your arm?
The water feels cool.

3. **Observe** What happens to the water after a few minutes?
Some of the water evaporates.

Draw Conclusions
4. Which have more energy, the particles of water that entered the air, or the particles of water that remain on your arm?
The particles of water in the air have more energy.

Activity Lab Book, p. 169

2 Teach

Read and Learn

Main Idea Have students examine the visuals in the lesson, and ask them to discuss what they think they will learn about.

Vocabulary Have students examine the vocabulary terms in the lesson. Have them divide the terms into those that they know and those that they don't. Have them give examples of the terms they do know.

Reading Skill Predict
Graphic Organizer 3 Have students fill in a Predict graphic organizer as they read each two pages of the lesson. They can use the Quick Check questions to identify each prediction to make.

My Prediction	What Happens

What happens when matter is heated?

FAST TRACK **Discuss the Main Idea**

Have students discuss how adding energy to matter can change it. Ask:

- **What are some examples of changing matter by adding energy to it?** Possible answers: boiling water by heating it, melting a freezer pop by leaving it out of the freezer

- **What two ways can adding energy affect matter?** Possible answer: It can increase the temperature or change its state.

RESOURCES and TECHNOLOGY

- **Reading and Writing,** pp. 188–190
- **Visual Literacy,** p. 61
- **PuzzleMaker CD-ROM**
- **Presentation Toolkit CD-ROM**
- *Physical and Chemical Changes*
- **e-Glossary**

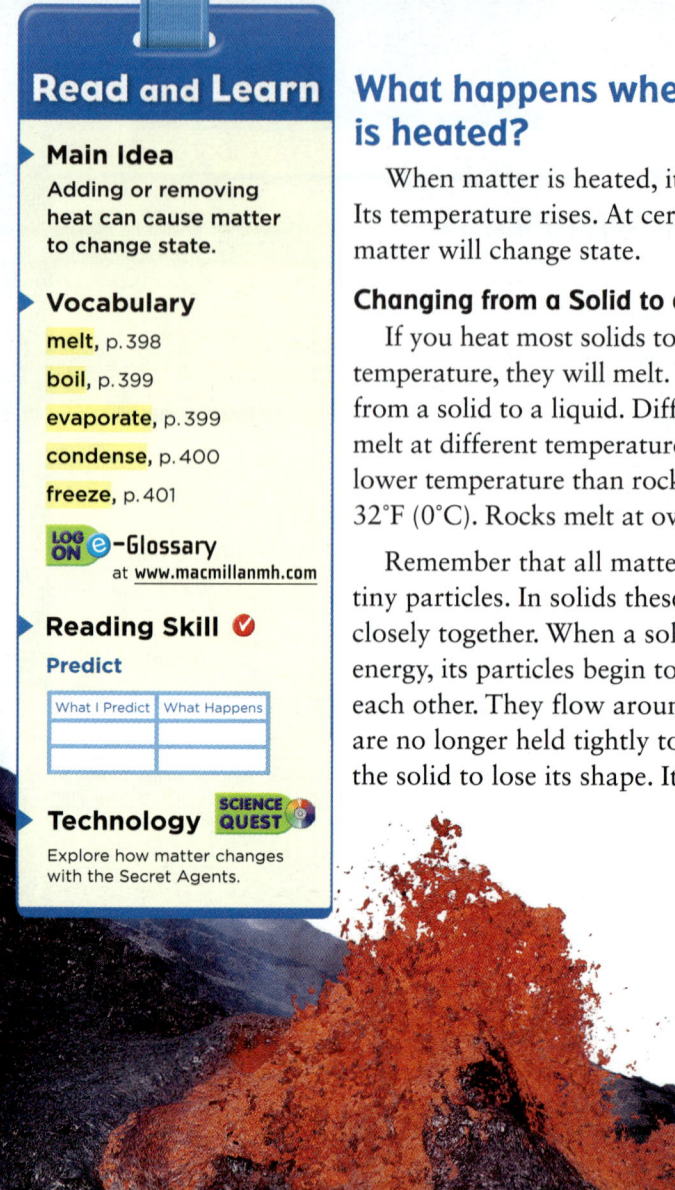

Read and Learn

▶ **Main Idea**
Adding or removing heat can cause matter to change state.

▶ **Vocabulary**
melt, p. 398
boil, p. 399
evaporate, p. 399
condense, p. 400
freeze, p. 401

LOG ON e-Glossary
at www.macmillanmh.com

▶ **Reading Skill** ✓
Predict

What I Predict	What Happens

▶ **Technology** **SCIENCE QUEST**
Explore how matter changes with the Secret Agents.

What happens when matter is heated?

When matter is heated, it gains energy. Its temperature rises. At certain temperatures, matter will change state.

Changing from a Solid to a Liquid

If you heat most solids to a high enough temperature, they will melt. To melt is to change from a solid to a liquid. Different kinds of matter melt at different temperatures. Ice melts at a lower temperature than rocks. Ice melts at 32°F (0°C). Rocks melt at over 1,100°F (593°C)!

Remember that all matter is made up of tiny particles. In solids these particles are held closely together. When a solid is heated and gains energy, its particles begin to move away from each other. They flow around each other and are no longer held tightly together. This causes the solid to lose its shape. It becomes a liquid.

The lava flowing from this volcano is rock that melted deep beneath Earth's surface.

398 EXPLAIN

Science Background

The temperature at which a material changes state depends partly on the mass of the particles in it and partly on the amount of attraction among them. For example, both water and bromine are liquids at room temperature. Because bromine particles are more massive than water's, the boiling point of bromine might be expected to be higher than water's boiling point. However, stronger electrical attraction exists among water molecules. The boiling point of water is higher because of the stronger attraction among water molecules.

See **Science Yellow Pages** in the Teacher Resource section for background information.

LOG ON Professional Development For more Science Background and resources from **NSDL** visit **www.macmillanmh.com/nsdl/**

Changing from a Liquid to a Gas

If you heat a liquid to a high enough temperature, it will boil. To **boil** is to change from a liquid to a gas. Energy from heat causes the particles in a liquid to move faster. They spread apart. The liquid turns into a gas.

Liquids can also **evaporate**, or change into a gas, slowly. When wet clothes are placed in the Sun, the water in the clothes evaporates. The Sun heats water droplets in the clothes. The water turns into a gas and your clothes dry. The gas state of water is called *water vapor*. You cannot see water vapor, but it is part of the air.

▲ These clothes will dry when the liquid water changes into a gas.

Heating Water

| solid | liquid | gas |

Read a Diagram

What happens when you heat ice?

Clue: Arrows show a sequence.

Science in Motion Watch how matter changes at www.macmillanmh.com

✓ *Quick Check*

Predict What will happen to cheese when it is heated?

Critical Thinking How does a blow dryer get your hair dry?

Differentiated Instruction

Leveled Activities

EXTRA SUPPORT Have students draw what happens to particles of liquid water when the water boils.

ENRICHMENT Ask students to research the difference between evaporation and boiling. Have them explain the difference in their own words.

▶ **Use the Visuals**

Refer students to the visuals at the bottom of page 399. Ask:

■ **What supplies energy in these photographs?** the gas flames

■ **What happens to the particles that absorb the energy?** They move faster. If they move fast enough, the material changes state.

■ **What change of state occurs?** Possible answers: photo 1: melting; photo 3: boiling

▶ **Develop Vocabulary**

melt *Word Origin* The word *melt* comes from the Proto–Indo-European prefix **meld-,** which means "softness." When solids melt, they become softer.

boil *Word Origin* The word *boil* comes from the Latin verb **bullire,** which means "to bubble, to seethe." Bubbles form when a liquid boils.

evaporate *Word Origin* The word *evaporate* comes probably from the Vulgar Latin (the everyday language of the Roman people) *extufare,* from *ex-,* meaning "out" and *vapor,* meaning "steam, vapor." When liquids evaporate, they change to a vapor.

Read a Diagram

Answer The water molecules absorb energy and start to move faster. When they move enough that they slide past each other, the water is a liquid.

 Science in Motion, *Heating Water*

✓ **Quick Check Answers**

• **Predict** It will melt.

• **Critical Thinking** Heat from the blow dryer is added to the water particles in the hair, causing them to move faster. When the particles move fast enough, they form a gas, leaving your hair and moving into the air.

What happens when matter is cooled?

▶ Discuss the Main Idea

Ask students to compare and contrast the terms that describe changes of state. Ask:

- During what processes is energy absorbed?
  boiling, evaporation, melting

- During what processes is energy given off?
  freezing, condensing

- What happens to the particles in liquid water when water freezes? They give off energy, slow down, and are locked in position.

- How is this different from what happens to the particles in liquid water when it boils? They absorb energy, speed up, and become a gas.

FAST TRACK Develop Vocabulary

condense *Word Origin* The word *condense* comes from the Latin verb **condensare,** which means "to make dense."

freeze *Scientific vs. Common Use* The word *freeze* often is used as meaning "to stop moving." When a liquid freezes, its particles do not stop moving, but they slow down and only move in place.

What happens when matter is cooled?

When matter is cooled, it loses energy. Its temperature drops. At certain temperatures, matter will change state.

Changing from a Gas to a Liquid

If you cool a gas to the right temperature, it will condense (kuhn•DENS). To **condense** is to change from a gas to a liquid. For example, on cool mornings, small droplets of water called dew can appear on grass and windows. This happens when water vapor in the air touches cool objects and loses energy. Particles of water vapor come closer together. They change into drops of liquid water.

Dew forms when water vapor in the air cools and condenses.

Water vapor in this tiger's breath condenses on a cold day.

400
EXPLAIN

Differentiated Instruction

Leveled Questions

EXTRA SUPPORT What process explains why dew forms on grass? condensation

ENRICHMENT Why should you wrap something that heat does not easily move through around water pipes in a cold location? Wrapping the pipes keeps the water from losing so much energy that it freezes.

RESOURCES and TECHNOLOGY

▶ **Activity Lab Book,** p. 170

◆ **Presentation Toolkit CD-ROM**

SCIENCE QUEST *Physical and Chemical Changes*

Changing from a Liquid to a Solid

If you cool a liquid to the right temperature, it will freeze. To **freeze** is to change from a liquid to a solid. The particles in the liquid lose energy and move slower and closer together. They get locked into position and form a solid. For example, when you put liquid water into the freezer, it loses energy. It cools to a certain temperature and turns into ice.

◄ When juice is cooled enough, it will freeze and become a solid.

Quick Lab

Condense Water Vapor

1 **Observe** Feel an empty plastic cup. Does it feel wet or dry? Does it feel hot or cold? Record your observations.

2 Fill your cup with ice cubes. Next add cold water to the cup.

3 **Observe** Feel your cup again. Does the cup feel wet or dry? Does the cup feel hot or cold? Record your observations.

4 **Observe** Look at your cup after five minutes. What do you notice about the outside of the cup? Is it wet or dry?

5 **Infer** Where did the water on the cup come from?

✓ Quick Check

Predict What will happen to water vapor when it is cooled?

Critical Thinking How could you make an ice pop?

401
EXPLAIN

ELL Support

Model/Pantomime Write the words *solid, liquid,* and *gas* on the board, and have students repeat them after you. Have students use small balls or other objects to model the motion of the particles during changes of state. For instance, when changing from solid to liquid, have them show the balls held together at first and then moving away from each other. Have them say the terms for each change of state as they model.

BEGINNING Ask students questions that require one-word answers such as, *Which liquid can also be solid and gas? What do you call frozen water?*

INTERMEDIATE Students can use phrases and short sentences to describe changes of state.

ADVANCED Students can describe changes of state by using complete sentences.

Quick Lab
 pairs   15 minutes

Objective Show how a gas condenses into a liquid.

Materials plastic cup, ice cubes, water

1 This activity will not work well if the relative humidity is low. If the humidity is too low, boiling water in the room before class will increase the humidity. Possible answer: The cup feels dry and cool.

3 Possible answer: The cup feels slightly damp and cold.

4 Possible answer: Small droplets of water are on the outside of the cup.

5 Be sure students understand that the water on the outside of the cup did not come from the water inside the cup. Possible answer: The water came from water vapor in the air when it condensed on the outside of the cold cup.

▶ Use the Visuals

Tell students that sometimes materials change directly from a solid to a gas without first becoming a liquid. Ask:

■ **What do you think happens when a piece of meat becomes "freezer burned"?** Possible answer: The ice in the meat changes to water vapor and leaves the meat, drying it out.

✓ Quick Check Answers

● **Predict** It will condense into a liquid if it is cooled enough.

● **Critical Thinking** Put some juice into an ice-pop form. Attach a thin, wooden stick. Put the shape into a freezer. The particles that make up the juice lose heat to the cold air in the freezer and begin to move more slowly. Eventually, the particles become fixed in place, and the juice becomes frozen.

How is water different from other kinds of matter?

▶ Discuss the Main Idea

To help in the discussion, draw a Main Idea graphic organizer on the board. It should show that water exists as ice, liquid water, and water vapor. Under each of these physical states for water, list details that students mention about them. These details could include descriptions of the states and where and how the states are used.

▶ Use the Visuals

Refer students to the photo on page 402. Using an overhead projector, show them six pennies arranged as closely together as possible. Tell them that those pennies represent the particles in liquid water. Next to those six pennies, show six more pennies that still touch and are in a pattern but are farther apart. These pennies represent the particles in ice. Tell students that ice floats on liquid water because the same number of particles of water take up more space in ice.

▶ Develop Vocabulary

Tell students that when a certain amount of matter takes up more space, it expands. When it takes up less space, it contracts. When water freezes, it expands, and when it melts, it contracts.

Ask students to give other examples of materials expanding and contracting. If possible, show students a picture of the expansion grates in a large bridge that allow the bridge to expand and contract with temperature changes without weakening the structure.

✔ Quick Check Answers

- **Predict** When the water in the bottle freezes, the bottle may crack because water increases in volume when it freezes.

- **Critical Thinking** Water's particles begin to take up less space because the particles are losing the empty spaces between them they had when the water was frozen.

How is water different from other kinds of matter?

Most kinds of matter shrink as they freeze. Their particles get packed more closely together. They take up less space. Yet water gets larger when it freezes.

As water freezes, its particles rearrange themselves. They make a special pattern. Empty spaces form between the particles. The frozen water takes up more space than the liquid water. This is why freezing a glass of water cracks the glass.

Ice floats in liquid water. This keeps lakes and ponds from freezing from the bottom up. Living things can survive under the ice.

 Quick Check

Predict What would happen if you put a plastic bottle filled with liquid water in the freezer? Why does this happen?

Critical Thinking Describe how water changes when it melts.

The particles in ice are more spread out than the particles in liquid water. This is why ice floats.

solid water

liquid water

FACT Ice, liquid water, and water vapor are all forms of water.

402 EXPLAIN

Homework Activity

Experiment with Freezing

Under the supervision of an adult, have students fill a round balloon with water and tie it off. Caution students to not fill the balloon to its maximum volume. The balloon must be able to stretch more as the water freezes. Have them measure the distance around the balloon. Then have them place the balloon in the freezer. When the water has frozen, have them measure the distance around the balloon again. Have them write their observations and their conclusions.

Lesson Review

Visual Summary

 When most **solids** are heated, they melt into **liquids**. When a **liquid** is heated, it changes into a **gas**.

 When a **gas** cools it usually condenses into a **liquid**. When a **liquid** cools it freezes into a **solid**.

 **Water** is a special kind of matter. It gets larger when it freezes.

Make a FOLDABLES Study Guide

Make a Layered-Look Book. Use it to summarize what you learned about changes of state.

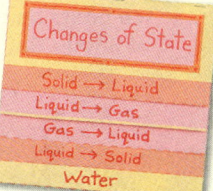

Changes of State
Solid → Liquid
Liquid → Gas
Gas → Liquid
Liquid → Solid
Water

Think, Talk, and Write

1 **Main Idea** How can heat change matter?

2 **Vocabulary** What happens when a gas condenses?

3 **Predict** After a rainstorm the Sun comes out and shines brightly. What will happen to puddles of rainwater?

What I Predict	What Happens

4 **Critical Thinking** You see drops of water on the bathroom mirror after a shower. What caused the water drops to form?

5 **Test Prep** How is water different from other liquids?
- **A** Water gets larger as it freezes.
- **B** Water gets smaller as it freezes.
- **C** Water stays the same as it freezes.
- **D** Water never freezes.

 Writing Link

Write a Story
Describe how your life would be different if liquids changed into solids when heated. For example, it could snow when the temperature outside was very hot.

Math Link

Find the Difference
Ice melts at 32°F. Water boils at 212°F. How many degrees are there between water's melting and boiling temperatures?

LOG ON **e-Review** Summaries and quizzes online at www.macmillanmh.com

403
EVALUATE

Formative Assessment

Approaching Have students find a picture in a magazine that shows a change of state, such as a liquid boiling or ice melting.

On-Level Have students write a paragraph explaining how water is different from other materials when it changes state.

Challenge Have students draw an Events Chain concept map. Starting with a solid, have them show the changes of state as more energy is added.

 **Key Concept Cards** For student intervention, see the prescribed routine on **Key Concept Cards 61 and 62.**

---

3 Close
Lesson Review

▶ **Summarize the Main Idea**

Have students save their lesson Study Guides. They will use them at the end of the chapter for review.

▶ **Make a** FOLDABLES **Study Guide**

See pages R27–R28 for instructions.

FAST TRACK

Think, Talk, and Write

1 **Main Idea** Adding heat can change solids to liquids and liquids to gases. Removing heat can change gases to liquids and liquids to solids.

2 **Vocabulary** Its particles lose energy and slow down until they form a liquid.

3 **Predict**

What I Predict	What Happens
The Sun will cause the water in the puddle to evaporate.	The water evaporates.

4 **Critical Thinking** Possible answer: Heat from the shower/bath water caused the air temperature to rise. Water vapor in the air condensed onto the mirror.

5 **Test Prep** A

Writing Link Before writing, have students choose solid objects around them and think about what effect it would have on students if those objects changed shape like liquids do. Have them also consider that their bodies could change shape.

Math Link 180° F

RESOURCES and TECHNOLOGY

▶ **Reading and Writing,** p. 191

▶ **School to Home Activities,** pp. 81–82

SCIENCE QUEST *Physical and Chemical Changes*

LOG ON **Science in Motion,** *Heating Water*

LOG ON **e-Review** Narrated Summary and Quiz

LOG ON **Progress Reporter** Assessments

Focus on Skills

Objective

- Determine whether salt water or fresh water freezes faster.

Materials graduated cylinder or metric measuring cup, water, two identical plastic cups, salt, measuring spoon, labels, freezer

Plan Ahead Acquire enough freezer space for each small group. If freezer space is not readily available, this activity can be done as a class activity or at home. Set aside time for checking cups every 15 minutes until the contents of one cup freezes.

EXTEND This activity will teach students how to make a prediction and then perform an experiment to determine whether the prediction is correct or not.

Inquiry Skill: Predict

▶ Learn It

- Explain to students that predictions are not expected to always be correct. Neither are they expected to be only guesses. Predictions are guesses about what will happen based on what is known about the topic.

- Point out that it is important to compare a prediction to actual results to determine whether the prediction is correct or not. If it is not correct, the prediction should be examined to determine how other predictions could be more accurate.

▶ Try It

1. Any freezer-safe cup can be used, but the cups should be identical so that the type of cup does not affect the results. Be sure students label the cups and do not taste the contents.

3. Results should show that fresh water freezes faster. The accuracy of the predictions depends on the predictions made.

Focus on Skills

Inquiry Skill: Predict

You just learned about how liquids change to solids. Which do you think freezes faster, salt water or fresh water? To find answers to questions like this, scientists first **predict** what they think will happen. Next, they experiment to find out what does happen. Then, they compare their results with their prediction.

▶ Learn It

When you **predict**, you state the possible results of an event or experiment. It is important to record your prediction before you do an experiment. Next, you record your observations as you experiment and record the final results. Then you have enough data to figure out if your prediction was correct.

Ice floats on the salt water of Shoup Bay in Alaska.

Integrate Reading

Read About Salt on Ice

Ask students to use reference books or the Internet, if available, to read about why salt is placed on icy roads and walkways.

- **Why does salt cause ice to melt?**

- **Is there a temperature at which salt water freezes?**

Have students compare the results of this activity to what they read. Their reading should tell them that salt melts icy surfaces because it makes water freeze at a lower temperature than the temperature at which pure water freezes.

Skill Builder

▶ Try It

Which do you think freezes faster, salt water or fresh water? **Predict** what will happen when you freeze fresh water and salt water. Write your prediction on a chart like the one shown. Then do an experiment to test your prediction.

Materials measuring cup, water, two plastic cups, salt, measuring spoon

1. Pour 125 mL of water into a plastic cup. Label this cup *Fresh Water*.

2. Pour 125 mL of water into another plastic cup. Add 1 tablespoon of salt and stir with a spoon. Label this cup *Salt Water*.

3. Place both cups into the freezer. Check them every 15 minutes. Draw or write your observations.

Now answer these questions. Which freezes faster, fresh water or salt water? Was your prediction correct?

▶ Apply It

Now that you have learned to think like a scientist, make another prediction. Do you **predict** that salt water or fresh water will evaporate faster? Plan an experiment to find out if your prediction is correct.

Which Freezes Faster?	
My Predictions	
Observations of fresh water	
Observations of salt water	
Results	

Fresh Water

Salt Water

405
EXTEND

▶ Apply It

Have students set up an experiment to find out whether fresh water or salt water evaporates more readily. Be sure students again use identical cups so the same surface area is exposed to the air.

Set the open cups in the same area. Both of them must be exposed to the same conditions so factors such as sunlight or breezes do not affect the results. Student results should show that fresh water evaporates faster.

Interested students might inquire as to why the fresh water evaporates more quickly. If all other factors are the same, relative rates of evaporation depend on the surface area of water. In salt water, some of the surface has particles from the salt on it, so the amount of water on the surface is less. The surface area is less, so the rate of evaporation is slower.

Name _____ Date _____   Focus on Skills

Predict

You just learned about how liquids change to solids. Which do you think freezes faster, salt water or fresh water? To find answers to questions like this, scientists predict what they think will happen. Next, they experiment to find out what does happen. Then, they compare their results with their prediction.

Learn It

When you predict, you state the possible results of an event or experiment. It is important to record your prediction before you do an experiment. Next, you record your observations as you experiment and record the final results. Then you have enough data to figure out if your prediction was correct.

Ice floats on the salt water of Shoup Bay in Alaska.

▶ Activity Lab Book, pp. 171–173

Plan Your Lesson

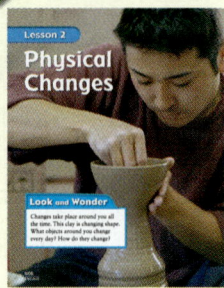

Lesson 2 Physical Changes

Objective

- Define physical changes as those that do not change the identity of a material.
- Describe how to make and separate mixtures.

Reading Skill Draw Conclusions

Text Clues	Conclusions

Graphic Organizer 13, p. TR15

Professional Development Look for **NSDL** to find recommended Science Background materials from the National Science Digital Library

FAST TRACK

Lesson Plan When time is short, follow the Fast Track and use the essential resources.

1 Introduce
Look and Wonder, p. 406
Resource Alternative Explore, p. 176

2 Teach
Discuss the Main Idea, p. 408
Discuss the Main Idea, p. 410
Resource Visual Literacy, p. 63

3 Close
Think, Talk, and Write, p. 413
Resource Assessment, p. 126

▶ Reading and Writing

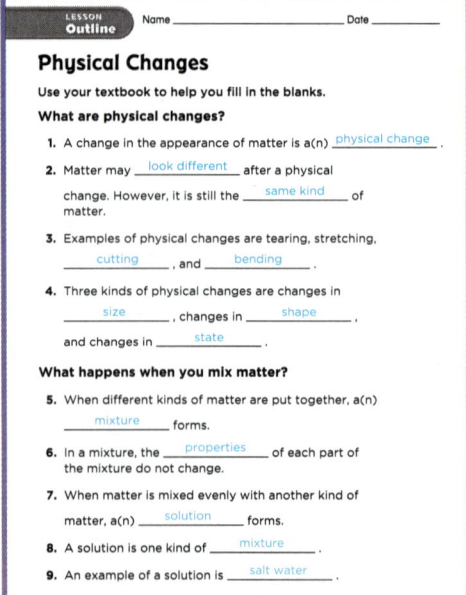

Outline, pp. 192–193
Also available as a student workbook

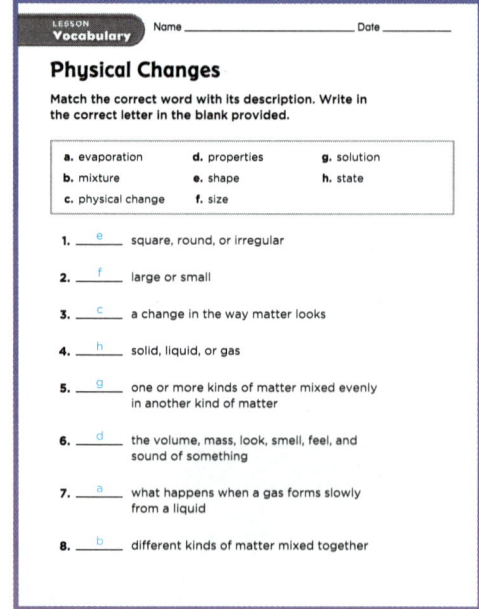

Vocabulary, p. 194
Also available as a student workbook

▶ Visual Literacy

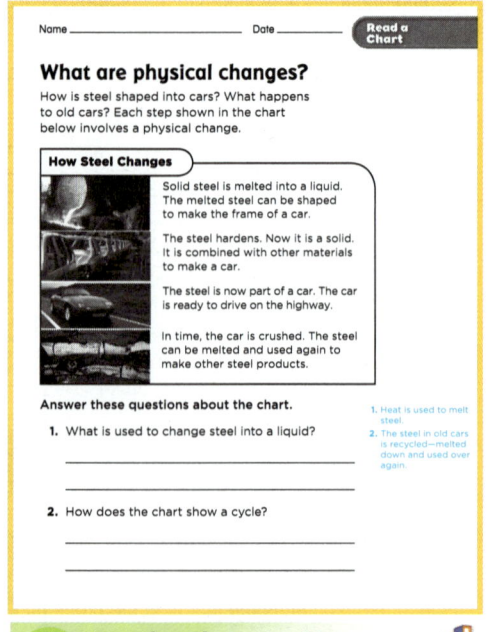

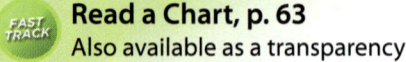

Read a Chart, p. 63
Also available as a transparency

▶ Activity Lab Book

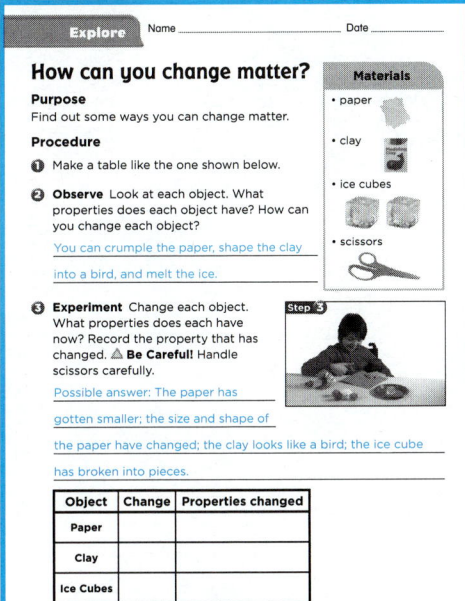

Explore   Name _____   Date _____

How can you change matter?

Purpose
Find out some ways you can change matter.

Procedure
❶ Make a table like the one shown below.

❷ **Observe** Look at each object. What properties does each object have? How can you change each object?
You can crumple the paper, shape the clay
into a bird, and melt the ice.

❸ **Experiment** Change each object. What properties do they have now? Record the property that has changed. ⚠ **Be Careful!** Handle scissors carefully.
Possible answer: The paper has
gotten smaller; the size and shape of
the paper have changed; the clay looks like a bird; the ice cube
has broken into pieces.

Materials
- paper
- clay
- ice cubes
- scissors

Step ❸

Object	Change	Properties changed
Paper		
Clay		
Ice Cubes		

Explore, pp. 174–175
Also available as a student workbook

Name _____   Date _____   **Quick Lab**

Separating Mixtures

❶ Mix some sand, marbles, and paper clips together in a bowl.

❷ **Experiment** Design an experiment to separate this mixture.
Possible answer: I can pick out the marbles and use a magnet to
separate out the paper clips.

❸ **Observe** Were you able to completely separate the mixture? How do you know?
Possible answer: I had some difficulty separating the paper clips
from the sand.

❹ **Predict** How could you separate a mixture of sugar and water?
Possible answer: I could put the cup of sugar and water in a warm
place and wait for the water to evaporate.

Quick Lab, p. 177
Also available as a student workbook

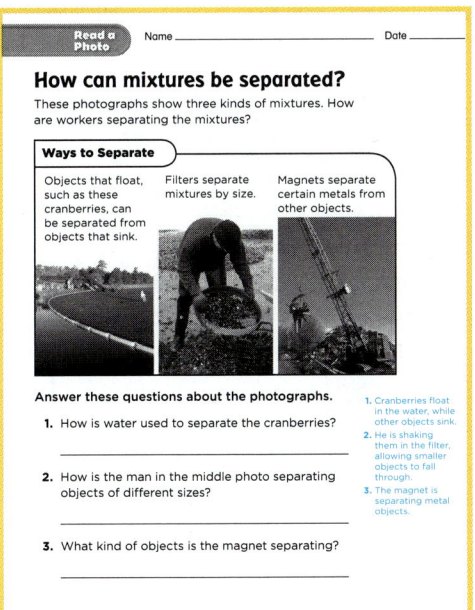

▶ Assessment

Read a Photo   Name _____   Date _____

How can mixtures be separated?

These photographs show three kinds of mixtures. How are workers separating the mixtures?

Ways to Separate

Objects that float, such as these cranberries, can be separated from objects that sink.

Filters separate mixtures by size.

Magnets separate certain metals from other objects.

Answer these questions about the photographs.

1. How is water used to separate the cranberries?
_____

2. How is the man in the middle photo separating objects of different sizes?
_____

3. What kind of objects is the magnet separating?
_____

1. Cranberries float in the water, while other objects sink.
2. He is shaking them in the filter, allowing smaller objects to fall through.
3. The magnet is separating metal objects.

Read a Photo, p. 64
Also available as a transparency

Lesson 2 Test   Name _____   Date _____

Circle the letter of the best answer for each question.

1. All of the following represent physical changes in paper <u>except</u>
  (A) burning.
  B cutting.
  C folding.
  D crumbling.

2. How is a solution formed?
  A by mixing different kinds of matter
  B by evenly mixing two of the same kinds of matter
  (C) by mixing one or more kinds of matter evenly into another kind of matter
  D by changing the state of two or more kinds of matter

3. All of the following are ways mixtures can be separated <u>except</u>
  A by evaporation
  B by its properties
  C by filters
  (D) by weight

4. Study the table below.

Mixtures
vegetable soup
salad dressing
clouds

What belongs in the empty box?
  A brass
  B salt water
  C chocolate milk
  (D) salad

Critical Thinking How can magnets be used to separate mixtures?
A magnet will attract any object that is magnetic. Therefore, any
objects with magnetic properties can be separated and collected
by the magnet.

FAST TRACK   **Lesson Test, p. 126**
Also available as a student workbook

ADDITIONAL RESOURCES

Activity Flipchart
for Science Centers and Workstations

p. 47

ELL
Teacher's Guide

p. 116

Transparencies for Visual Literacy

pp. 63–64

School to Home Activities

pp. 83–84

Key Concept Cards

63–64

Vocabulary Cards

151–153

Technology
 **Science Activity DVD**
 **Instructional Navigator CD-ROM**
 Presentation Toolkit CD-ROM
 **e-Review**
 **NSDL**

Lesson 2 Physical Changes

Objectives
- Define physical changes as those that do not change the identity of a material.
- Describe how to make and separate mixtures.

1 Introduce

▶ **Assess Prior Knowledge**

Show students a pencil. Ask them how many different ways they can change the pencil. Possible answers: sharpen it, break it

- **When you change the pencil in these ways, are the materials from the pencil still the same materials?**
  Possible answer: When I sharpen the pencil, the materials in the pencil remain the same.

Tell students that they will learn how to classify changes in a material according to whether or not the identity of the material changes.

 Look and Wonder

Invite students to share their responses to the Look and Wonder question:

- **How do they [the objects around you] change?**
  Possible answers: Water puddles dry up by evaporation; food is cooked by heating.

Write ideas on the board and note any misconceptions that students may have. Address these misconceptions as you teach the lesson.

RESOURCES and TECHNOLOGY
- Activity Lab Book, pp. 174–176
- Activity Flipchart, p. 47
- Science Activity DVD

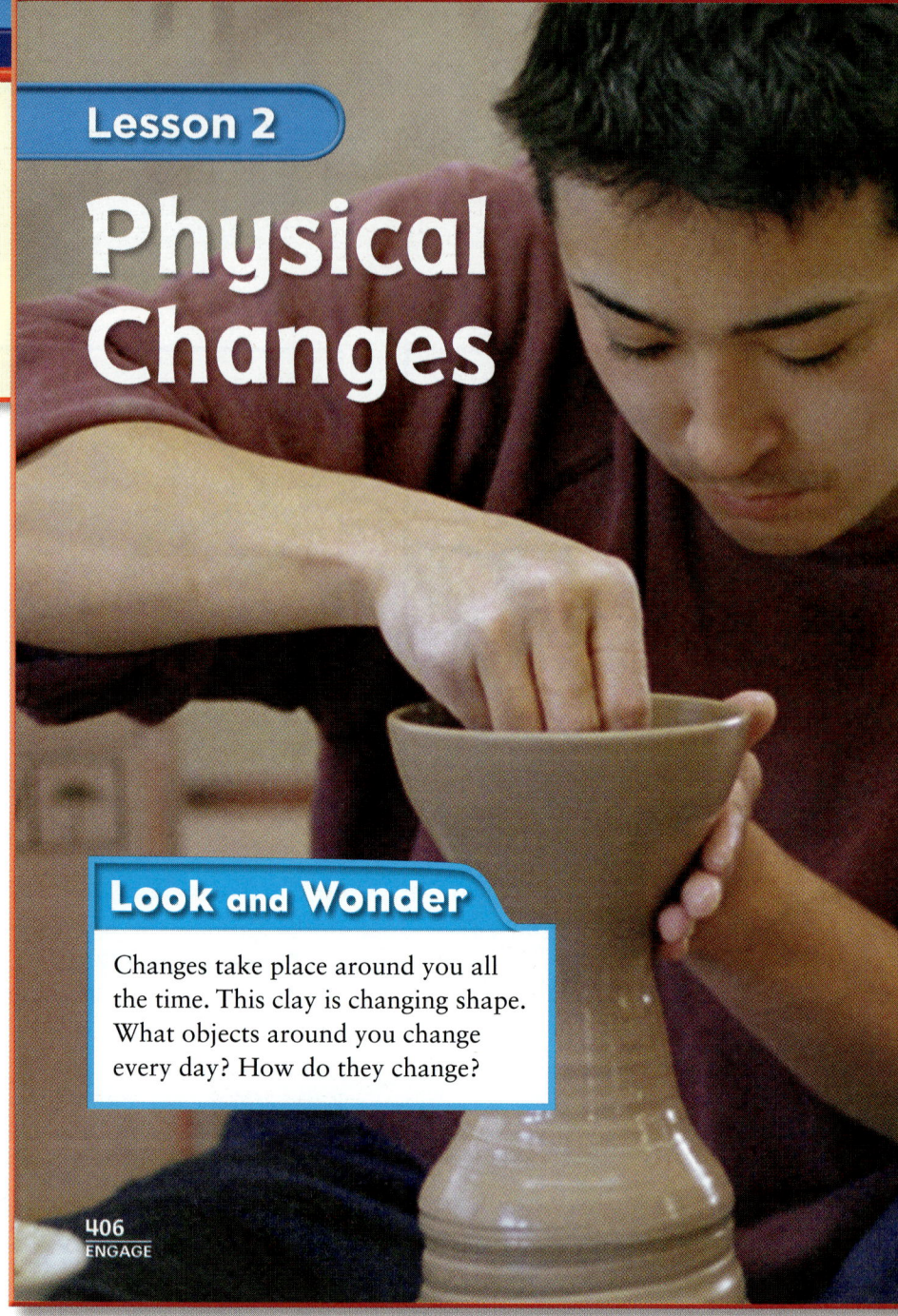

Lesson 2

Physical Changes

Look and Wonder

Changes take place around you all the time. This clay is changing shape. What objects around you change every day? How do they change?

406
ENGAGE

Warm Up

Start with a Model

Give each student two grapes, a paper towel, and rubber gloves. Place a grape on a paper towel and cut it in half. Have students peel one of their grapes. Have them crush a third grape with their fingers. Caution students to never eat or drink anything in a laboratory. Ask:

- **What do all these changes have in common?**
  Possible answer: The materials that make up the grapes remain the same.

Explore

Inquiry Activity

How can you change matter?

Purpose
Find out some ways you can change matter.

Procedure
1. Make a table like the one shown below.

Object	Change	Properties changed
Paper		
Clay		
Ice cubes		

2. **Observe** Look at each object. What properties does each object have? How can you change each object? Record your plan.

3. **Experiment** Change each object. What properties does each have now? Record the property that has changed. ⚠ **Be Careful.** Handle scissors carefully.

Draw Conclusions
4. How are the objects different after you made the changes?

5. **Infer** Do you think you changed the kind of matter making up the object? Explain.

Explore More
Experiment What would happen if you added a spoon of salt to a cup of water? How would the salt and water change? How could you remove the salt from the water?

Materials

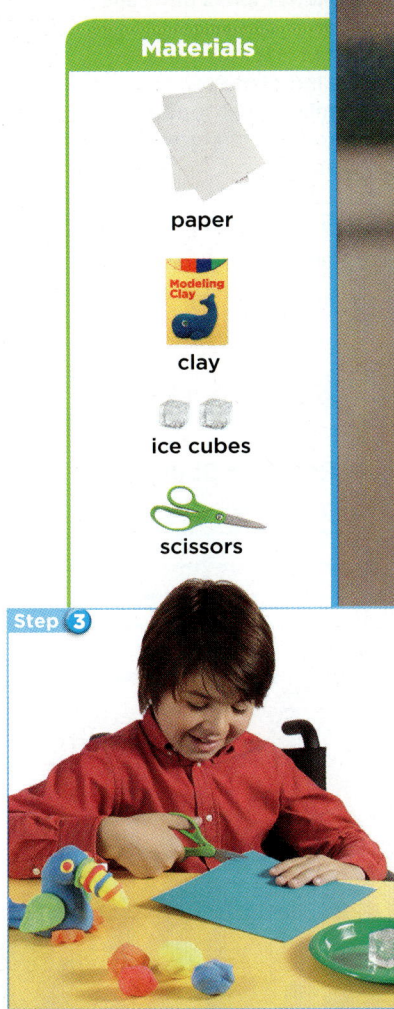

paper

clay

ice cubes

scissors

Step 3

407
EXPLORE

Explore

small groups   20 minutes

Plan Ahead Have a cooler available to store ice cubes until they are needed. Have paper or cloth towels available to clean up any water spills. Paper, clay, and ice cubes are used because they are easy for students to change. Other objects can be substituted for these items.

Purpose This activity helps students identify changes in matter and analyze whether the matter changed in identity.

Structured Inquiry

2. **Observe** Be sure students choose a property they can realistically change. Possible answer: Change the size or shape of the paper and clay using the scissors; change the size and shape of the ice by letting it melt.

3. **Experiment** **Be Careful!** Instruct students to handle the scissors with care. Students can list more than one property if more than one property changed. For example, if they cut the paper into two pieces, both size and shape change.

4. Possible answer: Changes are made to the size, shape, and state of the object.

5. **Infer** Accept all reasonable answers. Students may begin to see that certain changes in an object do not change the matter that makes it up.

Guided Inquiry Explore More

Experiment The salt seems to disappear in the water. The water looks the same but tastes salty. The salt and water can be separated by evaporating the water.

Open Inquiry

Ask students whether they think that a change of state changes the identity of matter. Have them think of their own question about changes to matter, and then ask them to make a plan and carry out an experiment to answer the question. Ask:

When ice melts, does it change what makes up the matter?

Alternative Explore

How does changing state change matter?

Materials narrow-mouthed jar, ice cube, hot tap water

For each group, fill the jar with hot water. Ensure the water is not hot enough to burn. Have students hold an ice cube over the top of the jar, being careful that the ice does not fall into the water. Have them observe that water droplets form near the jar's top. Discuss the changes of state that occurred and that water did not change identity during any particular change.

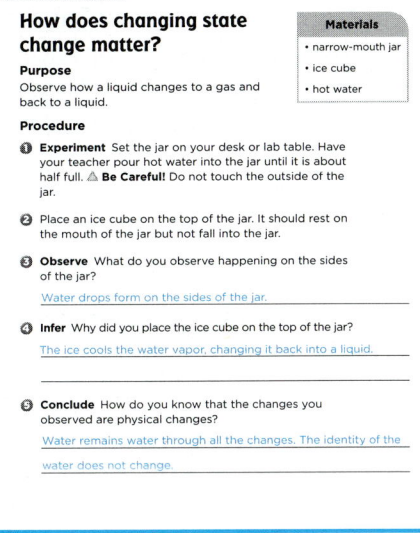

Alternative Explore Name _____ Date _____

How does changing state change matter?

Purpose
Observe how a liquid changes to a gas and back to a liquid.

Procedure
1. **Experiment** Set the jar on your desk or lab table. Have your teacher pour hot water into the jar until it is about half full. ⚠ **Be Careful!** Do not touch the outside of the jar.

2. Place an ice cube on the top of the jar. It should rest on the mouth of the jar but not fall into the jar.

3. **Observe** What do you observe happening on the sides of the jar?
   Water drops form on the sides of the jar.

4. **Infer** Why did you place the ice cube on the top of the jar?
   The ice cools the water vapor, changing it back into a liquid.

5. **Conclude** How do you know that the changes you observed are physical changes?
   Water remains water through all the changes. The identity of the water does not change.

Materials
- narrow-mouth jar
- ice cube
- hot water

Activity Lab Book, p. 176

2 Teach

Read and Learn

Main Idea Have students do a picture tour of the lesson and ask them to discuss what they think they will learn about.

Vocabulary Have students read aloud the vocabulary words. Ask students to read the definitions and then restate them in their own words.

Reading Skill **Draw Conclusions**
Graphic Organizer 13 Have students fill in a Draw Conclusions graphic organizer as they read each two pages of the lesson. They can use the Quick Check questions to identify each conclusion to draw.

Text Clues	Conclusions

What are physical changes?

🟢 **FAST TRACK** **Discuss the Main Idea**

Have students discuss changes in the way matter looks. Hold up a piece of fabric that will wrinkle when crushed. Ask:

■ **How can you change the way this fabric looks?** Possible answers: crush it, cut it, stretch it

■ **Does the identity of this material change during any of these changes?** No; it remains the same fabric.

Read a Chart

Answer The steel changes from a liquid to a solid. The steel is melted back into a liquid.

RESOURCES and TECHNOLOGY

▶ **Reading and Writing,** pp. 192–194

▷ **Visual Literacy,** pp. 63

💿 **PuzzleMaker CD-ROM**

💿 **Presentation Toolkit CD-ROM**

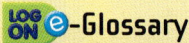 **e-Glossary**

Read and Learn

▶ **Main Idea**
Matter looks different after a physical change, but it is still the same kind of matter. You can mix matter together to form mixtures and solutions.

▶ **Vocabulary**
physical change, p. 408
mixture, p. 410
solution, p. 411

 **e-Glossary**
at www.macmillanmh.com

▶ **Reading Skill** ✅
Draw Conclusions

Text Clues	Conclusions

What are physical changes?

Matter can change. A **physical change** (FIZ•i•kuhl CHAYNJ) is a change in the way matter looks. Tearing a sheet of paper is a physical change. The size and shape of the paper are different, but the paper is still paper. Matter looks different after a physical change, but it is still made of the same kind of matter.

A change of state is also a physical change. When liquid water freezes, its state changes from a liquid to a solid. The water looks different, but it is still water.

Not all types of matter change in the same way. If you pull on a rubber band it stretches. When you let go, it returns to its original size. If you pull on a metal spoon, nothing happens. If you pull on a piece of thread, it might break.

Painting an object does not change what the object is made of. ▼

408
EXPLAIN

Science Background

How are properties used to separate mixtures? A mixture contains different kinds of matter that can be separated by physical means. The physical properties of the mixture's components can be used to separate them. In addition to the physical properties discussed in the lesson, physical properties include melting point, boiling point, density, and electrical conductivity. For example, sugar and water can be separated by boiling off the water because the boiling point of water is much less than the boiling point of sugar.

See **Science Yellow Pages** in the Teacher Resource section for background information.

🔵 **Professional Development** For more Science Background and resources from **NSDL** visit **www.macmillanmh.com/nsdl/**

How Steel Changes

Solid steel is melted into a liquid. The melted steel can be shaped to make the frame of a car.

The steel hardens. Now it is a solid. It is combined with other materials to make a car.

The steel is now part of a car. The car is ready to drive on the highway.

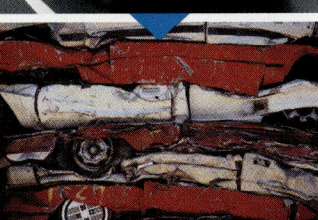

In time, the car is crushed. The steel can be melted and used again to make other steel products.

Read a Chart

What physical changes have happened to the steel?

Clue: Look at how the steel has changed in each photograph. Use the captions to help.

 Quick Check

Draw Conclusions Why is a change of state a physical change?

Critical Thinking Make a list of three physical changes you could make to a piece of paper.

409
EXPLAIN

▶ Use the Visuals

Refer students to the visuals on page 409. Ask:

■ **Which photo shows a change of state?** The first photo shows a change from solid to liquid metal.

■ **What physical change took place before the last photo was taken?** The car was crushed.

▶ Develop Vocabulary

physical change Have students give examples of physical changes to matter. Possible answers: tearing a sheet of paper, water freezing to ice, water evaporating to water vapor

✔ *Quick Check Answers*

• **Draw Conclusions** Even though the matter looks different when it changes state, it is still the same kind of matter.

• **Critical Thinking** Possible answers: cut, fold, crumple

Differentiated Instruction

Leveled Activities

EXTRA SUPPORT   Have students draw cause-and-effect pictures of two physical changes.

ENRICHMENT   Ask students to research what physical changes occur when food is eaten. Tell them that these changes include those that occur as food is chewed.

What happens when you mix matter?

What happens when you mix matter?

FAST TRACK ### Discuss the Main Idea

Tell students that there are two different types of mixtures. Point out the examples of the fruit mixture and the saltwater ocean solution. Ask:

- **How are the two types of mixtures alike?** Possible answers: Both types of mixtures contain at least two parts, each with its own properties.

- **How are the two types of mixtures different?** Possible answers: The different parts of the fruit mixture can be seen. The parts are so evenly mixed in the saltwater ocean solution that the parts can not be seen.

▶ **Develop Vocabulary**

mixture *Word Origin* The term *mixture* is named for the way a mixture is formed. The term comes from the Latin verb *mixtus,* which means "to mix." This meaning has been in use since around A.D. 1460.

solution *Word Origin* The word *solution* comes from the Latin *solutonem,* which means "a loosening or unfastening."

▶ **Address Misconceptions**

Some students may think that all solutions are liquid. Just as mixtures can be solid or liquid, the kind of mixture known as a solution can also appear in solid form.

FACT ▶ **Solutions can be solid.** Show examples of brass, gold, or steel or alloys made up of several solids.

RESOURCES and TECHNOLOGY
💿 **Presentation Toolkit CD-ROM**

Another kind of physical change is a mixture (MIKS•chuhr). A **mixture** is different kinds of matter mixed together. When you pour milk on your cereal, you are making a mixture. In a mixture the properties of each kind of matter do not change. The milk is still milk, and the cereal is still cereal.

A mixture can be a combination of solids, liquids, and gases. Vegetable soup is a mixture of liquids and solids. Salad dressing can be a mixture of different liquids. Clouds are a mixture of air, dust, and water droplets.

▲ What makes up this mixture?

410 EXPLAIN
FACT ▶ Solutions can be solid.

Classroom Equity

Encourage all students to get involved. To encourage more students to volunteer answers in this lesson, wait three to five seconds before calling on someone to answer a question. As an experiment, have students write their names on index cards. Then, take the deck, shuffle it, and turn up the cards one by one until you have called on each student.

Solutions

There are many kinds of mixtures. One kind of mixture is a solution (suh•LEW•shuhn). A **solution** forms when one or more kinds of matter are mixed evenly into another kind of matter.

Salt water is a solution. If you add salt to water, the salt mixes evenly with the water. You cannot see the salt, but it is still there. If the water evaporates, the salt will be left behind.

Not all solids form solutions in liquids. Try to mix sand with water. The sand will just sink to the bottom. Some things will not form solutions no matter how long you stir.

Some solutions contain no liquids at all. Air is a solution of different gases. Brass is a solution of several solids, including copper and zinc.

Brass is a solution of metals. It is used to make musical instruments.

Ocean water is a mixture. It contains many different types of matter, including salt, water, and oxygen.

 **Quick Check**

Draw Conclusions Do all kinds of matter form solutions with water? Explain your answer.

Critical Thinking You cannot see that salt is in salt water. How do you know it is there?

411
EXPLAIN

▶ Discuss the Main Idea

Have students write down what they had for breakfast or lunch today. Have them determine whether it is a mixture or not. Ask:

- **Did you eat or drink anything that was not a mixture?** Possible answer: No; everything was a mixture.

- **Were any of the mixtures solutions?** Possible answer: Yes; if they had a soft drink, it was a solution.

Point out that some unclear liquids such as milk are in between being solutions and mixtures. Emphasize that almost all things that someone eats or drinks is a mixture.

▶ Explore the Main Idea

ACTIVITY Have students use reference books or the Internet, if available, to research alloys. Ask:

- **What is an alloy?** An alloy is a solid solution of a metal and at least one other element, usually another metal.

- **What are some examples of alloys?** Possible answers: steel, brass, bronze

Point out that alloys can contain several different elements. For example, some types of steel contain iron and several other elements. Show students several examples of alloys, such as 14K gold, a brass musical instrument, or a steel paper clip. Have student volunteers examine these alloys with a hand lens and report that they cannot see different parts in the mixture.

✓ Quick Check Answers

- **Draw Conclusions** No; some things form solutions when mixed with water, such as sugar or salt. Other things, such as sand, remain a mixture.

- **Critical Thinking** Possible answers: The water tastes salty.

How can mixtures be separated?

 **Quick Lab**  small groups 15 minutes

Objective Show how to separate a mixture.

Materials marbles, sand, metal paper clips, bowl, colander, magnet

2 Designs might include using a magnet to remove the paper clips, and separating the sand and marbles by using the colander.

3 Answers will vary. The mixture is completely separated when no pieces of one item are mixed with pieces of another item.

4 Evaporate the water, leaving the sugar behind.

▶ Discuss the Main Idea

Tell students that the physical properties of the parts of a mixture are used to separate the mixture. Have student volunteers name mixtures and other students tell how the mixtures could be separated.

✔ Quick Check Answers

- **Draw Conclusions** Possible answer: Pick out each by hand.

- **Critical Thinking** Place the mixture in water, filter out the sand, then evaporate the water.

Quick Lab

Separating Mixtures

1 Mix some sand, marbles, and paper clips together in a bowl.

2 Experiment Design an experiment to separate this mixture.

3 Observe Were you able to completely separate the mixture? How do you know?

4 Experiment How could you separate a mixture of sugar and water?

How can mixtures be separated?

Some properties help you separate mixtures. These properties include size, shape, and color. One way to separate a mixture is to pick out each different type of matter. In a mixture of spaghetti and meatballs, you can pick out the meatballs.

Another way to separate a mixture is by evaporation. Leave a solution of salt and water in a warm place. As the water evaporates, the salt is left behind. The photos below show some other ways to separate mixtures.

✔ Quick Check

Draw Conclusions How can you separate peas from carrots?

Critical Thinking List some ways to separate sand from salt.

Objects that float, such as these cranberries, can be separated from objects that sink. ▼

Filters separate mixtures by size. ▼

Magnets separate certain metals from other objects. ▼

412
EXPLAIN

Homework Activity

Separate a Mixture

Have students make a booklet showing how to separate a mixture of sand, salt, and paper clips. Each page should show a step in the separation. Students should illustrate with drawings and explain in words each step in the separation.

Lesson Review

Visual Summary

Physical changes cause matter to look different. The kind of matter stays the same.

A **mixture** is a combination of two or more types of matter.

Some properties can help you **separate a mixture**.

Think, Talk, and Write

1. **Main Idea** Describe and explain three types of physical changes.

2. **Vocabulary** What is a mixture?

3. **Draw Conclusions** A sculptor carves a statue out of a rock. Is this a physical change? How do you know?

Text Clues	Conclusions

4. **Critical Thinking** How could you separate plastic paper clips from metal paper clips?

5. **Test Prep** Noodles and broth could be separated by
   A heating in an oven.
   B boiling in a pot.
   C filtering.
   D freezing.

Make a FOLDABLES Study Guide

Make a Trifold Book. Use it to summarize what you learned about physical changes.

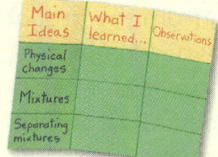

Math Link

Sort and Classify Materials
What materials dissolve in water? Try mixing several materials, such as salt, flour, sugar, soil, and cooking oil, with water. Then classify the materials into groups to show which dissolve and which do not. Make a chart to show your results.

Art Link

Experiment with Color
Cut a circle out of a coffee filter. Use a black marker to draw a spot in the center of the filter. Put the filter on a plate. Add a few drops of water to the spot. Watch what happens. Why do you think this happens? What does this tell you about ink?

 e-Review Summaries and quizzes online at www.macmillanmh.com

413
EVALUATE

Formative Assessment

Approaching Have students list four physical changes that could be made to an apple.

On-Level Have students explain how to separate a mixture of water, pebbles, and sugar.

Challenge Tell students that water contains both hydrogen and oxygen. Have students explain why pure water is not a mixture.

Key Concept Cards For student intervention, see the prescribed routine on **Key Concept Cards 63 and 64.**

---

3 Close

Lesson Review

▶ **Visual Summary**

Have students save their lesson Study Guides. They will use them at the end of the chapter for review.

▶ **Make a FOLDABLES Study Guide**

See pages R27–R28 for instructions.

FAST TRACK Think, Talk, and Write

1. **Main Idea** Possible answers: cutting paper, molding clay, bending wire

2. **Vocabulary** Two or more types of matter are mixed together, but the properties of each part stay the same.

3. **Draw Conclusions**

Text Clues	Conclusions
How the rock looks is changed.	The rock remains rock.

4. **Critical Thinking** Possible answer: by using a magnet

5. **Test Prep** C

Math Link dissolves in water: salt, flour, sugar; does not dissolve in water: soil, cooking oil

Art Link Students must use a water soluble marker. When the water touches the marker, the black ink separates into different colors. Encourage students to try different markers to see if the results differ. Black ink is a mixture of inks of different colors.

RESOURCES and TECHNOLOGY

▶ **Reading and Writing,** p. 195

▶ **Activity Lab Book,** p. 177

▶ **School to Home Activities,** pp. 83–84

e-Review Narrated Summary and Quiz

Progress Reporter Assessments

Lesson 2   **413**

Reading in Science

Objective
■ Make an inference about how metal is taken from rocks.

Mining Ores

Genre: Nonfiction

■ **What question do you think might be answered by reading this story?** Possible answer: How are metal ores mined?

Before Reading

Give students an object made of metal, such as a key, to pass around. Ask them to describe how the object looks and feels. Direct students to the photo of the keys. Ask:

■ **Are the keys solid, liquid, or gas?** solid

■ **What do you think the keys are made from?** Possible answer: metal, stainless steel

Explain that the keys are made of a metal. Discuss what your students have already learned about metal. Ask:

■ **Where do you think the metal that made these keys was found?** Possible answers: in the ground, in a mine, in a factory

Have students think about how the metal is mined. Ask:

■ **How do you think people get metal out of these rocks?** Possible answers: They dig for metal, they crush the rocks to get metal.

List what students already know about mining metals in the middle boxes of Graphic Organizer 14.

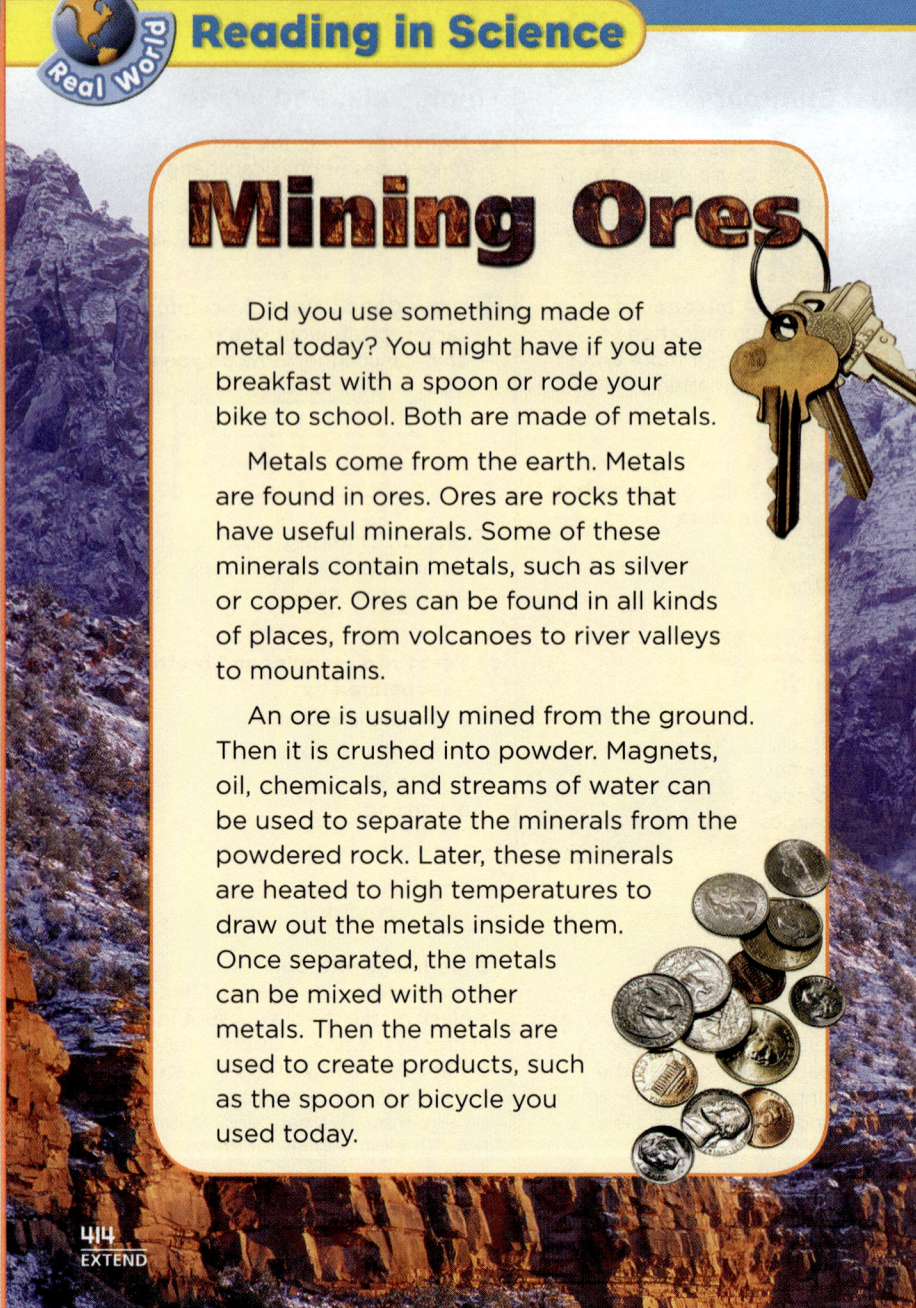

Real World Reading in Science

Mining Ores

Did you use something made of metal today? You might have if you ate breakfast with a spoon or rode your bike to school. Both are made of metals.

Metals come from the earth. Metals are found in ores. Ores are rocks that have useful minerals. Some of these minerals contain metals, such as silver or copper. Ores can be found in all kinds of places, from volcanoes to river valleys to mountains.

An ore is usually mined from the ground. Then it is crushed into powder. Magnets, oil, chemicals, and streams of water can be used to separate the minerals from the powdered rock. Later, these minerals are heated to high temperatures to draw out the metals inside them. Once separated, the metals can be mixed with other metals. Then the metals are used to create products, such as the spoon or bicycle you used today.

414
EXTEND

ELL Support

Paraphrase Highlight words in the article and captions, such as *crushed*. Point to the photos and ask: **Where is metal ore found? What happens to the rock in this picture?**

BEGINNING   Have students point to or say a word in the article that describes the pictures.

INTERMEDIATE   Have students use short phrases to paraphrase the pictures.

ADVANCED   Have students say complete sentences to paraphrase the article by describing the pictures.

Science, Technology, and Society

rock crusher

Infer

When you infer
▶ you use what you already know
▶ you use facts in the article
▶ you form new ideas

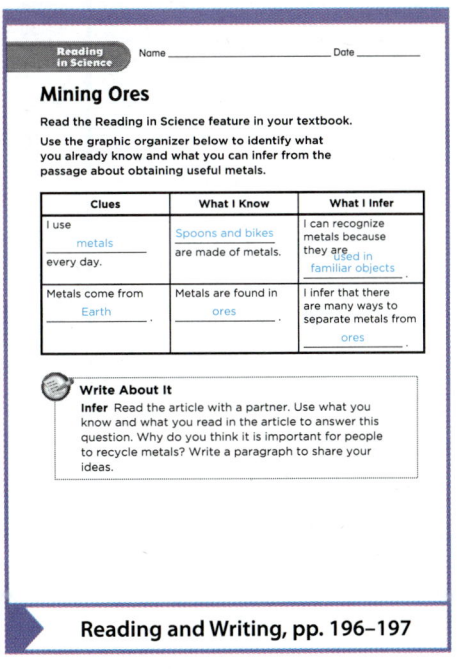

Write About It

Infer Read the article with a partner. Use what you know and what you read in the article to answer this question. Why do you think it is important for people to recycle metals? Write a paragraph to share your ideas.

LOG ON e-Journal Write about it online at www.macmillanmh.com

AMERICAN MUSEUM & NATURAL HISTORY

415
EXTEND

During Reading

Explain to students that they are reading about the process by which metals are taken from Earth. Explain that they should think about the steps taken to extract the metals and how they are made into everyday objects. Ask:

- **What does the metal ore look like in the ground?** The metal looks like part of the rocks.

- **What happens after the metal-rich minerals are mined?** The rocks are crushed.

Ask students to discuss the reasons the rocks might be crushed.

After Reading

Display Graphic Organizer 14 and read aloud what students already knew before reading the article. Ask:

- **Is getting metals from ore an easy or difficult process? Support your answer with clues from the article.** It's a difficult process. The ore must be mined and crushed into powder. The minerals must be separated from the ore and then heated to draw out the metals.

Write students' responses in the clues column.

- **What can we infer about metals from what we know and from the clues in the article?** Possible answers: Metals are used for many things and mining them is a difficult process.

Write About It

It is important for people to recycle metals because they are a nonrenewable natural resource. Once we take metal ore out of the ground, we cannot replace it.

Integrate Reading

Make a Design

Ask students to retell the steps of the process in which metal is mined and used to create products. Ask: **What are some of the jobs required to take metal from the Earth and turn it into a product?** Possible answers: scientist, miners, factory workers, product designers

Have students choose the job that is most interesting to them. Ask them to write a paragraph that tells about that job and why they would be good at doing it.

Reading in Science

Name _____ Date _____

Mining Ores

Read the Reading in Science feature in your textbook.

Use the graphic organizer below to identify what you already know and what you can infer from the passage about obtaining useful metals.

Clues	What I Know	What I Infer
I use _metals_ every day.	Spoons and bikes are made of metals.	I can recognize metals because they are used in familiar objects.
Metals come from _Earth_.	Metals are found in _ores_.	I infer that there are many ways to separate metals from _ores_.

Write About It

Infer Read the article with a partner. Use what you know and what you read in the article to answer this question. Why do you think it is important for people to recycle metals? Write a paragraph to share your ideas.

Reading and Writing, pp. 196–197

RESOURCES and TECHNOLOGY

📖 **Reading and Writing,** pp. 196–197

📗 **Technology: A Closer Look,** Lesson 1

LOG ON e-Journal Online research and writing

p. TR26 **Writing Rubric**

Reading in Science   **415**

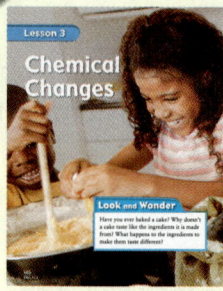

Lesson 3 Chemical Changes

Objective

- Describe chemical changes.
- Understand that chemical changes are part of our everyday life.

Reading Skill Infer

Clues	What I Know	What I Infer

Graphic Organizer 14, p. TR16,

LOG ON **Professional Development** Look for **NSDL** to find recommended Science Background materials from the National Science Digital Library

FAST TRACK

Lesson Plan When time is short, follow the Fast Track and use the essential resources.

1 Introduce
Look and Wonder, p. 416
Resource Alternative Explore, p. 180

2 Teach
Discuss the Main Idea, p. 418
Resource Visual Literacy, p. 65

3 Close
Think, Talk, and Write, p. 421
Resource Assessment, p. 127

► Reading and Writing

LESSON Outline Name _____ Date _____

Chemical Changes

Use your textbook to help you fill in the blanks.

What are chemical changes?

1. Rust and ash are results of _chemical changes_.

2. A chemical change takes place when a material forms a(n) _new kind_ of matter.

3. The _properties_ of a new material will be different from the original material.

What are the signs of a chemical change?

4. There are often _signs_ that a chemical change has taken place.

5. Signs of chemical change are _light_, _heat_, or _color_ change.

6. When a log burns, _heat and light_ are released and _new kinds of matter_ form.

7. Some chemical changes also make _bubbles_.

8. When baking soda and vinegar are mixed, bubbles of _carbon dioxide_ gas form.

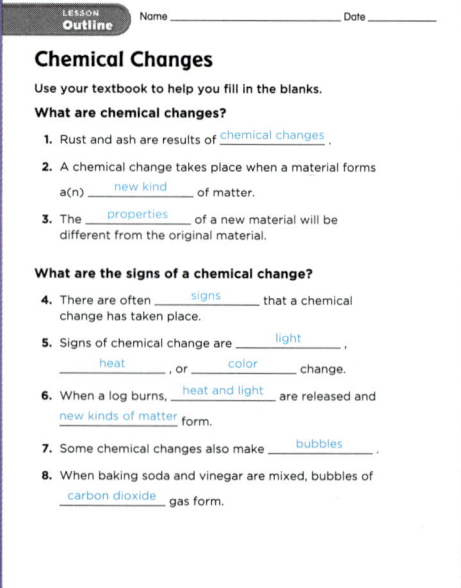

Outline, pp. 198–199
Also available as a student workbook

LESSON Vocabulary Name _____ Date _____

Chemical Changes

Match the correct word with its description below. Then write its letter in the blanks below.

a. baking	d. chemical change	g. properties
b. bubbles	e. green plants	h. rust
c. carbon dioxide	f. light and heat	

1. ___f___ These are two signs that a chemical change has taken place.

2. ___c___ Plants change this material into food and oxygen.

3. ___d___ This process goes on inside you every day and creates a new type of matter.

4. ___g___ These are the characteristics of a certain type of matter.

5. ___a___ During this activity, cake batter is changed chemically.

6. ___h___ When iron is chemically changed, this is made.

7. ___e___ These use the Sun's energy to make chemical changes.

8. ___b___ These show that carbon dioxide gas forms when baking soda is added to vinegar.

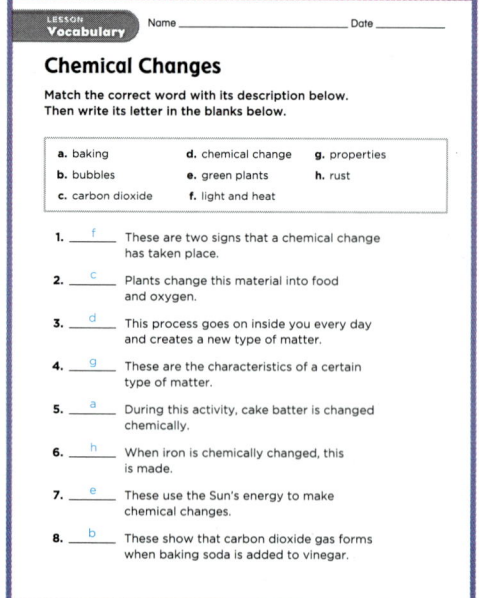

Vocabulary, p. 200
Also available as a student workbook

► **Visual Literacy**

Name _____ Date _____ Read a Diagram

What are chemical changes?

This diagram shows bananas in sequence, or in order. Notice how the bananas change.

A Chemical Change

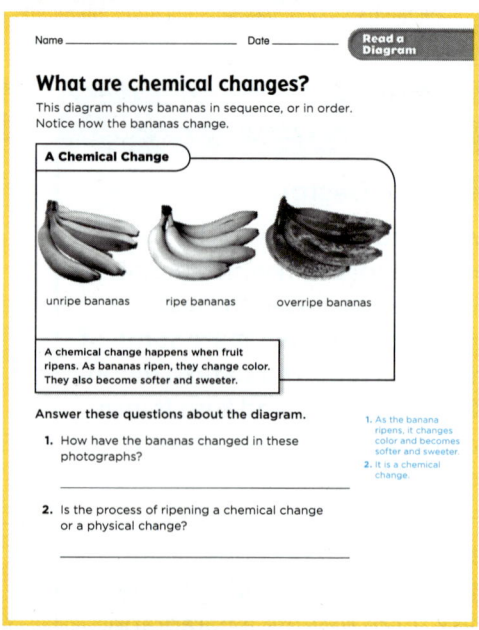

unripe bananas   ripe bananas   overripe bananas

A chemical change happens when fruit ripens. As bananas ripen, they change color. They also become softer and sweeter.

Answer these questions about the diagram.

1. How have the bananas changed in these photographs?

2. Is the process of ripening a chemical change or a physical change?

1. As the banana ripens, it changes color and becomes softer and sweeter.

2. It is a chemical change

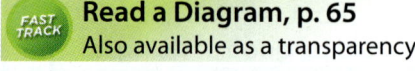

Read a Diagram, p. 65
Also available as a transparency

▶ Activity Lab Book

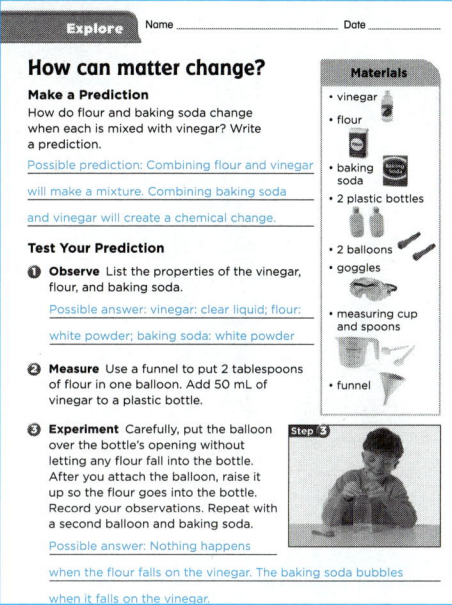

Explore   Name _____ Date _____

How can matter change?

Make a Prediction

How do flour and baking soda change when each is mixed with vinegar? Write a prediction.

Possible prediction: Combining flour and vinegar will make a mixture. Combining baking soda and vinegar will create a chemical change.

Test Your Prediction

❶ **Observe** List the properties of the vinegar, flour, and baking soda.

Possible answer: vinegar: clear liquid; flour: white powder; baking soda: white powder

❷ **Measure** Use a funnel to put 2 tablespoons of flour in one balloon. Add 50 mL of vinegar to a plastic bottle.

❸ **Experiment** Carefully, put the balloon over the bottle's opening without letting any flour fall into the bottle. After you attach the balloon, raise it up so the flour goes into the bottle. Record your observations. Repeat with a second balloon and baking soda.

Possible answer: Nothing happens when the flour falls on the vinegar. The baking soda bubbles when it falls on the vinegar.

Materials
- vinegar
- flour
- baking soda
- 2 plastic bottles
- 2 balloons
- goggles
- measuring cup and spoons
- funnel

Explore, pp. 178–179
Also available as a student workbook

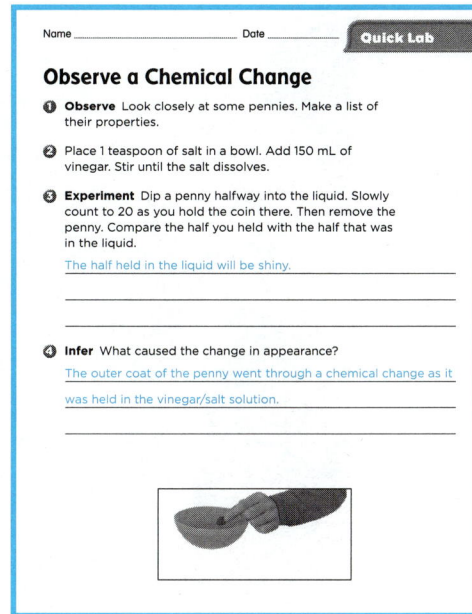

Name _____ Date _____   **Quick Lab**

Observe a Chemical Change

❶ **Observe** Look closely at some pennies. Make a list of their properties.

❷ Place 1 teaspoon of salt in a bowl. Add 150 mL of vinegar. Stir until the salt dissolves.

❸ **Experiment** Dip a penny halfway into the liquid. Slowly count to 20 as you hold the coin there. Then remove the penny. Compare the half you held with the half that was in the liquid.

The half held in the liquid will be shiny.

_____

_____

❹ **Infer** What caused the change in appearance?

The outer coat of the penny went through a chemical change as it was held in the vinegar/salt solution.

_____

Quick Lab, p. 181
Also available as a student workbook

▶ Assessment

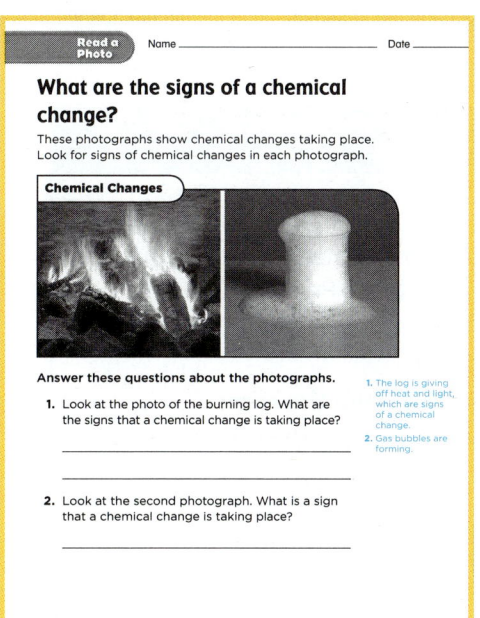

Read a Photo   Name _____ Date _____

What are the signs of a chemical change?

These photographs show chemical changes taking place. Look for signs of chemical changes in each photograph.

Chemical Changes

Answer these questions about the photographs.

1. Look at the photo of the burning log. What are the signs that a chemical change is taking place?

_____

_____

2. Look at the second photograph. What is a sign that a chemical change is taking place?

_____

_____

1. The log is giving off heat and light, which are signs of a chemical change.
2. Gas bubbles are forming.

Read a Photo, p. 66
Also available as a transparency

Name _____ Date _____   **Lesson 3 Test**

Circle the letter of the best answer for each question.

1. What is one way you can tell that a chemical change has taken place?

   A the object changes size
   B the object changes color
   C the object changes weight
   D the object changes shape

2. How does your body go through chemical changes?

   A by growing
   B by sleeping
   C by breaking down food
   D by learning new skills

3. What has to occur for cake batter to undergo a chemical change?

   A it has to be heated
   B it has to be cooled
   C it has to be mixed
   D it has to be frozen

4. Which of the following represents a chemical change of an apple?

   A peeling it
   B cooking it
   C turning brown
   D falling off a tree

Critical Thinking How is rust the result of a chemical change?

Iron becomes rust as a result of being exposed to elements like oxygen and water. The iron then becomes weaker and changes color, becoming rust.

_____

Lesson Test, p. 127
Also available as a student workbook

ADDITIONAL RESOURCES

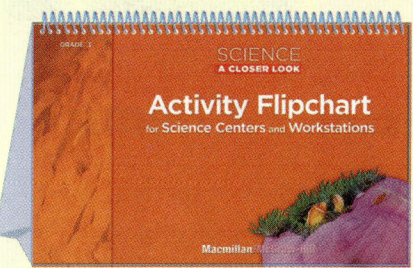

Activity Flipchart for Science Centers and Workstations

p. 48

ELL Teacher's Guide

p. 117

Transparencies for Visual Literacy

pp. 65–66

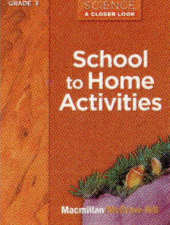

School to Home Activities

pp. 85–86

Key Concept Cards

65–66

Vocabulary Cards

154

Technology

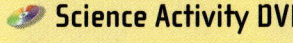 **Science Activity DVD**

 **Instructional Navigator CD-ROM**

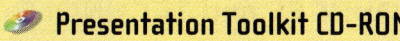 **Presentation Toolkit CD-ROM**

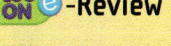 **e-Review**

 **NSDL**

Lesson 3 Chemical Changes

Objectives
- Describe chemical changes.
- Understand that chemical changes are part of our everyday life.

1 Introduce

▶ **Assess Prior Knowledge**

Show students a burning match. Allow most of the match to burn. Ask:

- **Is this change a physical change?** No.

- **How do you know that the match's burning is not a physical change?** The final substances have different properties than the original substances.

Look and Wonder

Invite students to share their responses to the Look and Wonder question:

- **What happens to the [cake] ingredients to make them taste different?** The ingredients undergo chemical changes that change their properties.

Write ideas on the board and note any misconceptions that students may have. Address these misconceptions as you teach the lesson.

RESOURCES and TECHNOLOGY
▶ **Activity Lab Book,** pp. 178–180
▶ **Activity Flipchart,** p. 48
💿 **Science Activity DVD**

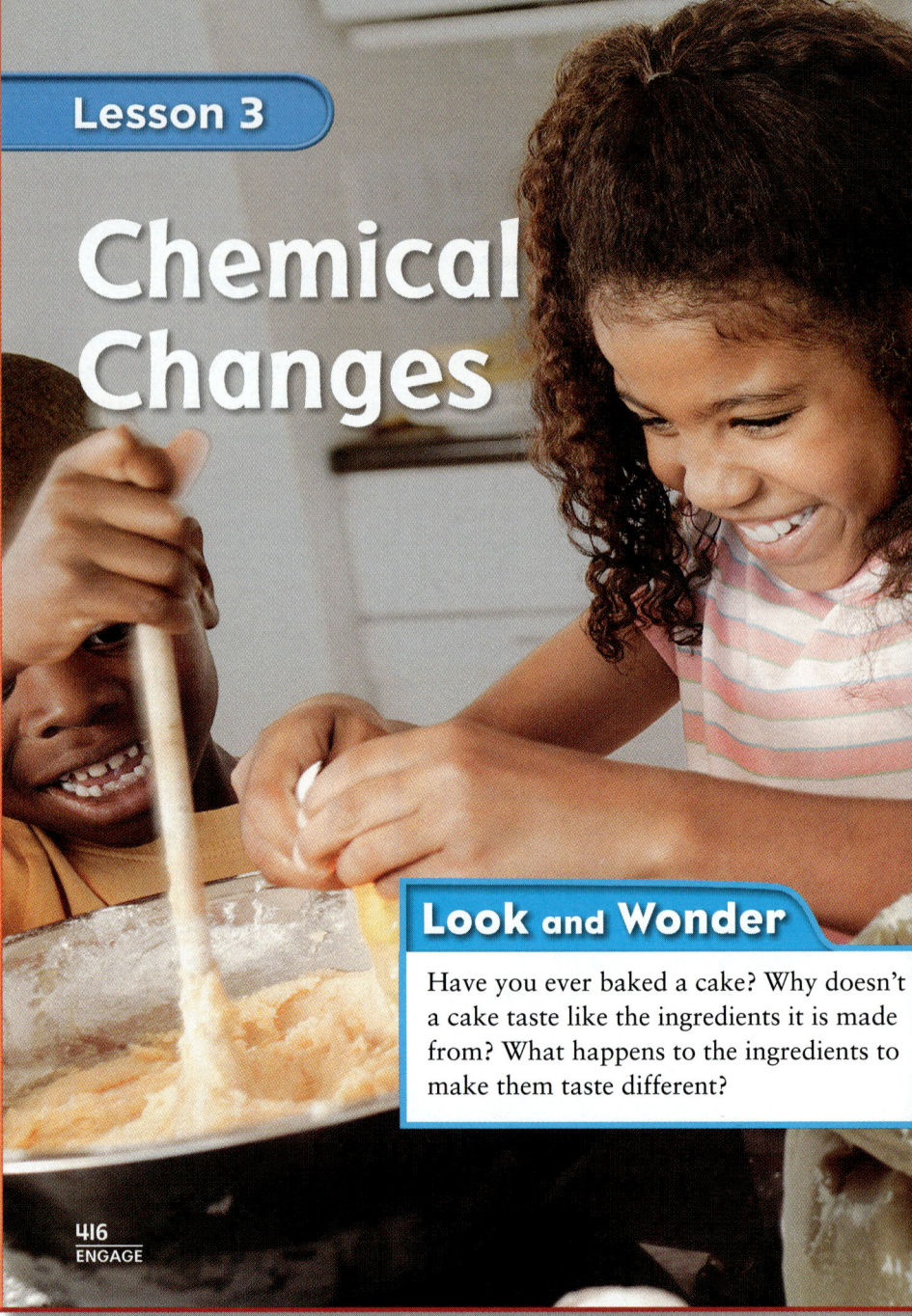

Lesson 3

Chemical Changes

Look and Wonder

Have you ever baked a cake? Why doesn't a cake taste like the ingredients it is made from? What happens to the ingredients to make them taste different?

416
ENGAGE

Warm Up

Start with a Book

Have students examine the visuals and read excerpts from *Forest Fires* by Luke Thompson (Children's Press, ©2000, ISBN 0-51-623370-X). Have them look closely at pictures of the forests before and after fires. Ask:

- **Do objects in a forest before a fire have the same properties as objects that form during a forest fire?** No.

- **What kind of changes occur during a fire?** chemical changes

Explore

Inquiry Activity

How can matter change?

Make a Prediction

How do flour and baking soda change when each is mixed with vinegar? Write a prediction.

Test Your Prediction

⚠️ **Be Careful.** Wear goggles.

1. **Observe** List the properties of the vinegar, flour, and baking soda.

2. **Measure** Use a funnel to put 2 tablespoons of flour in one balloon. Add 50 mL of vinegar to a plastic bottle.

3. **Experiment** Carefully, put the balloon over the bottle's opening without letting any flour fall into the bottle. After you attach the balloon, raise it up so the flour goes into the bottle. Record your observations.

4. Repeat steps 2 and 3 using the second balloon and baking soda instead of flour.

Draw Conclusions

5. Did your results match your prediction? Explain your answer.

6. **Infer** What do you think caused the differences in the balloons?

Explore More

Experiment What might happen to the balloon if you add two tablespoons of baking soda and 50 mL of water to a container? Try it and find out.

Materials

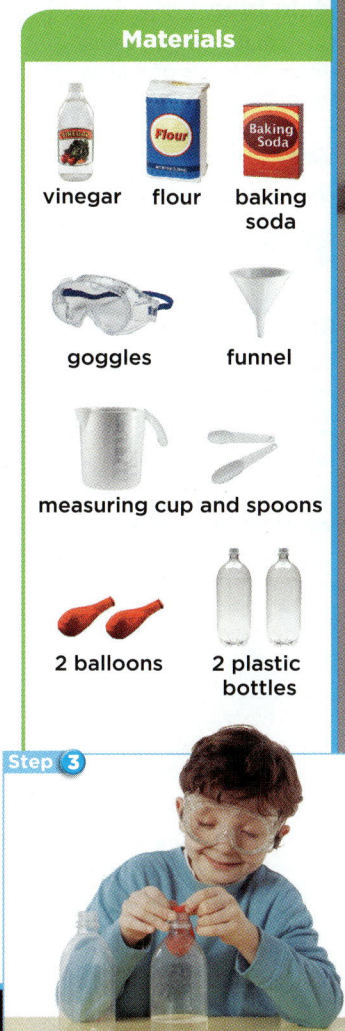

vinegar   flour   baking soda

goggles   funnel

measuring cup and spoons

2 balloons   2 plastic bottles

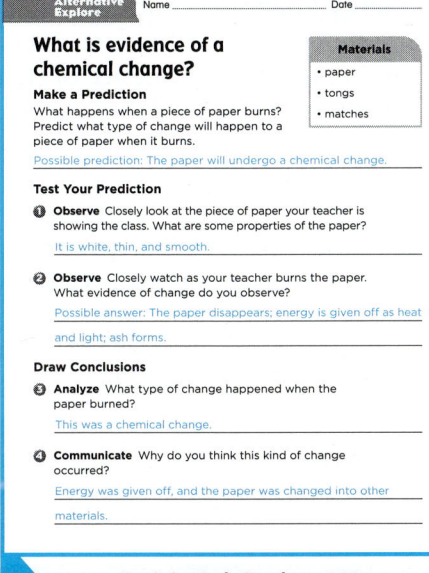

Step 3

417
EXPLORE

Alternative Explore

What is evidence of a chemical change?

Materials paper, tongs, matches

As a demonstration, hold a piece of paper with tongs and burn it in front of the class. Have students observe and record the evidence of this chemical change. The release of energy as light and heat are evidence of a chemical change. The identity of the material changes. Students will not observe them, but they may know that gases form.

Alternative Explore

Name _____   Date _____

What is evidence of a chemical change?

Materials
• paper
• tongs
• matches

Make a Prediction
What happens when a piece of paper burns? Predict what type of change will happen to a piece of paper when it burns.

Possible prediction: The paper will undergo a chemical change.

Test Your Prediction

1. **Observe** Closely look at the piece of paper your teacher is showing the class. What are some properties of the paper?

It is white, thin, and smooth.

2. **Observe** Closely watch as your teacher burns the paper. What evidence of change do you observe?

Possible answer: The paper disappears; energy is given off as heat and light; ash forms.

Draw Conclusions

3. **Analyze** What type of change happened when the paper burned?

This was a chemical change.

4. **Communicate** Why do you think this kind of change occurred?

Energy was given off, and the paper was changed into other materials.

Activity Lab Book, p. 180

Explore

 small groups    30 minutes

Plan Ahead Collect clear plastic bottles that have narrow openings and that are about 0.5 L in volume. Water bottles are suitable. Two bottles per student group are needed.

Purpose This activity helps students observe a chemical change. They will observe production of a gas, which is one example of evidence that a chemical change has occurred.

Structured Inquiry

Make a Prediction Possible prediction: The baking soda reacts with vinegar, but the flour does not.

1. **Observe** Observations should include that the physical properties of flour and baking soda are similar.

3. **Experiment** Students should observe that the flour just fell into the vinegar, and no other change occurred.

5. A gas that inflated the balloon formed when baking soda mixed with vinegar.

6. **Infer** A chemical change occurred between the vinegar and the baking soda. The chemical change produced a gas that inflated the balloon. No chemical change occurred between the vinegar and the flour.

Guided Inquiry   Explore More

Experiment The chemical change that occurs is not between baking powder and water. Baking powder contains two substances that react with each other when they are in solution. Bubbles again show that a chemical reaction occurs.

Open Inquiry

Ask students what would happen if baking powder were left out of the batter for dough. Have students think of their own question about the presence of baking powder in dough. Then have them make a plan and carry out an experiment to answer the question. Ask:

What happens when baking powder is left out of a recipe for baking a cake or cookies?

2 Teach

Read and Learn

Main Idea Have students skim the lesson and discuss what they think they will learn about chemical changes.

Vocabulary From what they know about physical changes, have students infer what a chemical change is.

Reading Skill Infer

Graphic Organizer 14 Have students fill in an Infer graphic organizer as they read each two pages of the lesson. They can use the Quick Check questions to identify each inference to make.

Clues	What I Know	What I Infer

What are chemical changes?

FAST TRACK Discuss the Main Idea

Have students review that in a physical change, the materials involved do not change identity. Ask:

- **If changes are either physical or chemical, what do you think a chemical change is?** Possible answer: A chemical change is a change in which the materials involved change identity.

- **Why is evaporating gasoline a physical change, and burning gasoline a chemical change?** When gasoline evaporates, it remains gasoline; when it burns, it changes into other materials.

▶ Develop Vocabulary

chemical change Have students draw a Main Idea concept map that lists examples of chemical changes.

RESOURCES and TECHNOLOGY

- ▶ **Reading and Writing,** pp. 198–200
- ▷ **Visual Literacy,** p. 65
- ▶ **Activity Lab Book,** p. 181
- ◉ **PuzzleMaker CD-ROM**
- ◉ **Presentation Toolkit CD-ROM**
- **e-Glossary**

Read and Learn

▶ **Main Idea**
Chemical changes cause different kinds of matter to form.

▶ **Vocabulary**
chemical change, p. 418

LOG ON e-Glossary
at www.macmillanmh.com

▶ **Reading Skill** ✓
Infer

Clues	What I Know	What I Infer

What are chemical changes?

You may have seen an apple turn brown or a burning log change into ash and smoke. Both are examples of a chemical change (KEM•i•kuhl CHAYNJ). A **chemical change** is a change that causes different kinds of matter to form. The properties of the new matter are different from those of the original materials.

Chemical changes happen every day. Your body uses chemical changes to break down the food you eat. Green plants use the Sun's energy to change carbon dioxide and water into food and oxygen. Cooking also uses chemical changes. Cake batter changes when you bake it. You know that it has changed because it feels and tastes different.

A Chemical Change

unripe → ripe → overripe

A chemical change happens when fruit ripens. As bananas ripen, they change color. They also become softer and sweeter.

Read a Diagram

How have the bananas in these photographs changed?

Clue: Compare the three photographs to find differences.

Science Background

Energy Changes Indications of a chemical change, such as production of a gas, are present in some chemical changes but not in others. The evidence of chemical change that is always present, regardless of the type of chemical change, is energy change. Sometimes this is obvious, such as when something burns. During other chemical changes, the energy change may be so slight that it is difficult to detect. For example, rusting is a chemical change that gives off energy, but the amount is so small that it cannot be observed without special equipment.

See **Science Yellow Pages** in the Teacher Resource section for background information.

LOG ON Professional Development For more Science Background and resources from **NSDL** visit **www.macmillanmh.com/nsdl/**

Lesson Review

Visual Summary

Chemical changes cause different kinds of matter to form.

You observe chemical changes every day.

Light and heat, formation of a gas, and a color change are **signs of a chemical change**.

Make a  **Study Guide**

Make a Trifold Book. Use it to summarize what you learned about chemical changes.

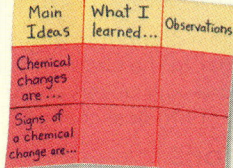

Think, Talk, and Write

1. **Main Idea** List three things that may tell you a chemical change has occurred.

2. **Vocabulary** What is a chemical change? Give an example.

3. **Infer** Two clear liquids are combined. Bubbles form. What kind of change might have happened? Explain.

Clues	What I Know	What I Infer

4. **Critical Thinking** Mrs. Hall wiped a discolored pot with a special cleaner. The pot returned to its original color. What happened?

5. **Test Prep** Which is a chemical change to a piece of paper?
   - **A** folding
   - **B** cutting
   - **C** tearing
   - **D** burning

 Math Link

Solve a Problem
A log takes one hour to burn down to ash. A banana turns brown and mushy in four days. How many hours did the longer chemical change take?

Social Studies Link

Conduct Research
Bread is made differently in other countries. Different ingredients make different chemical changes. Research how bread is made in other countries.

e-Review Summaries and quizzes online at www.macmillanmh.com

421
EVALUATE

---

3 Close

Lesson Review

▶ **Visual Summary**

Have students save their lesson Study Guides. They will use them at the end of the chapter for review.

▶ **Make a** FOLDABLES **Study Guide**

See pages R27–R28 for instructions.

FAST TRACK

Think, Talk, and Write

1. **Main Idea** Possible answers: heat, color change, bubbles

2. **Vocabulary** when a material changes to a new kind of material or materials; a burning match

3. **Infer**

Clues	What I Know	What I Infer
Bubbles formed.	Bubbles = chemical change.	Chemical change happened.

4. **Critical Thinking** Possible answer: The pot became tarnished as the result of chemical changes, and the cleaner caused a chemical change in the material on the pot that allowed the tarnish to be removed.

5. **Test Prep** D

Math Link 4 days $\times$ 24 hours/day = 96 hours

Social Studies Link Answers will vary depending on the countries chosen.

RESOURCES and TECHNOLOGY

▶ **Reading and Writing,** p. 201

▶ **School to Home Activities,** pp. 85–86

e-Review Narrated Summary and Quiz

Progress Reporter Assessments

Be a Scientist

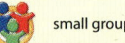 small groups   30 minutes

Skills observe, experiment, interpret data, infer, communicate

Objective

- Observe physical and chemical changes in a piece of chalk.

Materials chalk, hand lens, black construction paper, vinegar, dropper

Plan Ahead Make available a means for students to wash their hands and the droppers at the end of the activity. Have paper towels available to clean up any vinegar spills. Provide paper towels or newspaper for students to place on their work surface to protect the surface from chalk dust and vinegar.

EXTEND This activity will show students how both physical and chemical changes affect matter.

Structured Inquiry

How can physical and chemical changes affect matter?

Form a Hypothesis Possible hypothesis: If I break the chalk, it will change shape and size, but the matter will not change. If I add vinegar, the chalk will undergo a chemical change.

Test Your Hypothesis

1. **Observe** The broken end of the chalk shows a rough texture, but all the chalk appears to be the same material. This is a physical change because the identity of the chalk stays the same.

2. **Experiment** Observations should show that the chalk on the paper still looks the same as other chalk, just in much smaller particles. This is a physical change because the chalk's identity remains the same.

3. **Experiment** Bubbles form, which can be evidence of a chemical change.

Be a Scientist

Materials

chalk

hand lens

black construction paper

vinegar

dropper

Structured Inquiry

How can physical and chemical changes affect matter?

Form a Hypothesis

How will breaking chalk change the chalk? How will adding vinegar to the chalk change it? Write a hypothesis.

Test Your Hypothesis

1. **Observe** Break a piece of chalk in half. Use a hand lens to look at the broken end of the chalk. Record your observations. Is this a chemical or physical change?

2. **Experiment** Rub one of the chalk pieces on a piece of black paper. Using the hand lens look at the chalk on the paper. Record your observations. Is this a chemical or physical change?

3. **Experiment** Use a dropper to add one drop of vinegar to the chalk on the black paper. Record your observations. Is this a chemical or physical change?

Draw Conclusions

4. **Interpret Data** What did you observe? Which changes were physical changes? Was there a chemical change?

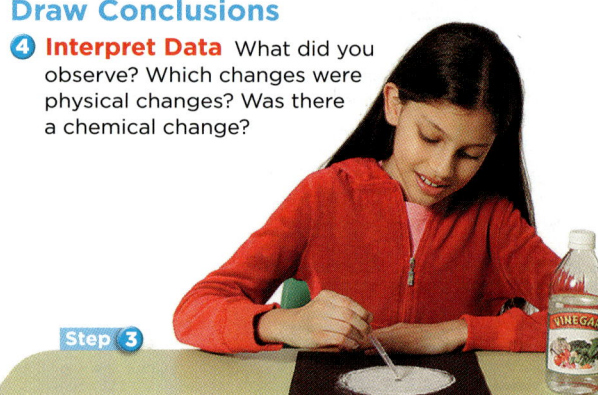

Integrate Reading

Acid Rain and Rocks

- Tell students that some rain contains materials similar to vinegar. This rain is called "acid rain." Point out that some rocks are similar to chalk in composition.

- Using research materials or the Internet, if available, have students find out what kind of change occurs to rocks when acid rain falls on them. Remind students that many buildings and sculptures are made from rock and can be affected by acid rain.

- Have students present their findings in an oral report using models or pictures as visual aids.

Inquiry Investigation

Infer Describe what happened to the chalk when you added the vinegar. What caused this to happen?

Communicate Use your observations to write your own definitions of chemical and physical change.

What are the signs of a chemical change?

Form a Hypothesis

How can you tell a chemical change has happened? Write a hypothesis.

Test Your Hypothesis

Design an experiment to investigate chemical changes. Use the materials shown. Write the steps you plan to follow. Record your results and observations.

Materials

plastic cups spoon milk

iron wool vinegar baking soda

Draw Conclusions

What changes did you observe? Did your experiment support your hypothesis? Why or why not?

Open Inquiry

What else would you like to know about physical and chemical changes? Come up with a question to investigate. For example, how does iron rust? Design an experiment to answer your question.

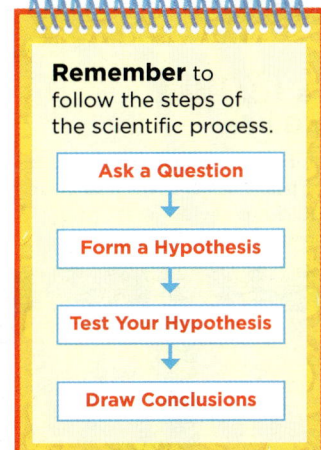

Remember to follow the steps of the scientific process.

Ask a Question
↓
Form a Hypothesis
↓
Test Your Hypothesis
↓
Draw Conclusions

423
EXTEND

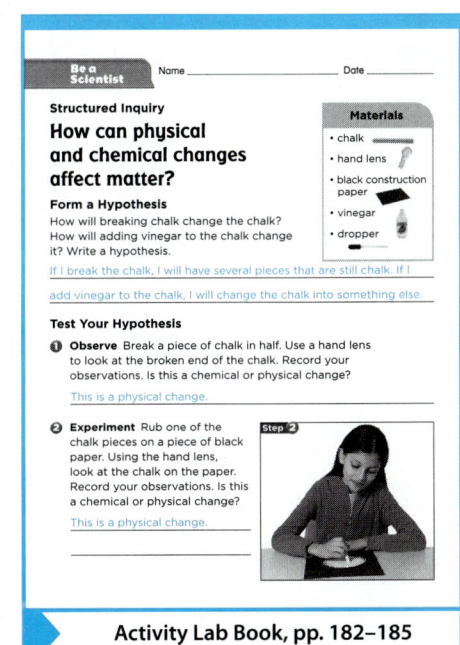

Be a Scientist   Name _____ Date _____

Structured Inquiry

How can physical and chemical changes affect matter?

Materials
• chalk
• hand lens
• black construction paper
• vinegar
• dropper

Form a Hypothesis
How will breaking chalk change the chalk? How will adding vinegar to the chalk change it? Write a hypothesis.

If I break the chalk, I will have several pieces that are still chalk. If I add vinegar to the chalk, I will change the chalk into something else.

Test Your Hypothesis

❶ **Observe** Break a piece of chalk in half. Use a hand lens to look at the broken end of the chalk. Record your observations. Is this a chemical or physical change?

This is a physical change.

❷ **Experiment** Rub one of the chalk pieces on a piece of black paper. Using the hand lens, look at the chalk on the paper. Record your observations. Is this a chemical or physical change?

This is a physical change.

▶ Activity Lab Book, pp. 182–185

Draw Conclusions

Interpret Data Breaking the chalk in half was a physical change. Rubbing the chalk on the paper was a physical change because the chalk kept its identity. The bubbles observed show that the chalk underwent a chemical change.

Infer The vinegar caused the chalk to undergo a chemical change, forming new materials with new properties.

Communicate Possible answer: The properties of a material stay the same during a physical change; new materials with new properties form during a chemical change.

What are the signs of a chemical change?

Form a Hypothesis Possible hypothesis: A chemical change occurs when new substances are formed during the change.

Test Your Hypothesis Encourage students to experiment adding two materials together until they find out which ones react with each other. Final designs may reflect that vinegar reacts with milk, iron wool, and baking soda to bring about chemical changes.

Draw Conclusions Vinegar causes milk to curdle and iron wool to rust, and bubbles form when it is added to baking soda. Answers will vary depending on the hypothesis and how it compares to the results.

Advise students that they should think about physical and chemical changes that they wonder about. Tell them that they should choose a question that is simple and safe to investigate. Encourage students to extend what they know about changes. Possible question: How does temperature affect the rate of chemical change?

RESOURCES and TECHNOLOGY

▶ **Activity Lab Book,** pp. 182–185

▶ **Activity Flipchart,** p. 49

▶ **Instructional Navigator CD-ROM**

CHAPTER 10 Review

▶ Visual Summary

Have students look at the pictures to review the main ideas of the chapter.

1. mixture

2. physical change

3. freeze

4. chemical change

5. evaporate

6. solution

7. melt

8. condense

▶ Make a **FOLDABLES** Study Guide

See pages R27–R28 in the back of the Teacher's Edition for more information on Foldables.

RESOURCES and TECHNOLOGY

▶ **Reading and Writing,** pp. 202–203

▶ **Assessment,** pp. 121–124, 128–131

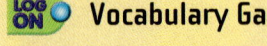 **Progress Reporter** Assessments

💿 **PuzzleMaker CD-ROM**

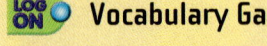 **Vocabulary Games**

CHAPTER 10 Review

Visual Summary

 **Lesson 1** Adding or removing heat can cause matter to change state.

 **Lesson 2** Matter looks different after a physical change, but it is still the same kind of matter.

 **Lesson 3** Chemical changes cause different kinds of matter to form.

Make a **FOLDABLES** Study Guide

Glue your lesson study guides to a piece of paper as shown. Use your study guide to review what you have learned in this chapter.

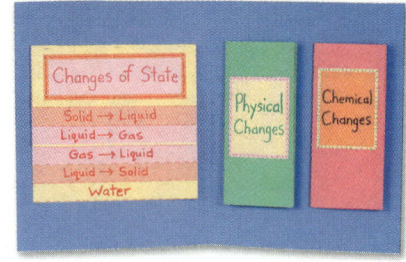

Vocabulary

Fill each blank with the best term from the list.

chemical change, p. 418

condense, p. 400

evaporate, p. 399

freeze, p. 401

mixture, p. 410

melt, p. 398

physical change, p. 408

solution, p. 411

1. When you mix spaghetti and meatballs together, you make a _____.

2. Tearing a sheet of paper is a _____.

3. If you _____ a liquid, it becomes a solid.

4. A change that causes different kinds of matter to form is a _____.

5. To change from a liquid to a gas slowly is to _____.

6. When you mix salt with water you make a _____.

7. To change from a solid to a liquid is to _____.

8. If you cool a gas to the right temperature, it will _____, or turn into a liquid.

  -**Review** Summaries and quizzes online at **www.macmillanmh.com**

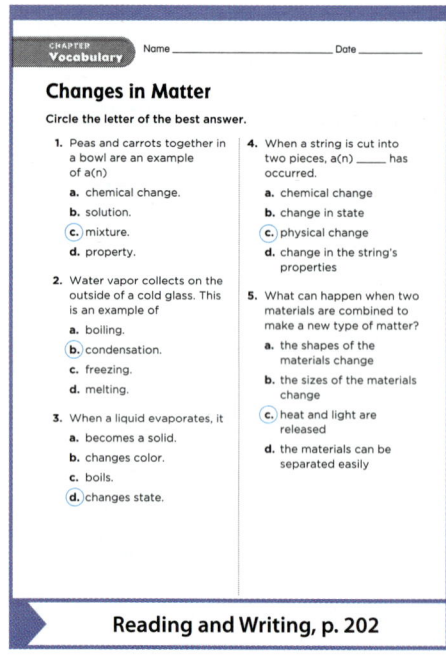

Reading and Writing, p. 202

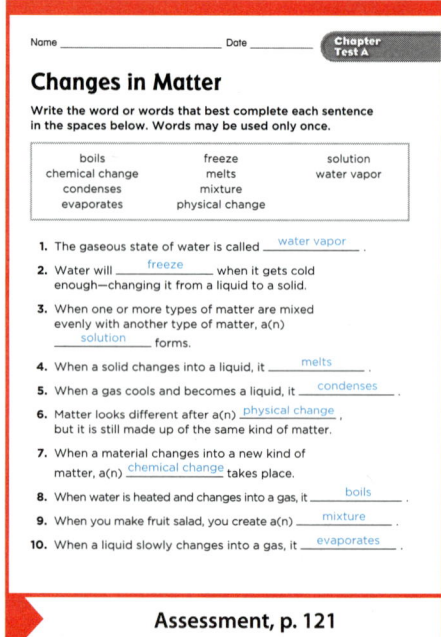

Assessment, p. 121

Skills and Concepts

Answer each of the following in complete sentences.

9. **Infer** What kind of change occurs when you toast bread? What kind of change occurs when butter melts on a piece of toast? Explain your answer.

10. **Expository Writing** Describe what happens to water as it freezes.

11. **Predict** It is a warm, sunny day. You leave a bar of chocolate on the window sill. How do you think it will change? Can you change it back?

12. **Critical Thinking** You add sugar to a glass of lemonade and stir it. You cannot see the sugar anymore. The lemonade tastes sweet now. What kind of mixture is this? How do you know?

13. Study the photograph. In what two states of matter is water shown? Describe how they are different.

14. In what ways can matter change?

Act It Out!

▶ With a partner, act out one important term or idea from this chapter. For example, you may choose a term such as melt. You may not speak during your skit.

▶ Present your skit to the class. Then let other students guess the term.

▶ What information about your term or idea did you show? How did you show it?

▶ What details helped you guess other pairs' terms and ideas?

Test Prep

1. **Which of the following BEST describes what happens when a log burns?**

   A A chemical change is taking place.

   B The logs are becoming a liquid.

   C The logs are getting bigger.

   D Smoke is boiling.

425

Summative Assessment and Intervention

Assessment Book provides a test for Chapter 10.

Leveled Readers may be used to reteach lesson content in an alternative format. The leveled readers deliver chapter content in different readability levels. The back cover of each reader provides comprehension building activities specific to the book content. (see p. 395)

Key Concept Cards 61–66 contain prescribed routines for student intervention.

Skills and Concepts

9. **Infer** Chemical change: the toast changed color; physical change: the butter changed state. Changing state is a physical change.

10. **Expository Writing** As the water freezes, its particles rearrange. They make a special pattern. They create empty spaces between themselves. Because of this pattern, solid water takes up more space than liquid water.

11. **Predict** The heat will cause the chocolate to melt. You can change the chocolate back to a solid if you put the chocolate somewhere cold, such as in a refrigerator. As it cools, it will become a solid again.

12. **Critical Thinking** This is a solution. The sugar is mixed evenly in the lemonade. That is why every sip tastes sweet.

13. Water is shown as a solid and as a liquid. Frozen water takes up more space than liquid water.

 14. Students should use information from the chapter to answer.

Performance Assessment

Scoring Rubric

Act It Out! (4-point rubric)

4 Points Student has (1) chosen a vocabulary term or main idea from this chapter; (2) prepared and presented a skit showing details about the term or idea; (3) successfully performed the skit without using words; (4) participated in guessing other pairs' terms and ideas.

3 Points Student has correctly completed three of the four tasks described.

2 Points Student has correctly completed two of the four tasks described.

1 Point Student has correctly completed one of the four tasks described.

Test Prep

1. A

Careers in Science

Objective

- Learn how environmental chemists help to reduce pollution.

Environmental Chemist

Genre: Nonfiction Mention that this career focuses on how product use affects the environment and how environmental problems can be solved. Ask: **What real parts of the environment are being tested in the pictures?** the water and the beach

Talk About It

- Ask: **What kind of tests might an environmental chemist perform on water to make sure it is safe to drink?** Possible answer: Check for bacteria and other living things that make drinking water unsafe; check for materials and chemicals that might be poisonous.

Learn About It

- Ask: **Do you think an environmental chemist would find more air pollution in the city or in the country? Why?** Possible answer: More air pollution would probably be found in the city because there are more people, vehicles, and businesses.

✏ Write About It

Show students pictures of the effects of an oil spill on the ocean. Have them write what they think caused the problem and what could be done to clean it up. After students write their ideas individually, have them work in small groups to compile a group solution to the oil spill. The group solution should incorporate several ideas from each student.

RESOURCES and TECHNOLOGY

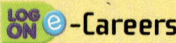 e-Careers

 **Writing Rubric** P. TR26

Careers in Science

Environmental Chemist

Do you like helping keep plants and animals healthy? Are you concerned with keeping the environment clean? If so, then you might like to be an environmental chemist.

An environmental chemist is a kind of scientist. These scientists help keep the water, land, and air free of pollution. Pollution can hurt plants, animals, and people. Environmental chemists protect living things by helping clean up pollution. Environmental chemists also show people how to reduce pollution.

To become an environmental chemist, you should begin learning about the environment where you live. Start a recycling program in your home or at school to reduce waste. You could also join a group that helps protect the environment.

▲ This scientist is collecting data on water pollution.

▲ This chemist is testing to see how much oil is left on a beach after a spill.

> **Here are some other Physical Science careers:**
> - carpenter
> - lab technician
> - chemical engineer
> - pharmacist

426

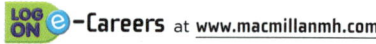 **e-Careers** at www.macmillanmh.com

---

Integrate Writing

Design a Recycling Program

Have students work in small groups to plan a recycling program for the school. Tell students to conduct research and present their strategies and findings in a written report. Have students answer the following questions in their proposals:

- How would this program benefit our school and our community?
- What materials can be recycled, and where can they be taken?
- What does our school need to do to participate?

After their plans are approved, have students draw and write on posters to promote the program and show how recycling helps the environment.

National Science Education Standards

The following Fundamental Concepts and Principles for Grades K–4 Content Standard B are covered in this unit:

- The position of an object can be described by locating it relative to another object.
- The position and motion of objects can be changed by pushing or pulling.
- Sound is produced by vibrating objects.
- Light travels in a straight line until it strikes an object.
- Heat can be produced in many ways.
- Electricity in circuits can produce light, heat, sound, and magnetic effects.

CHAPTER 11

Forces and Motion

Main Idea Motion is defined as an object's change in position. An object is moved by forces acting on it. When a force moves an object, work is done. Simple machines make work easier.

Lesson 1 Position and Motion
An object is in motion when its position changes..

Lesson 2 Forces
Forces can change an object's motion.

Lesson 3 Work and Energy
Work is done when a force changes an object's motion. Energy is the ability to do work.

Lesson 4 Using Simple Machines
Simple machines make work easier to do.

CHAPTER 12

Forms of Energy

Main Idea Energy takes many forms, including heat, sound, light, and electricity.

Lesson 1 Heat
Heat affects matter in many ways. Heat always moves from warmer objects to cooler objects.

Lesson 2 Sound
Sounds are made when objects vibrate. Pitch and volume can be used to compare sounds.

Lesson 3 Light
Light is a form of energy that allows you to see objects. Light moves in straight paths.

Lesson 4 Electricity
Electricity is made up of charged particles . These charges can flow through a circuit.

CHAPTER 11
Forces and Motion

 CD-ROM

Instructional Navigator CD-ROM
Interactive Lesson Planner, Teacher's Edition, Worksheets, and Links to Online Resources

Presentation Toolkit CD-ROM

Test Generator

PuzzleMaker CD-ROM

 DVD

Science: Master Teacher DVD Set

Science Activity DVD

 **www.macmillanmh.com**

 Science in Motion How a Lever Works

Online Teacher's Edition

Leveled Reader Database

Translated Concept Summaries

Progress Reporter Assessments

Professional Development

e-Glossary

e-Journal

e-Review

NSDL National Science Digital Library

CHAPTER 12
Forms of Energy

 CD-ROM

Instructional Navigator CD-ROM
Interactive Lesson Planner, Teacher's Edition, Worksheets, and Links to Online Resources

Presentation Toolkit CD-ROM

Test Generator

PuzzleMaker CD-ROM

 DVD

Science: Master Teacher DVD Set

Science Activity DVD

 **SCIENCE QUEST** *Light Energy*

 **www.macmillanmh.com**

 Science in Motion How You Hear Sounds and *Seeing Colors*

Online Teacher's Edition

Leveled Reader Database

Translated Concept Summaries

Progress Reporter Assessments

Professional Development

e-Glossary

e-Journal

e-Review

NSDL National Science Digital Library

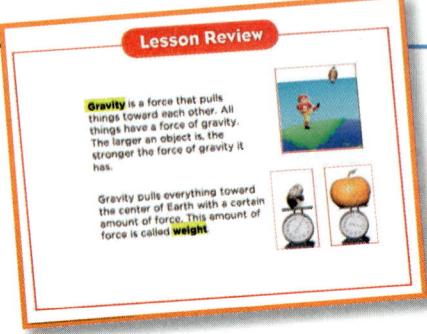

◄ Presentation Toolkit CD-ROM

CHAPTER 11
Forces and Motion

 CD-ROM

Student Navigator CD-ROM

PuzzleMaker CD-ROM

 DVD

Science Activity DVD

 **www.macmillanmh.com**

 Science in Motion How a Lever Works

Online Student Edition

Online Vocabulary Games

Translated Concept Summaries

e-Careers

e-Glossary

e-Journal

e-Review

CHAPTER 12
Forms of Energy

 CD-ROM

Student Navigator CD-ROM

PuzzleMaker CD-ROM

 DVD

Science Activity DVD

 **SCIENCE QUEST** *Light Energy*

 www.macmillanmh.com

 Science in Motion How You Hear Sounds and *Seeing Colors*

Online Student Edition

Online Vocabulary Games

Translated Concept Summaries

e-Careers

e-Glossary

e-Journal

e-Review

Operation: Science Quest ▶

Literature
Poem

Objective
- Describe how something moves and sounds.

Literature

Poem

428

Jump Rope

Genre: Poem A poem uses rhythm and rhyme to convey the characteristics of an everyday object. Ask:

- **What is an example of words that rhyme?**
  Answers will vary. Possible answers: *line* and *fine; back* and *black; grow* and *slow*

Before Reading

Before students read the poem "Jump Rope," take them outside and have them jump rope, either individually or in pairs. Ask:

- **What kinds of words would you use to describe how a jump rope moves?** Possible answers: *hop, skip, swing, swoop*

- **What kinds of words would you use to describe the sound a jump rope makes?** Possible answers: *slap, whoosh*

During Reading

Ask students to make a list of all the descriptive words they find as they read the poem. Ask:

- **What descriptive words did you find in the poem?**
  Possible answers: *quick feet; slap; slip; scissor skip*

- **Why do you think the author of the poem used these words to describe a jump rope?** Possible answer: She used these words because they describe what she feels about a jump rope.

RESOURCES and TECHNOLOGY
▶ **Reading and Writing,** p. 204

e-Journal Online research and writing

Writing Rubric P. TR26

ELL Support

Play Games Discuss with students the jump-rope games they are familiar with. Reread the poem with students. Explain that "double Dutch" is a jump rope game using two different ropes. Tell them that other jump rope games can be played individually with a shorter rope or with one long rope. Take students outside to play jump-rope games, asking them to pay attention to the sound the jump rope makes and how it moves. Have them orally describe the sounds.

BEGINNING Students can point to or name descriptive words and/or rhyming words in the poem.

INTERMEDIATE Students describe how a jump rope moves or sounds by using short sentences and phrases.

ADVANCED Students can use complete sentences to describe the sound a jump rope makes as it moves.

Jump Rope

by Rebecca Kai Dotlich

S w i n g s
        up,
whirls around,
brushes ground
beneath quick feet.
Sweeps walks,
slap, slip,
double Dutch,
scissor skip.
Flip, flap,
LOOPS around,
slip, slap,
swoops down.
Slides and swirls,
twirls and twists,
song for a jump rope
sounds like this:
Tiger leap,
spider spin,
your turn next,
jump on in!

 Write About It

Response to Literature This poet uses rhythm and rhyme to describe how a jump rope moves. How else do things move on the playground? Write a poem about another movement game.

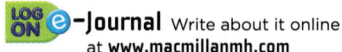 **-Journal** Write about it online at www.macmillanmh.com

429

After Reading

Explain to students that this poem describes what one person, the author, thinks about a jump rope. Encourage students to write a poem, either rhyming or non-rhyming, to describe what they think about a jump rope. Ask:

- **How does your poem compare with the poem in the textbook?** Accept all reasonable answers.

- **Do you think you and the author of the poem in the textbook think the same about a jump rope?** Accept all reasonable responses.

Write About It

If students have trouble describing the game, ask:

- **How do you move when you do this activity?**

- **How do your friends move when they do this activity?**

Before they begin their poems, encourage students to look back at the list of words they created while reading the poem "Jump Rope."

More to Read

Animals in Motion: How Animals Swim, Jump, Slither, and Glide, by Pamela Hickman and Pat Stephens (Kids Can Press, 2000)

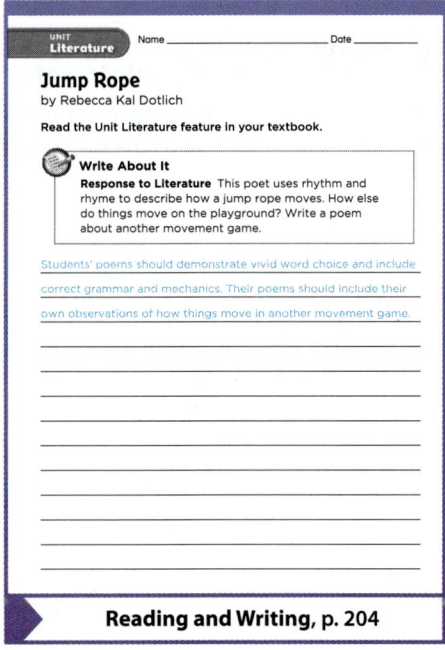

Presentation Toolkit CD-ROM Lesson Presentations

 Instructional Navigator CD-ROM Interactive Lesson Planner, Teacher's Edition, Worksheets, and Online Resources

Lesson	OBJECTIVES AND READING SKILLS	VOCABULARY	RESOURCES AND TECHNOLOGY
1 Position and Motion PAGES 432–439 PACING: 2 days FAST TRACK: 1 day	■ Describe and relate position and motion. ■ Define speed using distance and time. **Reading Skill** Compare and Contrast *Graphic Organizer 10*	position distance motion speed	▶ Reading and Writing, pp. 206–209 ▶ Visual Literacy, pp. 67–68 ▶ Activity Lab Book, pp. 186–189 ▶ Transparencies, pp. 67–68
2 Forces PAGES 442–449 PACING: 2 days FAST TRACK: 1 day	■ Identify a force as a push or a pull, and relate force to motion. ■ Define common forces, such as friction, gravity, and magnetism. **Reading Skill** Cause and Effect *Graphic Organizer 8*	force magnet gravity weight friction	▶ Reading and Writing, pp. 212–215 ▶ Visual Literacy, pp. 69–70 ▶ Activity Lab Book, pp. 190–193 ▶ Transparencies, pp. 69–70
3 Work and Energy PAGES 452–459 PACING: 2 days FAST TRACK: 1 day	■ Define energy and work. ■ Discuss the forms of energy and how energy changes from one form to another. **Reading Skill** Summarize *Graphic Organizer 5*	work energy kinetic energy potential energy	▶ Reading and Writing, pp. 216–219 ▶ Visual Literacy, pp. 71–72 ▶ Activity Lab Book, pp. 198–201 ▶ Transparencies, pp. 71–72
4 Using Simple Machines PAGES 462–471 PACING: 2 days FAST TRACK: 1 day	■ Identify and describe simple machines, and apply their use to real-world tasks. ■ Define what a compound machine is and give several examples. **Reading Skill** Problem and Solution *Graphic Organizer 12*	simple machine lever pulley wheel and axle inclined plane screw wedge compound machine	▶ Reading and Writing, pp. 220–223 ▶ Math, pp. 17–18 ▶ Visual Literacy, pp. 73–74 ▶ Activity Lab Book, pp. 205–208 ▶ Transparencies, pp. 73–74 ▶ **Science in Motion,** *How a Lever Works*

Chapter II Review
PAGES 474–475

■ Summarize chapter concepts.

Resources
▶ Assessment, pp. 134–137, 142–145
▶ Reading and Writing, pp. 226–227

Technology
▶ **Progress Reporter** Assessments
▶ **e-Review**

PACING Assumes a day is a 25- to 35-minute session

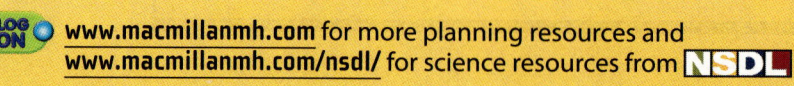 **LOG ON** www.macmillanmh.com for more planning resources and www.macmillanmh.com/nsdl/ for science resources from **NSDL**

Activity Planner

 **Science Activity DVD** Explore Activity demos

Materials included in the Activity Kit are listed in *italics*.

EXPLORE Activities

Explore p. 433 | PACING: 30 minutes

Objective Describe the positions of cubes in relation to a reference point.

Skills communicate, observe, infer

Materials notebook, two sets of 10 colored blocks

⭐ **PLAN AHEAD** Make the needed cubes out of paper if blocks are not available.

Explore p. 443 | PACING: 30 minutes

Objective Determine the relationship between force and motion.

Skills observe, measure, use variables, infer, interpret data, experiment

Materials six books, cardboard, *masking tape, toy car,* tennis ball, ruler

⭐ **PLAN AHEAD** Arrange classroom furniture so students have room to work.

Explore p. 453 | PACING: 20 minutes

Objective Develop an operational definition of work.

Skills classify, communicate, infer, experiment

Materials book, chair

⭐ **PLAN AHEAD** Prepare a data table for students to use.

Explore p. 463 | PACING: 30 minutes

Objective Determine how a machine affects work.

Skills experiment, use variables, communicate, interpret data

Materials *clay,* thick marker, ruler, 2 small *cups,* large blocks, one-gram cubes

⭐ **PLAN AHEAD** Use large metal paper clips in place of one-gram cubes if needed.

QUICK LAB Activities

Quick Lab p. 438 | PACING: 15 minutes

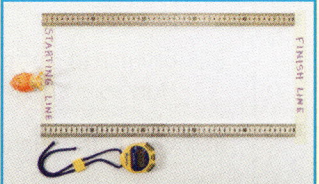

Objective Measure distance and time to determine a wind-up toy's speed.

Skills measure, communicate, use numbers

Materials 2 metersticks, *masking tape,* wind-up toy, *stopwatch*

⭐ **PLAN AHEAD** Screen wind-up toys to select only those that travel in a straight line.

Quick Lab p. 447 | PACING: 10 minutes

Objective Show how gravity depends on mass.

Skills predict, observe, infer

Materials empty plastic bottle, full plastic bottle

⭐ **PLAN AHEAD** Fill bottles with water before beginning the activity.

Quick Lab p. 457 | PACING: 15 minutes

Objective Plan a meal that provides enough energy or an activity.

Skills use numbers

Food	Calories of Energy
1 cup of apple juice	120
1 slice of wheat bread	75
1 slice of turkey	30
1 slice of cheese	60
1 cup of lettuce	7

⭐ **PLAN AHEAD** Explain that students can use more than one serving of a particular food in planning a meal.

Quick Lab p. 469 | PACING: 15 minutes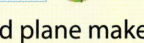

Objective Demonstrate how an inclined plane makes work easier.

Skills measure, interpret data

Materials cardboard, 4 books, spring scale, bag of 25 *marbles*

⭐ **PLAN AHEAD** Use smooth cardboard to reduce the effect of friction.

FOR MORE ACTIVITIES

Activity Flipchart for Science Centers and Workstations

Focus on Skills	Be a Scientist	Everyday Science
Teach the inquiry skill: Infer, p. 54	Magnetic force, p. 52	Energy, p. 67

Use the Activities in your work station. See **Activity Lab Book** for more support.

For a comprehensive list of consumable and non-consumable materials, see the back of the unit tab.

Chapter 11 Planner **430B**

Technology

For additional language support and vocabulary development, go to **www.macmillanmh.com**

Science in Motion
How a Lever Works

Vocabulary Games

Academic Language

English language learners need help in building their understanding of the academic language used in daily instruction and science activities. The following strategies will help to increase students' language proficiency and comprehension of content and instruction words.

Strategies to Reinforce Academic Language

- **Use Context** Academic language should be explained in the context of the task. Use gestures, expressions, and visuals to support meaning.

- **Use Visuals** Use charts, transparencies, and graphic organizers to explain key labels to help students understand classroom language.

- **Model** Use academic language as you demonstrate the task to help students understand instruction.

Academic Language Vocabulary Chart

The following chart shows chapter vocabulary and inquiry skills as well as some Spanish cognates. **Vocabulary** words help students comprehend the main ideas. **Inquiry Skills** help students develop questions and perform investigations. **Cognates** are words that are similar in English and Spanish.

Vocabulary	Inquiry Skills	Cognates	
		English	**Spanish**
position, p. 434	communicate, p. 433	communicate, p. 433	*comunicar*
distance, p. 435	observe, p. 433		
motion, p. 436	infer, p. 433	observe, p. 433	*observar*
speed, p. 438	measure, p. 443	infer, p. 433	*inferir*
force, p. 444	use variables, p. 443	position, p. 434	*posición*
magnet, p 446	interpret data, p. 443	distance, p. 435	*distancia*
gravity, p. 447	experiment, p. 443	interpret data, p. 443	*interpretar datos*
weight, p. 447	classify, p. 453	experiment, p. 443	*experimentar*
friction, p. 448		force, p. 444	*fuerza*
work, p. 454		gravity, p. 447	*gravedad*
energy, p. 456		friction, p. 448	*fricción*
potential energy, p. 456		classify, p. 453	*clasificar*
kinetic energy, p. 456		energy, p. 456	*energía*
simple machine, p. 465		potential energy, p. 456	*energía potencial*
lever, p. 466		simple machine, p. 465	*máquina simple*
pulley, p. 467		kinetic energy, p. 456	*energía cinética*
wheel and axle, p. 467			
inclined plane, p. 468			
screw, p. 468			
wedge, p. 469			
compound machine, p. 470			

Vocabulary Routine

Use the routine below to discuss the meaning of each word on the vocabulary chart. Use gestures and visuals to model all words.

Define *Position* is the location of an object.

Example This desk is next to that one. That is the *position* of this desk.

Ask How would you describe your *position* in this room?

Students may respond to questions according to proficiency level with gestures, one-word answers, or phrases.

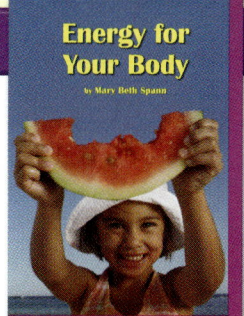

ELL Leveled Reader

Energy for Your Body
by Mary Beth Spann

Summary Read about how the five basic food groups help our bodies work.

Reading Skill
Summarize

Vocabulary Activities

Help students understand the concepts of **position** and **motion.**

BEGINNING Have three volunteers stand next to each other facing the class. Students describe their positions using phrases such as *next to*, *between*, and *to the right of*. Have volunteers hop in place. Students identify that motion as *up and down*. Repeat with other positions and motions.

INTERMEDIATE Direct students' attention to the photos on page 437. Elicit the position of each person. Make a five-column chart on the board. Draw arrows representing the following: *straight line, round and round, zigzag, back and forth,* and *up and down*. Brainstorm games that students play. Using the following model, students state the motions used in the games: *When I jump rope, I move up and down.* List the games in the appropriate column.

ADVANCED Students play "Simon Says," issuing commands such as: *Put your hand on your head. Jump up and down. Put your hand on the shoulder of the person on your right.*

Language Transfers

Grammar Transfer
In Cantonese, Haitian Creole, Hmong, and Vietnamese, the linking verb *be* is omitted:
The desk next to that one.

Phonics Transfer
Native speakers of Spanish and Cantonese may have difficulty with the /sh/ sound as in *position*.

Forces and Motion

 THE BIG IDEA What makes something move?

Chapter Preview Read the headings in the lesson. Predict what the lessons will be about.

▶ Assess Prior Knowledge

Before reading the chapter, create a **KWL** chart with students. Read the Big Idea question and then ask:

- How is force related to motion?
- What are the different types of energy?
- How does a machine help you work?

Forces and Motion

What We **K**now	What We **W**ant to Know	What We **L**earned
An object moves when it changes position.	How does force change motion?	
When something moves, it has energy.	Can energy change form?	
A machine does work.		

Answers shown represent sample student responses.

Follow the **Instructional Plan** at right after assessing students' prior knowledge of chapter content.

RESOURCES and TECHNOLOGY

▶ **School to Home Activities,** pp. 89–96

▶ **Reading and Writing,** pp. 205–227

💿 **Presentation Toolkit CD-ROM**

💿 **Puzzlemaker CD-ROM**

🔵 **www.macmillanmh.com**

🔵 **e-Journal**

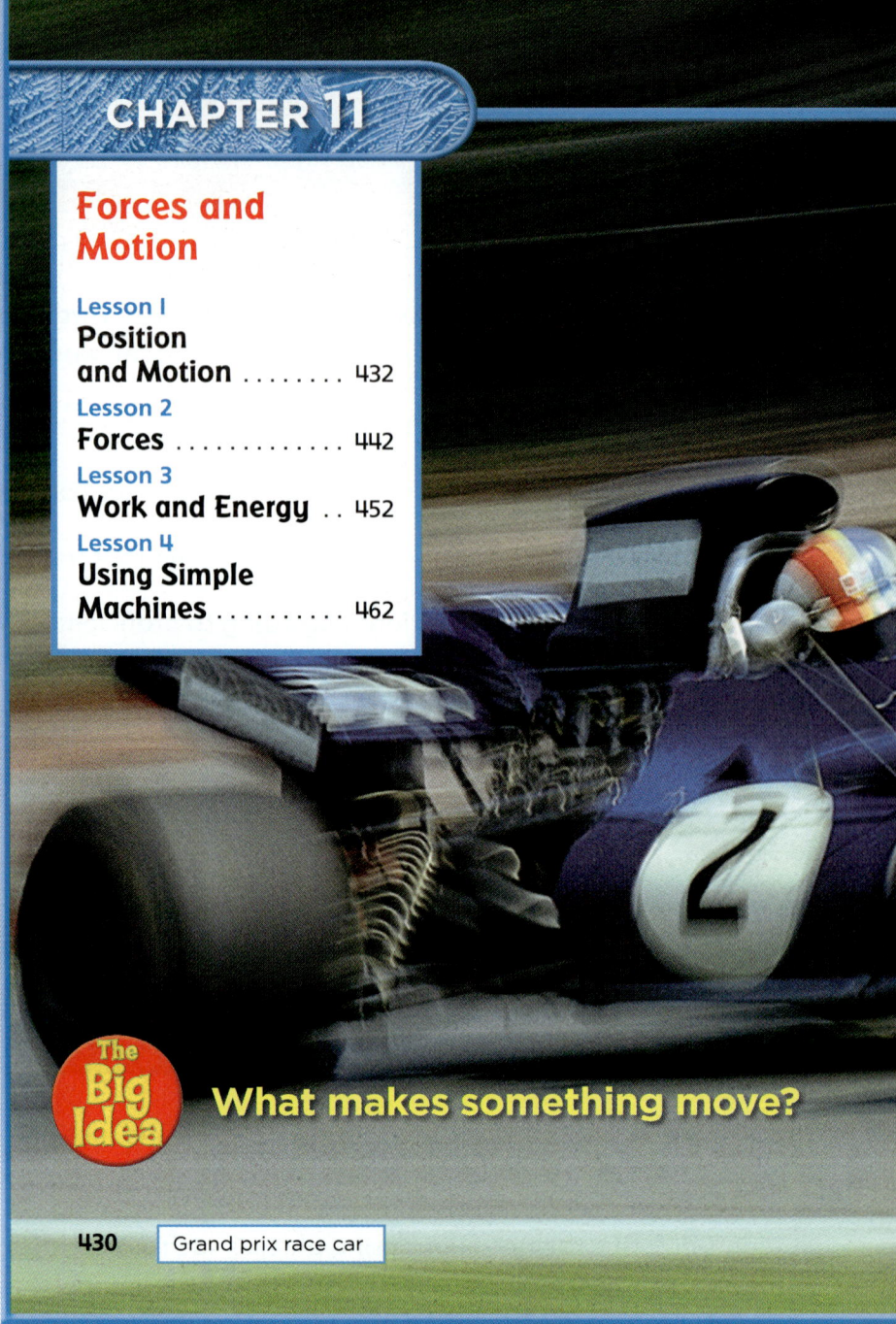

Forces and Motion

Lesson 1
Position and Motion 432

Lesson 2
Forces 442

Lesson 3
Work and Energy . . 452

Lesson 4
Using Simple Machines 462

 What makes something move?

430 Grand prix race car

Differentiated Instruction

Instructional Plan

Chapter Concept Forces cause many kinds of change.

EXTRA SUPPORT Students who need to describe the motion of objects should cover all of **Lesson 1,** pp. 432–441, before continuing with the rest of the chapter.

ON LEVEL Students who can describe the motion of objects can cover just the topic of speed, **Lesson 1,** pp. 438–439, and then go to **Lesson 2,** pp. 442–451, and **Lesson 3,** pp. 452–461, to focus on the Chapter 11 concept.

ENRICHMENT Students who are ready to enrich their understanding of the Chapter 11 concept may investigate forces with simple machines, **Lesson 4,** pp. 462–473.

Key Vocabulary

position
the location of an object
(p. 434)

motion
a change in position
(p. 436)

force
a push or pull; a force
can make an object
move (p. 444)

work
what is done when a force
changes an object's motion
(p. 454)

energy
the ability to do work;
energy is what makes
motion possible (p. 456)

simple machine
a machine with few or no
moving parts
(p. 465)

More Vocabulary

distance, p. 435
speed, p. 438
magnet, p. 446
gravity, p. 447
weight, p. 447
friction, p. 448
kinetic energy, p. 456
potential energy, p. 456
lever, p. 466
pulley, p. 467
wheel and axle, p. 467
inclined plane, p. 468
screw, p. 468
wedge, p. 469
compound machine,
p. 470

Vocabulary Preview

- Have a volunteer read the **Key Vocabulary** words aloud to the class. Ask students to find one or two words in the chapter by using the given page references. Add these words and their definitions to a class "Word Wall."

- Encourage students to use the illustrated glossary in the student edition's reference section. Guide students to explore the **e-Glossary**, which offers audio pronunciations, definitions, and sentences using the vocabulary words.

431

Science Leveled Readers

ALSO ON AUDIO CD
Leveled Reader Library

APPROACHING

Moving Fast Get amazing facts on some fast-moving animals and machines.
ISBN: 978-0-02-286171-1

ON LEVEL

Energy for Your Body How the five basic food groups help our bodies work.
ISBN: 0-02-284673-5

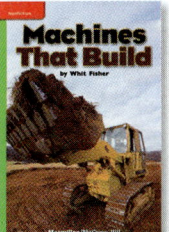

BEYOND

Machines That Build Simple levers, ramps, and pulleys help big machines operate.
ISBN: 0-02-285945-4

ELL

Energy for Your Body Uses sheltered language of On-Level Reader.
ISBN: 0-02-283487-7

See teaching strategies in the Leveled Reader Teacher's Guide. To order, call 1-800-442-9685.

 Leveled Reader Database Online Readers, searchable by topic, reading level, and keywords

Plan Your Lesson

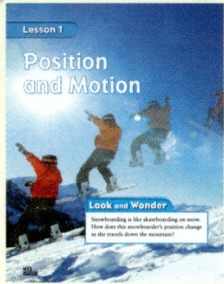

Lesson 1 Position and Motion

Objective

- Describe and relate position and motion.
- Define speed using distance and time.

Reading Skill Compare and Contrast

Different   Alike   Different

Graphic Organizer 10, p. TR12

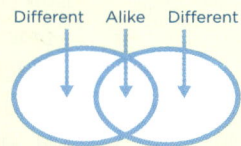

LOG ON **Professional Development** Look for **NSDL** to find recommended Science Background materials from the National Science Digital Library

FAST TRACK

Lesson Plan When time is short, follow the Fast Track and use the essential resources.

1 Introduce
Look and Wonder, p. 432

Resource Alternative Explore, p. 188

2 Teach
Use the Visuals, p. 435

Discuss the Main Idea, p. 436

Discuss the Main Idea, pl 438

Resource Visual Literacy, p. 67

3 Close
Think, Talk, and Write, p. 439

Resource Assessment, p. 138

▶ Reading and Writing

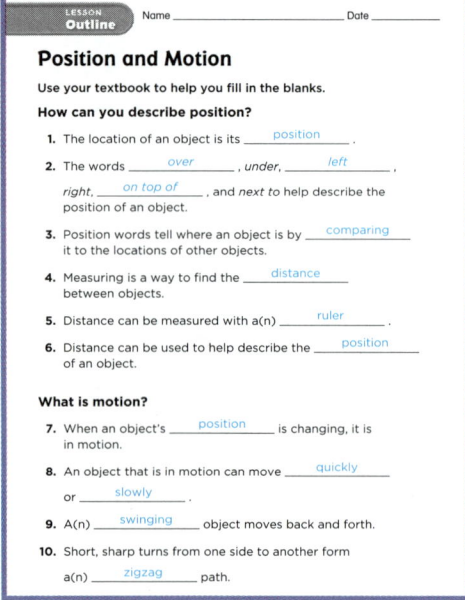

Outline, pp. 206–207
Also available as a student workbook

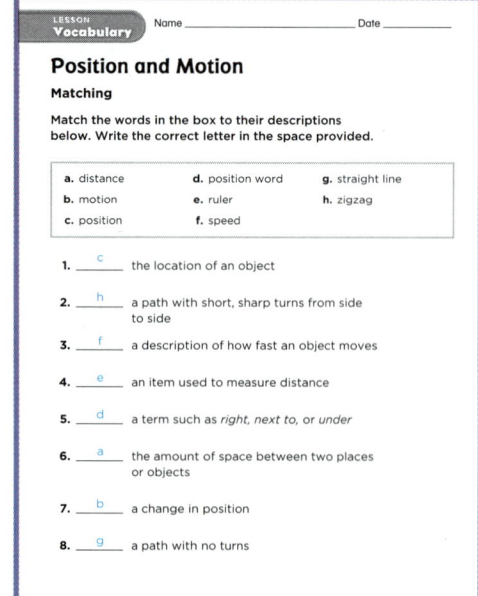

Vocabulary, p. 208
Also available as a student workbook

▶ Visual Literacy

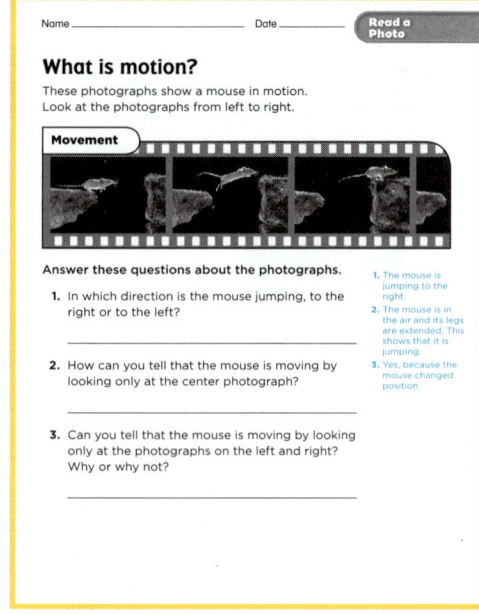

Read a Photo, p. 67
Also available as a transparency

▶ Activity Lab Book

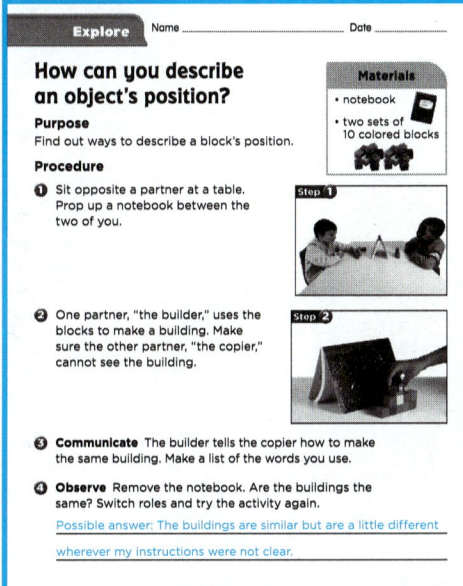

Explore Name _____ Date _____

How can you describe an object's position?

Purpose
Find out ways to describe a block's position.

Procedure

Materials
- notebook
- two sets of 10 colored blocks

❶ Sit opposite a partner at a table. Prop up a notebook between the two of you.

Step 1

❷ One partner, "the builder," uses the blocks to make a building. Make sure the other partner, "the copier," cannot see the building.

Step 2

❸ **Communicate** The builder tells the copier how to make the same building. Make a list of the words you use.

❹ **Observe** Remove the notebook. Are the buildings the same? Switch roles and try the activity again.

Possible answer: The buildings are similar but are a little different wherever my instructions were not clear.

Explore, pp. 186–187
Also available as a student workbook

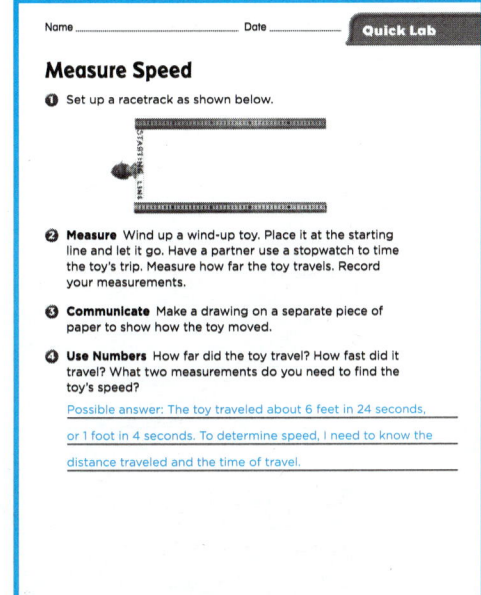

Name _____ Date _____   **Quick Lab**

Measure Speed

❶ Set up a racetrack as shown below.

❷ **Measure** Wind up a wind-up toy. Place it at the starting line and let it go. Have a partner use a stopwatch to time the toy's trip. Measure how far the toy travels. Record your measurements.

❸ **Communicate** Make a drawing on a separate piece of paper to show how the toy moved.

❹ **Use Numbers** How far did the toy travel? How fast did it travel? What two measurements do you need to find the toy's speed?

Possible answer: The toy traveled about 6 feet in 24 seconds, or 1 foot in 4 seconds. To determine speed, I need to know the distance traveled and the time of travel.

Quick Lab, p. 189
Also available as a student workbook

▶ Assessment

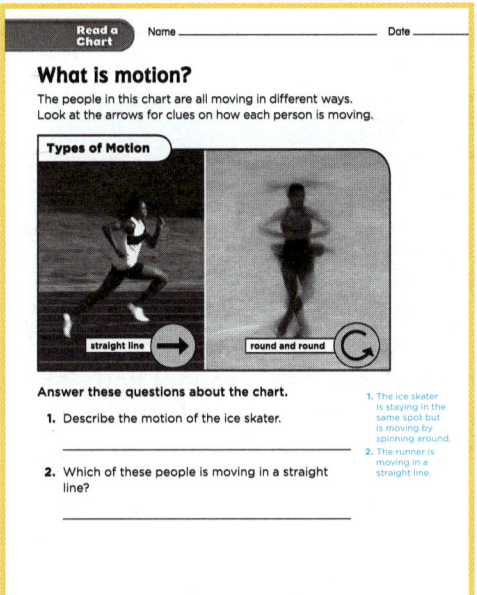

Read a Chart Name _____ Date _____

What is motion?

The people in this chart are all moving in different ways. Look at the arrows for clues on how each person is moving.

Types of Motion

straight line   round and round

Answer these questions about the chart.

1. Describe the motion of the ice skater.

_____

2. Which of these people is moving in a straight line?

_____

1. The ice skater is staying in the same spot but is moving by spinning around.
2. The runner is moving in a straight line.

Read a Chart, p. 68
Also available as a transparency

Lesson 1 Test Name _____ Date _____

Circle the letter of the best answer for each question.

1. What does the word *position* mean?
   A the speed of an object
   B the location of an object
   C the amount of space between objects
   D how an object moves

2. What two things must be known to measure speed?
   A how far an object traveled and how long it took to go that distance
   B how much an object weighs and how far it traveled
   C how dense an object is and how far it traveled
   D how long an object took to go a certain distance

3. What do the words *over, under, left,* and *right* give clues to?
   A speed
   B distance
   C position
   D balance

4. At the speed shown, which vehicle would travel 10 miles in the least amount of time?

   Vehicles

   A train
   B car
   C plane
   D boat

Critical Thinking A student is asked to run 50 yards. He can choose to run it in a straight or zigzag course. Which course would allow him to reach his destination first?
He would reach his destination first by using the straight course because the fastest distance between two points is a straight line.

FAST TRACK **Lesson Test, p. 138**
Also available as a student workbook

ADDITIONAL RESOURCES

Activity Flipchart
for Science Centers and Workstations

p. 50

ELL Teacher's Guide

p. 127

Transparencies for Visual Literacy

pp. 67–68

School to Home Activities

pp. 89–90

Key Concept Cards

67–68

Vocabulary Cards

155–158

Technology

🌀 **Science Activity DVD**

🌀 **Instructional Navigator CD-ROM**

🌀 **Presentation Toolkit CD-ROM**

 **e-Review**

 **NSDL**

Lesson 1 Position and Motion

Objectives
- Describe and relate position and motion.
- Define speed using distance and time.

1 Introduce

▶ Assess Prior Knowledge

Ask each student to describe his or her position. Ask:

- **What kinds of words do you need to use?** Possible answer: words that show my location compared to the location of something else

- **If you move somewhere else, how will you describe that position?** Possible answer: I will give my new location compared to the reference position.

- **What is motion?** Possible answers: when something moves; a change of position

FAST TRACK

Look and Wonder

Invite students to share their responses to the Look and Wonder statement and question.

- **How does this snowboarder's position change as she travels down the mountain?** Answers should include the skier's position at various times in reference to other objects.

Write ideas on the board and note any misconceptions that students may have. Address these misconceptions as you teach the lesson.

RESOURCES and TECHNOLOGY
- ▶ Activity Lab Book, pp. 186–188
- ▶ Activity Flipchart, p. 50
- ▶ Science Activity DVD

432 Unit F Chapter 11

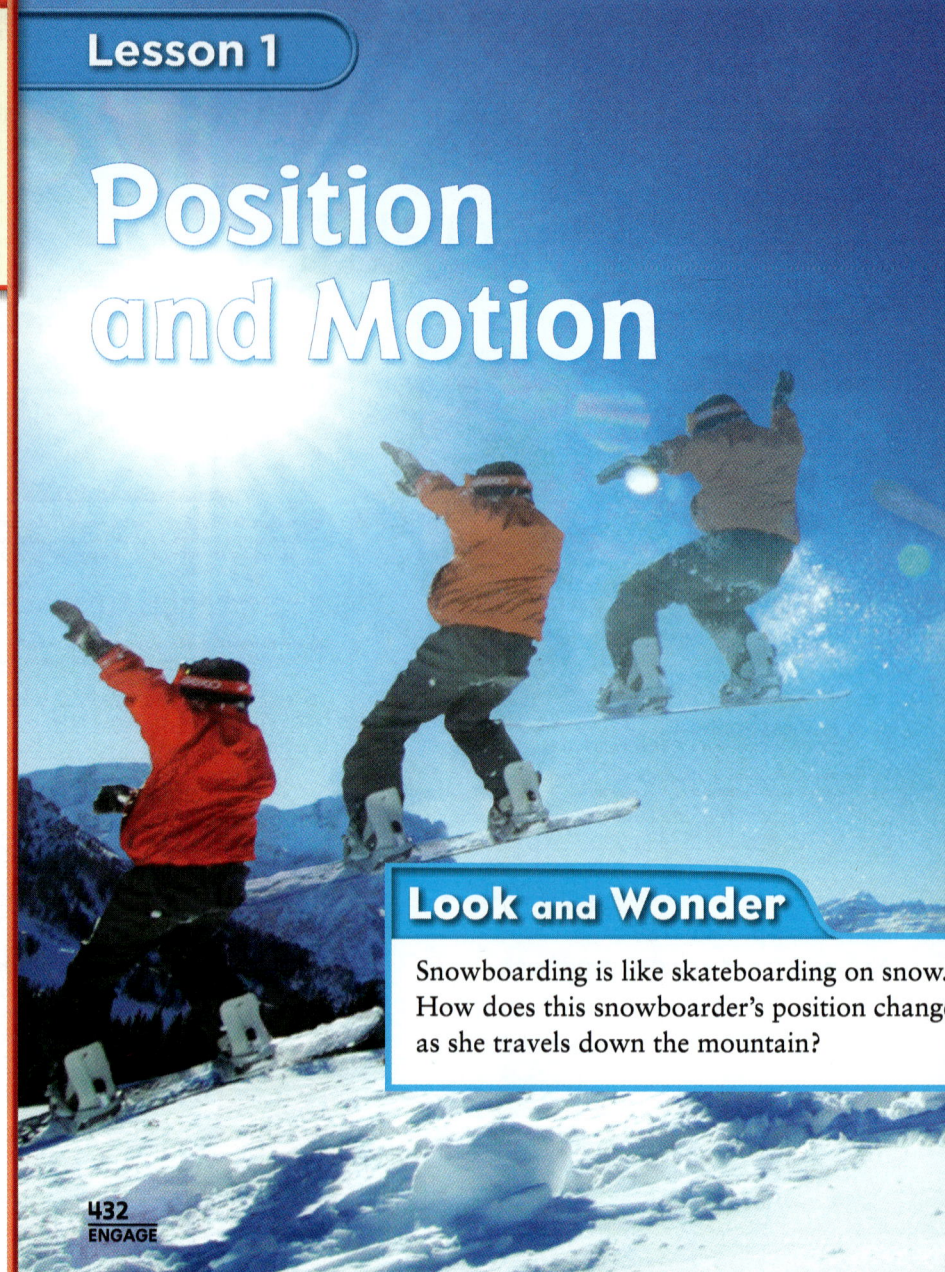

Lesson 1

Position and Motion

Look and Wonder

Snowboarding is like skateboarding on snow. How does this snowboarder's position change as she travels down the mountain?

432 ENGAGE

Warm Up

Start with a Book

Have students use a book or a magazine of their choice to identify references to position or motion. Point out that words such as *jumped* and *ran* refer to motion. Have students copy a section of text and identify references to position or motion. They should be able to justify their choices. If they cannot find references in the text, allow them to choose examples from the visuals in the book or magazine.

Explore

Inquiry Activity

How can you describe an object's position?

Purpose
Find out ways to describe a block's position.

Procedure
1. Sit opposite a partner at a table. Prop up a notebook between the two of you.
2. One partner, "the builder," uses the blocks to make a building. Make sure the other partner, "the copier," cannot see the building.
3. **Communicate** The builder tells the copier how to make the same building. Make a list of the words you use.
4. **Observe** Remove the notebook. Are the buildings the same? Switch roles and try the activity again.

Draw Conclusions
5. What words did you use to describe your building?
6. **Infer** Could you describe the position of each block without comparing it to other blocks around it?

Explore More
Communicate
How could you direct someone from your home to your school?

Materials

notebook

two sets of 10 colored blocks

Step 2

Step 3

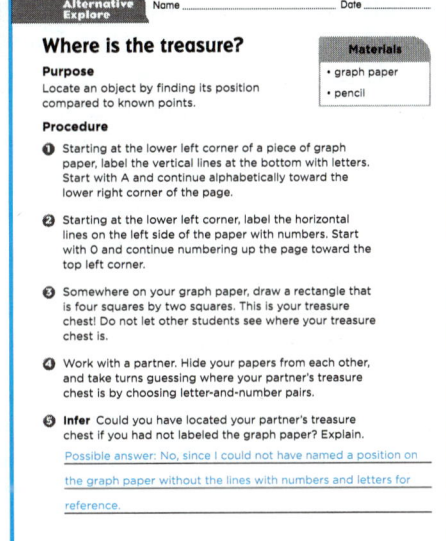

433
EXPLORE

Explore

pairs   30 minutes

Plan Ahead If block sets are not available, make cubes from a pattern that consists of four identical squares in a row and one square of the same size on either side of the row. Cut, fold, and tape the cubes together.

Purpose This activity helps students describe position in terms of a reference point.

Structured Inquiry

Procedure Help students position their notebooks so that they cannot see each other's buildings.

3. **Communicate** Caution students to use only words and not gestures.

4. **Observe** If the buildings are different, have students use the instructions from Step 2 for error analysis.

5. Possible answers: *up, down, over, under, next to, on top of*

6. **Infer** Partners would be unable to give clear directions without comparing each block's location to the location of other blocks around it.

Guided Inquiry Explore More

Communicate Students should conclude that certain terms help describe location. These terms include *up, down, over, under, next to, to the right of,* and *to the left of.*

Open Inquiry

Ask students to think of questions they could ask to find an object in the classroom. Then have student partners choose an object. Have students ask yes-or-no questions of the partners about the object's position until they find the item.

Possible question: Where is the teacher's desk?

Alternative Explore

Where's the treasure?

Materials graph paper, pencil

Tell students that a treasure is buried, and they need to find its position. Have students label one axis on graph paper with letters and the other axis with numbers. Have each student secretly draw a treasure chest that is four blocks by two blocks on the graph paper. Have them work with a partner and take turns guessing letter-and-number pairs until each locates the other's treasure chest.

Alternative Explore   Name _____ Date _____

Where is the treasure?

Purpose
Locate an object by finding its position compared to known points.

Materials
• graph paper
• pencil

Procedure
1. Starting at the lower left corner of a piece of graph paper, label the vertical lines at the bottom with letters. Start with A and continue alphabetically toward the lower right corner of the page.
2. Starting at the lower left corner, label the horizontal lines on the left side of the paper with numbers. Start with 0 and continue numbering up the page toward the top left corner.
3. Somewhere on your graph paper, draw a rectangle that is four squares by two squares. This is your treasure chest! Do not let other students see where your treasure chest is.
4. Work with a partner. Hide your papers from each other, and take turns guessing where your partner's treasure chest is by choosing letter-and-number pairs.
5. **Infer** Could you have located your partner's treasure chest if you had not labeled the graph paper? Explain.
   Possible answer: No, since I could not have named a position on the graph paper without the lines with numbers and letters for reference.

Activity Lab Book, p. 188

2 Teach

Read and Learn

Main Idea Have students choose a visual in the lesson and infer what its picture content has to do with position or motion.

Vocabulary Have students read aloud the vocabulary words. Have them define them in their own words, then compare these definitions to those in the text.

Reading Skill Compare and Contrast
Graphic Organizer 10 Have students fill in a Compare and Contrast graphic organizer as they read each two pages of the lesson. They can use the Quick Check questions to identify each comparison and contrast.

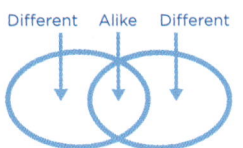

Different   Alike   Different

How can you describe position?

▶ Discuss the Main Idea

Have students brainstorm ideas of how they could describe the position of the classroom door. Ask:

- **What words describe this position?** Possible answers: *next to the wall; in the wall across from the windows*

- **How could you find the distance from the door to the trash can?** Possible answer: use a ruler or a meter stick

- **Why might this distance change from day to day?** The trash can might change in its position from day to day.

Read and Learn

Main Idea
An object is in motion when its position changes.

Vocabulary
position, p. 434
distance, p. 435
motion, p. 436
speed, p. 438

[LOG ON] **e-Glossary**
at www.macmillanmh.com

Reading Skill ✓
Compare and Contrast

Different   Alike   Different

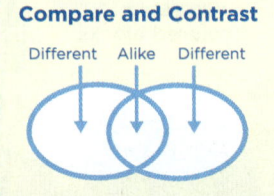

How can you describe position?

Look at the children below. Where is the boy in the red shirt? He is next to the girl in the pink shirt. He is under the girl wearing the blue overalls. When you describe where something is, you describe its position (puh·ZISH·uhn). **Position** is the location of an object.

You can describe something's position by comparing it to the position of other things. Words such as *over*, *under*, *left*, *right*, *on top of*, *beneath*, and *next to* give clues about position. You could say that a mouse is under a table or that a cat is on top of a shelf. When we describe the position of something, we compare it with objects around it.

How can you describe the position of the girl in the pink shirt?

434
EXPLAIN

Science Background

Tools used to measure distance depend on the accuracy needed in the measurement. For most measurements, tools such as rulers, meter sticks, and measuring tapes are adequate. If extremely accurate measurements are needed for mathematical or scientific purposes, tools are used that do not change when conditions, such as temperature, change. Some extremely accurate measurements of length are made using electromagnetic radiation of a specific wavelength.

See **Science Yellow Pages** in the Teacher Resource section for background information.

Distance

You can also describe something's position by measuring its distance (DIS•tuhns) from other objects. **Distance** is the amount of space between two objects or places. Distance can be measured in inches, yards, or miles. In the metric system, distance is often measured in centimeters, meters, or kilometers. You can use a ruler or meterstick to measure distances. The distance between the two toys shown below is 10 centimeters.

Quick Check

Compare and Contrast What must you compare an object to in order to describe its position?

Critical Thinking Use position words to describe the location of your classroom.

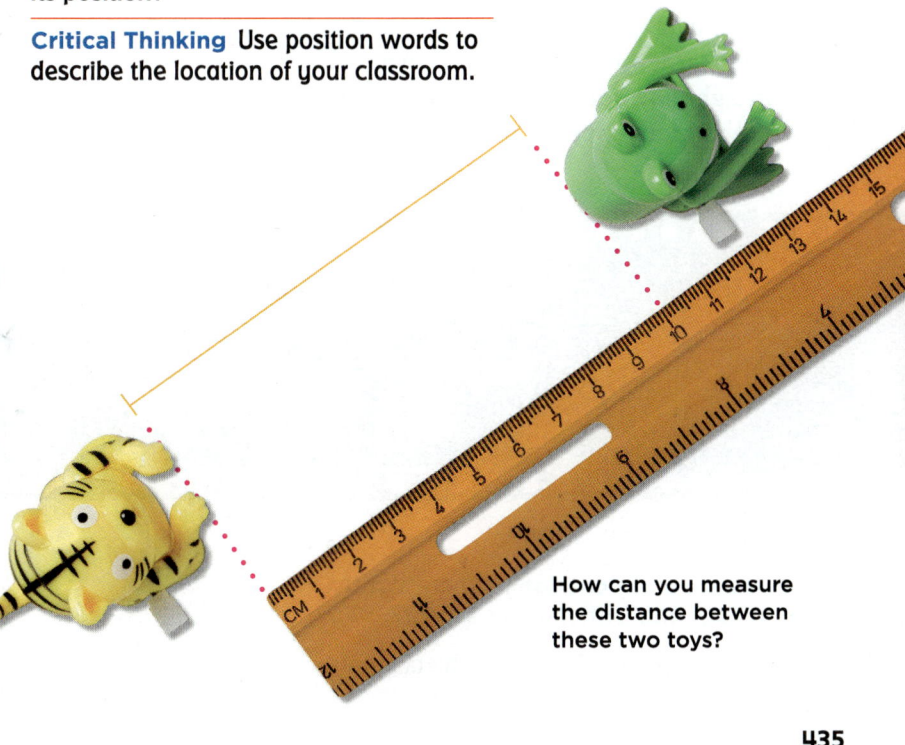

How can you measure the distance between these two toys?

435
EXPLAIN

FAST TRACK Use the Visuals

Refer students to the visual on page 435. Emphasize that two positions are needed to find distance. Ask:

- **What other distances could you find in the picture?** Possible answers: the length or height of each toy; the height of the ruler

- **Your height is one measurement of distance. What are other measurements of distance?** Possible answers: length, width, depth

▶ Develop Vocabulary

position *Word Origin* The word *position* comes from the Latin verb **positionem,** which means "act or fact of placing." When an object is placed somewhere, it is in a certain position.

distance *Word Origin* The word *distance* comes from the Latin verb **distantia,** which means "a standing apart." The amount of distance, or space, between two objects is the amount the objects are apart.

▶ Explore the Main Idea

ACTIVITY Provide students with rulers and have them measure distances between objects. Show students how to read a ruler, and have them measure only to the nearest inch or centimeter. Point out that a distance measurement contains both a number and a unit. Have students brainstorm different units of distance that they have used. Possible answers: inch, centimeter, meter, foot, yard

✓ Quick Check Answers

- **Compare and Contrast** to the positions of other objects near it

- **Critical Thinking** Possible answers: *beside, left, right, next to, opposite*

Differentiated Instruction

Leveled Activities

EXTRA SUPPORT Have students make cards with one position term written on each card. Have them take turns drawing a card and using blocks to model the term on the card. Then have them say sentences that describe the position. For example, the sentence could state, *The red block is above the blue block.*

ENRICHMENT Provide students with the position of an object in the classroom relative to a reference point. Give them a distance, such as 10 cm, between that object and a second object. Have them locate the second object and describe its position compared to the original reference point.

What is motion?

FAST TRACK Discuss the Main Idea

Have students explain in their own words what is meant by motion. Ask:

- **How do you know something moves?** It changes position.

- **What are some different types of motion?** Possible answers: straight, circular, zigzag, back and forth

▶ Use the Visuals

Have students look at the photos on page 437. Ask students to describe each type of motion in their own words. Then have them move a pencil to model each type of motion.

▶ Develop Vocabulary

motion *Word Origin* Point out that the word *motion* has its origin in the Latin word *motionem,* meaning "a moving, an emotion." When an object has motion, it has movement, or is moving.

✓ Quick Check Answers

- **Compare and Contrast** Both types of motion include a change in direction as well as a change in position.

- **Critical Thinking** Possible answers: a bike or car wheel, pinwheel, fan, skater, carousel, spinning top, flying disk

What is motion?

Look at the pictures of the mouse below. In the first box, the mouse is on a rock. In the second box, it is between the two rocks. What happened to the mouse? It moved. You know that the mouse moved because its position changed. While an object is changing position, it is in motion (MOH•shuhn). **Motion** is a change in position.

Objects can move in different ways. Look at the chart on the next page. The runner moves forward in a straight line. The figure skater spins round and round on the ice. The snowboarder moves down the hill in a zigzag. A zigzag is a path with short, sharp turns from one side to another. The skateboarder moves back and forth in the pipe. Straight line, round and round, zigzag, and back and forth are types of motion.

▲ A swing moves back and forth.

▼ How can you tell that the mouse has moved?

 Quick Check

Compare and Contrast How are zigzag and back and forth motions similar?

Critical Thinking List some objects that move round and round.

436
EXPLAIN

Types of Motion

straight line →

round and round ↻

zigzag ⚡

back and forth ⟲

Read a Chart

What are some ways objects can move?

Clue: Arrows can show directions.

437
EXPLAIN

▶ **Explore the Main Idea**

ACTIVITY   In small groups or in pairs, have students write a short story that includes each of the types of motion mentioned in the text. For example, a student might be walking home from school. He or she might walk in a straight line down the sidewalk, and then walk in a zigzag pattern to avoid getting wet from a sprinkler.

Read a Chart

Answer straight line, round and round, zigzag, back and forth

▶ **Address Misconceptions**

A common misconception is that motion is fully described by a change in position. Because motion results in a change of position, and it takes time to change position, motion also involves distance and time. Therefore, motion involves speed: the distance an object travels over a given time.

Differentiated Instruction

Leveled Questions

EXTRA SUPPORT   How do you know a bicycle has motion when you ride it? It changes position.

ENRICHMENT   How does straight motion differ from zigzag motion? Straight motion does not change direction, but zigzag motion does change direction several times.

What is speed?

Quick Lab

pairs   15 minutes

Objective Measure distance and time for a moving object and calculate its speed.

Materials 2 metersticks, masking tape, toy car, stopwatch

2 Check the toys before class to make sure they move in a straight line. Have students measure distance in centimeters and time in seconds.

4 Using data from two student pairs at a time, calculate speeds of the toy cars by dividing distance by time. The speeds will have units of centimeters per second. Have students compare the speeds to determine which toy had the greater speed.

▶ **Discuss the Main Idea**

Discuss speed with students. Point out that speed is a measurement, so it has both a number and a unit. Ask:

- **What is speed?** how fast something moves

- **What two kinds of units make up a unit of speed?** distance and time

▶ **Develop Vocabulary**

speed *Word Origin* This word's meaning as "quickness of motion" comes from the Old English word *spedum*. Old English is English as it was written and spoken from about A.D. 450 to A.D. 1100.

✓ *Quick Check Answers*

- **Compare and Contrast** An airplane is faster because it moves a certain distance in less time.

- **Critical Thinking** The red car; objects that move faster cover greater distances in the same time.

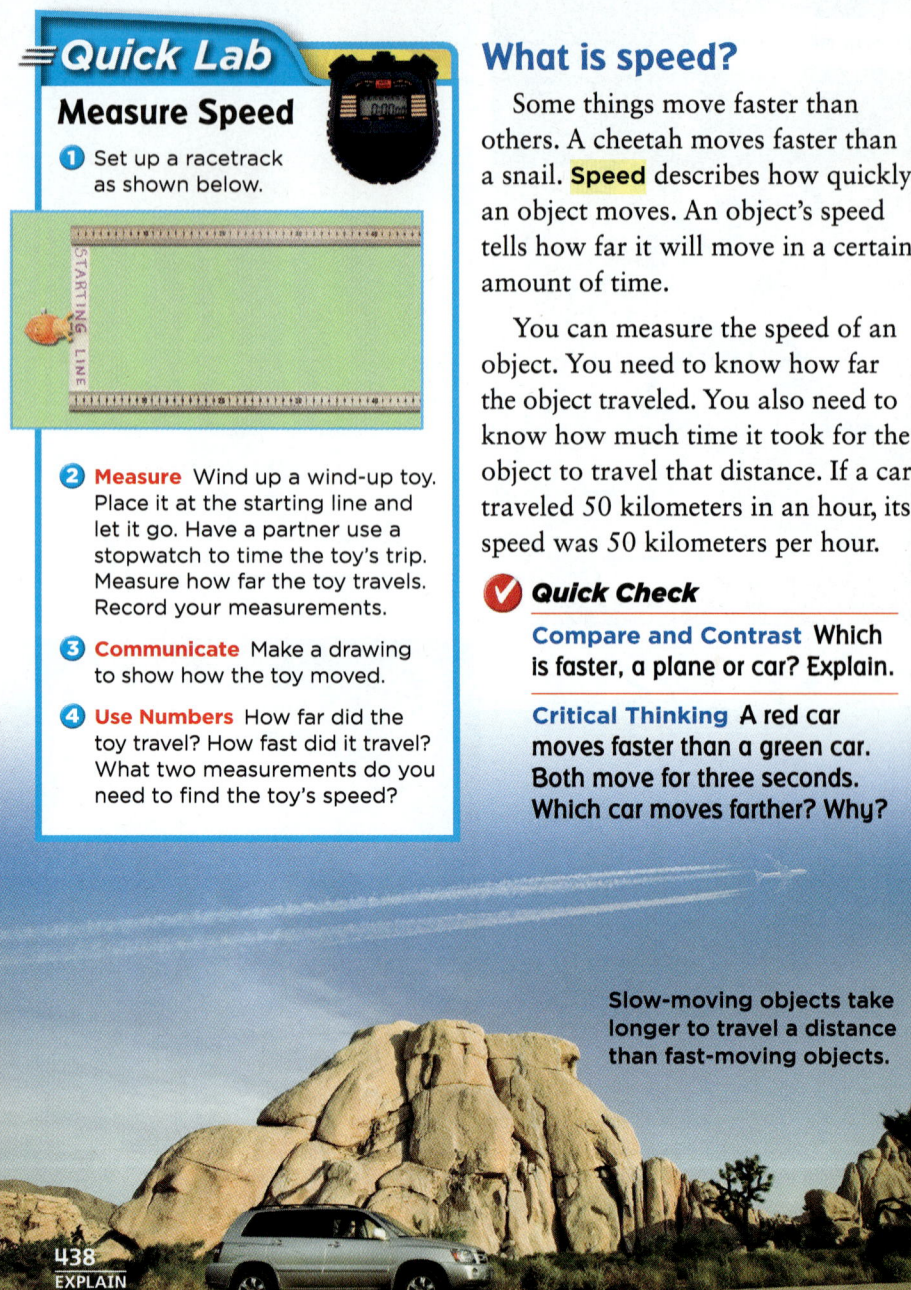

Quick Lab

Measure Speed

1 Set up a racetrack as shown below.

STARTING LINE

2 **Measure** Wind up a wind-up toy. Place it at the starting line and let it go. Have a partner use a stopwatch to time the toy's trip. Measure how far the toy travels. Record your measurements.

3 **Communicate** Make a drawing to show how the toy moved.

4 **Use Numbers** How far did the toy travel? How fast did it travel? What two measurements do you need to find the toy's speed?

What is speed?

Some things move faster than others. A cheetah moves faster than a snail. **Speed** describes how quickly an object moves. An object's speed tells how far it will move in a certain amount of time.

You can measure the speed of an object. You need to know how far the object traveled. You also need to know how much time it took for the object to travel that distance. If a car traveled 50 kilometers in an hour, its speed was 50 kilometers per hour.

✓ *Quick Check*

Compare and Contrast Which is faster, a plane or car? Explain.

Critical Thinking A red car moves faster than a green car. Both move for three seconds. Which car moves farther? Why?

Slow-moving objects take longer to travel a distance than fast-moving objects.

438
EXPLAIN

Homework Activity

Comparing Speeds

Have students examine the speedometers on any vehicles they have at home. If students do not have vehicles at home, provide each student with a picture of a speedometer. Have them explain what two units of speed are shown on the speedometer. mi/h and km/h Ask them to look at the speedometer and determine which is going faster, a car moving at 50 mi/h or one going 50 km/h. 50 mi/h

Lesson Review

Visual Summary

Position is the location of an object.

When an object's position changes, the object is in **motion**. Objects can move in different ways.

Speed describes how quickly an object moves.

Make a FOLDABLES **Study Guide**

Make a Three-Tab Book. Use it to summarize what you learned about position and motion.

Think, Talk, and Write

❶ **Main Idea** How can you tell if an object has moved?

❷ **Vocabulary** What is the position of an object?

❸ **Compare and Contrast** How is zigzag motion like back and forth motion? How are they different?

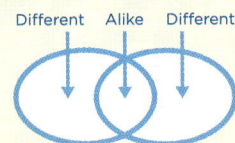
Different   Alike   Different

❹ **Critical Thinking** Suppose you rode a bike at 10 kilometers per hour for three hours. How far would you travel?

❺ **Test Prep** Which tool measures distance?
A stopwatch
B thermometer
C pan balance
D meterstick

 **Writing Link**

Write a Description
Hold a ball in your hand. Drop it. How does the ball move? Then toss the ball to a friend. How does the ball move? Describe the different ways the ball moves.

 Math Link

Make a Graph
Use research materials to find the speed of five objects. Organize this information into a chart. Then make a bar graph. Is it easier to compare data using a chart or a bar graph? Explain your answer.

LOG ON **e-Review** Summaries and quizzes online at www.macmillanmh.com

439
EVALUATE

Formative Assessment

Approaching Have students use an object of their choice to model changing position and changing speed.

On-Level Have students write a paragraph that explains how to find the speed of a moving bicycle.

Challenge Have students draw a picture that shows the starting and ending positions of a car moving slowly and a similar picture for a car moving quickly for one second. The picture of the faster car should show that its starting and ending positions are farther apart.

Key Concept Cards For student intervention, see the prescribed routine on **Key Concept Cards 67** and **68.**

3 Close

Visual Summary

▶ **Summarize the Main Idea**

Have students save their lesson Study Guides. They will use them at the end of the chapter for review.

▶ **Make a** FOLDABLES **Study Guide**

See pages R27–R28 for instructions.

FAST TRACK **Think, Talk, and Write**

❶ **Main Idea** Compare its position at different times to the positions of nearby objects that have not moved.

❷ **Vocabulary** the location of the object

❸ **Compare and Contrast**

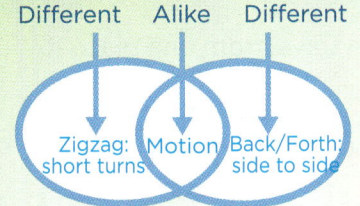
Different   Alike   Different
Zigzag: short turns   Motion   Back/Forth: side to side

❹ **Critical Thinking** 30 kilometers

❺ **Test Prep** D

Writing Link Encourage students to also find examples of each type of motion listed.

Math Link Students may want to find related speeds, such as the maximum speed five different animals can run.

RESOURCES and TECHNOLOGY

▶ **Reading and Writing,** p. 209

▶ **Activity Lab Book,** p. 189

▶ **School to Home Activities,** pp. 89–90

LOG ON **e-Review** Narrated Summary and Quiz

LOG ON **Progress Reporter** Assessments

Reading in Science

Objective
- Identify problem and solution in an article.

Travel Through Time

Genre: Nonfiction

Have students read the title and skim the photos on the time line. Ask:

- **What do you think this time line will show you?** when different forms of transportation were built

- **Why do you think people invented machines for travel?** to get somewhere in a shorter amount of time

Before Reading

Discuss how people can travel from one place to another. Ask:

- **How do you travel to school?** Possible answers: bus, car, bike, walk, train

- **When you and your family travel long distances, what kind of transportation do you take?** Possible answers: train, plane, car, bus

- **How do you think people traveled before these machines were invented?** Possible answers: by boat, on horseback, wagon, stagecoach, on foot

During Reading

Read the introductory paragraph and photo captions aloud with students. Ask:

- **Why was the first steam engine a helpful invention?** It helped people travel long distances quickly.

- **When the first car was invented, did everyone travel by car? Why or why not?** No; cars were not widely available and were too expensive for the average person.

- **Are the machines on the time line like cars and planes today? How are they different?** Possible answer: Cars used to have three wheels, and now they have four.

Real World **Reading in Science**

Travel Through Time

People have always wanted to travel. They found ways to travel within their state, across the country, and throughout the world. People have even traveled into space. The time line below shows some of the first machines that helped people travel to distant places.

1804

In England Richard Trevithick built the first steam engine for a train. The steam engine helped people travel great distances. It also helped them get to their destinations more quickly.

1884

In Germany Karl Friedrich Benz built the first car to run on gasoline. It worked similarly to the cars you see on the road today. However, his car had only three wheels!

440 EXTEND

AMERICAN MUSEUM & NATURAL HISTORY

ELL Support

Discuss Have students describe the photographs and then read the captions to them. Have them work together to describe each machine, using their own words. Ask: **Which machine do you like the best?**

BEGINNING Students can point to and say the name of their favorite machine. Students can use short phrases to explain why they like it the best.

INTERMEDIATE Students can use short sentences and phrases to explain why their machine is the best.

ADVANCED Students can use comparative language to explain why their machine is better than each of the other three.

1961

Russian astronaut Yuri Gagarin was the first person in space. His spaceship had special engines. They produced a force that was stronger than the pull of Earth's gravity. These engines helped the spaceship leave Earth's surface and orbit the planet.

1903

Wilbur and Orville Wright constructed the first motorized airplane that flew and landed safely. Their airplane's engine ran on gasoline. It flew for 12 seconds over 36 meters (120 feet).

Problem and Solution

A problem and solution
► gives a problem
► tells how to solve the problem

Write About It

Problem and Solution How have machines helped people learn about distant places? Read the article again. Then write about ways machines have helped people solve problems.

 e-Journal Write about it online at www.macmillanmh.com

441
EXTEND

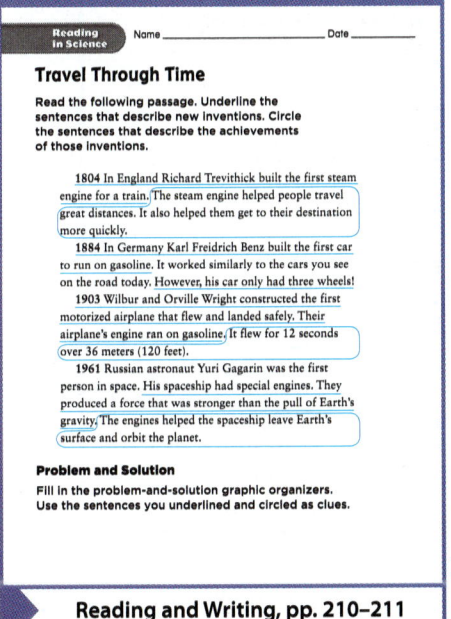

After Reading

Point out to students that the inventors of these machines often came up against problems that needed to be solved in order to get their machine to work. Display Graphic Organizer 12. Discuss with students the challenge engineers had to face when developing the first space rocket. Ask:

■ **What is the force that keeps us on Earth?** gravity

■ **When the Russians wanted to send an astronaut into space, what is one problem they had?** Possible answer: They needed to get the spaceship through Earth's gravity.

■ **What steps do you think they took to solve this problem?** Possible answers: study Earth's gravity, invent engines that were strong enough to overcome Earth's gravity

■ **What was their final solution?** They made special engines that produced a force that was stronger than the pull of Earth's gravity.

Use students' responses to complete the Problem and Solution Graphic Organizer.

Write About It

Possible answer: Various machines have allowed people to travel to and learn more about distant places, even, eventually, outer space.; written examples will differ.

Integrate Reading

Make a Time Line

Have students make their own time lines using three small sheets and one large sheet of construction paper. Ask them to think about three vehicles in which they have ridden within the past month. Have them draw a picture of each on the smaller sheets. Then have students place the vehicles in order of their use on the large sheet. Encourage them to date each drawing and write a caption.

Reading in Science   Name _____ Date _____

Travel Through Time

Read the following passage. Underline the sentences that describe new inventions. Circle the sentences that describe the achievements of those inventions.

1804 In England Richard Trevithick built the first steam engine for a train. The steam engine helped people travel great distances. It also helped them get to their destination more quickly.

1884 In Germany Karl Freidrich Benz built the first car to run on gasoline. It worked similarly to the cars you see on the road today. However, his car only had three wheels!

1903 Wilbur and Orville Wright constructed the first motorized airplane that flew and landed safely. Their airplane's engine ran on gasoline. It flew for 12 seconds over 36 meters (120 feet).

1961 Russian astronaut Yuri Gagarin was the first person in space. His spaceship had special engines. They produced a force that was stronger than the pull of Earth's gravity. The engines helped the spaceship leave Earth's surface and orbit the planet.

Problem and Solution
Fill in the problem-and-solution graphic organizers. Use the sentences you underlined and circled as clues.

► Reading and Writing, pp. 210–211

RESOURCES and TECHNOLOGY

► **Reading and Writing,** pp. 210–211

► **Technology: A Closer Look,** Lesson 1

e-Journal Online research and writing

p. TR26 **Writing Rubric**

Reading in Science   **441**

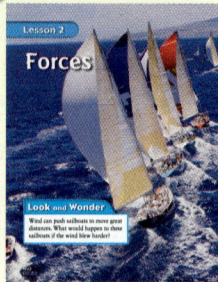

Stop Here to Plan Your Lesson

Lesson 2 Forces

Objective
- Identify a force as a push or a pull, and relate force to motion.
- Define common forces, such as friction, gravity, and magnetism.

Reading Skill Cause and Effect

Cause → Effect

Graphic Organizer 8, p. TR10

Professional Development Look for 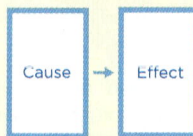 to find recommended Science Background materials from the National Science Digital Library

 **FAST TRACK**

Lesson Plan When time is short, follow the Fast Track and use the essential resources.

1 Introduce
Look and Wonder, p. 442
Resource Alternative Explore, p. 192

2 Teach
Discuss the Main Idea, p. 444
Discuss the Main Idea, p. 446
Discuss the Main Idea, p. 448
Resource Visual Literacy, p. 69

3 Close
Think, Talk, and Write, p. 449
Resource Assessment, p. 139

▶ Reading and Writing

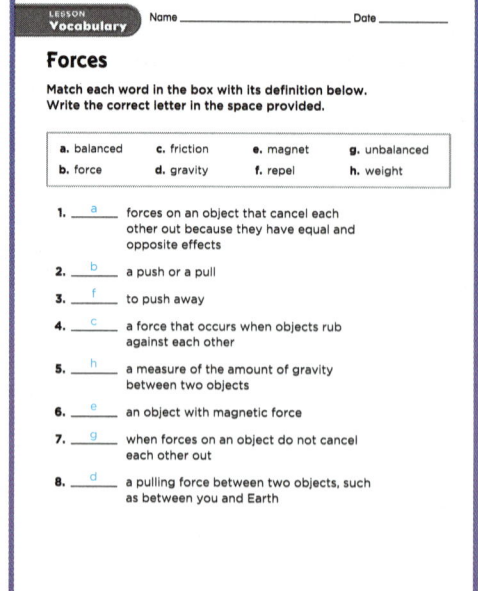

Outline, pp. 212–213
Also available as a student workbook

Vocabulary, p. 214
Also available as a student workbook

 **▶ Visual Literacy**

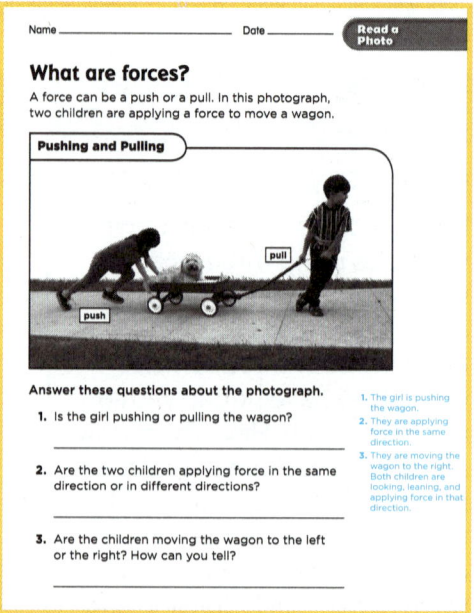

Read a Photo, p. 69
Also available as a transparency

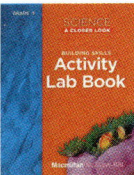

▶ Activity Lab Book

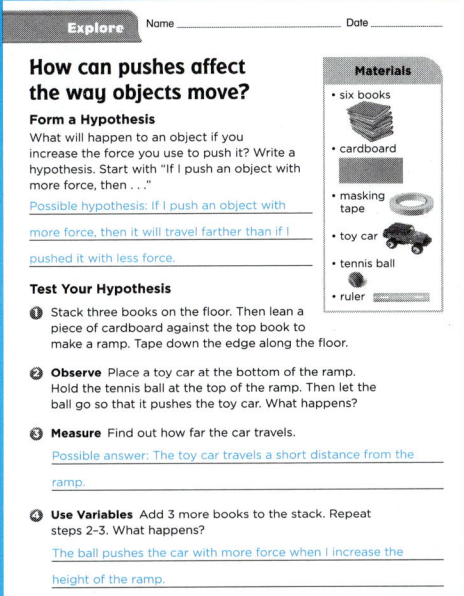

| Explore | Name _____ Date _____ |

How can pushes affect the way objects move?

Form a Hypothesis
What will happen to an object if you increase the force you use to push it? Write a hypothesis. Start with "If I push an object with more force, then . . ."

Possible hypothesis: If I push an object with
more force, then it will travel farther than if I
pushed it with less force.

Test Your Hypothesis

❶ Stack three books on the floor. Then lean a piece of cardboard against the top book to make a ramp. Tape down the edge along the floor.

❷ Observe Place a toy car at the bottom of the ramp. Hold the tennis ball at the top of the ramp. Then let the ball go so that it pushes the toy car. What happens?

❸ Measure Find out how far the car travels.

Possible answer: The toy car travels a short distance from the
ramp.

❹ Use Variables Add 3 more books to the stack. Repeat steps 2–3. What happens?

The ball pushes the car with more force when I increase the
height of the ramp.

Materials
- six books
- cardboard
- masking tape
- toy car
- tennis ball
- ruler

Explore, pp. 190–191
Also available as a student workbook

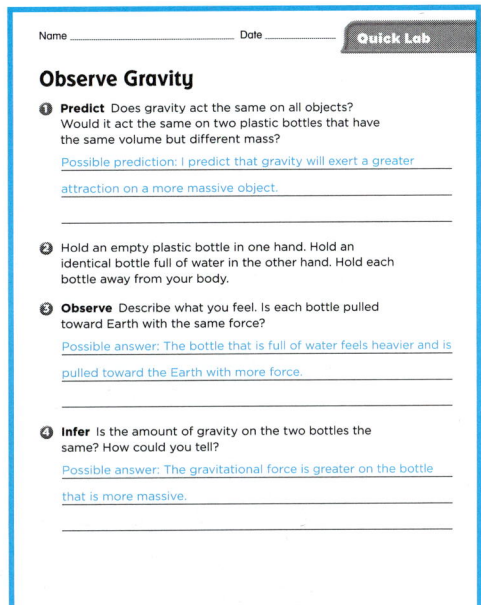

| Name _____ Date _____ | Quick Lab |

Observe Gravity

❶ Predict Does gravity act the same on all objects? Would it act the same on two plastic bottles that have the same volume but different mass?

Possible prediction: I predict that gravity will exert a greater
attraction on a more massive object.

❷ Hold an empty plastic bottle in one hand. Hold an identical bottle full of water in the other hand. Hold each bottle away from your body.

❸ Observe Describe what you feel. Is each bottle pulled toward Earth with the same force?

Possible answer: The bottle that is full of water feels heavier and is
pulled toward the Earth with more force.

❹ Infer Is the amount of gravity on the two bottles the same? How could you tell?

Possible answer: The gravitational force is greater on the bottle
that is more massive.

Quick Lab, p. 193
Also available as a student workbook

▶ Assessment

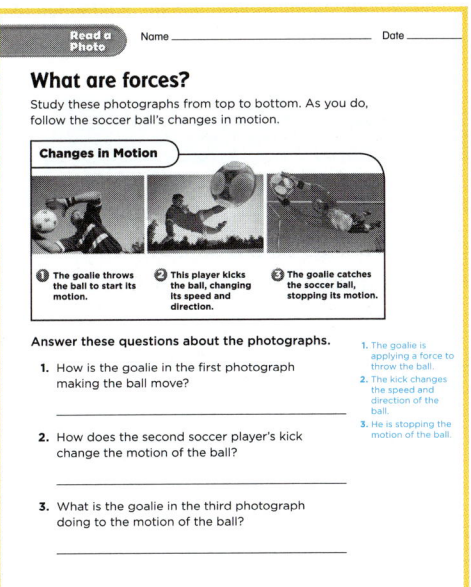

| Read a Photo | Name _____ Date _____ |

What are forces?

Study these photographs from top to bottom. As you do, follow the soccer ball's changes in motion.

Changes in Motion

❶ The goalie throws the ball to start its motion.
❷ This player kicks the ball, changing its speed and direction.
❸ The goalie catches the soccer ball, stopping its motion.

Answer these questions about the photographs.

1. How is the goalie in the first photograph making the ball move?

2. How does the second soccer player's kick change the motion of the ball?

3. What is the goalie in the third photograph doing to the motion of the ball?

1. The goalie is applying a force to throw the ball.
2. The kick changes the speed and direction of the ball.
3. He is stopping the motion of the ball.

Read a Photo, p. 70
Also available as a transparency

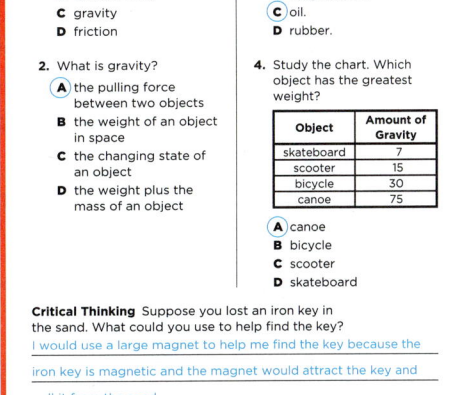

| Name _____ Date _____ | Lesson 2 Test |

Circle the letter of the best answer for each question.

1. What causes a magnet to attract and repel?
   A magnetic force
   B contact force
   C gravity
   D friction

2. What is gravity?
   A the pulling force between two objects
   B the weight of an object in space
   C the changing state of an object
   D the weight plus the mass of an object

3. All of the following will increase friction except
   A sandpaper.
   B rough stones.
   C oil.
   D rubber.

4. Study the chart. Which object has the greatest weight?

Object	Amount of Gravity
skateboard	7
scooter	15
bicycle	30
canoe	75

   A canoe
   B bicycle
   C scooter
   D skateboard

Critical Thinking Suppose you lost an iron key in the sand. What could you use to help find the key?
I would use a large magnet to help me find the key because the
iron key is magnetic and the magnet would attract the key and
pull it from the sand.

FAST TRACK **Lesson Test, p. 139**
Also available as a student workbook

ADDITIONAL RESOURCES

Activity Flipchart
for Science Centers and Workstations

p. 51

ELL Teacher's Guide

p. 128

Transparencies for Visual Literacy

pp. 69–70

School to Home Activities

pp. 91–92

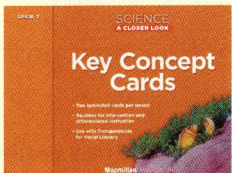

Key Concept Cards

69–70

Vocabulary Cards

159–163

Technology

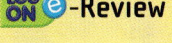 **Science Activity DVD**

Instructional Navigator CD-ROM

Presentation Toolkit CD-ROM

LOG ON e-Review

 **LOG ON NSDL**

Lesson 2 Forces

Objectives

- Identify a force as a push or a pull, and relate force to motion.
- Define common forces, such as friction, gravity, and magnetism.

1 Introduce

▶ Assess Prior Knowledge

Show students a book lying on a desk. Have a student push and then pull the book. Ask:

- **How does pushing on the book affect it?** It makes it move away from the source of the push.

- **How does pulling on the book affect it?** It makes it move toward the source of the pull.

Have students choose an activity, such as playing a musical instrument or a sport, and discuss how different objects react to pushes and pulls during the activity.

FAST TRACK Look and Wonder

Invite students to share their responses to the Look and Wonder statement and question.

- **What would happen to these sailboats if the wind blew harder?** The sailboats would move faster.

Write ideas on the board and note any misconceptions that students may have. Address these misconceptions as you teach the lesson.

RESOURCES and TECHNOLOGY

▶ **Activity Lab Book,** pp. 190–192

▶ **Activity Flipchart,** p. 51

● **Science Activity DVD**

Lesson 2

Forces

Look and Wonder

Wind can push sailboats to move great distances. What would happen to these sailboats if the wind blew harder?

442
ENGAGE

Warm Up

Start with a Demonstration

Tell students that this lesson explains pushes and pulls. Show them a toy car with a bar magnet attached to the front or back of it. Ask:

- **What will happen when a push or pull is given to the car?** The car will move.

- **What can give a push or pull to the car?** Possible answer: a hand

Hold another magnet near the magnet on the car and have students observe what happens. Turn over the magnet in your hand and again hold it near the car. Point out that pushes and pulls have many sources.

Explore

Inquiry Activity

How can pushes affect the way objects move?

Form a Hypothesis

What will happen to an object if you increase the force you use to push it? Write a hypothesis. Start with "If I push an object with more force, then . . ."

Test Your Hypothesis

1. Stack three books on the floor. Then lean a piece of cardboard against the top book to make a ramp. Tape down the edge along the floor.

2. **Observe** Place a toy car at the bottom of the ramp. Hold a tennis ball at the top of the ramp. Then let the ball go so that it pushes the toy car. What happens?

3. **Measure** Find out how far the car travels.

4. **Use Variables** Add 3 more books to the stack. The ball pushes the car with more force when you increase the height of the ramp. Repeat steps 2 and 3.

Draw Conclusions

5. **Infer** What caused the car to move?

6. **Interpret Data** When did the car travel farther?

7. **Infer** How does the amount of force you use to push an object affect how far the object travels?

Explore More

Experiment What would happen if you added a weight to the toy car and repeated the activity?

Materials

six books   cardboard

masking tape   toy car

tennis ball

ruler

Step 1

Step 2

443
EXPLORE

Alternative Explore

How does mass affect push?

Materials 3 books, cardboard, toy car, tennis ball, baseball, tape, ruler

Have students set up a ramp by leaning a piece of cardboard against a stack of three books. Tape the other end to the table. Place a toy car at the end of the ramp. Roll a tennis ball down the ramp so that it hits the car. Measure how far the car moves. Repeat using a baseball instead of a tennis ball. Students relate the ball's mass to the size of the push.

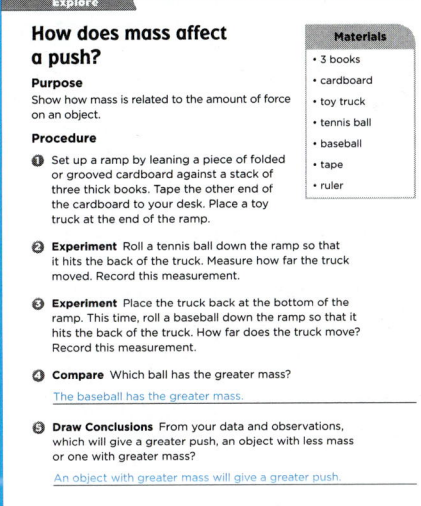

Alternative Explore Name _____ Date _____

How does mass affect a push?

Purpose
Show how mass is related to the amount of force on an object.

Procedure
1. Set up a ramp by leaning a piece of folded or grooved cardboard against a stack of three thick books. Tape the other end of the cardboard to your desk. Place a toy truck at the end of the ramp.

2. **Experiment** Roll a tennis ball down the ramp so that it hits the back of the truck. Measure how far the truck moved. Record this measurement.

3. **Experiment** Place the truck back at the bottom of the ramp. This time, roll a baseball down the ramp so that it hits the back of the truck. How far does the truck move? Record this measurement.

4. **Compare** Which ball has the greater mass?
The baseball has the greater mass.

5. **Draw Conclusions** From your data and observations, which will give a greater push, an object with less mass or one with greater mass?
An object with greater mass will give a greater push.

Materials
- 3 books
- cardboard
- toy truck
- tennis ball
- baseball
- tape
- ruler

Activity Lab Book, p. 192

Explore

 small groups   30 minutes

Plan Ahead Move room furniture so students have adequate room for the activity. Check the cardboard to make sure it is long enough to form a slope when using up to six books.

Purpose This activity helps students determine the relationship between force and motion.

Structured Inquiry

Form a Hypothesis Possible hypothesis: If I push an object with more force, then it will move farther.

1. Be sure the top of the cardboard is at the edge of the top book.

2. **Observe** Be sure the toy car's long dimension is perpendicular to the bottom edge of the ramp and that the ball is rolled so that it hits the truck at the same place each time.

4. **Use Variables** Ask students why they are repeating the activity. Point out that the data are more reliable if more trials are done because there could be mistakes in a single trial. If students' experiments did not support their hypotheses, encourage students to develop new hypotheses.

5. **Infer** The tennis ball pushed the car.

6. **Interpret Data** Possible answers: when the ramp was steeper; when it was pushed harder

7. **Infer** The greater the force used on an object, the greater the distance the object will travel.

Guided Inquiry   Explore More

Experiment The car with the weight will not travel as far.

Open Inquiry

Ask students how the surface affects a push. Have them think of a question about how a push is affected by the surface an object moves over. Ask them to make a plan and carry out an experiment to answer the question. Ask:

What is the difference between pushing an object over a painted surface and over a rough surface?

2 Teach

Read and Learn

Main Idea Have students do a picture tour of the lesson and ask them to discuss what they think they will learn about pushes and pulls.

Vocabulary Have students make a small picture dictionary of the vocabulary words and their definitions.

Reading Skill **Cause and Effect**
Graphic Organizer 8 Have students fill in a Cause and Effect graphic organizer as they read each two pages of the lesson. They can use the Quick Check questions to identify each cause and effect.

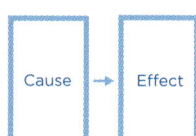

What are forces?

Discuss the Main Idea

Have students list common pushes and pulls. Tell them that a force is a push or a pull. Ask:

- **What effect does a force have on an object?** It changes its motion.

- **In what ways can motion be changed by a force?** Motion can change in speed or direction.

- **Can forces act on an object and not change its motion?** Yes; if all the forces acting on an object are equal in opposite directions, the forces are balanced, and motion does not change.

RESOURCES and TECHNOLOGY

- Reading and Writing, pp. 212–214
- Visual Literacy, p. 70
- Presentation Toolkit CD-ROM
- e-Glossary

Read and Learn

- **Main Idea**
  Forces can change an object's motion.

- **Vocabulary**
  force, p. 444
  magnet, p. 446
  gravity, p. 447
  weight, p. 447
  friction, p. 448

  LOG ON e-Glossary
  at www.macmillanmh.com

- **Reading Skill** ✓
  Cause and Effect

  Cause → Effect

What are forces?

Objects do not move by themselves. You have to apply a force (FAWRS) to make them start moving. A **force** is a push or a pull. You use forces to move things all the time. When you pull on a door handle or push a wagon, you apply a force to make something move.

Forces can be large or small. The force a crane uses to lift a truck is huge. The force your hand uses to lift a feather is tiny. It takes more force to move heavy objects than light objects. Forces also affect an object's speed. The more force you use, the faster an object will move.

A push and a pull make this red wagon move.

push

pull

Science Background

Sir Isaac Newton stated three laws that related motion and forces. His first law states that an object in motion remains in motion unless acted upon by a force, and an object at rest remains at rest unless a force acts on it. The second law states that unbalanced forces result in movement in the direction of the net force. The third states that forces occur in equal and opposite pairs. For example, if a hand pushes a book, the book exerts a force equal in strength and opposite in direction.

See **Science Yellow Pages** in the Teacher Resource section for background information.

LOG ON **Professional Development** For more Science Background and resources from **NSDL** visit **www.macmillanmh.com/nsdl/**

Changes in Motion

Forces can change the motion of objects. They can make objects start moving, speed up, slow down, or stop moving. They can make objects change direction, too.

Forces can change a soccer ball's motion. A goalie applies a force to throw the ball to a teammate. The ball starts to move. The teammate applies another force when he kicks the ball. The ball changes direction. Each time a force is applied the motion of the ball changes. When a goalie catches the ball, the ball's motion is stopped.

A change in an object's motion is the result of all the forces that are acting on the object. Think of the game tug of war. When both sides pull equally on the rope, the forces are balanced. Nothing moves. If one side pulls harder, the forces become unbalanced. Now the rope moves. There is a change of motion.

 Quick Check

Cause and Effect How can forces affect an object's motion?

Critical Thinking What happens when you kick a moving ball?

Changes in Motion

① The goalie throws the ball to start its motion.

② This player kicks the ball, changing its speed and direction.

③ The goalie catches the soccer ball, stopping its motion.

Read a Photo

How have forces changed the motion of this soccer ball?

Clue: Captions give information.

445
EXPLAIN

▶ **Use the Visuals**

Refer students to the visuals on page 445. Ask:

■ **What part of motion changes in each of the photos?** Speed changes in the first and last photos; in the second photo, both direction and speed change.

▶ **Explore the Main Idea**

ACTIVITY Have students use diagrams to illustrate balanced and unbalanced forces. Have them draw two diagrams of a student pushing on a large box. Have them use one of these drawings to illustrate balanced forces by showing another student exerting an equal force on the other side of the box. Have them use equal-sized arrows to show that the forces are equal in size and opposite in direction. Then have them use the other drawing to illustrate unbalanced forces by showing another student exerting a greater force on the other side of the box. Have them use different-sized arrows to show that the forces are not equal in size and that the box moves in the direction of the larger force.

Read a Photo

Answer Forces cause the ball to start moving, change direction and speed, and stop moving.

▶ **Develop Vocabulary**

force *Word Origin* The word *force* comes from the Latin word *fortis,* which means "strong." *Force* and *strength* have similar connotations, as both terms indicate an ability to bring about change.

✔ **Quick Check Answers**

• **Cause and Effect** Forces can cause an object to start moving, change direction, change speed, or stop moving.

• **Critical Thinking** It might move at a faster speed or in a different direction.

ELL Support

Discuss/Restate To review forces, ask students how forces change the motion of things. Elicit that they can make things start moving, speed up, slow down, stop moving, or change direction. Ask students what happens when a football player receives the ball and then makes a pass. Elicit that the ball changes direction.

BEGINNING Give students sentence frames such as, *A goalie throws a ball to start its _____.* Give them choices such as, *a. game, b. team, c. motion.* Elicit that the correct answer is c. *motion.*

INTERMEDIATE Students can use phrases and short sentences to describe how forces change the motion of things

ADVANCED Students can use complete sentences to describe how forces change the motion of things.

What are types of forces?

FAST TRACK

Discuss the Main Idea

Lead a discussion on what students know about magnetism and gravity. Ask:

- Does a magnet exert a force on all objects? no
- Upon what objects does a magnet exert a force? objects that contain iron
- Does gravity exert a force on all objects? yes

▶ Develop Vocabulary

magnet *Word Origin* Point out that the word *magnet* has its origin in the Latin word *magnetum,* "lodestone." A lodestone is a magnetite rock that possesses polarity.

gravity *Word Origin* The word *gravity* is derived from the Latin word *gravis,* meaning "heavy."

weight *Word Origin* The word *weight* comes from the Old English word *gewiht.* The more weight an object has, the more difficult it is to carry or bear from one place to another.

▶ Explore the Main Idea

ACTIVITY Use several disc magnets with holes in the middle and a pencil to illustrate how magnets attract and repel each other. Place the magnets on an upright pencil so that they repel each other. The repulsion causes a space between the magnets. Take the magnets off the pencil and turn over alternate magnets. The magnets will now attract each other. There will be no space between the magnets, and lifting the top magnet will lift all of the magnets off of the pencil.

RESOURCES and TECHNOLOGY

▶ Activity Lab Book, p. 193

💿 Presentation Toolkit CD-ROM

What are types of forces?

There are many types of forces. The forces you are probably most familiar with are contact forces. *Contact forces* happen between objects that touch. Think about a baseball game. The pitcher must touch the ball to throw it to home plate. A bat must touch the ball to change its direction. Some forces can act on an object without touching the object. Magnetism and gravity are examples.

When the bat hits the ball, the ball changes direction.

Magnetism

Have you ever used magnets? What did you notice? When you bring two magnets together, they may *attract,* or pull on, each other. They may also *repel,* or push away from, each other. Magnets can attract or repel each other without touching. The force that causes this to happen is called *magnetic force.* A **magnet** is any object with a magnetic force.

Magnets can attract or repel each other. They can also attract things made of certain metals like iron. They cannot attract things made of wood, glass, plastic, or rubber. Magnets can attract or repel objects through solids, liquids, or gases.

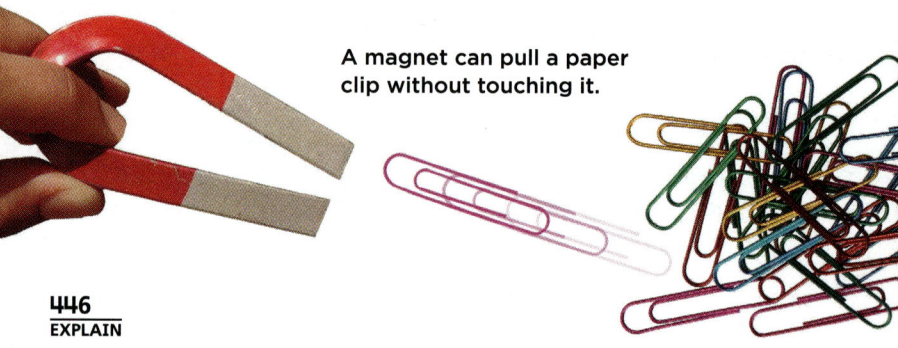

A magnet can pull a paper clip without touching it.

446
EXPLAIN

ELL Support

Use Visual Information Refer students to the visual on p. 446. Ask: *Which picture shows an example of a contact force?* Remind them that there are many types of forces, and that a contact force occurs when two objects touch each other. Also discuss the forces of magnetism and gravity, emphasizing that these forces can act on an object without touching it.

BEGINNING Have students look at the photos and point to and verbally identify the photo that shows a contact force.

INTERMEDIATE Have students look at the photo that shows the bat and ball. Have them use phrases to list other examples of contact forces.

ADVANCED Have students identify the photo that shows magnetic force. Have them use complete sentences to explain how magnetism and gravity are alike.

Gravity

You cannot see gravity, but it is what keeps you on Earth. **Gravity** is a pulling force between two objects, such as you and Earth. Gravity pulls objects together. When you jump up, Earth's gravity pulls you down. Gravity pulls through solids, liquids, or gases.

How much gravity does it take to keep you on Earth? The answer is your weight (WAYT). An object's **weight** is a measure of the pull of gravity on it. The more mass an object has, the more gravity pulls on it.

✔ Quick Check

Cause and Effect What effect does gravity have on objects?

Critical Thinking How can you pick up metal paper clips without touching them?

Gravity is pulling these skydivers to Earth.

447
EXPLAIN

≡ Quick Lab

Observe Gravity

① **Predict** Does gravity act the same on all objects? Would it act the same on two plastic bottles that have the same volume but different mass?

② Hold an empty plastic bottle in one hand. Hold an identical bottle full of water in the other hand. Hold each bottle away from your body.

③ **Observe** Describe what you feel. Is each bottle pulled toward Earth with the same force?

④ **Infer** Is the amount of gravity on the two bottles the same? How could you tell?

≡ Quick Lab

👤 individual   ⏲ 10 minutes

Objective Show how gravity depends on mass.

Materials plastic empty water bottle, plastic full water bottle

① Possible prediction: The force of gravity is greater on an object that has greater mass.

③ The forces are different.

④ The bottle with the water in it was pulled toward Earth with a greater force. This bottle felt heavier.

Have students compare their results with their predictions. Ask them to explain what they learned in the Quick Lab that might help them make a better prediction about the effect of gravity on an object.

▶ Explore the Main Idea

ACTIVITY Have students research units used for weight in the metric and English systems of measurement. Point out that weight is a measurement, and therefore must include a number and a unit. The metric unit for weight is the newton (N), which is the unit for any force. A medium-sized apple weighs about one newton. The pound is the unit for weight in the English system of measurement.

✔ Quick Check Answers

- **Cause and Effect** Gravity pulls objects toward each other.

- **Critical Thinking** If they contain iron, you could use a magnet to pick them up.

Differentiated Instruction

Leveled Questions

EXTRA SUPPORT What is happening between two magnets when they push away from each other? They are repelling each other.

ENRICHMENT Gravity on the Moon is much less than it is on Earth. Where would you weigh more, on Earth or on the Moon? on Earth

What is friction?

▶ Discuss the Main Idea

Explain what friction is and how it slows or stops motion. Ask:

- **Why do you think a car stops when the brakes are applied?** The brakes cause friction.

- **How does friction help you walk?** Friction between your feet and the ground keeps you from slipping.

- **Why is ice on a sidewalk dangerous?** The ice reduces friction between your feet and the ground, and you could slip.

▶ Develop Vocabulary

friction *Word Origin* The word *friction* comes from the Latin word *frictionem,* which means "a rubbing, rubbing down." Friction is the force that resists motion between two surfaces that rub against each other.

▶ Explore the Main Idea

ACTIVITY Have students work in small groups. Ask them to make lists of examples of instances when lots of friction is helpful, lots of friction is harmful, little friction is helpful, and little friction is harmful. Possible answers, in order: Friction between tires and the road keep a bicycle from sliding. Machine parts rub against each other and wear out. A slick slide is easy to slide down. Ice on a road causes a car to slide.

✓ *Quick Check Answers*

- **Cause and Effect** The brake pad presses against the wheel. The friction between the pad and the wheel stops the bicycle.

- **Critical Thinking** Friction pushes against a moving object.

▲ This water slide is smooth and has little friction.

Friction between the brake pad and the bike rim stops the bike. ▼

bike rim

brake pad

448
EXPLAIN

What is friction?

A block slides on the floor. It then slows down and stops. Why does this happen? A force called friction (FRIK•shuhn) is acting on the block. **Friction** is a force that occurs when one object rubs against another. It pushes against moving objects and causes them to slow down.

Different surfaces produce different amounts of friction. Rough surfaces, such as sandpaper, usually produce a lot of friction. Smooth surfaces, such as ice, usually produce less friction.

People use slippery things to reduce friction. Oil is often put on moving parts of machines to reduce friction. People use rough or sticky things to increase friction. The brakes on a bike use rubber pads to increase friction. When you squeeze the brake handles, the brake pads press against the rim of the wheel. The friction between the pads and rim cause the bike to stop.

✓ *Quick Check*

Cause and Effect What happens when you squeeze a hand brake on a bicycle?

Critical Thinking How can you tell that friction is a force?

Homework Activity

Friction and Other Forces

Have students design and conduct an experiment to show how friction influences how pushes affect objects. For example, students could release a ball down a ramp to strike a toy car sitting on waxed paper. They could then repeat the experiment using sandpaper instead of waxed paper. The results would show that the car would go farther on the waxed paper because there is less friction between the wheels and the surface.

Lesson Review

Visual Summary

A force is a push or a pull. Forces can change the motion of objects.

Contact, magnetism, and gravity are different types of forces.

Friction is a force that occurs when one object rubs against another.

Make a **FOLDABLES** Study Guide

Make a Trifold book. Use it to summarize what you learned about forces.

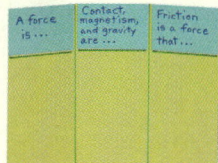

Think, Talk, and Write

1 **Main Idea** How can forces affect an object?

2 **Vocabulary** What is friction? Talk about it.

3 **Cause and Effect** You are swinging on a swing. What force causes you to slow down as you go up?

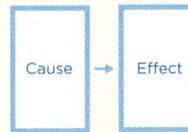

Cause → Effect

4 **Critical Thinking** How can friction help keep you safe?

5 **Test Prep** Which is an example of a contact force?
  A a magnet attracting a paper clip
  B two magnets repelling each other
  C a bat hitting a ball
  D gravity pulling on a leaf

 **Math Link**

Order Numbers
Weigh five objects on a spring scale. Measure their weight in newtons, the unit of force in the metric system. Organize your data in a bar graph from least weight to greatest weight.

 Health Link

Use Your Muscles
You use your muscles when you push or pull things. Find out about some of the muscles in your body. What do your muscles do? How do your muscles help you move?

 e-Review Summaries and quizzes online at www.macmillanmh.com

449
EVALUATE

Formative Assessment

Approaching Have students explain why motor oil is added to car engines.

On-Level Have students draw examples of balanced and unbalanced forces.

Challenge Have students research what two factors determine how much gravity is present between two objects.

Key Concept Cards For student intervention, see the prescribed routine on **Key Concept Cards 69** and **70.**

---

3 Close

Lesson Review

▶ **Visual Summary**

Have students save their lesson Study Guides. They will use them at the end of the chapter for review.

▶ **Make a FOLDABLES Study Guide**

See pages R27–R28 for instructions.

FAST TRACK **Think, Talk, and Write**

1 **Main Idea** Forces push or pull an object. They can cause objects to start moving, change speed, change direction, or stop moving. Magnetic forces cause objects to repel or attract each other. Gravity pulls objects toward Earth's center, which tends to hold objects on Earth's surface.

2 **Vocabulary** Friction is a force that resists motion between two objects that rub against each other.

3 **Cause and Effect**

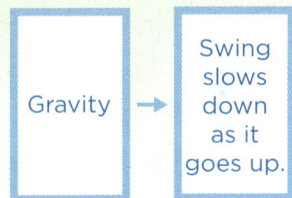

Gravity → Swing slows down as it goes up.

4 **Critical Thinking** Possible answers: Applying bicycle brakes keeps you from going too fast. You can walk without slipping because of the friction between your feet and the ground.

5 **Test Prep** C

Math Link If an object will not hang from the hook on the scale, it can be placed in a mesh bag.

Health Link Accept all reasonable answers based on accurate information.

RESOURCES and TECHNOLOGY

▶ **Reading and Writing,** p. 219

▶ **School to Home Activities,** pp. 93–94

▶ **e-Review** Narrated Summary and Quiz

▶ **Progress Reporter** Assessments

Be a Scientist

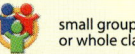

small groups or whole class

30 minutes

Skills experiment, measure, use numbers, interpret data

Objective

- Experiment to determine how the distance between a magnet and an object affects magnetic force.

Materials magnet, paper clips, ruler

Plan Ahead To save time, you can prepare a data table before class and provide it to students.

EXTEND This activity determines the relationship between magnetic force and the distance between the object and the magnet.

Structured Inquiry

How does distance affect the pull of a magnet on metal objects?

Form a Hypothesis Possible hypothesis: If you move a magnet closer to a pile of paper clips, then more paper clips will be attracted to the magnet.

Test Your Hypothesis

1. Be sure students measure the distance above the paper clips for each trial. The pile of paper clips could be a different height for each trial.

3. **Measure** If any paper clips stuck to other paper clips by magnetism, have students count those paper clips as being stuck to the magnet. Have students return the paper clips that were picked up by the magnet to the pile before completing another trial.

Be a Scientist

Materials

magnet

paper clips

ruler

Structured Inquiry

How does distance affect the pull of a magnet on metal objects?

Form a Hypothesis

You know that some metal objects, such as paper clips, are attracted to magnets. What happens when you change the distance between a magnet and a pile of paper clips? How does this affect the magnet's pull on the paper clips? Write a hypothesis. "If you move a magnet closer to a pile of paper clips, then..."

Test Your Hypothesis

1. Gather a pile of paper clips on your desk. Stand up a ruler near the paper clips.

2. **Experiment** Hold a magnet as shown below. Slowly lower the magnet until it is only 1 cm above the pile.

450
EXTEND

Integrate Math

Graphing Data

Have students create a line graph for the data from the activity.

- Have them label the horizontal axis with numbers that show distances between the magnet and the paper clips.

- Have them label the vertical axis with numbers to show the number of paper clips picked up in each trial.

- Have students work in their lab groups to graph their data and write a statement that interprets the graph.

Inquiry Investigation

3 **Measure** Move the magnet away from the pile. Remove the paper clips and count how many stuck to the magnet. Record the data in a table.

4 Repeat steps 1–3, holding the magnet 2 cm and 3 cm away from the pile of paper clips. Record your data.

Draw Conclusions

5 **Use Numbers** At what distance did the magnet pick up the most paper clips?

6 **Interpret Data** Does a magnet's pull on objects get greater or smaller as the magnet moves away from the objects?

Step 3

Distance	Number of Paper Clips
1 cm	
2 cm	
3 cm	

Draw Conclusions

5 at 1 cm

6 **Interpret Data** The magnetic force gets weaker as the magnet gets farther away.

Guided Inquiry

Can magnetic force pass through an object?

Form a Hypothesis

Can a magnetic force pass through different objects, such as wood, plastic, paper, or foil? Write a hypothesis.

Test Your Hypothesis

Design a plan to test your hypothesis. List the materials you will use. Write down the steps you plan to follow.

Draw Conclusions

Did any of the objects block magnetic force? Did any of the objects make the magnetic force stronger or weaker? Share your results with your classmates.

Open Inquiry

What other questions do you have about magnets? For example, what common objects are attracted to magnets? Design an experiment to find out.

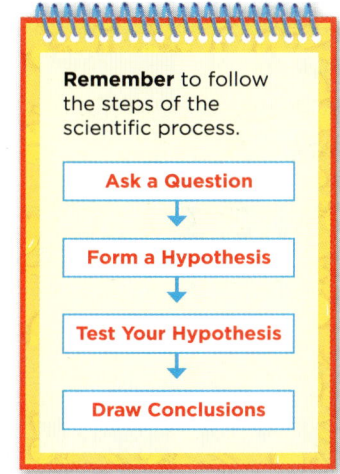

Remember to follow the steps of the scientific process.

Ask a Question
↓
Form a Hypothesis
↓
Test Your Hypothesis
↓
Draw Conclusions

451
EXTEND

Guided Inquiry

Can magnetic force pass through an object?

Form a Hypothesis Possible hypothesis: Magnetic force can pass through thin materials.

Test Your Hypothesis Check students' designs for accuracy and safety. Be sure their plans include the resources they need and steps they will follow. If designs are approved and time allows, have students perform their experiments. Be sure they record their results and observations as they follow their plans.

Draw Conclusions Have students evaluate whether or not their results support their hypotheses. Have them write a summary statement that shows their conclusions. Then have them share their conclusions with their classmates.

Open Inquiry

Encourage students to group with other students with similar questions. Have them design an experiment to answer one of their questions. Allow them to use reference books and the Internet, if available, to find the answers to questions that do not require experimentation to answer.

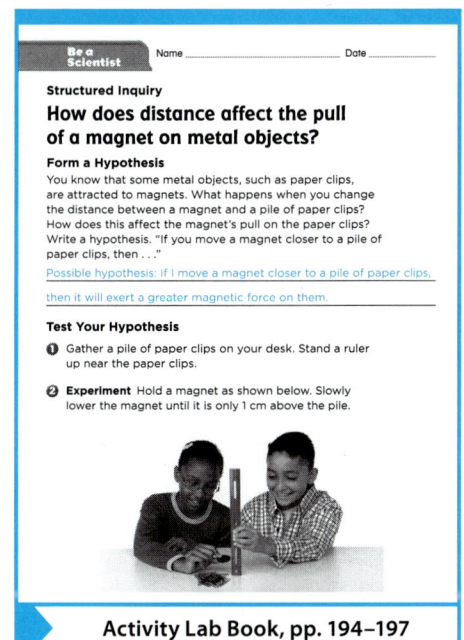

Be a Scientist   Name _____ Date _____

Structured Inquiry

How does distance affect the pull of a magnet on metal objects?

Form a Hypothesis

You know that some metal objects, such as paper clips, are attracted to magnets. What happens when you change the distance between a magnet and a pile of paper clips? How does this affect the magnet's pull on the paper clips? Write a hypothesis. "If you move a magnet closer to a pile of paper clips, then . . ."

Possible hypothesis: if I move a magnet closer to a pile of paper clips, then it will exert a greater magnetic force on them.

Test Your Hypothesis

1 Gather a pile of paper clips on your desk. Stand a ruler up near the paper clips.

2 **Experiment** Hold a magnet as shown below. Slowly lower the magnet until it is only 1 cm above the pile.

Activity Lab Book, pp. 194–197

RESOURCES and TECHNOLOGY

▶ **Activity Lab Book,** pp. 194–197

▶ **Activity Flipchart,** p. 52

▶ **Instructional Navigator CD-ROM**

Plan Your Lesson

Lesson 3 Work and Energy

Objective

- Define energy and work.
- Discuss the forms of energy and how energy changes from one form to another.

Reading Skill Summarize

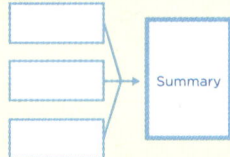

Summary

Graphic Organizer 5, p. TR7

LOG ON **Professional Development** Look for **NSDL** to find recommended Science Background materials from the National Science Digital Library

FAST TRACK

Lesson Plan When time is short, follow the Fast Track and use the essential resources.

1 Introduce
Look and Wonder, p. 452
Resource Alternative Explore, p. 200

2 Teach
Discuss the Main Idea, p. 454
Discuss the Main Idea, p. 456
Resource Visual Literacy, p. 71

3 Close
Think, Talk, and Write, p. 459
Resource Assessment, p. 140

▶ Reading and Writing

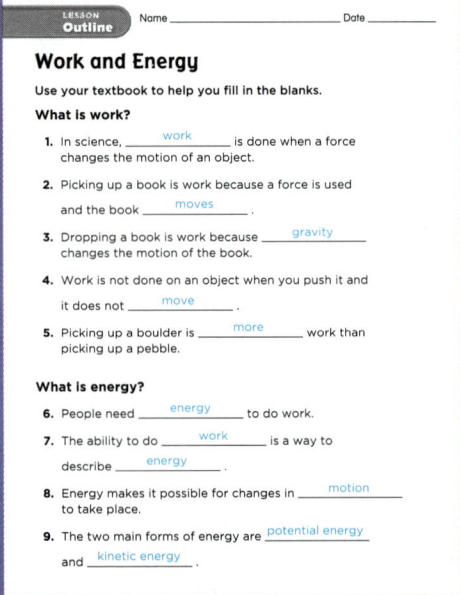

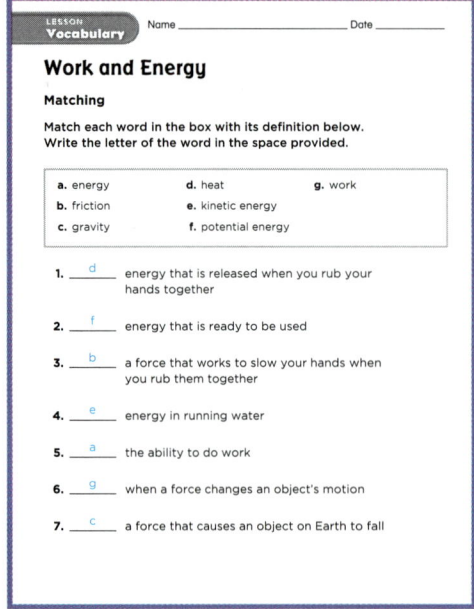

Outline, pp. 216–217
Also available as a student workbook

Vocabulary, p. 218
Also available as a student workbook

 # Visual Literacy

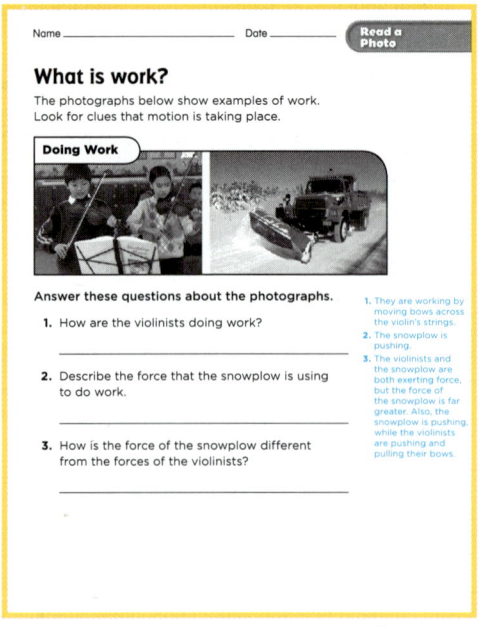

Read a Photo, p. 71
Also available as a transparency

▶ Activity Lab Book

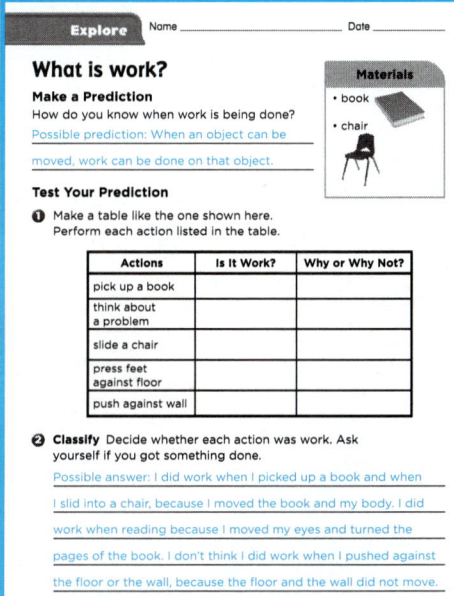

Explore — Name _____ Date _____

What is work?

Make a Prediction
How do you know when work is being done?
Possible prediction: When an object can be moved, work can be done on that object.

Materials
• book
• chair

Test Your Prediction
1. Make a table like the one shown here. Perform each action listed in the table.

Actions	Is It Work?	Why or Why Not?
pick up a book		
think about a problem		
slide a chair		
press feet against floor		
push against wall		

2. **Classify** Decide whether each action was work. Ask yourself if you got something done.
Possible answer: I did work when I picked up a book and when I slid into a chair, because I moved the book and my body. I did work when reading because I moved my eyes and turned the pages of the book. I don't think I did work when I pushed against the floor or the wall, because the floor and the wall did not move.

Explore, pp. 198–199
Also available as a student workbook

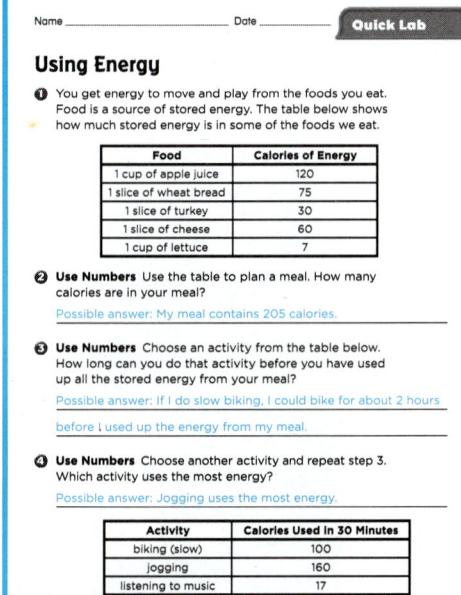

Quick Lab — Name _____ Date _____

Using Energy

1. You get energy to move and play from the foods you eat. Food is a source of stored energy. The table below shows how much stored energy is in some of the foods we eat.

Food	Calories of Energy
1 cup of apple juice	120
1 slice of wheat bread	75
1 slice of turkey	30
1 slice of cheese	60
1 cup of lettuce	7

2. **Use Numbers** Use the table to plan a meal. How many calories are in your meal?
Possible answer: My meal contains 205 calories.

3. **Use Numbers** Choose an activity from the table below. How long can you do that activity before you have used up all the stored energy from your meal?
Possible answer: If I do slow biking, I could bike for about 2 hours before I used up the energy from my meal.

4. **Use Numbers** Choose another activity and repeat step 3. Which activity uses the most energy?
Possible answer: Jogging uses the most energy.

Activity	Calories Used in 30 Minutes
biking (slow)	100
jogging	160
listening to music	17

Quick Lab, p. 201
Also available as a student workbook

▶ Assessment

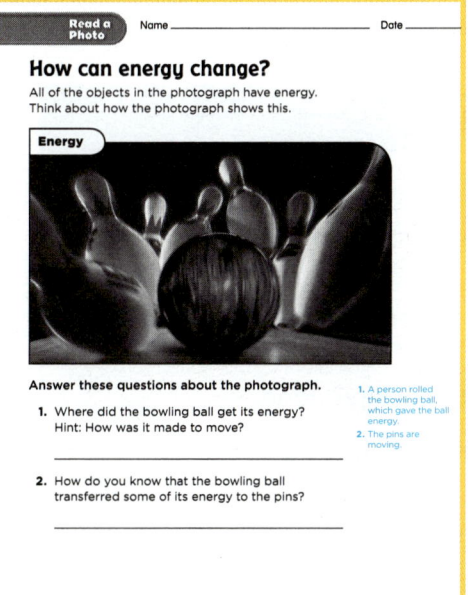

Read a Photo — Name _____ Date _____

How can energy change?

All of the objects in the photograph have energy. Think about how the photograph shows this.

Energy

Answer these questions about the photograph.

1. Where did the bowling ball get its energy? Hint: How was it made to move?
_____

2. How do you know that the bowling ball transferred some of its energy to the pins?
_____

1. A person rolled the bowling ball, which gave the ball energy.
2. The pins are moving.

Read a Photo, p. 72
Also available as a transparency

Assessment — Lesson 3 Test — Name _____ Date _____

Circle the letter of the best answer for each question.

1. What makes changes in motion possible?
   A energy
   B work
   C gravity
   D weight

2. Which activity requires the most work?

Activity	Amount of Energy Used
Jumping rope	moderate
Lifting a book	light
Hammering a nail	heavy
Reading a book	none

   A jumping rope
   B lifting a book
   C hammering a nail
   D reading a book

3. What happens when a force changes an object's motion?
   A the object changes state
   B work is done
   C gravity is interrupted
   D a machine is being used

4. What are the two main forms of energy?
   A energy of speed and energy of sound
   B energy of motion and potential energy
   C energy of force and energy of motion
   D energy of speed and potential energy

Critical Thinking Explain why kicking a soccer ball is work.
Work is done only when a force changes an object's motion. Work is being done when a ball is kicked because the object's motion is changed.

Lesson Test, p. 140
Also available as a student workbook

Lesson 3 Work and Energy

Objectives

- Define energy and work.
- Discuss the forms of energy and how energy changes from one form to another.

1 Introduce

▶ **Assess Prior Knowledge**

Have students list different things they consider to be work. Student responses will probably reflect common examples including chores and jobs that people do to earn money. Tell students that some of these items meet the scientific definition of work, and some do not. Keep the list until the end of the lesson, and review it after students learn the scientific definition of work.

Look and Wonder

Invite students to share their responses to the Look and Wonder statement and question.

- **Is this tugboat doing work? Why or why not?**

   Possible answer: The tugboat is doing work pulling the ship. It is using force to change the ship's motion.

Write ideas on the board and note any misconceptions that students may have. Address these misconceptions as you teach the lesson.

RESOURCES and TECHNOLOGY

▶ **Activity Lab Book,** pp. 198–200

▶ **Activity Flipchart,** p. 53

🖸 **Science Activity DVD**

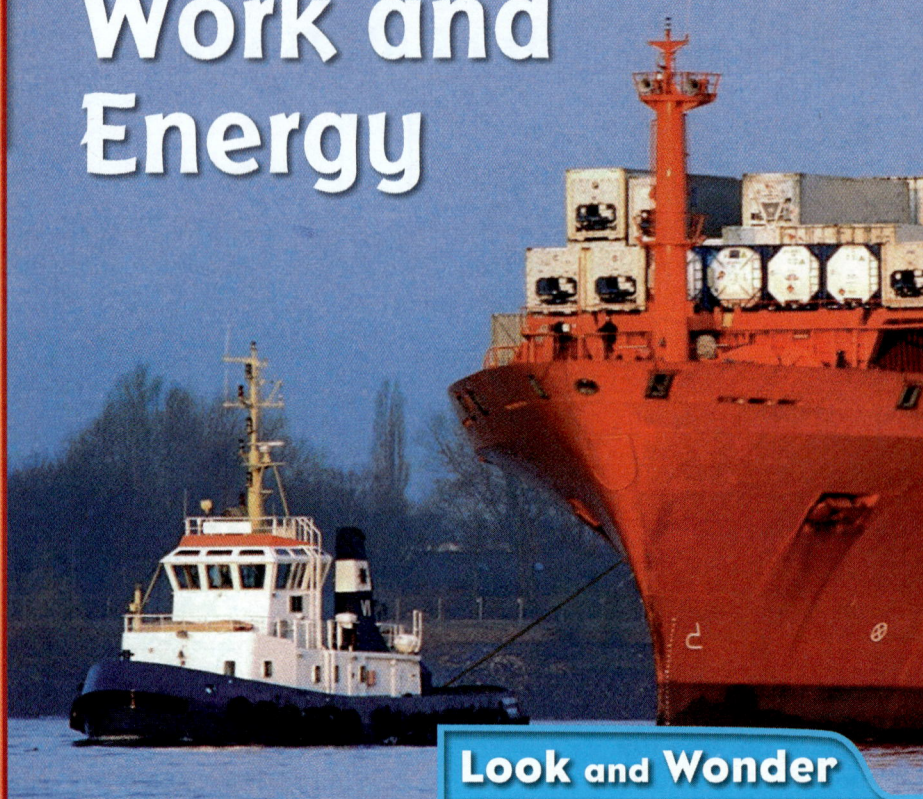

Lesson 3

Work and Energy

Look and Wonder

This tugboat is pulling a large container ship to the dock. Is this tugboat doing work? Why or why not?

452
ENGAGE

Warm Up

Start with a Discussion

Discuss energy and work with students. Have them brainstorm first. Ask:

- **What is energy?** Possible answer: something needed for activity

- **What supplies your body with energy?** food

- **How are energy and work related?** It takes energy to do work.

Keep a copy of the answers for review at the end of the lesson.

Explore

Inquiry Activity

What is work?

Make a Prediction

How do you know when work is being done?

Test Your Prediction

1. Make a table like the one shown below. Perform each action listed in the table.

Actions	Is It Work?	Why or Why Not?
pick up a book		
think about a problem		
slide a chair		
press feet against floor		
push against wall		

Materials

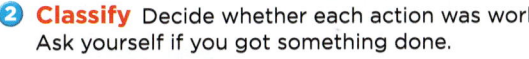

book

chair

2. **Classify** Decide whether each action was work. Ask yourself if you got something done.

Draw Conclusions

3. **Communicate** Explain why you classified each action the way you did. Record this information in the table.

4. **Infer** What do you think work is?

Explore More

Experiment Perform other actions at home. Try to find actions where you do different amounts of work.

453
EXPLORE

Alternative Explore

What work is done while cooking?

Provide students with a list of actions that are done when baking cookies. Have them consider which of the actions are work, and list the actions as either *Work* or *Not Work*. Then have them list actions that occur when they are getting ready for school in the morning. Have them list three actions that are work and three actions that are not work.

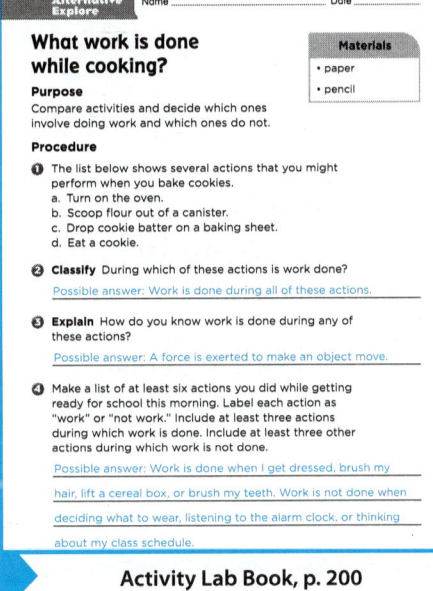

Activity Lab Book, p. 200

Explore

small groups • 20 minutes

Plan Ahead If time is short, prepare a data table for students to use.

Purpose This activity helps students develop an operational definition of work.

Structured Inquiry

Make a Prediction Possible prediction: Work is done when something moves.

1. Each student should participate by doing at least one action. Each student can do as many of the actions as time allows.

2. **Classify** Students might categorize activities as *work* if they are difficult or require significant force, such as pushing against the wall. Accept all reasonable explanations.

4. **Infer** Students should infer that picking up a book is work because a force is necessary to move the book.

Have students compare their predictions to their conclusions.

Guided Inquiry | Explore More

Experiment Students should explain how they decided whether a lot of work or a little work was done.

Open Inquiry

Ask students if work is done when they play sports. Have them think of a question about the activities done during a sport. Have students make a plan and carry out an experiment to answer their question. Ask: **Is work done when a soccer ball is kicked?**

Lesson 3 **453**

2 Teach

Read and Learn

Main Idea Have students read the titles in the lesson and ask them to discuss what they think they will learn about work and energy.

Vocabulary Have students read aloud the vocabulary words. Ask students to share and write each other's definitions on the board.

Reading Skill Summarize
Graphic Organizer 6 Have students fill in a Summarize graphic organizer as they read each page of the lesson. They can use the Quick Check questions to help them summarize.

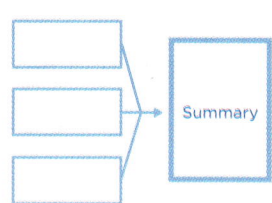

What is work?

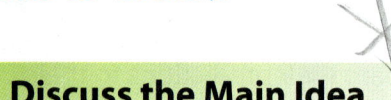

FAST TRACK Discuss the Main Idea

Discuss with students what work they think they have done today. Ask:

- **Why do you think these actions were work?**
  Possible answers: It was difficult to do. I earned money.

Discuss with students the scientific definition of *work*. Have them revisit the actions they thought were work and decide whether the actions fit the scientific definition of work.

Read and Learn

▶ **Main Idea**
Work is done when a force changes an object's motion. Energy is the ability to do work.

▶ **Vocabulary**
work, p. 454
energy, p. 456
kinetic energy, p. 456
potential energy, p. 456

e-Glossary
at www.macmillanmh.com

▶ **Reading Skill** ✓
Summarize

[Summarize graphic organizer: boxes leading to Summary]

What is work?

Do you know what work is? You might say that you do work every day at school. In science work has a special meaning. Work is done when a force moves an object or changes an object's motion. This means that picking up a book is work. A force changes the book's motion. Work is done when a book falls to the floor. Gravity changes the book's motion. Gravity does the work. Pushing on a wall is not work. No matter how hard you push, the wall does not move.

Work can be easy or hard. Picking up a small pebble is work. Lifting a large boulder is, too. In both examples a force is used to move an object.

✓ **Quick Check**

Summarize How can you tell if an action is work?

Critical Thinking Can play be work? Why or why not?

When you paint at an easel, you are doing work. Your hand moves the brush. ▶

454
EXPLAIN

Science Background

When Is It Work? An action is considered work based on two requirements. An object must be moved by a force, and the movement must be in the same direction as the force. For example, work is done when a rock is picked up; both the force and the motion are upward. If the rock is carried across the yard, no work is done; motion and force are present, but the force on the rock is upward, and the motion is horizontal.

See **Science Yellow Pages** in the Teacher Resource section for background information.

Professional Development For more Science Background and resources from **NSDL** visit **www.macmillanmh.com/nsdl/**

What is work?

The photographs below show examples of work.
Look for clues that motion is taking place.

Doing Work

Answer these questions about the photographs.

1. How are the violinists doing work?

   _____

2. Describe the force that the snowplow is using
   to do work.

   _____

3. How is the force of the snowplow different
   from the forces of the violinists?

   _____

Doing Work

Read a Photo

How is work being done in each picture?

Clue: Look for a force and a change in an object's motion in each picture.

455
EXPLAIN

Differentiated Instruction

Leveled Activities

EXTRA SUPPORT   Have students draw pictures of work being done. Have them label force and motion using two separate arrows in their pictures.

ENRICHMENT   Ask students to research how the amount of work done is calculated from the amount of force and the distance the object moves. Work = force × distance

---

▶ Use the Visuals

Have students sequence, from smallest to largest, the amount of work done in the photos on page 455. Remind students that work involves both the amount of force required and the distance the object is moved. Greater force and greater distance indicate that more work is done. For some photos, it will be difficult to compare the amount of work done. Accept any sequence that students can justify. Possible sequence: pencil, violin, jump rope, garden work, crane work, plowing

▶ Develop Vocabulary

work *Scientific vs. Common Use* The common use of the word *work* usually means a chore or a job that earns money. The scientific meaning of *work* is more specific. It requires using a force to bring about a change in the motion of an object.

▶ Address Misconceptions

A common misconception is that work involves any action during which something moves. Work is done if the force on the object is in the same direction as the movement. Lifting a book is work because the upward force of the hand results in upward movement of the book. When carrying the book across the room, no work is done on the book. The force on the book is upward and is not in the direction of its motion. However, work is done on your body because a horizontal force results in horizontal motion of your body.

Read a Photo

Answer Force from the hand moves the pencil. Force from the hand moves the bow. Force from the pitchfork moves the soil. Force from the plow moves the snow. Force from the crane moves debris. Force from arms move the rope.

✓ *Quick Check Answers*

- **Summarize** It is work if it involves a force, an object, and a change in the object's motion.
- **Critical Thinking** Yes, if it involves a force, an object, and a change in motion of the object.

What is energy?

FAST TRACK ## Discuss the Main Idea

Lead a discussion on what energy is and what types of energy there are. Ask:

- **What type of energy is in a baseball bat when it is held by the batter?** stored energy

- **What type of energy is in a baseball bat when it is swung?** energy of motion

▶ Use the Visuals

Have students look at the photo on pages 456–457. Ask students to identify when the roller coaster has stored energy and when it has energy of motion.
The roller coaster has stored energy at the top of the hill and energy of motion moving down the hill.

▶ Develop Vocabulary

energy *Word Origin* Point out that *energy* has its origin in the Late Latin word **energia,** meaning "activity, operation." Energy is the ability to do work. Energy of motion is associated with activity.

kinetic energy *Scientific vs. Common Use* The word *kinetic* is used to describe something active and energetic. For example, *She is a kinetic dancer.* In science, all active or moving objects have kinetic energy.

potential energy *Scientific vs. Common Use* When something has potential, it has the ability to do something. Explain to students that the potential energy from food gives us the ability to work and play.

RESOURCES and TECHNOLOGY

▶ **Activity Lab Book,** p. 201

💿 **Presentation Toolkit CD-ROM**

What is energy?

Work cannot be done without energy (EN•uhr•jee). **Energy** is the ability to do work. Energy is what makes motion possible. An object needs energy to move. When you do work on an object, you give it that energy.

Kinds of Energy

When you throw a paper airplane, you do work. You give the airplane energy. The airplane starts to move. Energy of motion is called **kinetic energy** (ki•NET•ik EN•uhr•jee). All moving objects—roller coasters, cars, even people—have kinetic energy.

When you pull a sled to the top of a hill, you do work. You give the sled potential energy (puh•TEN•shuhl EN•uhr•jee). **Potential energy** is stored energy that is ready to be used. As the sled moves down the hill, its potential energy changes into kinetic energy.

How do you get energy to move, live, and grow? You get most of your energy from food. Food is a source of stored energy. Gasoline, wood, and food all have stored energy that is ready to be used.

 Quick Check

Summarize What can energy do?

Critical Thinking When does a roller coaster have the most potential energy?

456
EXPLAIN

ELL Support

Paraphrase Provide students with practice in stating the definition of energy in their own words. Ask students if they can see energy. Ask them when they know they have energy and when they know they do not. Ask them how they can see what energy does. Discuss the two types of energy with students.

BEGINNING   Students can point to and say "energy of motion" or "stored energy" as they observe the visuals.

INTERMEDIATE   Read aloud the first two paragraphs on page 456. Students should then use descriptive phrases or short sentences to give other examples of energy.

ADVANCED   Students can use complete sentences to explain the difference between energy of motion and stored energy.

Quick Lab

Using Energy

1 You get energy to move and play from the foods you eat. Food is a source of stored energy. The table below shows how much stored energy is in some of the foods we eat.

Food	Calories of Energy
1 cup of apple juice	120
1 slice of wheat bread	75
1 slice of turkey	30
1 slice of cheese	60
1 cup of lettuce	7

2 Use Numbers Use the table to plan a meal. How many calories are in your meal?

3 Use Numbers Choose an activity from the table below. How long can you do that activity before you have used up all the stored energy from your meal?

4 Use Numbers Choose another activity and repeat step 3. Which activity uses the most energy?

Activity	Calories Used in 30 Minutes
biking (slow)	100
jogging	160
listening to music	17

457
EXPLAIN

Quick Lab

individual   15 minutes

Objective Show how much energy is in different foods.

1 Advise students to write *Calories* with a capital *C*. The word *calorie* with a lower case *c* is a very small unit of energy. One Calorie equals 1,000 calories. Food Calories are the larger unit.

2 Answers will vary. Emphasize that the meals can contain more than one serving of the listed foods. The total number of Calories in a given food choice is the sum of the Calories in each serving of that type of food.

3 If students cannot divide the total number of Calories available by the Calories used, show them how to group and subtract numbers to find the answer.

4 jogging

✔ Quick Check Answers

- **Summarize** Energy can make matter move and change.

- **Critical Thinking** When it is at the top of a hill, it's energy is ready to be used.

Differentiated Instruction

Leveled Questions

EXTRA SUPPORT **What type of energy is in a log?** stored energy

ENRICHMENT **What is kinetic energy? What is potential energy? Use a dictionary to help you answer the questions.** Kinetic energy is energy of motion. Potential energy is stored energy.

How can energy change?

▶ Discuss the Main Idea

Discuss with students common examples of energy's changing form. Ask:

- **How does energy move?** It is transferred from one object to another.

- **How can rubbing your hands together cause heat?** Rubbing uses energy and causes friction, which turns that energy into heat.

▶ Develop Vocabulary

Remind students that they discussed conservation when they learned about protecting natural resources to keep them from being used up. Conservation is also a term that refers to energy. The law of conservation of energy states that energy is not used up, it just changes to another form of energy.

▶ Use the Visuals

Point out to students that the photo on page 458 shows how energy can be transferred from one object to another. Have them draw a picture of their own that shows energy transferred among three objects. An example might be a ball thrown, hit by a bat, and caught by a player.

▶ Address Misconceptions

Students may think that energy is finite.

FACT ▶ Energy does not get used up. It changes form.

✔ *Quick Check Answers*

- **Summarize** Energy can change from one form to another, as when energy of motion is changed by friction into heat energy. Energy can also move from one object to another.

- **Critical Thinking** The energy in the bowling ball is transferred to the pin, causing the ball to slow down.

Energy from the ball makes the pins move.

How can energy change?

Energy can move from one object to another. When you roll a bowling ball, you transfer energy from your body to the ball. When the ball hits the pins, it transfers energy to the pins. The pins move.

Energy can also change from one form into another. Rub your hands together. What do you notice? They get warmer. Your moving hands have energy. As friction slows your hands down, some of that energy is changed into heat, a kind of energy.

✔ *Quick Check*

Summarize How can energy change?

Critical Thinking Why does a bowling ball slow down when it hits a pin?

458
EXPLAIN

FACT ▶ Energy does not get used up. It changes form.

Homework Activity

Activities and Energy

Have students choose one activity they like to do. Have them do the activity, thinking about the energy changes and transfers involved. Have them list the energy changes and transfers. For example, when playing the piano, stored energy from food changes to energy of motion in your fingers. Your fingers transfer the energy by hitting the keys.

Lesson Review

Visual Summary

Work is done when a force moves an object or changes an object's motion.

Energy is the ability to do work.

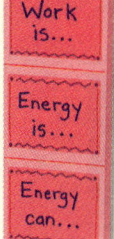

Energy can move from one object to another. Energy can also change form.

Make a FOLDABLES Study Guide

Make a Three-Tab book. Use it to summarize what you learned about work and energy.

Work is...

Energy is...

Energy can...

Think, Talk, and Write

❶ **Main Idea** What is energy? How can it change?

❷ **Vocabulary** What is work? Give two examples.

❸ **Summarize** A soccer ball is at your feet. You kick the ball and it travels across the field. Use the terms work and energy to describe what happens.

[diagram: three boxes → Summary]

❹ **Critical Thinking** How is an apple like gasoline in a car?

❺ **Test Prep** Which is an example of work being done?
A studying for a test
B picking up a feather
C holding a heavy box over your head
D pushing on a wall

 **Writing Link**

Explanatory Writing
A rock on a hill has potential energy. What happens to this energy as the rock rolls down the hill? Write about it. Then make a drawing to illustrate your writing.

 Art Link

Make a Collage
Cut pictures from magazines of objects with kinetic energy. Paste your pictures onto a poster. Write how you think each object got its energy.

LOG ON e-Review Summaries and quizzes online at **www.macmillanmh.com**

459
EVALUATE

Formative Assessment

Approaching Have students demonstrate one example of energy changing from one form to another.

On-Level Have students write a paragraph that explains the relationship between energy and work with examples.

Challenge Have students model an energy transfer and share the model with the rest of the class.

Key Concept Cards For student intervention, see the prescribed routine on **Key Concept Cards 71** and **72**.

3 Close

Lesson Review

▶ **Visual Summary**

Have students save their lesson Study Guides. They will use them at the end of the chapter for review.

▶ **Make a FOLDABLES Study Guide**

See pages R27–R28 for instructions.

FAST TRACK **Think, Talk, and Write**

❶ **Main Idea** Energy is the ability to do work. It can change from one form to another. For example, stored energy can be changed into energy of motion.

❷ **Vocabulary** Work is done when a force moves an object; lifting a pencil and peddling a bicycle.

❸ **Summarize**

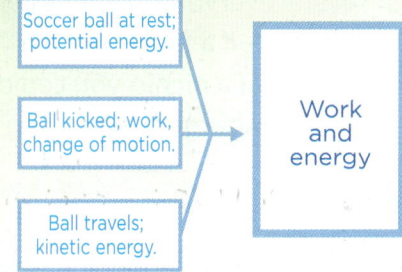

Soccer ball at rest; potential energy.

Ball kicked; work, change of motion.

Ball travels; kinetic energy.

→ Work and energy

❹ **Critical Thinking** Both objects contain stored energy that can change to energy of motion.

❺ **Test Prep** B

Writing Link Most potential: not moving; most kinetic: at its highest speed going down the hill

Art Link Check posters for accuracy.

RESOURCES and TECHNOLOGY

▶ **Reading and Writing,** p. 219

▶ **School to Home Activities,** pp. 93–94

LOG ON e-Review Narrated Summary and Quiz

LOG ON Progress Reporter Assessments

Focus on Skills

Objective
■ Relate the speed of water and the energy it has.

Materials paper plate, ruler, scissors, pencil, thread, paper clip, tape, faucet

Plan Ahead Have paper or cloth towels available for water spills. For safety, you might prefer to punch small holes in the middle of the plates ahead of time.

EXTEND This activity will teach students how to use their observations of work done by running water to make inferences about the relationship between work and energy.

Inquiry Skill: Infer

▶ Learn It
Provide students with several everyday examples of making inferences. For example, if a family ate ten hamburgers and five hot dogs at a picnic, you could infer that the family prefers hamburgers to hot dogs.

▶ Try It

Answer Yes; water caused the plate to spin and the paper clip to move. If something moves an object, it is doing work.

① Have students make four 3-cm marks on the plate, 90 degrees apart. Caution them to cut carefully.

② **Be Careful!** If students punch the holes themselves, caution them to be careful. The pencils are sharp.

⑤ The plate turns and lifts the paper clip.

⑥ The plate lifts the paper clip faster.

▶ The energy from the running water moves the wheel. The moving wheel lifts the paper clip.

▶ Fast-moving water has more energy because it can lift the paper clip faster.

▶ It would move more slowly because it takes more energy to move a heavier object.

Focus on Skills

Inquiry Skill: Infer
When you do an experiment, you are trying to answer a question. Sometimes you can answer a question from the data you collect. Other times, you must **infer** the answer using facts you know.

▶ Learn It
When you **infer**, you form an idea based on observations and facts. As you make observations, it is important to record your data. The more data you collect, the better you will be able to infer.

▶ Try It
Can running water do work? To answer this question, make a water wheel. Then observe what happens to it under running water. Use your observations and what you know about work to **infer** if water can do work.

Materials paper plate, ruler, scissors, pencil, thread, paper clip, tape, faucet

◀ A water wheel is a machine that uses the energy of moving water to power mills and factories.

460
EXTEND

Integrate Reading

Write About Energy
Have students write a short story about life in a pioneer town where the two energy sources are wind and running water.

• Allow them to use references if they need to find out more about these energy sources.

• This story can be done as a class project, with all students contributing to it, or it can be done individually or in small groups.

• The stories might include energy transfers involved in using energy from running water to grind flour in a mill, or using wind and a windmill to pump water.

• If individual or group stories are written, they can be compiled into a book.

Skill Builder

1. Cut four 3-cm slits into a plastic plate. Then bend the slits to create a pinwheel.

2. Gently push a pencil through the center of the plastic plate. ⚠ **Be Careful.** Point the pencil away from your body. Ask an adult for help.

3. Tie one end of a piece of thread to a paper clip. Tape the other end to the pencil, near the hole in the plate.

4. Turn on the faucet so that a little water flows out.

5. Rest the pencil across the palms of your hands. Then hold the edge of the plate 2 cm under the water. Record your observations.

6. Repeat with a larger stream of water. Record what you observe.

Now use observations and facts you know to answer these questions.

▸ What makes the water wheel move?

▸ Does using more water give the wheel more energy? How can you tell?

▸ Can running water do work? Explain your answer.

▸ Apply It

You have learned to **infer** the answer to a question from the data you collect and the facts you know. Now you can infer answers to new questions. For example, can wind do work? How might you use your wheel to infer the answer?

Step 3

Step 5

461
EXTEND

▶ Apply It

As an alternative to the suggestion in the Pupil book, have students set up a similar experiment using a metal teaspoon instead of 25 paper clips.

- Have students include a hypothesis, steps to follow, and a data table in their plans.

- Students will determine that more energy is required to move a heavier object. A slow stream of water will not lift the spoon. A faster stream of water might lift the spoon, depending on the mass of the spoon and the speed of the water.

- Student inferences should include that for more weight to be lifted, more energy must be used. Faster-moving water has more energy than slower-moving water.

Focus on Skills

Name _____ Date _____

Infer

When you do an experiment, you are trying to answer a question. Sometimes you can answer a question from the data you collect. Other times, you must infer the answer using facts you know.

Learn It

When you **infer**, you form an idea based on observations and facts. As you make observations, it is important to record your data. The more data you collect, the better you will be able to infer.

◀ A water wheel is a machine that uses the energy of moving water to power mills and factories.

▸ Activity Lab Book, pp. 202–204

RESOURCES and TECHNOLOGY

▸ **Activity Lab Book,** pp. 202–204

▸ **Activity Flipchart,** p. 54

💿 **Instructional Navigator CD-ROM**

Plan Your Lesson

Lesson 4 Using Simple Machines

Objective
- Identify and describe simple machines, and apply their use to real-world tasks.
- Define what a compound machine is and give several examples.

Reading Skill Problem and Solution

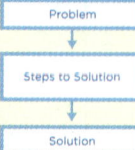

Problem

Steps to Solution

Solution

Graphic Organizer 12, p. TR14

LOG ON **Professional Development** Look for **NSDL** to find recommended Science Background materials from the National Science Digital Library

FAST TRACK

Lesson Plan When time is short, follow the Fast Track and use the essential resources.

1 Introduce
Look and Wonder, p. 462

Resource Alternative Explore, p. 207

2 Teach
Discuss the Main Idea, p. 464

Discuss the Main Idea, p. 466

Discuss the Main Idea, p. 468

Discuss the Main Idea, p. 470

Resource Visual Literacy, p. 73

3 Close
Think, Talk, and Write, p. 471

Resource Assessment, p. 141

▶ Reading and Writing

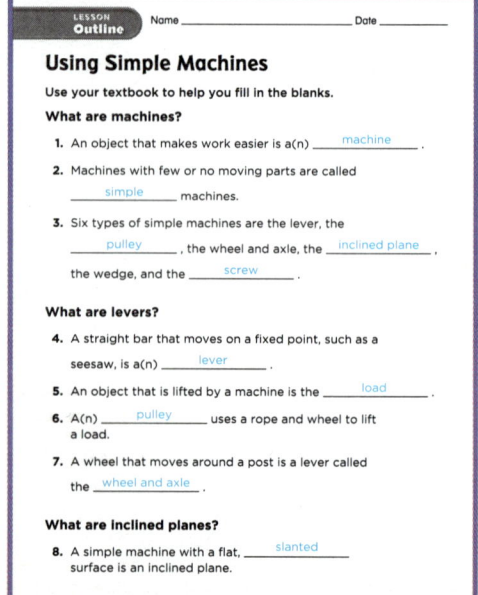

Outline, pp. 220–221
Also available as a student workbook

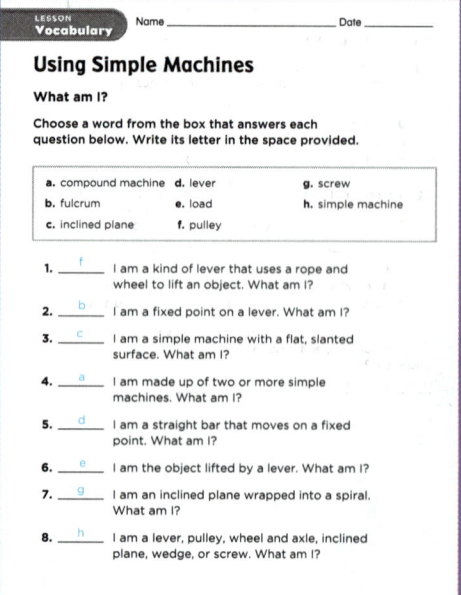

Vocabulary, p. 222
Also available as a student workbook

▶ Visual Literacy

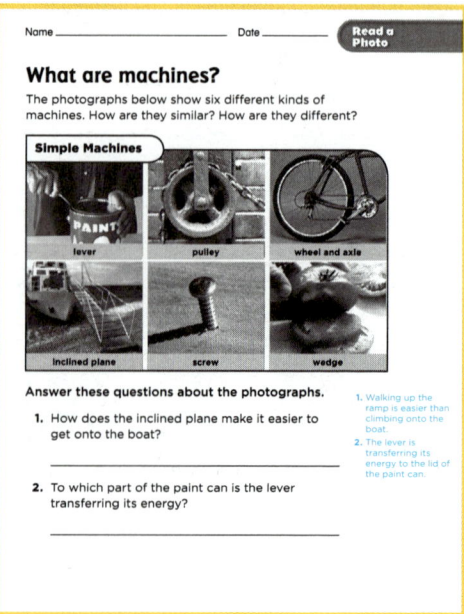

Read a Photo, p. 73
Also available as a transparency

▶ Activity Lab Book

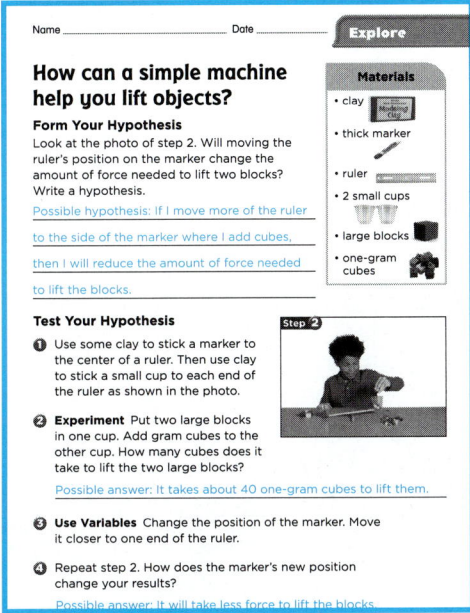

Explore, pp. 205–206
Also available as a student workbook

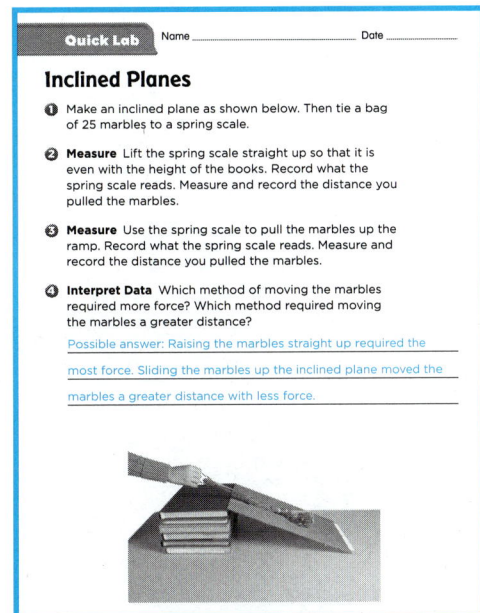

Quick Lab, p. 208
Also available as a student workbook

▶ Assessment

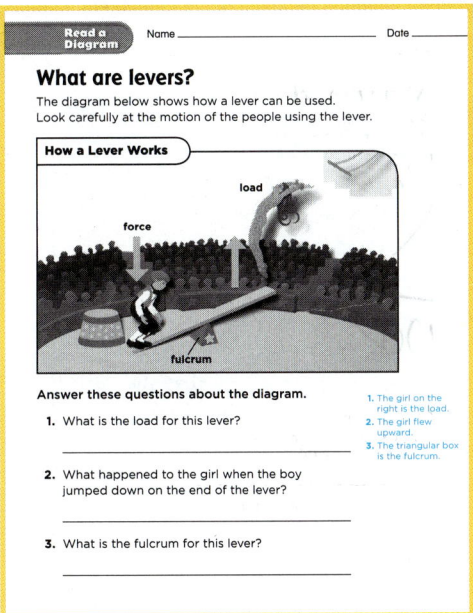

Read a Diagram, p. 74
Also available as a transparency

Lesson Test, p. 141
Also available as a student workbook

ADDITIONAL RESOURCES

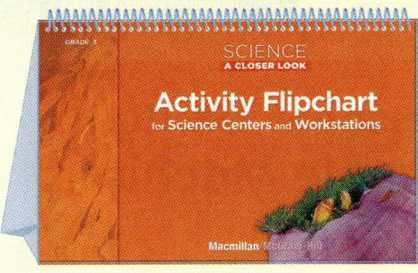

p. 55

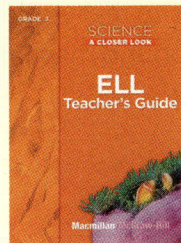

p. 130   **pp. 73–74**

pp. 95–96   **73–74**

168–175

Technology

- **Science Activity DVD**
- **Instructional Navigator CD-ROM**
- **Presentation Toolkit CD-ROM**
- **Science in Motion**
  How a Lever Works
- **e-Review**
- **NSDL**

Lesson 4 Using Simple Machines

Objectives

- Identify and describe simple machines and apply their uses to real-world tasks.
- Define what a compound machine is and give several examples.

1 Introduce

▶ **Assess Prior Knowledge**

Question students about prior use of machines. Ask the following questions and discuss student answers. Accept all reasonable student responses, but save them for later discussion. Ask:

- **Why is it easier to put a nail into a board using a hammer?** Possible answer: The hammer helps concentrate force on the nail.

- **Which is easier, pulling the lid off a paint can with your fingers or prying it off with a screwdriver?** with a screwdriver

- **Is it easier to lift and carry a heavy box or push it up a ramp?** push it up a ramp

 Look and Wonder

Invite students to share their responses to the Look and Wonder statement and question.

- **How can this wheelbarrow make harvesting a garden easier?** It makes moving soil, tools, and plants easier.

Write ideas on the board and note any misconceptions that students may have. Address these misconceptions as you teach the lesson.

RESOURCES and TECHNOLOGY

▶ **Activity Lab Book,** pp. 205–207

▶ **Activity Flipchart,** p. 55

◉ **Science Activity DVD**

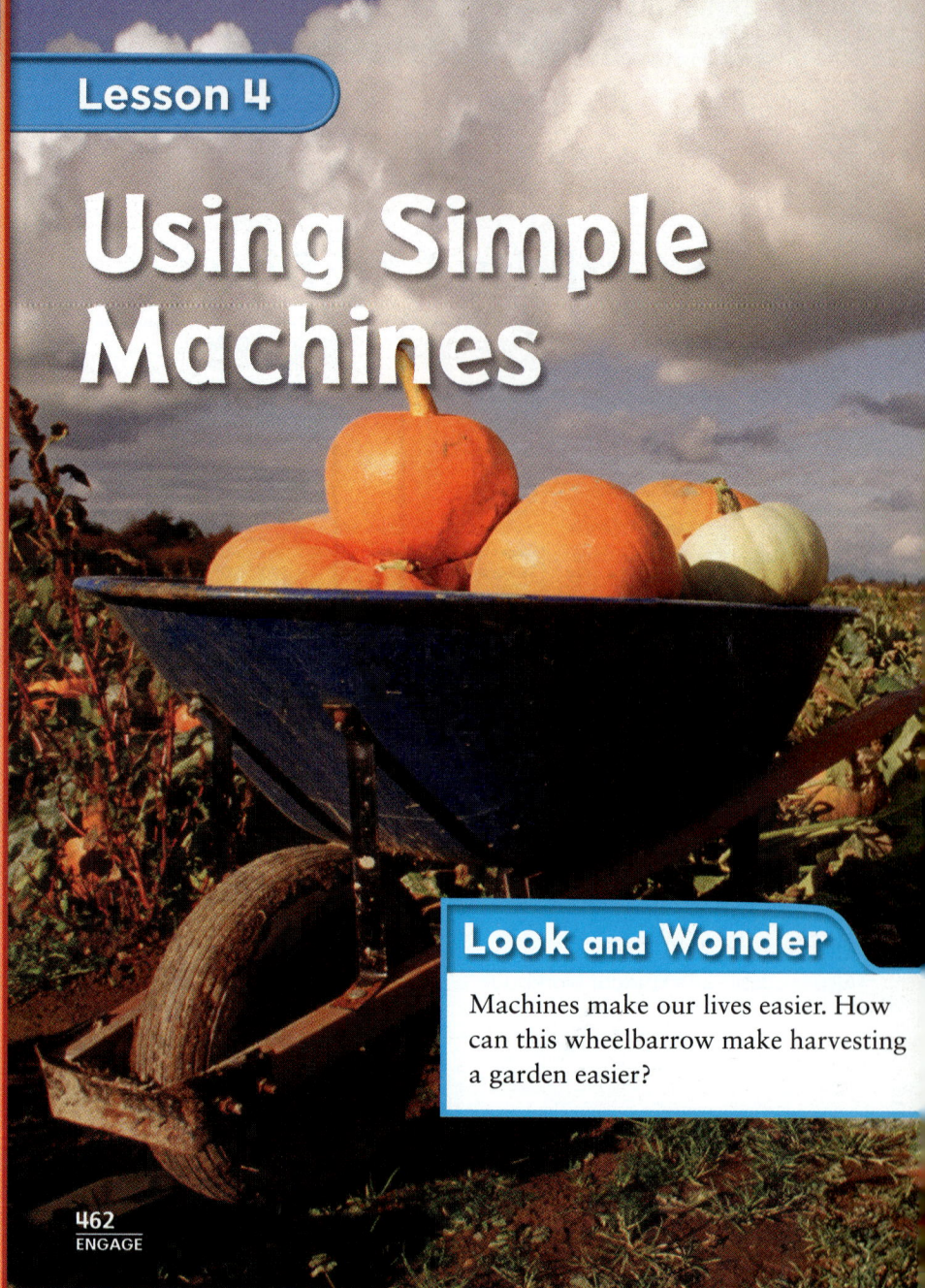

Lesson 4

Using Simple Machines

Look and Wonder

Machines make our lives easier. How can this wheelbarrow make harvesting a garden easier?

462
ENGAGE

Warm Up

Start with a Book

Divide the class into six small groups. Assign each group a simple machine. Have them read about this machine in the book *Simple Machines* by Adrienne Mason (Kids Can Press, ISBN 1-55-074399-6). Tell students that they will be class "experts" for the machine they studied. As each simple machine is discussed, encourage students who studied that particular simple machine to contribute details about it.

Explore

Inquiry Activity

How can a simple machine help you lift objects?

Form a Hypothesis

Look at the photos of steps 2 and 4. Will moving the ruler's position on the marker change the amount of force needed to lift two blocks? Write a hypothesis.

Test Your Hypothesis

1. Use some clay to stick a marker to the center of a ruler. Then use clay to stick a small cup to each end of the ruler as shown below.

2. **Experiment** Put two large blocks in one cup. Add gram cubes to the other cup. How many cubes does it take to lift the two large blocks?

3. **Use Variables** Change the position of the marker. Move it closer to one end of the ruler.

4. **Experiment** Repeat step 2. How does the marker's new position change your results?

Draw Conclusions

5. **Communicate** How does this machine lift objects?

6. **Interpret Data** How does the position of the marker change the number of gram cubes you need to lift the two large blocks?

Explore More

Experiment When are the two blocks lifted higher in the air—when the marker is near the two large blocks or when it is near the gram cubes? Try it to find out.

Materials

clay   thick marker

ruler

2 small cups

large blocks   one-gram cubes

Step 2

Step 4

463
EXPLORE

Alternative Explore

How does a machine make work easier?

Materials book, metric ruler

Have students place a ruler on the edge of a desk with half of the ruler hanging over. Place a book on the end of the ruler on the desk. Have students find how hard they must push and how far the ruler's end is hanging off the desk to raise the book 4 cm off the desk. Have them repeat with different ruler positions and relate the length of the hanging end to the amount of force needed.

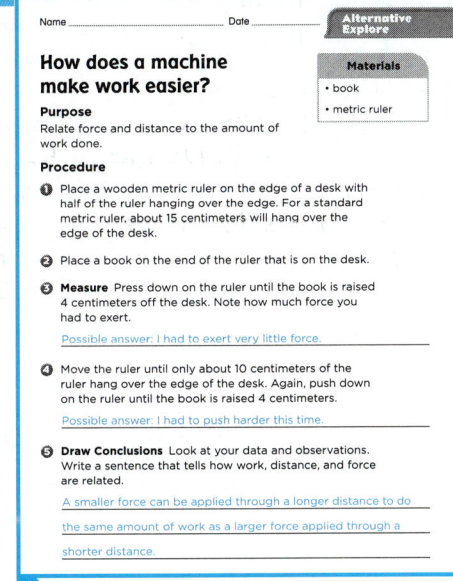

Name _____ Date _____

Alternative Explore

How does a machine make work easier?

Materials
- book
- metric ruler

Purpose
Relate force and distance to the amount of work done.

Procedure

1. Place a wooden metric ruler on the edge of a desk with half of the ruler hanging over the edge. For a standard metric ruler, about 15 centimeters will hang over the edge of the desk.

2. Place a book on the end of the ruler that is on the desk.

3. **Measure** Press down on the ruler until the book is raised 4 centimeters off the desk. Note how much force you had to exert.

   Possible answer: I had to exert very little force.

4. Move the ruler until only about 10 centimeters of the ruler hang over the edge of the desk. Again, push down on the ruler until the book is raised 4 centimeters.

   Possible answer: I had to push harder this time.

5. **Draw Conclusions** Look at your data and observations. Write a sentence that tells how work, distance, and force are related.

   A smaller force can be applied through a longer distance to do the same amount of work as a larger force applied through a shorter distance.

Activity Lab Book, p. 207

Explore

small groups   30 minutes

Plan Ahead Large, metal paper clips can be used in place of gram cubes. Explain to students that gravity acts on the mass cubes and that this force, which is weight, is what moves the blocks.

Purpose This activity helps students relate ease of work and the use of a simple machine.

Structured Inquiry

Form a Hypothesis Possible hypothesis: If I change the position of the ruler, then the force needed to move the blocks will change.

1. Be sure equal amounts of clay are used to attach the cups so that the mass of the clay does not affect results.

2. **Experiment** Possible answer: It takes 40 gram cubes to lift two blocks.

4. Answers will vary depending on where the student places the marker. If the student places the marker close to the end where they are adding the 1-gram blocks, more blocks are required to raise the large blocks. Less blocks will be required if they place the marker on the other end.

5. **Communicate** The cubes push down on one end of the ruler, and they raise the end with the two large blocks. This machine changes the force needed to lift an object.

6. **Interpret Data** When the marker is closer to the two large blocks, fewer cubes are required to lift the load. When the marker is farther from the two large blocks, more cubes are required.

Guided Inquiry   Explore More

Experiment when the marker is near the gram cubes

Open Inquiry

Ask students how the distance a force moves is related to the amount of force. Have them think of their own question about this relationship. Have them make a plan and carry out an experiment to answer the question. Ask:

If the force used to move two blocks is less, does this force move through a longer or a shorter distance?

Lesson 4   **463**

2 Teach

Read and Learn

Main Idea Have students look at the different machines they see in the visuals in the lesson. Have them infer what they have in common.

Vocabulary Have students create a main event concept map that relates the vocabulary terms and adds details for each.

Reading Skill Problem and Solution **Graphic Organizer 12** Have students fill in a Problem and Solution **graphic organizer** as they read each two pages of the lesson. They can use the Quick Check questions to identify each problem and its solution.

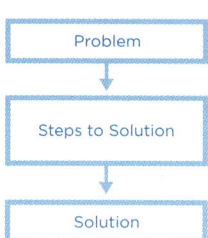

Problem
↓
Steps to Solution
↓
Solution

What are machines?

 **Discuss the Main Idea**

Discuss with students what a machine is in terms of its purpose. Ask:

- **Does a machine change the amount of work done?** no

- **What is the purpose of a machine?** It makes work easier.

- **How is a simple machine different from other machines?** A simple machine has few or no moving parts; it does work in one motion.

RESOURCES and TECHNOLOGY

▶ **Reading and Writing**, pp. 220–222

📀 **Presentation Toolkit CD-ROM**

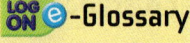 **e-Glossary**

Read and Learn

▶ **Main Idea**
Simple machines make work easier to do.

▶ **Vocabulary**
simple machine, p. 465
lever, p. 466
pulley, p. 467
wheel and axle, p. 467
inclined plane, p. 468
screw, p. 468
wedge, p. 469
compound machine, p. 470

e-Glossary
at **www.macmillanmh.com**

▶ **Reading Skill** ✓
Problem and Solution

Problem
↓
Steps to Solution
↓
Solution

What are machines?

You use machines every day. You might use a machine to travel to school. You might use a machine to sharpen your pencil. How would you describe a machine? A *machine* is something that makes work easier to do. Machines do not change the amount of work done. They simply change the way you do the work. For example, it is easier to lift and carry a heavy rock with a wheelbarrow than with your hands.

Some machines help you use less force to do work. Other machines change the direction in which you push or pull.

How is this backhoe making work easier? ▼

464
EXPLAIN

Science Background

Levers There are three types of levers. The type is determined by the relative locations of the resistance force, or load, the effort force, and the fulcrum. In a first-class lever, such as scissors, the fulcrum is between the load and the effort force. In a second-class lever, such as a wheelbarrow, the load is between the fulcrum and the effort force. In a third-class lever, such as a broom, the effort force is between the load and the fulcrum.

See **Science Yellow Pages** in the Teacher Resource section for background information.

 Professional Development For more Science Background and resources from **NSDL** visit **www.macmillanmh.com/nsdl/**

Machines such as cars, giant cranes, and bulldozers have lots of parts. They help us do a lot of work in a short amount of time. Yet the first machines were simple ones. **Simple machines** are machines with few or no moving parts. There are six types of simple machines. They are the lever, the pulley, the wheel and axle, the inclined plane, the screw, and the wedge.

Simple Machines

lever | pulley | wheel and axle

inclined plane | screw | wedge

 Quick Check

Problem and Solution How do machines help people solve problems?

Critical Thinking What are three simple machines you use every day?

465
EXPLAIN

▶ ## Use the Visuals

Refer students to the visuals on pages 464–465. Have students use the pictures to infer how each simple machine makes work easier. Have them draw or list other examples of each of these simple machines.

▶ ## Address Misconceptions

A common misconception is that machines reduce the amount of work done. Work is done by multiplying force by distance. Simple machines often use the same amount of force through a longer distance. The inclined plane is probably the easiest example used to illustrate this concept. Have students lift a box containing several books. Then have them push the box up a ramp to the same height they lifted it. Ask:

- **Which method of raising the box required you to use more force?** lifting the box

- **In which method did you move the box a greater distance?** using the ramp

- **In which method was more work done?** Neither. The same amount of work was done in both methods.

▶ ## Develop Vocabulary

simple machine *Scientific vs. Common Use* This vocabulary term is based on a noun, *machine,* which refers to an object that makes work easier, and an adjective, *simple,* which means that it is not complex. Therefore, a simple machine operates in an uncomplicated way to make work easier.

✔ Quick Check Answers

- **Problem and Solution** Possible answers: They make work easier. They change the way people do work.

- **Critical Thinking** Possible answers: screwdriver; fishing pole; pulley to raise a flag; baseball bat; entrance ramp

What are levers?

FAST TRACK **Discuss the Main Idea**

A lever is a bar free to move around a fixed point, the fulcrum. Using diagrams, lead a discussion on how three of the simple machines can be considered levers. For example, a pulley can be considered a lever with the axle acting as a fulcrum, and it has a rope instead of a bar. A wheel and axle acts as a lever attached to a shaft. Although simple machines fall into two main categories, there are enough differences in the six simple machines for them to be considered separate types. Ask:

- **What are the two main categories of simple machines?** levers and inclined planes

▶ Develop Vocabulary

lever *Word Origin* Point out that the word *lever* has its origin in the Old French word *levier,* meaning "to raise." Levers commonly use less force to raise an object a greater distance.

pulley *Word Origin* This word probably comes from the Greek *polos,* meaning "pole," or "axis."

wheel and axle *Scientific vs Common Use* The scientific meaning of *wheel and axle* comes from the common meanings of the root words. A wheel and axle contains a wheel fastened to a central shaft, called an *axle.*

What are levers?

How are a wheelbarrow and a seesaw alike? They are both levers (LEV•uhrz). A **lever** is a straight bar that moves on a fixed point. The fixed point is the *fulcrum.*

A lever can be used to lift something. The object lifted is called the *load.* In the diagram below, the girl is the load. When the boy presses down on one end of the lever, the load is lifted. The closer the fulcrum is to the load, the less force you need to lift the load.

Levers can make it easier for people to lift objects. They can change how much force you need to move something. They can also change the direction of the force you use. Pressing down on a lever lifts up the load.

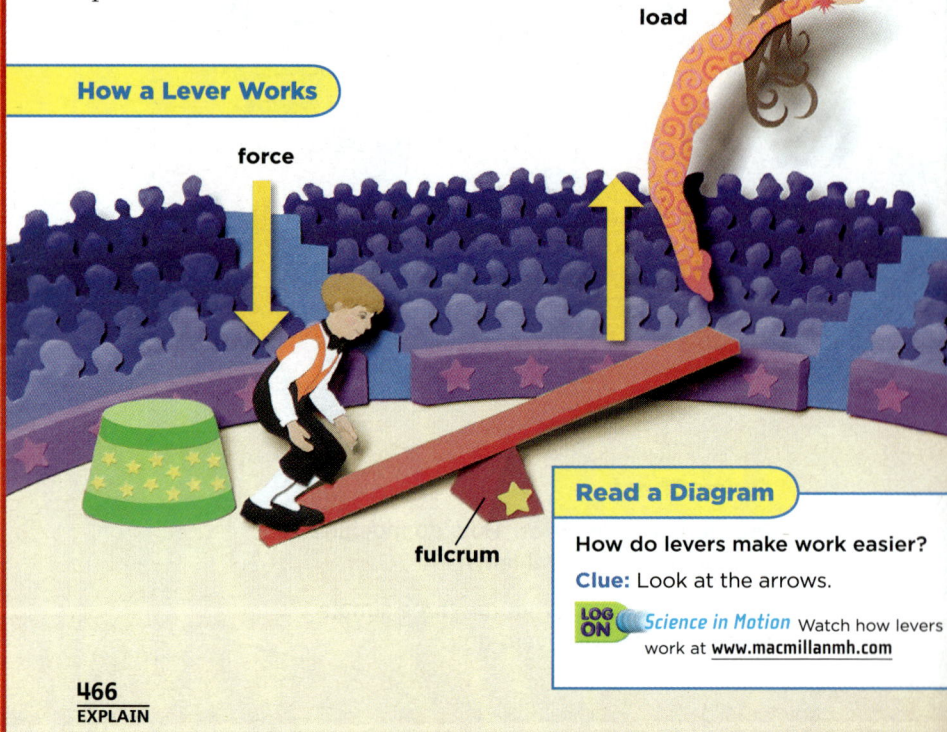

How a Lever Works

load

force

fulcrum

Read a Diagram

How do levers make work easier?

Clue: Look at the arrows.

LOG ON *Science in Motion* Watch how levers work at **www.macmillanmh.com**

466
EXPLAIN

Classroom Equity

Students benefit from role models they can relate to. Girls and minority students often get an unspoken message that science, especially the physical sciences, is not for them. Use this lesson on levers and pulleys to introduce students to some "real life" female or minority engineers who are involved in building machines. Have students research such a person and ask them to report their findings to the class.

Pulley

A <mark>pulley</mark> (PUL•ee) is a special kind of lever. It uses a rope and a wheel to lift an object. When you pull down on one end of the rope, the other end rises up. The pulley shown here makes work easier by changing the direction of the force you use to lift an object.

Wheel and Axle

A <mark>wheel and axle</mark> (AK•suhl) is another special lever made up of a wheel that moves around a post. The post is called an axle. Doorknobs and Ferris wheels are wheel and axles.

A wheel and axle can make work easier to do. Try opening a door by turning the knob. Now try it by turning the thin bar behind the knob. Which requires less force? Turning a wheel requires less force than turning an axle.

▲ A pulley makes it easier to lift this bucket.

The axle makes a smaller movement. The wheel makes a larger movement. ▼

wheel

axle

467
EXPLAIN

✔ Quick Check

Problem and Solution How could you move a heavy rock?

Critical Thinking Which simple machine could you use to raise a flag?

▶ Discuss the Main Idea

Show students the gears in a bicycle. Have them identify each of the wheels and axles in the gear system. Tell them that gears are sets of wheels and axles. Ask:

- **What does a wheel in a gear have that a wheel in another wheel and axle might not have?** teeth along the outer edge of the wheel

- **Why might you use a gear instead of a single wheel and axle?** Possible answer: A large wheel can be turned slowly and cause a smaller wheel to turn faster.

Read a Diagram

Answer by using a smaller force through a greater distance; by changing the direction of the force

 *Science in Motion* *How a Lever Works*

✔ *Quick Check Answers*

- **Problem and Solution** Possible answer: by placing one end of a bar under the rock and putting force on the other end of the bar to lift the rock

- **Critical Thinking** the wheel

Differentiated Instruction

Leveled Questions

EXTRA SUPPORT When a screwdriver is used as a lever to pry open the lid on a paint can, what is the fulcrum? the edge of the can

ENRICHMENT What simple machine might be used to pull the engine out of a car? a pulley Encourage students to investigate how a block and tackle, which is a system of pulleys, is used for such purposes.

What are inclined planes?

FAST TRACK ### Discuss the Main Idea

Use diagrams to show examples of inclined planes. Include the following information: An inclined plane is a simple machine that has a flat, slanted surface. A screw is an inclined plane wrapped around a shaft. A wedge has two sloping sides formed by two inclined planes. Ask:

- **What types of inclined planes are used to cut food?** knives, or wedges

- **What types of inclined planes are used to hang pictures on a wall?** screws

▶ Develop Vocabulary

inclined plane The meaning of *inclined* is "slanted," and *plane* refers to a flat surface. An inclined plane is a flat surface that is slanted.

screw *Scientific vs. Common Use* Students are probably familiar with common screws that hold objects together. This common screw is the same object as the screw that is a simple machine.

wedge *Scientific vs. Common Use* In common usage, a wedge is something that divides or comes between. For example, a wedge between two people might be caused by an argument between them. Scientifically, a wedge is a simple machine that splits things apart.

It takes less force to push a box up a ramp than to lift it straight up.

What are inclined planes?

You have probably seen ramps in buildings such as your school. A ramp is an inclined plane (in•KLINED PLAYN). An **inclined plane** is a simple machine with a flat, slanted surface.

Inclined planes can make work easier to do. They reduce the force you need to move an object. Think about moving a heavy box onto a truck. You could not lift it off the ground to put it in the truck. You could slide it up an inclined plane instead. Sliding a box up an inclined plane requires less force than lifting the box straight up. However, you must push the box a longer distance.

Screw

A **screw** is an inclined plane wrapped into a spiral. It takes less force to turn a screw than to pound a nail. A screw changes a turning force into a downward force.

This machine is called an auger. An auger is a giant screw.

468
EXPLAIN

Differentiated Instruction

Leveled Activities

EXTRA SUPPORT   Have students use models to show that a wedge is formed from two inclined planes.

ENRICHMENT   Have students identify, classify, and sketch the simple machines in playground equipment. Examples might include a wheel and axle in a merry-go-round and an inclined plane on a slide.

RESOURCES and TECHNOLOGY

▶ Activity Lab Book, p. 208

▶ Presentation Toolkit CD-ROM

Wedge

If you put two inclined planes back to back you get a wedge (WEJ). A **wedge** is a simple machine that splits objects apart.

The head of an ax is a wedge. When you swing an ax, the downward force is changed into a sideways force. The sideways force pushes, or splits, the wood apart.

Most cutting tools, such as knives, are wedges. As you press a knife into food, the knife pushes the food apart.

The downward force of the ax changes to the sideways force that splits the log.

≡ Quick Lab

Inclined Planes

1. Make an inclined plane as shown below. Then tie a bag of 25 marbles to a spring scale.

2. **Measure** Lift the spring scale straight up so that it is even with the height of the books. Record what the spring scale reads. Measure and record the distance you pulled the marbles.

3. **Measure** Use the spring scale to pull the marbles up the ramp. Record what the spring scale reads. Measure and record the distance you pulled the marbles.

4. **Interpret Data** Which method of moving the marbles required more force? Which method required moving the marbles a greater distance?

✓ Quick Check

Problem and Solution Which simple machine would you use to cut a banana?

Critical Thinking Where have you seen ramps used in your community?

469
EXPLAIN

≡ Quick Lab
 small groups   15 minutes

Objective Show that an inclined plane requires less force over a longer distance to do work.

Materials cardboard, 4 books, spring scale, 25 marbles

1. Use smooth cardboard so that friction does not affect the results.

2. Have students read through the lab and make a data table that can be used to record their observations.

3. Be sure students pull with a steady force.

4. Lifting the marbles required more force but less distance. Pulling the marbles up the inclined cardboard required moving them a greater distance, but with less effort force.

▶ Explore the Main Idea

ACTIVITY Have students wrap a strip of paper at a slant around a pencil to model a screw. Have them point out the inclined plane on the post of the screw.

✓ Quick Check Answers

- **Problem and Solution** You would use a knife, which is a wedge.

- **Critical Thinking** Possible answers: schools, loading platforms, restaurants, theaters

ELL Support

Practice Using Language Clarify the meaning of the word *wedge* for students. Write it on the board and have students repeat the word after you. Be sure you emphasize the voiced last consonant sound /je/ since many students may be unfamiliar with voiced consonants at the end of words or in final syllables. Explain that any tool that has a force that can split or push an object apart is a wedge. Ask students which wedges they are familiar with. Elicit knives, axes, and so on.

BEGINNING Students can point to or name wedges and inclined planes.

INTERMEDIATE Students can use phrases and short sentences to describe wedges, inclined planes, or other simple machines.

ADVANCED Students can use complete sentences to describe wedges, inclined planes, or other simple machines.

Lesson 4   **469**

How do machines work together?

▶ Discuss the Main Idea

Have students look up the word *compound* in the dictionary and infer the definition of a compound machine. Show them a classroom pencil sharpener. Ask:

■ **What are the simple machines that make up this compound machine?** wheel and axle in the handle; screw in the part that sharpens the pencil

▶ Use the Visuals

Emphasize to students that the photos on page 470 show machines that are obviously compound machines. Point out that the simple machines in some compound machines are not always clearly seen. Ask:

■ **What simple machines make up an axe?** The blade is a wedge, and the handle is a lever.

▶ Develop Vocabulary

compound machine Any compound is made up of two or more things. For example, a compound word, such as *blackboard,* is made up of two or more simpler words. Therefore, a compound machine is made up of two or more simple machines.

✔ Quick Check Answers

• **Problem and Solution** You get a compound machine.

• **Critical Thinking** levers

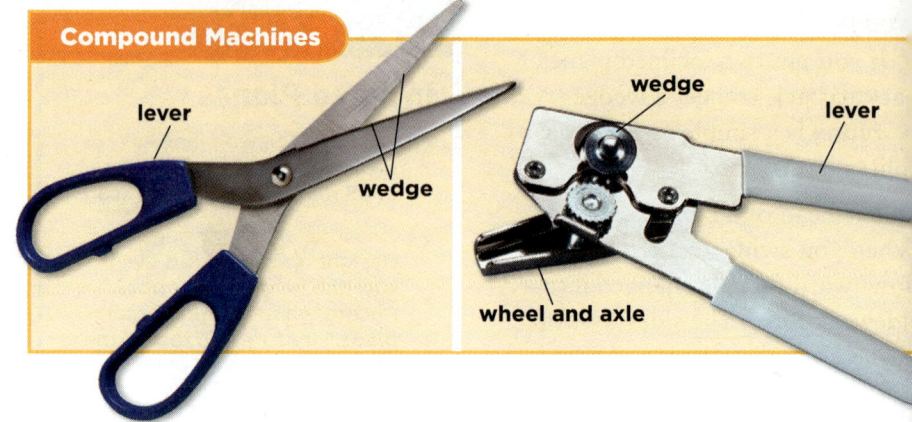

Compound Machines

lever   wedge   wedge   lever

wheel and axle

How do machines work together?

Most of the tools you use every day are compound machines. A **compound machine** is two or more simple machines put together.

A pair of scissors is a compound machine. Two wedges and two levers make an excellent cutting tool. The point where they are connected is the fulcrum. When the handles are pushed together, the edges cut through material.

A can opener is also a compound machine. It contains a wedge, a lever, and a wheel and axle acting as one machine.

Quick Check

Problem and Solution What do you get if you put two or more simple machines together?

Critical Thinking What simple machines are part of a bicycle?

470
EXPLAIN

Homework Activity

Machine Inventory

With adult permission and supervision, have students survey their homes and make a list of five simple machines they find. Have them identify the simple machine, the object that contains it, and how it is used. Also have them sketch the object, labeling the simple machine. Encourage them to find at least four different simple machines.

Lesson Review

Visual Summary

Machines help make work easier to do.

The lever, wheel and axle, pulley, inclined plane, screw, and wedge are types of **simple machines**.

A compound machine is made of two or more simple machines.

Make a FOLDABLES Study Guide

Make a Trifold Book. Use it to summarize what you learned about machines.

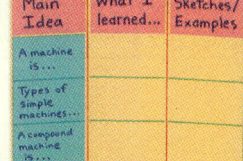

Think, Talk, and Write

1. **Main Idea** How do machines make work easier?

2. **Vocabulary** What is a simple machine? Talk about it.

3. **Problem and Solution** Suppose you were going to build a pyramid that was 10 meters high. How might you build it? What simple machines might you use?

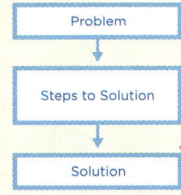

4. **Critical Thinking** How does a woodpecker use its beak as a simple machine?

5. **Test Prep** Which of the following is a compound machine?
   A lever
   B inclined plane
   C scissors
   D wheel and axle

 Writing Link

Write a Report
Make a list of the simple machines you find in your neighborhood. For example, you might see a ramp at the library or a seesaw at the playground. Write about how these machines are used.

Social Studies Link

Do Research
The giant pyramids in Egypt were built many years ago. There were no bulldozers, trucks, or tractors then. Use research materials to learn more about how people think these pyramids may have been built.

 e-Review Summaries and quizzes online at www.macmillanmh.com

471
EVALUATE

Formative Assessment

Approaching Have students draw a wheelbarrow and label the parts of this lever.

On-Level Have students draw each simple machine. Have them label the lever in a pulley and a wheel and axle and the inclined plane in a screw and a wedge.

Challenge Have students write a paragraph explaining why a screwdriver can act as a lever for some tasks and a wheel and axle for others.

Key Concept Cards For student intervention, see the prescribed routine on **Key Concept Cards 73** and **74.**

---

3 Close

Lesson Review

▶ **Visual Summary**

Have students save their lesson Study Guides. They will use them at the end of the chapter for review.

▶ **Make a FOLDABLES Study Guide**

See pages R27–R28 for instructions.

FAST TRACK

Think, Talk, and Write

1. **Main Idea** Machines make work easier by changing the size or direction of an applied force.

2. **Vocabulary** A machine with few or no moving parts; simple machines help make work easier.

3. **Problem and Solution**

   What simple machines might I use to build a 10-meter high pyramid?

   ↓

   Need a lever, a pulley for heavy objects, and inclined plane to push items up

   ↓

   The 10-meter high pyramid is built using these simple machines.

4. **Critical Thinking** The beak acts as a wedge, splitting the bark of the tree.

5. **Test Prep** C

Writing Link Check student lists for accuracy.

Social Studies Link Sources may vary as to whether the primary simple machine used was the inclined plane or the lever.

RESOURCES and TECHNOLOGY

▸ **Reading and Writing,** p. 223

▸ **School to Home Activities,** pp. 95–96

LOG ON **Science in Motion,** *How a Lever Works*

LOG ON **e-Review** Narrated Summary and Quiz

LOG ON **Progress Reporter** Assessments

Writing in Science

Objective
■ Write an explanatory paragraph about a compound machine.

A Very Useful Machine

Learn It

Discuss how an explanatory paragraph should be clear and easy to follow. Emphasize that the purpose of an explanatory paragraph is to provide the reader with necessary information.

Try It

■ Discuss with students times they have had to learn something new. Ask:

■ **How could a good explanatory paragraph have helped you?** A good explanatory paragraph would have explained what to do with easy-to-follow details and necessary time-order words.

Apply It

■ Have the class divide the explanatory paragraph about can openers into a series of steps. Ask:

■ **How is giving the operation of a can opener in a series of steps helpful?** It makes it easier for the reader to understand how the can opener works.

 Write About It

■ Provide a list of compound machines for students to use.

■ Students can work in pairs so partners can help each other figure out how the machine works and explain the steps in using it.

RESOURCES and TECHNOLOGY

▶ **Reading and Writing, pp. 224–225**

e-Journal Online research and writing

Writing Rubric P. TR26

Writing in Science

A Very Useful Machine

A can opener is a compound machine. It makes opening cans easy. How does it work? First, you attach the cutting wheel to the can's lid. Then, you press the two long handles together. This causes the cutting wheel to cut into the top of the can. Next, you turn the crank. This moves the wheel that cuts the can. The wheel continues to turn as long as you turn the crank. When the top of the can comes off, you can let go of the handles and remove the can opener.

Explanatory Writing

A good explanation

▶ explains how something works or gives information about how to do something

▶ gives details that are easy to follow

▶ uses time-order words such as *first*, *next*, and *after*

Three simple machines make up a can opener. They are the wedge, the lever, and the wheel and axle.

Write About It

Explanatory Writing Choose another compound machine. Find out how it works. Then write a paragraph that explains how to use it.

e-Journal Write about it online at www.macmillanmh.com

472
EXTEND

Integrate Writing

Order a Laboratory Procedure

• Provide a sample laboratory investigation to one student in a pair. It should have 3–5 steps. Have that student make an index card for each step, without using the numbers.

• Tell the partner to review the cards and place the steps in the proper order.

• Then have the pairs write a paragraph describing how to do the investigation, using details and time-order words.

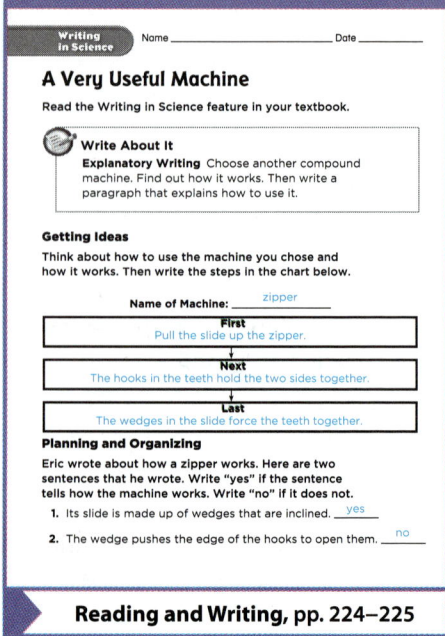

Writing in Science   Name _____ Date _____

A Very Useful Machine

Read the Writing in Science feature in your textbook.

Write About It
Explanatory Writing Choose another compound machine. Find out how it works. Then write a paragraph that explains how to use it.

Getting Ideas
Think about how to use the machine you chose and how it works. Then write the steps in the chart below.

Name of Machine: zipper

First
Pull the slide up the zipper.

Next
The hooks in the teeth hold the two sides together.

Last
The wedges in the slide force the teeth together.

Planning and Organizing
Eric wrote about how a zipper works. Here are two sentences that he wrote. Write "yes" if the sentence tells how the machine works. Write "no" if it does not.

1. Its slide is made up of wedges that are inclined. ___yes___
2. The wedge pushes the edge of the hooks to open them. ___no___

▶ **Reading and Writing, pp. 224–225**

Math in Science

Using Number Patterns

Levers can help people lift heavy loads. Suppose you know you need to use 60 newtons (N) of force to lift a bag of soil with a wheelbarrow. You need two times as much force, 120 newtons, to lift 2 bags of soil. How much force would you need to lift 3 bags of soil with this wheelbarrow? Find the number patterns in the chart below to solve this problem.

How to Use Patterns

► A number pattern is a series of numbers that follows a rule. To find a pattern, look at how the numbers change.

► To find the next number in this pattern, add 60 to the previous number.

| 1 bag | 2 bags | 3 bags |
| 60 N | 120 N | 180 N |

     +60   +60

Solve It

How many bags of soil could you lift using this wheelbarrow if you apply 240 newtons of force? How much force would you need to lift 5 bags of soil? Explain how you found your answer.

Bags of Soil	Force Needed (in newtons)
1	60
2	120
3	180
4	240

473
EXTEND

Using Number Patterns

Learn It

Review number patterns with students. Write on the board several number lists with simple patterns, such as *12-15-18-21* or *27-37-47-57*, and have students provide the next several numbers in the series. Ask:

■ **If a pinwheel costs 1 dollar, and two pinwheels cost 2 dollars, and three pinwheels cost 3 dollars, how much do four pinwheels cost?** 4 dollars

Try It

■ Tell students that one bowl of papier-mâché is enough to make three small puppets. Write the following on the board:

1 bowl—3 puppets

2 bowls—6 puppets

3 bowls—

Ask:

■ **How many puppets can you make with three bowls of papier-mâché?** 12 puppets

Apply It

■ Tell students to write a table similar to the one on the pupil page. Ask them to write a math problem based on their tables. Have students exchange papers and solve each other's problems.

Solve It

■ 4 bags; 300 newtons; the force needed to lift each bag increases by 60 each time. That is the number pattern.

Integrate Math

Patterns in Screws

● Bring in several different-size wood screws and distribute them to groups of students.

● Have students count the number of threads on the screw and measure the length of the threaded part.

● Tell student groups to make a table like the one on page 473. The table should tell how many threads a screw would have if it were two, three, and four times longer than the screw students have.

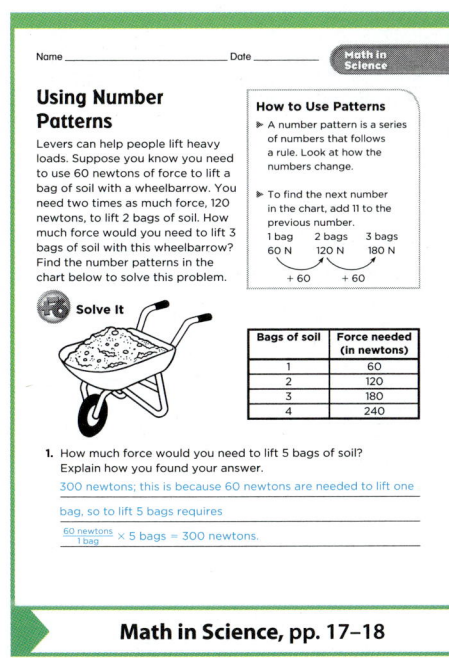

Name _____ Date _____

Math in Science

Using Number Patterns

Levers can help people lift heavy loads. Suppose you know you need to use 60 newtons of force to lift a bag of soil with a wheelbarrow. You need two times as much force, 120 newtons, to lift 2 bags of soil. How much force would you need to lift 3 bags of soil with this wheelbarrow? Find the number patterns in the chart below to solve this problem.

How to Use Patterns

► A number pattern is a series of numbers that follows a rule. Look at how the numbers change.

► To find the next number in the chart, add 11 to the previous number.

| 1 bag | 2 bags | 3 bags |
| 60 N | 120 N | 180 N |

  + 60   + 60

Solve It

Bags of soil	Force needed (in newtons)
1	60
2	120
3	180
4	240

1. How much force would you need to lift 5 bags of soil? Explain how you found your answer.

300 newtons; this is because 60 newtons are needed to lift one bag, so to lift 5 bags requires

$\frac{60 \text{ newtons}}{1 \text{ bag}}$ × 5 bags = 300 newtons.

Math in Science, pp. 17–18

RESOURCES and TECHNOLOGY

► **Math in Science,** pp. 17–18

Writing and Math in Science   **473**

CHAPTER 11 Review

▶ Visual Summary

Have students look at the pictures to review the main ideas of the chapter.

Vocabulary

1. motion
2. energy
3. speed
4. inclined plane
5. magnet
6. lever
7. force
8. wedge
9. friction
10. compound machine

▶ Make a FOLDABLES Study Guide

See pages R27–R28 in the back of the Teacher's Edition for more information on Foldables.

CHAPTER 11 Review

Visual Summary

Lesson 1 An object is in motion when its position changes.

Lesson 2 Forces can change an object's motion.

Lesson 3 Work is done when a force moves an object. Energy is the ability to do work.

Lesson 4 Simple machines make work easier to do.

Make a FOLDABLES Study Guide

Glue your lesson study guides to a piece of paper as shown. Use your study guide to review what you have learned in this chapter.

Vocabulary

Fill each blank with the best term from the list.

energy, p. 456	**lever**, p. 466
compound machine, p. 470	**magnet**, p. 446
force, p. 444	**motion**, p. 436
friction, p. 448	**speed**, p. 438
inclined plane, p. 468	**wedge**, p. 469

1. An object in _____ is changing its position.

2. The ability to do work is called _____.

3. How quickly an object moves is described by its _____.

4. A ramp is an example of an _____.

5. You can use a _____ to attract things made of iron.

6. A straight bar that moves on a fixed point is a _____.

7. A push or pull is called a _____.

8. A knife acts as a _____ when cutting food.

9. You squeeze the hand brakes on a bike. The force that slows down the bike is _____.

10. A machine made up of two or more simple machines is a _____.

 **e-Review** Summaries and quizzes online at **www.macmillanmh.com**

RESOURCES and TECHNOLOGY

▶ **Reading and Writing,** pp. 226–227

▶ **Assessment,** pp. 134–137, 142–145

🖫 **Progress Reporter** Assessments

🖫 **Puzzlemaker CD-ROM**

🖫 **Vocabulary Games**

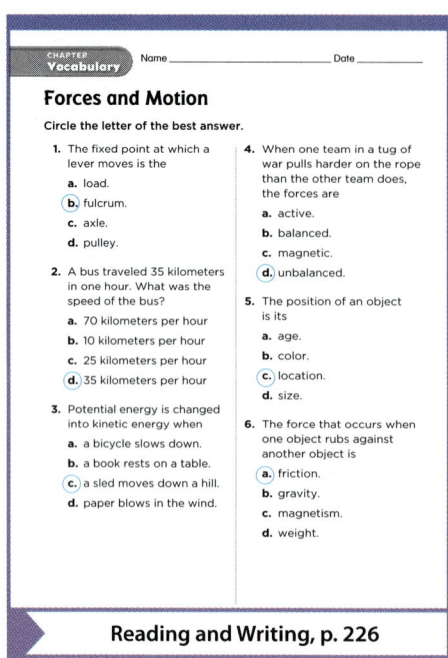

Forces and Motion

Circle the letter of the best answer.

1. The fixed point at which a lever moves is the
   a. load.
   b. fulcrum.
   c. axle.
   d. pulley.

2. A bus traveled 35 kilometers in one hour. What was the speed of the bus?
   a. 70 kilometers per hour
   b. 10 kilometers per hour
   c. 25 kilometers per hour
   d. 35 kilometers per hour

3. Potential energy is changed into kinetic energy when
   a. a bicycle slows down.
   b. a book rests on a table.
   c. a sled moves down a hill.
   d. paper blows in the wind.

4. When one team in a tug of war pulls harder on the rope than the other team does, the forces are
   a. active.
   b. balanced.
   c. magnetic.
   d. unbalanced.

5. The position of an object is its
   a. age.
   b. color.
   c. location.
   d. size.

6. The force that occurs when one object rubs against another object is
   a. friction.
   b. gravity.
   c. magnetism.
   d. weight.

Reading and Writing, p. 226

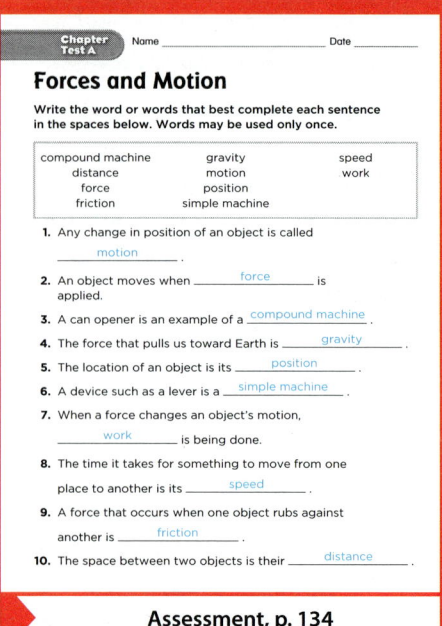

Forces and Motion

Write the word or words that best complete each sentence in the spaces below. Words may be used only once.

compound machine	gravity	speed
distance	motion	work
force	position	
friction	simple machine	

1. Any change in position of an object is called ___motion___.

2. An object moves when ___force___ is applied.

3. A can opener is an example of a ___compound machine___.

4. The force that pulls us toward Earth is ___gravity___.

5. The location of an object is its ___position___.

6. A device such as a lever is a ___simple machine___.

7. When a force changes an object's motion, ___work___ is being done.

8. The time it takes for something to move from one place to another is its ___speed___.

9. A force that occurs when one object rubs against another is ___friction___.

10. The space between two objects is their ___distance___.

Assessment, p. 134

Skills and Concepts

Answer each of the following in complete sentences.

11. **Problem and Solution** A car has just traveled 100 kilometers. What else do you need to know to figure out its average speed?

12. **Explanatory Writing** When does a roller coaster have the most potential energy? When does it have the most kinetic energy?

13. **Infer** Would you move faster down a water slide or a regular slide? Explain your answer.

14. **Critical Thinking** Both cars and people need energy. How is the way they get energy alike?

15. What simple machines are shown below? How are they the same? How are they different?

16. What makes something move?

Performance Assessment

Not-So-Simple Machines

Examples of Machines

▶ Machines help make work easier to do. Are there any tasks that you wish could be easier to do? Pick one such task and design an imaginary machine to help you do it.

▶ Write instructions that tell how the machine works. Draw a simple diagram of the machine.

▶ List some of the simple machines that make up your imaginary compound machine.

Test Prep

1. What two kinds of simple machines make up a pair of scissors?

   A wheels and axles
   B pulleys and wedges
   C levers and pulleys
   D levers and wedges

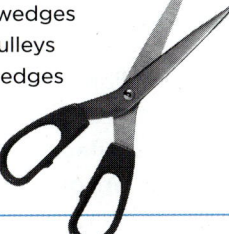

475

Summative Assessment and Intervention

Assessment Book provides a test for Chapter 11.

Leveled Readers may be used to reteach lesson content in an alternative format. The leveled readers deliver chapter content in different readability levels. The back cover of each reader provides comprehension building activities specific to the book content. (see p. 431)

Key Concept Cards 67–74 contain prescribed routines for student intervention.

Skills and Concepts

11. **Problem and Solution** You need to know the amount of time the car has been traveling.

12. **Explanatory Writing** When a roller-coaster moves down a hill it changes potential energy into kinetic energy. It has the most potential energy when it is at the top of the hill. It has the most kinetic energy when it is at the bottom of the hill.

13. **Infer** You would move faster down a water slide, because water reduces friction making the slide's surface less sticky and more slippery.

14. **Critical Thinking** Both need to get energy from another source. Cars need to take in gasoline and people need to take in food in order to move.

15. The axe is a wedge and the auger is a screw. Both are simple machines that use inclined planes. A wedge uses two inclined planes to change downward force into sideways force. A screw uses an inclined plane to change turning force into downward force.

The Big Idea **16.** Students should use information from the chapter to answer.

Performance Assessment

Scoring Rubric

4 Points Student has (1) selected a task and conceived of a machine that could make that task easier; (2) clearly described how the machine works; (3) created a simple diagram showing the basic operation of the machine; (4) identified at least one simple machine in their imaginary compound machine.

3 Points Student has met three of the four criteria.

2 Points Student has met two of the four criteria.

1 Point Student has met one of the four criteria.

Test Prep

1. D

Presentation Toolkit CD-ROM Lesson Presentations

 Instructional Navigator CD-ROM Interactive Lesson Planner, Teacher's Edition, Worksheets, and Online Resources

Lesson	OBJECTIVES AND READING SKILLS	VOCABULARY	RESOURCES AND TECHNOLOGY
1 Heat PAGES 478–485 **PACING:** 2 days **FAST TRACK:** 1 day	■ Describe how heat moves. ■ Compare insulators and conductors. 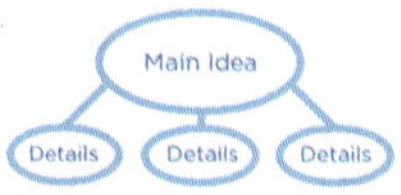 **Reading Skill** Main Idea and Details *Graphic Organizer 1*	heat thermal energy temperature thermometer conductor insulator	▶ Reading and Writing, pp. 229–232 ▶ Visual Literacy, pp. 75–76 ▶ Activity Lab Book, pp. 209–212 ▶ Transparencies, pp. 75–76
2 Sound PAGES 488–495 **PACING:** 2 days **FAST TRACK:** 1 day	■ Describe how vibrations produce sounds. ■ Compare the pitch and volume of a sound. 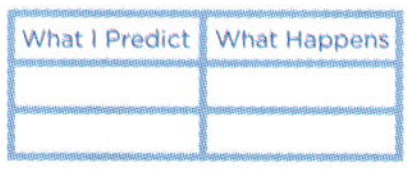 **Reading Skill** Predict *Graphic Organizer 3*	vibrate sound volume pitch	▶ Reading and Writing, pp. 233–236 ▶ Visual Literacy, pp. 77–78 ▶ Activity Lab Book, pp. 216–219 ▶ Transparencies, pp. 77–78 ▶ **Science in Motion,** *How You Hear Sounds*
3 Light PAGES 498–507 **PACING:** 2 days **FAST TRACK:** 1 day	■ Explore how light travels. ■ Describe how colors are seen. 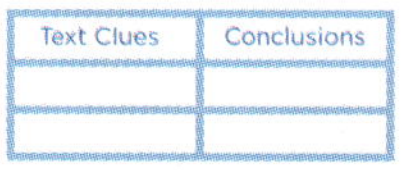 **Reading Skill** Draw Conclusions *Graphic Organizer 13*	light absorb reflect opaque shadow transparent translucent refract	▶ Reading and Writing, pp. 237–240 ▶ Visual Literacy, pp. 79–80 ▶ Activity Lab Book, pp. 224–227 ▶ Transparencies, pp. 79–80 ▶ **Operation: Science Quest,** *Light Energy* ▶ **Science in Motion,** *Seeing Colors*
4 Electricity PAGES 510–517 **PACING:** 2 days **FAST TRACK:** 1 day	■ Describe electrical charge. ■ Identify the parts of a circuit. 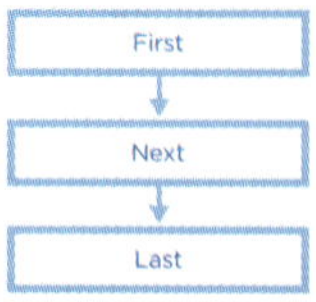 **Reading Skill** Sequence *Graphic Organizer 7*	electrical charge static electricity electric current circuit switch	▶ Reading and Writing, pp. 243–246 ▶ Math, pp. 19–20 ▶ Visual Literacy, pp. 81–82 ▶ Activity Lab Book, pp. 228–231 ▶ Transparencies, pp. 81–82

| **Chapter 12 Review** PAGES 520–521 | ■ Summarize chapter concepts. | **Resources**
 ▶ Assessment, pp. 148–151, 156–159
 ▶ Reading and Writing, pp. 249–250 | **Technology**
 ▶ **Progress Reporter** Assessments
 ▶ **e-Review** |

PACING Assumes a day is a 25- to 35-minute session

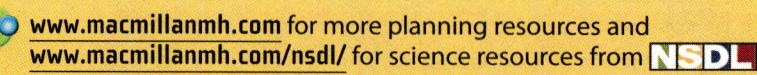

www.macmillanmh.com for more planning resources and
www.macmillanmh.com/nsdl/ for science resources from **NSDL**

Activity Planner

 Science Activity DVD Explore Activity demos

Materials included in the Activity Kit are listed in *italics*.

EXPLORE Activities

Explore p. 479 | PACING: 30 minutes

Objective Determine that air expands when heated.

Skills predict, observe, communicate, infer, experiment

Materials *dropper*, empty plastic bottle, water, *plastic disk*

⭐ **PLAN AHEAD** Chill bottles in a refrigerator or cooler until the activity is to begin.

Explore p. 489 | PACING: 30 minutes

Objective Demonstrate that movement causes sound.

Skills observe, infer, experiment

Materials goggles, paper, plastic ruler, *rubber band*, cardboard box

⭐ **PLAN AHEAD** Explain that students must wear goggles to prevent possible eye injuries caused by rubber bands or rulers.

Explore p. 499 | PACING: 30 minutes

Objective Demonstrate the reflection of light by a mirror.

Skills observe, experiment, communicate

Materials *mirror, flashlight*

⭐ **PLAN AHEAD** Acquire enough mirrors and flashlights before the class assembles.

Explore p. 511 | PACING: 30 minutes

Objective Demonstrate the flow of electricity in a simple circuit.

Skills experiment, communicate, infer

Materials *D-cell battery*, one 20-cm piece of insulated wire, light bulb

⭐ **PLAN AHEAD** Purchase flashlight light bulbs for the activity.

QUICK LAB Activities

Quick Lab p. 481 | PACING: 15 minutes

Objective Observe how different substances get hotter.

Skills predict, use variables, measure, experiment, use numbers, interpret data

Materials 2 *plastic cups*, water, *soil*, 2 *thermometers*, desk lamp

⭐ **PLAN AHEAD** Use a lamp with an incandescent light bulb as a heat source.

Quick Lab p. 493 | PACING: 15 minutes

Objective Observe how sounds change in pitch and volume.

Skills predict, experiment

Materials *plastic drinking straws*, scissors

⭐ **PLAN AHEAD** Remind students to use scissors carefully.

Quick Lab p. 505 | PACING: 15 minutes

Objective Observe how white light is made up of several colors.

Skills predict, observe

Materials *paper plates*, pencil, crayons

⭐ **PLAN AHEAD** Model how to fold the paper plate in half enough times to make eight equal sections.

Quick Lab p. 516 | PACING: 15 minutes

Objective Observe that only some materials are good conductors of electricity.

Skills experiment, observe, infer

Materials *battery*, battery holder, wire, light bulb, bulb socket, crayons, *paper clips*

⭐ **PLAN AHEAD** Make sure students correctly set up the circuit.

FOR MORE ACTIVITIES

Activity Flipchart for Science Centers and Workstations

Focus on Skills	Be a Scientist	Learning Lab	Everyday Science
Teach the inquiry skill: Experiment, p. 57	Sound movement, p. 59	Solar energy, pp. 74–75	Light, p. 68; Nocturnal Animals, p. 69

See **Activity Lab Book Teacher's Guide** for more support.

For a comprehensive list of consumable and non-consumable materials, see the back of the unit tab.

Technology

For additional language support and vocabulary development, go to **www.macmillanmh.com**

- **Science in Motion**
  How You Hear Sounds
  Seeing Colors
- **Vocabulary Games**

Academic Language

English language learners need help in building their understanding of the academic language used in daily instruction and science activities. The following strategies will help to increase students' language proficiency and comprehension of content and instruction words.

Strategies to Reinforce Academic Language

- **Use Context** Academic language should be explained in the context of the task. Use gestures, expressions, and visuals to support meaning.

- **Use Visuals** Use charts, transparencies, and graphic organizers to explain key labels to help students understand classroom language.

- **Model** Use academic language as you demonstrate the task to help students understand instruction.

Academic Language Vocabulary Chart

The following chart shows chapter vocabulary and inquiry skills as well as some Spanish cognates. Vocabulary words help students comprehend the main ideas. Inquiry Skills help students develop questions and perform investigations. Cognates are words that are similar in English and Spanish.

Vocabulary	Inquiry Skills	Cognates English	Spanish
heat, p. 480	predict, p. 479	predict, p. 479	*predecir*
temperature, p. 482	observe, p. 479	observe, p. 479	*observar*
thermal energy, p. 482	communicate, p. 479	communicate, p. 479	*comunicar*
thermometer, p. 483	infer, p. 479	infer, p. 479	*inferir*
conductor, p. 484	experiment, p. 479	experiment, p. 479	*experimentar*
insulator, p. 484		temperature, p. 482	*temperatura*
vibrate, p. 490		thermometer, p. 483	*termómetro*
sound, p. 490		conductor, p. 484	*conductor*
volume, p. 492		vibrate, p. 490	*vibrar*
pitch, p. 493		volume, p. 492	*volumen*
light, p. 500		absorb, p. 500	*absorber*
absorb, p. 500		opaque, p. 502	*opaco*
reflect, p. 501		transparent, p. 503	*transparente*
opaque, p. 502		translucent, p. 503	*translúcido*
shadow, p. 502		refract, p. 503	*refractar*
transparent, p. 503		static electricity, p. 513	*electricidad estática*
translucent, p. 503		electric current, p. 514	*corriente eléctrica*
refract, p. 503		circuit, p. 515	*circuito*
electrical charge, p 512			
static electricity, p. 513			
electric current, p. 514			
circuit, p. 515			
switch, p. 515			

Vocabulary Routine

Use the routine below to discuss the meaning of each word on the vocabulary chart. Use gestures and visuals to model all words.

Define: *Temperature* is a measure of how hot or cold something is.

Example: This water is cold. It has a *temperature* of 40°.

Ask: What tool would you use to check the *temperature* in a swimming pool?

Students may respond to questions according to proficiency level with gestures, one-word answers, or phrases.

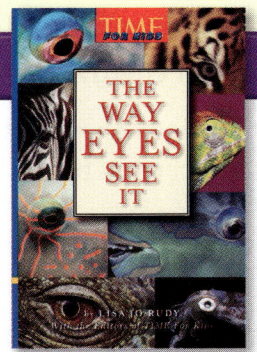

ELL Leveled Reader

The Way Eyes See It
by Lisa Jo Rudy

Summary Find out ways in which animals' eyes help them survive.

Reading Skill
Compare and Contrast

Vocabulary Activities

Help students understand thermal energy and temperature.

BEGINNING Refer to page 482. Point to the pictures and the labels. Ask: *Do you like iced tea or hot tea? Which has a higher temperature—a cup of hot water or a cup of cold water? Elicit examples and list them on a T-Chart.*

INTERMEDIATE Direct students' attention to page 482. On the board draw a cup of ice, a cup of water, and a cup of boiling water. Have students compare and contrast two of the cups, saying which one has more thermal energy or less thermal energy, and which one has a higher or lower temperature.

ADVANCED Direct students' attention to page 482. Have them draw two pictures. One item should have more thermal energy than the other. Students compare their pictures, explaining which one has more thermal energy. Ask: *Which one has a higher temperature?* Students pair one of their pictures with a picture from another student, and repeat the activity.

Language Transfers

Grammar Transfer
In Cantonese, Haitian Creole, Hmong, Korean, and Vietnamese, articles may be omitted:
Thermometer measures temperature.

Phonics Transfer
Cantonese, Hmong, and Korean do not have the /th/ sound in *thermal*.

CHAPTER 12

Forms of Energy

 THE BIG IDEA What are the main forms of energy? How are they used?

Chapter Preview Examine the graphics in the lesson. Predict what the lessons will be about.

▶ Assess Prior Knowledge

Before reading the chapter, create a **KWL** chart with students. Read the Big Idea question and then ask:

- How does heat move from one object to another?
- How are sounds made?
- How does light move?
- What is electricity?

Forms of Energy

What We **K**now	What We **W**ant to Know	What We **L**earned
Temperature is a measure of how hot or cold something is.	How does heat move?	
Sounds are made when an object vibrates.	What are pitch and volume?	
Light is a form of energy.		

Answers shown represent sample student responses.

Follow the **Instructional Plan** at right after assessing students' prior knowledge of chapter content.

RESOURCES and TECHNOLOGY

- **School to Home Activities,** pp. 99–106
- **Reading and Writing in Science,** pp. 228–250
- **Assessment,** pp. 148–161
- **Presentation Toolkit CD-ROM**
- **PuzzleMaker CD-ROM**
- **LOG ON** www.macmillanmh.com
- **LOG ON** e-Journal

Forms of Energy

Lesson 1
Heat 478

Lesson 2
Sound 488

Lesson 3
Light 498

Lesson 4
Electricity 510

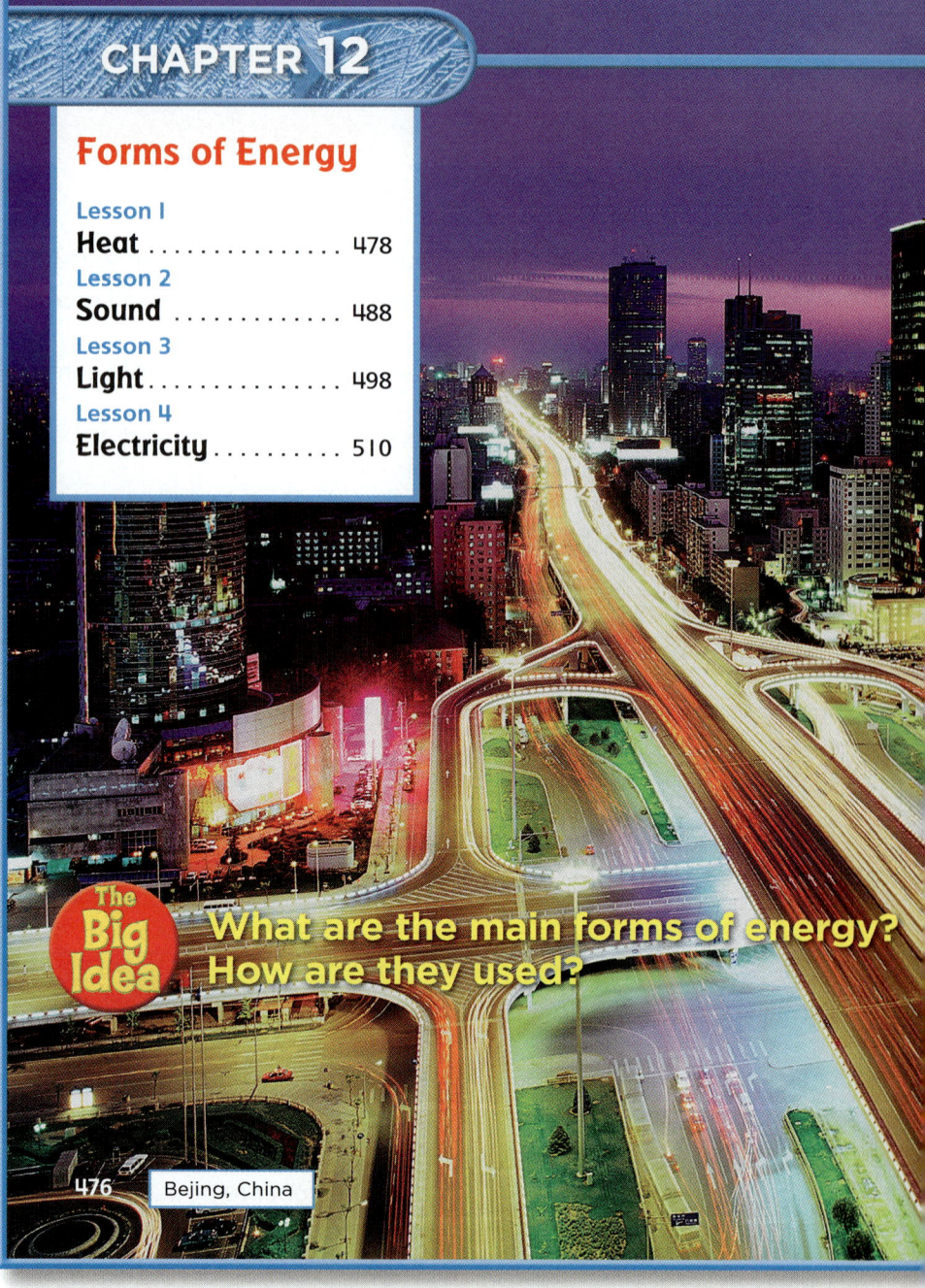

 **The Big Idea** What are the main forms of energy? How are they used?

476 Bejing, China

Differentiated Instruction

Instructional Plan

Chapter Concept Energy can change matter.

EXTRA SUPPORT Students who need to explain sounds as vibrations should cover all of **Lesson 2,** pp, 488–497.

ON LEVEL After exploring heat in **Lesson 1,** pp. 478–487, students who can explain sounds as vibrations can just cover what sounds move through, **Lesson 2,** pp, 494–495, and then go to **Lesson 3,** pp. 498–507, to describe interactions of light and matter, including colored light.

ENRICHMENT **Lesson 4,** pp. 508–517 builds on the topic of electricity from earlier grades by exploring how current electricity interacts with matter and produces heat and light.

Key Vocabulary

heat
energy that moves from a warmer object to a colder object (p. 480)

temperature
a measure of how hot or cold something is (p. 482)

vibrate
to move back and forth quickly (p. 490)

sound
the energy made by objects that vibrate (p. 490)

light
a form of energy that allows you to see objects (p. 500)

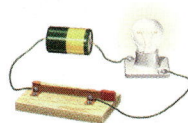
electric current
a flow of electrical charge (p. 514)

More Vocabulary

thermal energy, p. 482

thermometer, p. 483

conductor, p. 484

insulator, p. 484

volume, p. 492

pitch, p. 493

absorb, p. 500

reflect, p. 501

opaque, p. 502

shadow, p. 502

transparent, p. 503

translucent, p. 503

refract, p. 503

electrical charge, p. 512

static electricity, p. 513

circuit, p. 515

switch, p. 515

477

Vocabulary Preview

■ Have a volunteer read the **Key Vocabulary** words aloud to the class. Ask students to find one or two words in the chapter by using the given page references. Add these words and their definitions to a class "Word Wall."

■ Encourage students to use the illustrated glossary in the student edition's reference section. Guide students to explore the **e-Glossary**, which offers audio pronunciations, definitions, and sentences using the vocabulary words.

Science Leveled Readers

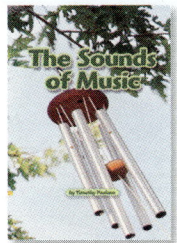
APPROACHING

The Sounds of Music Look at musical instruments and how they produce sounds.
ISBN: 0-02-284672-7

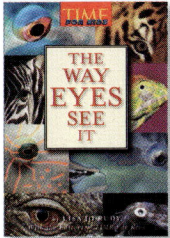
ON LEVEL

The Way Eyes See It Find out ways in which animals' eyes help them survive.
ISBN: 0-02-284676-X

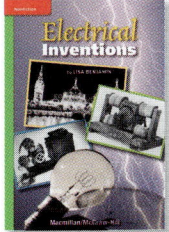
BEYOND

Electrical Inventions Meet great inventors and see their amazing inventions.
ISBN: 0-02-285939-X

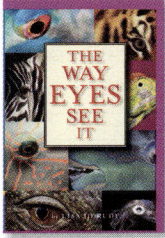
ELL

The Way Eyes See It Uses sheltered language of On-Level Reader.
ISBN: 0-02-283490-7

See teaching strategies in the Leveled Reader Teacher's Guide. To order, call 1-800-442-9685.

 Leveled Reader Database Online Readers, searchable by topic, reading level, and keywords

Plan Your Lesson

Lesson 1 Heat

Objective

- Describe how heat moves.
- Compare insulators and conductors.

Reading Skill Main Idea and Details

Main Idea
Details Details Details

Graphic Organizer 1, p. TR3

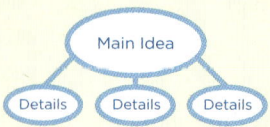
Professional Development Look for **NSDL** to find recommended Science Background materials from the National Science Digital Library

FAST TRACK

Lesson Plan When time is short, follow the Fast Track and use the essential resources.

1 Introduce
Look and Wonder, p. 478
Resource Alternative Explore, p. 211

2 Teach
Develop Vocabulary, p. 481
Discuss the Main Idea, p. 482
Resource Visual Literacy, p. 75

3 Close
Think, Talk, and Write, p. 485
Resource Assessment, p. 152

▶ Reading and Writing

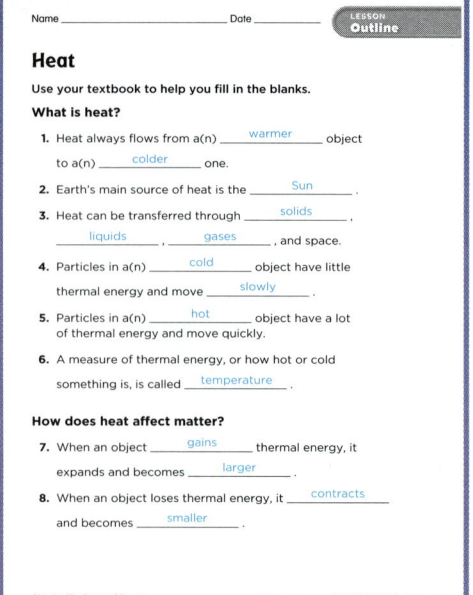

Outline, pp. 229–230
Also available as a student workbook

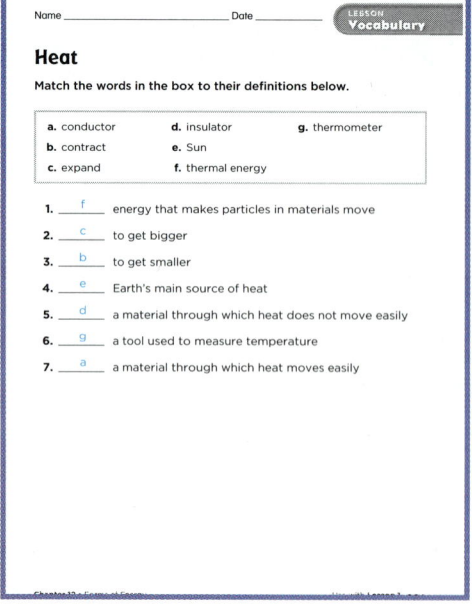

Vocabulary, p. 231
Also available as a student workbook

▷ Visual Literacy

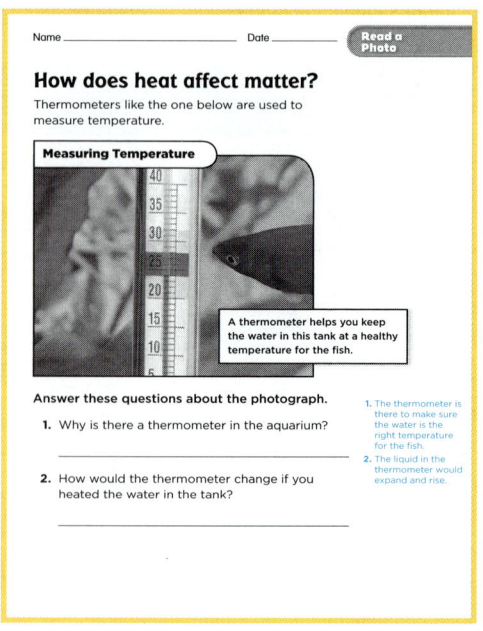

Read a Photo, p. 75
Also available as a transparency

▶ Activity Lab Book

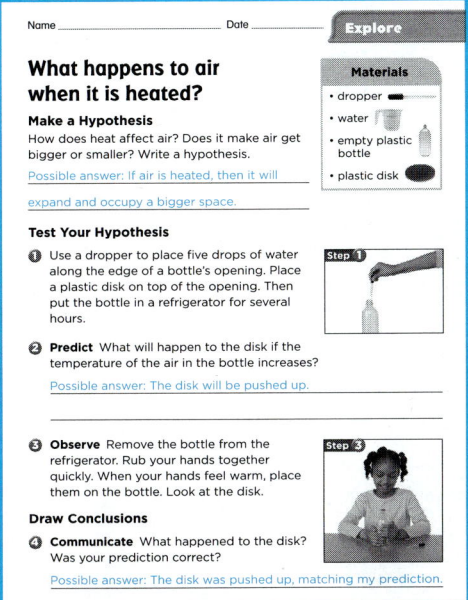

Explore

Name _____ Date _____

What happens to air when it is heated?

Make a Hypothesis

How does heat affect air? Does it make air get bigger or smaller? Write a hypothesis.

Possible answer: If air is heated, then it will expand and occupy a bigger space.

Materials
- dropper
- water
- empty plastic bottle
- plastic disk

Test Your Hypothesis

❶ Use a dropper to place five drops of water along the edge of a bottle's opening. Place a plastic disk on top of the opening. Then put the bottle in a refrigerator for several hours.

Step 1

❷ **Predict** What will happen to the disk if the temperature of the air in the bottle increases?

Possible answer: The disk will be pushed up.

❸ **Observe** Remove the bottle from the refrigerator. Rub your hands together quickly. When your hands feel warm, place them on the bottle. Look at the disk.

Step 3

Draw Conclusions

❹ **Communicate** What happened to the disk? Was your prediction correct?

Possible answer: The disk was pushed up, matching my prediction.

Explore, pp. 209–210
Also available as a student workbook

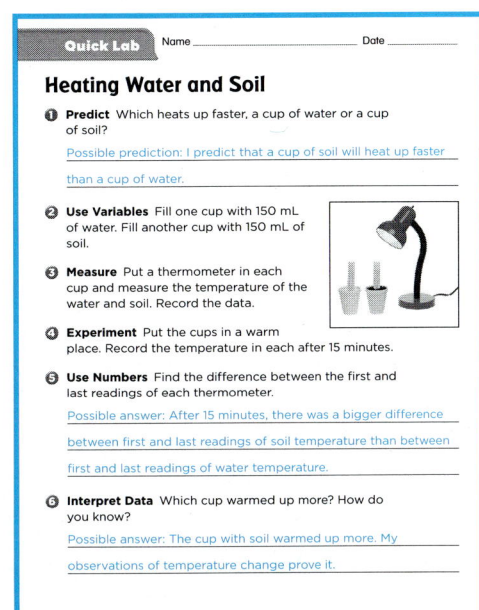

Quick Lab Name _____ Date _____

Heating Water and Soil

❶ **Predict** Which heats up faster, a cup of water or a cup of soil?

Possible prediction: I predict that a cup of soil will heat up faster than a cup of water.

❷ **Use Variables** Fill one cup with 150 mL of water. Fill another cup with 150 mL of soil.

❸ **Measure** Put a thermometer in each cup and measure the temperature of the water and soil. Record the data.

❹ **Experiment** Put the cups in a warm place. Record the temperature in each after 15 minutes.

❺ **Use Numbers** Find the difference between the first and last readings of each thermometer.

Possible answer: After 15 minutes, there was a bigger difference between first and last readings of soil temperature than between first and last readings of water temperature.

❻ **Interpret Data** Which cup warmed up more? How do you know?

Possible answer: The cup with soil warmed up more. My observations of temperature change prove it.

Quick Lab, p. 212
Also available as a student workbook

▶ Assessment

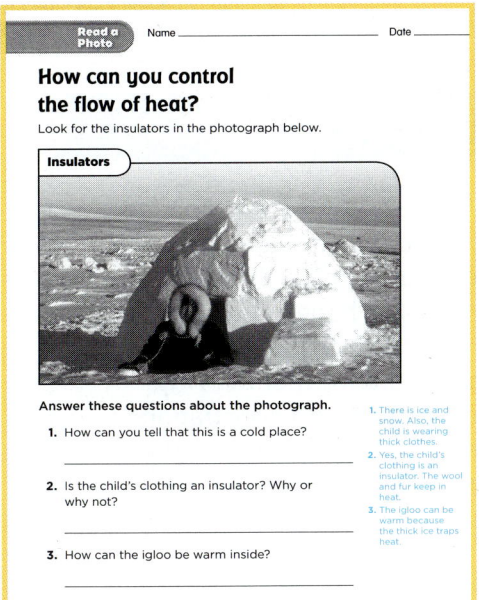

Read a Photo Name _____ Date _____

How can you control the flow of heat?

Look for the insulators in the photograph below.

Insulators

Answer these questions about the photograph.

1. How can you tell that this is a cold place?

2. Is the child's clothing an insulator? Why or why not?

3. How can the igloo be warm inside?

1. There is ice and snow. Also, the child is wearing thick clothes.
2. Yes, the child's clothing is an insulator. The wool and fur keep in heat.
3. The igloo can be warm because the thick ice traps heat.

Read a Photo, p. 76
Also available as a transparency

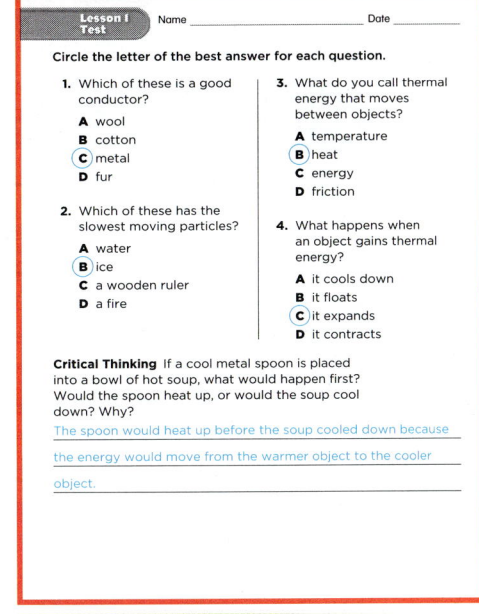

Lesson 1 Test Name _____ Date _____

Circle the letter of the best answer for each question.

1. Which of these is a good conductor?
   - **A** wool
   - **B** cotton
   - **C** metal ⬅
   - **D** fur

2. Which of these has the slowest moving particles?
   - **A** water
   - **B** ice ⬅
   - **C** a wooden ruler
   - **D** a fire

3. What do you call thermal energy that moves between objects?
   - **A** temperature
   - **B** heat ⬅
   - **C** energy
   - **D** friction

4. What happens when an object gains thermal energy?
   - **A** it cools down
   - **B** it floats
   - **C** it expands ⬅
   - **D** it contracts

Critical Thinking If a cool metal spoon is placed into a bowl of hot soup, what would happen first? Would the spoon heat up, or would the soup cool down? Why?

The spoon would heat up before the soup cooled down because the energy would move from the warmer object to the cooler object.

FAST TRACK **Lesson Test, p. 152**
Also available as a student workbook

ADDITIONAL RESOURCES

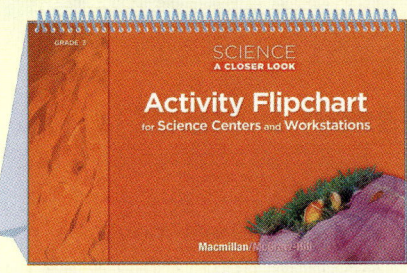

p. 56

p.139

pp. 75–76

pp. 99–100

75–76

176–181

Technology

- 💿 **Science Activity DVD**
- 💿 **Instructional Navigator CD-ROM**
- 💿 **Presentation Toolkit CD-ROM**
- **e-Review**
- **NSDL**

Lesson 1 Heat

Objectives
- Describe how heat moves.
- Compare insulators and conductors.

1 Introduce

▶ Assess Prior Knowledge

Have students observe the photo of the balloon on page 478. Explain that this type of balloon is a hot-air balloon. Ask:

- **Why do you think this type of balloon is called a hot-air balloon?** Possible answer: The heat from the fire makes the air inside the balloon hot.

- **Why do you think this balloon is able to get off the ground?** Possible answer: The warm air inside the balloon rises and carries the balloon with it.

Look and Wonder

Invite students to share their responses to the Look and Wonder statement and question:

- **What happens to air as it is heated?** Possible answer: The air inside the balloon expands, or takes up more space.

Write ideas on the board and note any misconceptions that students might have. Address these misconceptions as you teach the lesson.

RESOURCES and TECHNOLOGY
- ▶ **Activity Lab Book**, pp. 209–211
- ▶ **Activity Flipchart**, p. 56
- 💿 **Science Activity DVD**

Lesson 1

Heat

Look and Wonder

Hot air causes these balloons to rise into the sky. What happens to air as it is heated?

478
ENGAGE

Warm Up

Start with a Demonstration

Students should work in pairs. Have one student hold out his or her hands palms up. Have a partner place his or her hands palms down on the first student's hands. Ask students to describe the difference in temperature between the pairs of hands. Possible answer: My hands are warmer than my partner's hands. Ask students to repeat the demonstration after one of the them has rubbed his or her hands together briskly for five seconds. Students should observe that the rubbed hands are much warmer.

Explore

What happens to air when it is heated?

Form a Hypothesis

How does heat affect air? Does it make air get bigger or smaller? Write a hypothesis.

Test Your Hypothesis

1 Use a dropper to place five drops of water along the edge of a bottle's opening. Place a plastic disk on top of the opening. Then put the bottle in a refrigerator for several hours.

2 **Predict** What will happen to the disk if the temperature of the air in the bottle increases?

3 **Observe** Remove the bottle from the refrigerator. Rub your hands together quickly. When your hands feel warm, place them on the bottle. Look at the disk.

Draw Conclusions

4 **Communicate** What happened to the disk? Was your prediction correct?

5 **Infer** Think about what happened to the disk. What happens to air when it is heated?

Explore More

Experiment Place an empty plastic bottle in the refrigerator for several hours. Remove the bottle from the refrigerator and immediately stretch a balloon over the opening. What happens to the balloon?

Materials

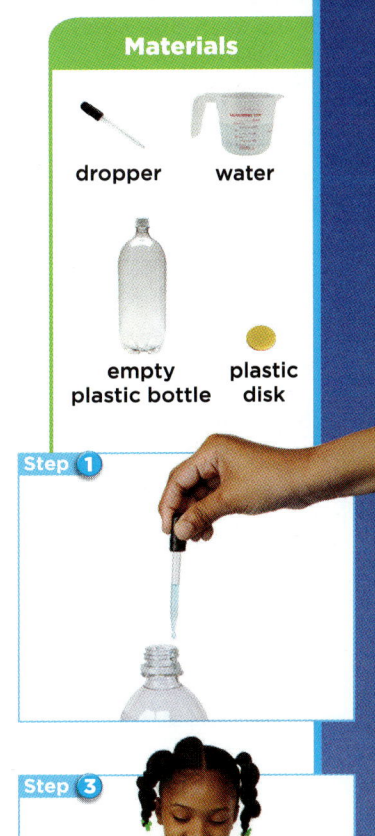

dropper   water

empty plastic bottle   plastic disk

Step **1**

Step **3**

479
EXPLORE

Explore

small groups   30 minutes

Plan Ahead Place the bottles in the refrigerator for a few hours before you begin the activity. Then the chilled air in the bottle will warm in the ambient room air, and the disk will bounce nicely.

Purpose In this activity, students observe that warm air expands and rises.

Structured Inquiry

Make a Hypothesis Possible hypothesis: If air is heated, then it expands and rises.

Test Your Hypothesis

2 **Predict** Possible prediction: The plastic disk will be pushed off of the bottle.

4 **Communicate** Students will observe that the disk is pushed up. This observation may or may not match student predictions.

5 **Infer** Possible answer: When air is heated, it expands, or gets bigger, and it rises.

If students' experiments did not support their hypotheses, encourage them to develop new hypotheses.

Guided Inquiry   Explore More

Experiment The balloon will expand as the temperature increases.

Open Inquiry

Ask students what would happen to the balloon if the plastic bottle were placed in a sunny window. Have them think of their own question about what would happen to the balloon. Then have them make a plan and carry out an experiment to answer the question. Ask:

Would the balloon expand if the plastic jar were placed by a sunny window?

Alternative Explore

Do your hands feel warm?

Materials 3 plastic bowls, tap water, ice, thermometer

Have students fill the first bowl with hot tap water, the second bowl with lukewarm tap water, and the third bowl with cool tap water and ice cubes. **Be Careful!** Caution students not to make the hot water so hot that it can burn their skin. Students will compare how their hands feel in different temperatures of water.

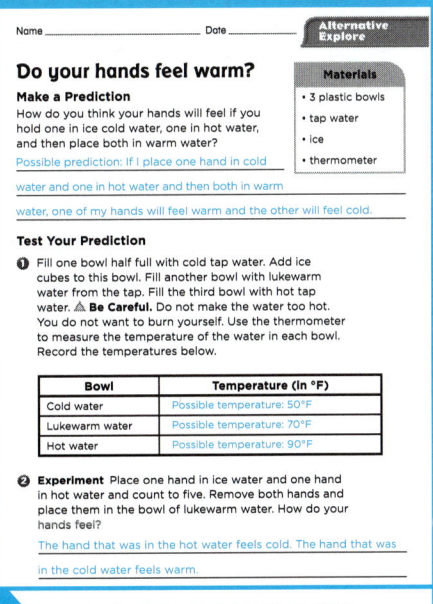

Name _____ Date _____   Alternative Explore

Do your hands feel warm?   Materials

Make a Prediction

How do you think your hands will feel if you hold one in ice cold water, one in hot water, and then place both in warm water?

Possible prediction: If I place one hand in cold water and one in hot water and then both in warm water, one of my hands will feel warm and the other will feel cold.

Test Your Prediction

1 Fill one bowl half full with cold tap water. Add ice cubes to this bowl. Fill another bowl with lukewarm water from the tap. Fill the third bowl with hot tap water. **Be Careful.** Do not make the water too hot. You do not want to burn yourself. Use the thermometer to measure the temperature of the water in each bowl. Record the temperatures below.

Bowl	Temperature (in °F)
Cold water	Possible temperature: 50°F
Lukewarm water	Possible temperature: 70°F
Hot water	Possible temperature: 90°F

2 **Experiment** Place one hand in ice water and one hand in hot water and count to five. Remove both hands and place them in the bowl of lukewarm water. How do your hands feel?

The hand that was in the hot water feels cold. The hand that was in the cold water feels warm.

• 3 plastic bowls
• tap water
• ice
• thermometer

Activity Lab Book, p. 211

2 Teach

Read and Learn

Main Idea Have students preview the visuals and then write one question they think will be answered in this lesson.

Vocabulary Have students read aloud the vocabulary words. Ask volunteers to give definitions. Make a list of vocabulary words and student definitions on the board.

Reading Skill **Main Idea and Details** Graphic Organizer 1

Have students fill in a Main Idea and Details graphic organizer as they read each two pages of the lesson. They can use the Quick Check questions to identify each main idea and its details.

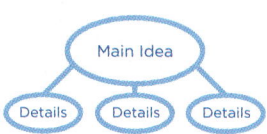

What is heat?

▶ Discuss the Main Idea

Show students a variety of thermometers, such as those used in refrigerators, ovens, gardens, and with humans. Ask:

- **What are these instruments called?** thermometers

- **What are they used for?** Possible answers: They tell how hot or cold something is. They tell the temperature of something.

FAST TRACK **Develop Vocabulary**

heat *Scientific vs. Common Use* Explain to students that the word *heat* has meanings other than the amount of thermal energy in an object. Among others, *heat* can mean a specific part of a race or preparing a horse for a race.

RESOURCES and TECHNOLOGY

▶ **Reading and Writing,** pp. 229–231

▷ **Visual Literacy,** p. 75

🖬 **PuzzleMaker CD-ROM**

🖬 **Presentation Toolkit CD-ROM**

 **e-Glossary**

Read and Learn

▶ Main Idea
Heat affects matter in many ways. Heat always moves from warmer objects to cooler objects.

▶ Vocabulary
heat, p. 480
thermal energy, p. 482
temperature, p. 482
thermometer, p. 483
conductor, p. 484
insulator, p. 484

LOG ON **e-Glossary**
at www.macmillanmh.com

▶ Reading Skill ✓
Main Idea and Details

The Sun's energy warms the air, land, and water.

What is heat?

Have you ever put your hands on a bowl of hot soup? What happened? Your hands got warm. Heat moved from the hot bowl to your cool hands. **Heat** is a form of energy that moves between objects. Heat can travel through solids, liquids, and gases. It can even travel through space. No matter what it travels through, heat always flows from a warmer object to a cooler one.

Sources of Heat

The Sun is Earth's main source of heat. A *source* is where something comes from. The Sun's heat warms the air, land, and water. Without the Sun's heat, it would be too cold on Earth for most living things to survive.

480
EXPLAIN

Science Background

Galileo is credited with inventing the first thermometer in 1593. That instrument, called a *thermoscope,* is a sealed glass tube filled with liquid and glass spheres. The glass spheres also are filled with liquid, and they rise and fall within the tube as the temperature changes. In 1612, the Italian scientist Santorio designed the first thermometer with a scale. During the mid-1600s, the Grand Duke of Tuscany invented the first thermometer using a liquid sealed within a glass tube.

See **Science Yellow Pages** in the Teacher Resource section for background information.

LOG ON **Professional Development** For more Science Background and resources from **NSDL** visit **www.macmillanmh.com/nsdl/**

How can a simple machine help you lift objects?

Materials

- clay
- thick marker
- ruler
- 2 small cups
- large blocks
- one-gram cubes

Form Your Hypothesis

Look at the photo of step 2. Will moving the ruler's position on the marker change the amount of force needed to lift two blocks? Write a hypothesis.

_____

_____

_____

_____

Test Your Hypothesis

Step 2

1. Use some clay to stick a marker to the center of a ruler. Then use clay to stick a small cup to each end of the ruler as shown in the photo.

2. **Experiment** Put two large blocks in one cup. Add gram cubes to the other cup. How many cubes does it take to lift the two large blocks?

_____

3. **Use Variables** Change the position of the marker. Move it closer to one end of the ruler.

4. Repeat step 2. How does the marker's new position change your results?

© Macmillan/McGraw-Hill

Name _____ Date _____

Draw Conclusions

⑤ Communicate How does this machine lift objects?

_____

_____

⑥ Interpret Data How does the position of the marker change the number of gram cubes you need to lift the two large blocks?

_____

_____

Explore More

Experiment When are the two blocks lifted higher in the air—when the marker is near the two large blocks or when it is near the gram cubes? Try to find out.

_____

_____

Open Inquiry

How is the distance a force moves an object related to the amount of force? Think of your own question about this relationship. Make a plan and carry out an experiment to answer the question.

My question is: _____

_____

How I can test it: _____

_____

_____

My results are: _____

_____

Quick Lab

Heating Water and Soil

1 **Predict** Which heats up faster, a cup of water or a cup of soil?

2 **Use Variables** Fill one cup with 150 mL of water. Fill another cup with 150 mL of soil.

3 **Measure** Put a thermometer in each cup and measure the temperature of the water and soil. Record the data.

4 **Experiment** Put the cups in a warm place. Record the temperature in each after 15 minutes.

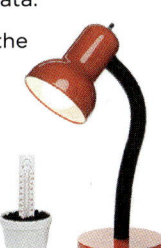

5 **Use Numbers** Find the difference between the first and last readings of each thermometer.

6 **Interpret Data** Which cup warmed up more? How do you know?

Fires, light bulbs, and stoves are some other sources of heat. Fires use chemical changes to produce heat. Light bulbs and some stoves use electricity to produce heat. Rubbing two objects together can also produce heat. This is why your hands get warm when you rub them together.

Heating Objects

Some objects heat up faster than others. For example, at the beach you will find sand and water. Both are warmed by the Sun. The sand gets very hot, but the water stays much cooler.

✓ Quick Check

Main Idea and Details Describe how heat flows.

Critical Thinking What are some ways people use heat?

481
EXPLAIN

Quick Lab

whole class   15 minutes

Objective Observe how different substances heat up.

Materials 2 plastic cups, water, soil, 2 thermometers

1 Possible prediction: A cup of soil will heat up faster.

3 Record the initial temperature of each cup in a data table on the board.

4 Be sure that the two cups are placed at an equal distance from the heat source. Record the temperature of each cup in the data table.

5 Review with students the correct way to set up the subtraction problems. Have volunteers perform the calculations for each set of temperature readings.

6 The soil, because its final temperature was higher than the final temperature of the water.

▶ Use the Visuals

Refer students to the visual shown on pages 480–481. Ask:

- **What is the source of energy that heats Earth's surface?** the Sun

- **What happens to particles of water and sand as they are warmed by the Sun?** They move faster.

✓ Quick Check Answers

- **Main Idea and Details** Heat travels through matter. It always flows from warmer objects to cooler objects.

- **Critical Thinking** Possible answer: to heat houses, to cook food, to run engines

Differentiated Instruction

Leveled Activities

EXTRA SUPPORT Provide students with a variety of thermometers. Have them take the temperature of several common classroom objects and record their findings in a data table or chart.

ENRICHMENT Have students use encyclopedias or the Internet, if available, to research the history of the thermometer. Students might want to research the work and scientific contributions of individuals such as Gabriel Fahrenheit, Anders Celsius, and Sir William Thomson Kelvin. Have students write and illustrate a brief report on their research to share with the class.

How does heat affect matter?

FAST TRACK **Discuss the Main Idea**

Lead a discussion with students about thermal contraction and expansion. Point out that when most matter is heated, it expands, or gets bigger, and when most matter cools, it contracts, or gets smaller. Ask:

- **Why do you think a balloon would shrink if you put it in a freezer?** Possible answer: The air in the balloon would contract.

- **A coach might tell you to store a basketball inside during cold weather. Explain.** Possible answer: If the ball is left outside, it will seem flat. The air inside the ball contracts more than the material the ball is made of.

▶ **Develop Vocabulary**

thermal energy *Word Origin* Explain to students that thermal energy is heat energy. Point out that the word *thermal* is from the Greek word **therme,** which means "heat."

temperature *Word Origin* Point out to students that the word *temperature* comes from the Latin word **temperatura,** which means "a tempering, a moderation." Ask students how this meaning might relate to the measure of how hot or cold something is.

thermometer *Word Origin* Tell students that the word *thermometer* comes from two Greek words: **thermos,** which means "hot," and **metron,** which means "measure." .

RESOURCES and TECHNOLOGY

▶ Activity Lab Book, p. 212

🖢 Presentation Toolkit CD-ROM

How does heat affect matter?

Remember that all matter is made of tiny particles. These particles are always moving. The energy that makes them move is called **thermal energy** (THUR•muhl EN•uhr•jee). Heating matter increases how much thermal energy the particles have. A hot object, such as hot soup, has a lot of thermal energy. Its particles move quickly. A cold object, such as an ice cube, has much less thermal energy. Its particles move slowly.

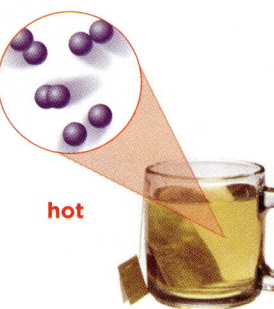
hot

cold

Thermal energy is what makes objects feel hot or cold. In fact, when you measure an object's temperature (TEM•puhr•uh•chuhr), you are really measuring its thermal energy. **Temperature** is a measure of how hot or cold something is. It describes how much thermal energy an object has. The more thermal energy an object has, the higher its temperature will be.

Measuring Temperature

A thermometer helps you keep the water in this tank at a healthy temperature for the fish.

482
EXPLAIN

ELL Support

Ask Questions Ask students questions such as the following to make sure they understand the concept of heat: **What is heat?** energy that moves from a warmer object to a colder object **What happens when a warm object touches a cool object?** Heat is transferred from the warm object to the cool object. **What will eventually happen to these two objects?** They will eventually be the same temperature.

BEGINNING Students use single words to answer, or point to the answers to the questions on pages 483–484.

INTERMEDIATE Students answer the questions using short phrases or simple sentences.

ADVANCED Students answer the questions using complete, grammatically correct sentences.

Roads can crack as they expand and contract with changing temperatures.

ENGAGE EXPLORE **EXPLAIN** EVALUATE EXTEND

Expanding and Contracting

When heat flows into an object, the object gains thermal energy. Its temperature increases. Its particles move faster and farther apart. The object gets bigger, or *expands*. When heat flows away from an object, the object loses thermal energy. Its temperature decreases. Its particles move more slowly. The object gets smaller, or *contracts*.

You can see matter expand or contract in a thermometer (thuhr•MOM•i•tuhr). A **thermometer** is a tool used to measure temperature. Some thermometers are made up of a clear tube filled with a liquid. When the temperature of the liquid increases, the liquid expands. It rises and fills more of the tube. When the temperature of the liquid decreases, the liquid contracts. It fills less of the tube.

Changing State

Heat can cause matter to change state. Solids, such as ice cream, can melt when they are heated. Liquids, such as water, can evaporate when they are heated. They can freeze when heat flows from them.

Read a Photo

What is the temperature shown on the thermometer? Give your answer in ˚C.

Clue: Line up the top of the red liquid with the black markings on the thermometer.

✓ Quick Check

Main Idea and Details List some ways heat affects matter.

Critical Thinking What happens to an ice pop when it is out of the freezer?

483
EXPLAIN

▶ Address Misconceptions

Some students might think that as heated particles move faster and farther apart, the particles themselves become bigger. Inform students that the particles stay the same size regardless of the amount of heat added to the matter. Only the spaces between the particles increase in size.

Read a Photo

Answer 24°C

✓ Quick Check Answers

• **Main Idea and Details** Possible answers: heat added to matter—particles in matter move faster and farther apart; heat removed from matter—particles in the matter slow down and contract; heat removed or added—matter may change state

• **Critical Thinking** It begins to change state, or melt, because of the addition of heat.

Differentiated Instruction

Leveled Questions

EXTRA SUPPORT How does heat move? Heat moves from a warmer object to a cooler object.

ENRICHMENT On a hot summer day, you put ice cubes into a glass of warm water. How does heat move to make the ice melt? Heat moves from the warm water to the cold ice cubes, causing them to change state by melting.

How can you control the flow of heat?

▶ Use the Visuals

Refer students to the visuals on page 484. Ask:

- **Why are cooking pots made of metal?** Metal is a good conductor of heat.

- **What kind of material is a blanket?** A blanket is an insulator.

- **Why would someone add insulation to a house when it is being built?** Insulation helps to keep warm air inside the house during the winter. It also helps to keep warm air outside the house during the summer.

Develop Vocabulary

conductor *Scientific vs. Common Use* Explain to students that the word *conductor* in common use refers to the person who directs an orchestra or the person who collects the fare or tickets on a bus or train.

insulator *Word Origin* Point out to students that the word *insulator* comes from the Latin word *insulatus,* which means "make into an island." Explain that an island is insulated, or separated, from other land by a body of water, and that an insulator separates heat from an object.

✔ Quick Check Answers

- **Main Idea and Details** Possible answers: A conductor is a material that heat moves through easily. Metal is a conductor. An insulator is a material that heat does not move through easily. Wool, cotton, and fur are examples of insulators.

- **Critical Thinking** The foam is a good insulator. It keeps the hot chocolate warm longer. The foam cup also keeps your hands from burning.

How can you control the flow of heat?

Heat moves more easily through some materials than others. That is why pots are often made of metal. Heat moves easily through metals. Heat moves from the stove to the metal pot. The whole pot gets warm. Materials such as metals are good conductors (kuhn•DUK•tuhrz). A **conductor** is a material that heat moves through easily.

When you are cold, you wrap yourself in a blanket to keep warm. A blanket is an insulator (IN•suh•lay•tuhr). An **insulator** is a material that heat does not move through easily. Wool, cotton, and fur are examples of insulators.

▲ Metal pots are conductors.

✔ Quick Check

Main Idea and Details What is a conductor? What is an insulator? Give an example of each.

Critical Thinking Why do people use foam cups when drinking hot chocolate?

Snow can be an insulator. Heat cannot flow easily through the walls of this igloo.

484
EXPLAIN

Homework Activity

Thermal Images

Ask students to use encyclopedias, magazines, or the Internet, if available, to find a picture of a thermal image of a house or another kind of building, a person, or the landscape. Have students write a brief report explaining what the thermal image is and what it shows. Encourage students to share their results with their classmates.

Lesson Review

Visual Summary

Energy that moves from a warmer object to a colder object is called heat.

Heat affects matter in many ways.

Heat moves easily through **conductors**. Heat does not move easily through **insulators**.

Make a  **Study Guide**

Make a Three-Tab Book. Use it to summarize what you learned about heat.

Temperature is...

Heat is...

Conductors and insulators are...

Think, Talk, and Write

1 Main Idea How does heat move?

2 Vocabulary How is temperature different from heat?

3 Main Idea and Details How does matter change when heat flows into it?

Main Idea

Details   Details   Details

4 Critical Thinking Sometimes when people have a fever, they put a cold wet cloth on their forehead. How does this cool them off?

5 Test Prep Most of Earth's heat comes from

A the Sun.
B water.
C batteries.
D electricity.

 Math Link

Solve a Problem
Yesterday the temperature outside was 21°C. Overnight the temperature dropped 8 degrees. What is the new temperature?

Health Link

Do Research
Fires are a source of heat. However, fires can be dangerous. Find out ways to prevent fires. Make a pamphlet to share this information.

 e-Review Summaries and quizzes online at www.macmillanmh.com

485
EVALUATE

Formative Assessment

Approaching Have students make flash cards of the vocabulary words in this lesson. Ask students to write the term on one side of a card and the definition on the other side. Ask them to save their cards until they finish the chapter.

On-Level Have students write an outline of each section of the lesson, making sure to include all the key ideas and vocabulary words from each section.

Challenge Ask students to make an illustrated poster that lists and describes different kinds of conductors and insulators found in their homes or in the school. Encourage students to share their findings with the class.

Key Concept Cards For student intervention, see the prescribed routine on **Key Concept Cards 75 and 76.**

---

3 Close

Lesson Review

▶ **Visual Summary**

Have students save their lesson Study Guides. They will use them at the end of the chapter for review.

▶ **Make a** FOLDABLES **Study Guide**

See pages R27–R28 for instructions.

Think, Talk, and Write

FAST TRACK

1 Main Idea Heat always moves from warmer objects to cooler objects.

2 Vocabulary Temperature is a measure of how hot or cold something is, whereas heat is the flow of thermal energy through matter.

3 Main Idea and Details

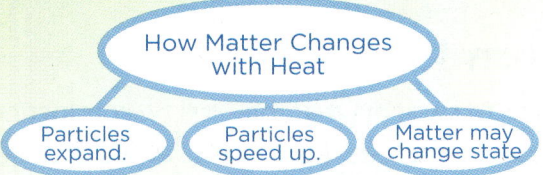

How Matter Changes with Heat

Particles expand.   Particles speed up.   Matter may change state.

4 Critical Thinking Heat moves from the hot forehead to the cooler cloth. The cloth gets warm and the forehead gets cooler.

5 Test Prep A

Math Link 13º C

Health Link Accept all reasonable pamphlets based on accurate information.

RESOURCES and TECHNOLOGY

▶ **Reading and Writing in Science,** p. 232

▶ **School to Home Activities,** pp. 99–100

e -Review Narrated Summary and Quiz

Progress Reporter Assessments

Lesson 1   **485**

Focus on Skills

Objective

- Experiment to compare the effectiveness of three different materials as insulators.

Materials paper cup, plastic cup, foam cup, 6 ice cubes, plastic wrap, 3 rubber bands, stopwatch

Plan Ahead Gather enough materials for each small group. Paper, plastic, and foam cups should be the same size. The six ice cubes also should be the same size. Allow enough time for students to make observations every 10 minutes for an hour.

EXTEND This activity will guide students through an experiment that tests different materials for their insulating qualities.

Inquiry Skill: Experiment

▶ Learn It

- Explain to students that they will be conducting an experiment to test the effectiveness of three materials used in cups as insulators. Ask:

- **What is the function of a cooler?** It keeps cold things cold.

- **Does a cooler function as an insulator or a conductor?** an insulator

- **Why do you think that it is important to use the same size cups when testing them for their insulating qualities?** to make sure that cup size isn't a factor in the results of the experiment

▶ Try It

1. Possible hypothesis: The foam cup will keep the ice cubes solid the longest.

2. Make sure that the ice cubes are the same size.

4. Cups can be placed on a sunny windowsill, under a lamp, or near a heater.

Focus on Skills

Inquiry Skill: Experiment

You just learned about heat. You read that an insulator is a material that does not allow heat to pass through it easily. How can you find out if something is an insulator? You can **experiment** to answer a question like this.

▶ Learn It

When you **experiment**, you perform tests to answer a question. You make observations and collect data. Then you interpret the data to answer a question. When you experiment, it is important to test only one thing at a time. This helps you know what caused your results.

▶ Try It

Experiment to find out which is the best insulator: paper, plastic, or foam.

Materials paper cup, plastic cup, foam cup, 6 ice cubes, plastic wrap, 3 rubber bands

1. Which material do you think will keep the ice cubes solid the longest: paper, plastic, or foam? Write a hypothesis.

2. Put two ice cubes into each cup.

3. Cover each cup with plastic wrap. Use a rubber band to seal the wrap to the cup.

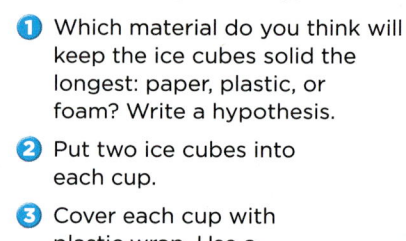

486
EXTEND

Integrate Reading

How Do People Stay Warm in Cold Climates?

Have students use an encyclopedia or the Internet, if available, to read about how Inuit people and others who live in cold areas stay warm. Ask students to read about what kinds of clothes these people wear, what kinds of homes or shelters they live in, and anything other ways they stay warm. Ask:

- **What kinds of clothes do Inuit people wear to keep them warm?** Possible answers: fur, wool, layers

- **What kinds of homes do people in cold climates live in to keep warm?** Possible answer: homes with good insulation

- **How does an igloo keep a person warm?** The ice of the igloo acts as an insulator and keeps heat inside the igloo.

Tell students to write and illustrate a report of their findings to share with the class.

④ Place the cups in a warm place.

⑤ Observe the ice in the cups every ten minutes for one hour. Record the changes that you observe.

Now use your results to draw conclusions.

▶ In which cup did the ice cubes melt the slowest?

▶ Which cup is the best insulator?

▶ Apply It

Now **experiment** to find out which is the best conductor of heat: aluminum, plastic, or wax paper. Remember that a conductor is a material that lets heat pass through it easily.

Repeat this experiment using three different types of wraps and three paper cups. Wrap aluminum foil around one cup, plastic wrap around the second cup, and wax paper around the third cup. Remember to record your observations.

◀ A cooler is an insulator. It keeps your food from getting warm.

487
EXTEND

Students should conclude that the ice cubes melted the slowest in foam cup. Therefore, the foam cup is the best insulator.

▶ Apply It

Tell students that conductors are materials that allow heat to pass through them easily. Have students design an experiment about conductors and collect the materials they will need for the experiment. Then have them work in small groups to conduct the experiment. Encourage students to make observations every 10 minutes for an hour and to record their observations in a data table of their own design.

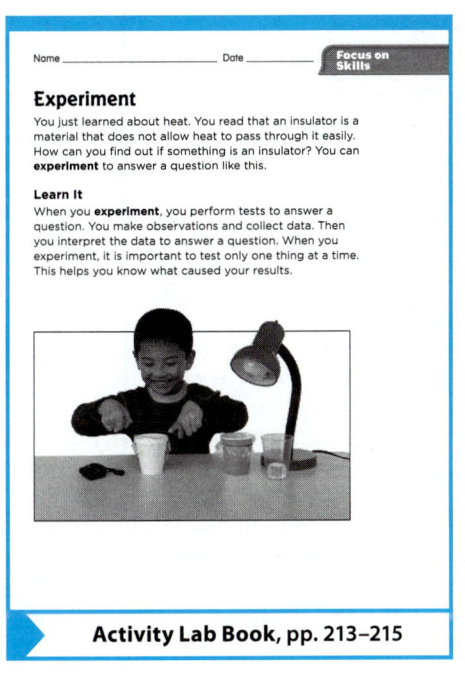

Name _____ Date _____ | Focus on Skills

Experiment

You just learned about heat. You read that an insulator is a material that does not allow heat to pass through it easily. How can you find out if something is an insulator? You can **experiment** to answer a question like this.

Learn It

When you **experiment**, you perform tests to answer a question. You make observations and collect data. Then you interpret the data to answer a question. When you experiment, it is important to test only one thing at a time. This helps you know what caused your results.

▶ **Activity Lab Book, pp. 213–215**

RESOURCES and TECHNOLOGY
▶ **Activity Lab Book,** pp. 213–215

▶ **Activity Flipchart,** p. 57

▶ **Instructional Navigator CD-ROM**

Focus on Skills   **487**

Plan Your Lesson

Lesson 2 Sound

Objective

- Describe how vibrations produce sounds.
- Compare the pitch and volume of a sound.

Reading Skill Predict

What I Predict	What Happens

Graphic Organizer 3, p. TR5

 Professional Development Look for **NSDL** to find recommended Science Background materials from the National Science Digital Library

FAST TRACK

Lesson Plan When time is short, follow the Fast Track and use the essential resources.

1 Introduce

Look and Wonder, p. 488

Resource Alternative Explore, p. 218

2 Teach

Develop Vocabulary, p. 490

Discuss the Main Idea, p. 492

Resource Visual Literacy, p. 77

3 Close

Think, Talk, and Write, p. 495

Resource Assessment, p. 153

▶ Reading and Writing

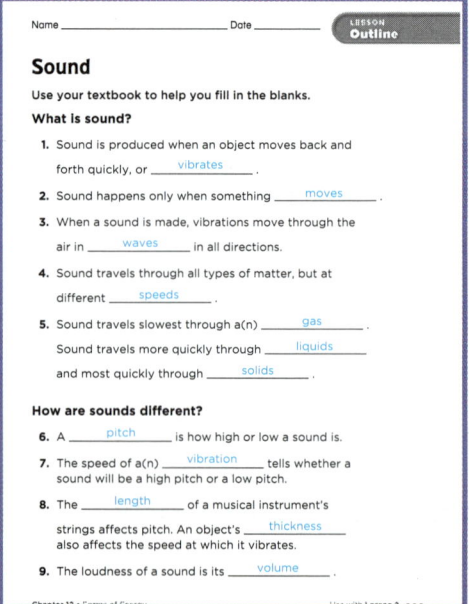

Outline, pp. 233–234
Also available as a student workbook

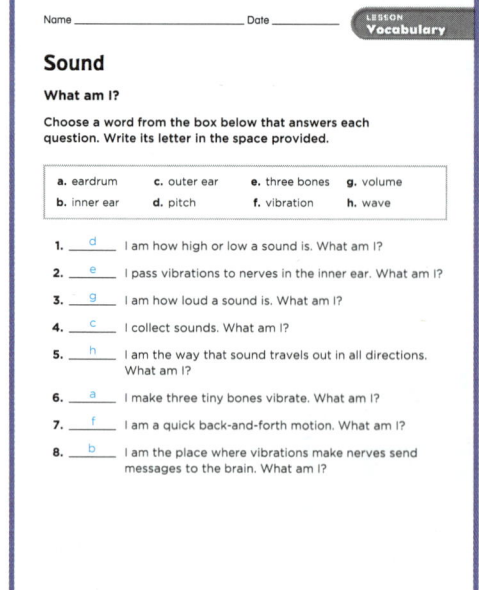

Vocabulary, p. 235
Also available as a student workbook

▷ Visual Literacy

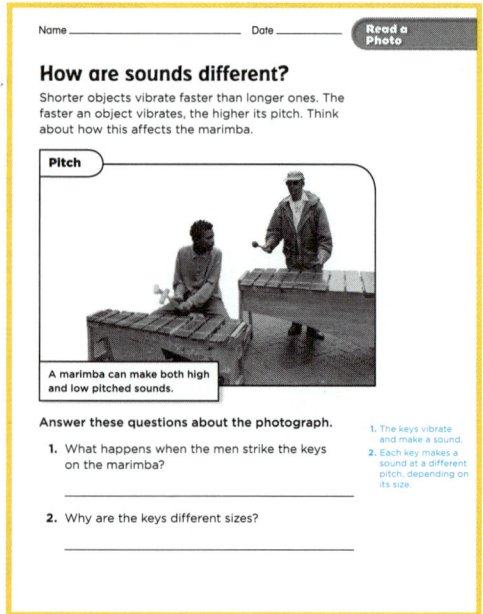

Read a Photo, p. 77
Also available as a transparency

▶ Activity Lab Book

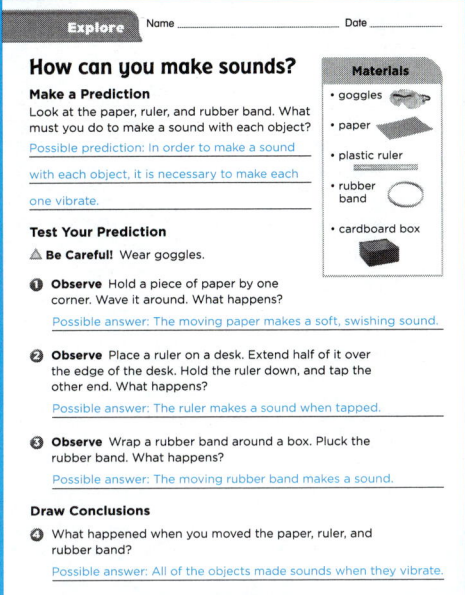

Explore Name _____ Date _____

How can you make sounds?

Make a Prediction
Look at the paper, ruler, and rubber band. What must you do to make a sound with each object?

Possible prediction: In order to make a sound with each object, it is necessary to make each one vibrate.

Test Your Prediction
△ **Be Careful!** Wear goggles.

❶ **Observe** Hold a piece of paper by one corner. Wave it around. What happens?

Possible answer: The moving paper makes a soft, swishing sound.

❷ **Observe** Place a ruler on a desk. Extend half of it over the edge of the desk. Hold the ruler down, and tap the other end. What happens?

Possible answer: The ruler makes a sound when tapped.

❸ **Observe** Wrap a rubber band around a box. Pluck the rubber band. What happens?

Possible answer: The moving rubber band makes a sound.

Draw Conclusions

❹ What happened when you moved the paper, ruler, and rubber band?

Possible answer: All of the objects made sounds when they vibrate.

Materials
- goggles
- paper
- plastic ruler
- rubber band
- cardboard box

Explore, pp. 216–217
Also available as a student workbook

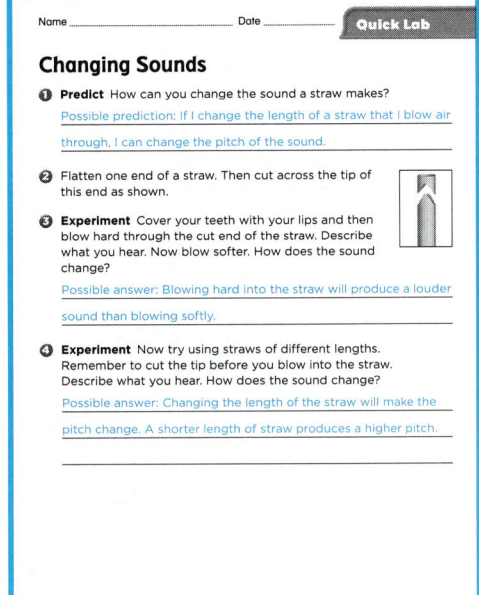

Name _____ Date _____ **Quick Lab**

Changing Sounds

❶ **Predict** How can you change the sound a straw makes?

Possible prediction: If I change the length of a straw that I blow air through, I can change the pitch of the sound.

❷ Flatten one end of a straw. Then cut across the tip of this end as shown.

❸ **Experiment** Cover your teeth with your lips and then blow hard through the cut end of the straw. Describe what you hear. Now blow softer. How does the sound change?

Possible answer: Blowing hard into the straw will produce a louder sound than blowing softly.

❹ **Experiment** Now try using straws of different lengths. Remember to cut the tip before you blow into the straw. Describe what you hear. How does the sound change?

Possible answer: Changing the length of the straw will make the pitch change. A shorter length of straw produces a higher pitch.

Quick Lab, p. 219
Also available as a student workbook

▶ Assessment

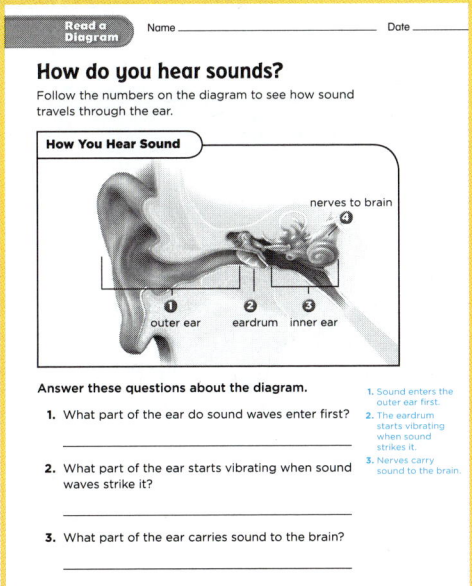

Read a Diagram Name _____ Date _____

How do you hear sounds?

Follow the numbers on the diagram to see how sound travels through the ear.

How You Hear Sound

nerves to brain ❹

❶ ❷ ❸
outer ear   eardrum   inner ear

Answer these questions about the diagram.

1. What part of the ear do sound waves enter first?
_____

2. What part of the ear starts vibrating when sound waves strike it?
_____

3. What part of the ear carries sound to the brain?
_____

1. Sound enters the outer ear first.
2. The eardrum starts vibrating when sound strikes it.
3. Nerves carry sound to the brain.

Read a Diagram, p. 78
Also available as a transparency

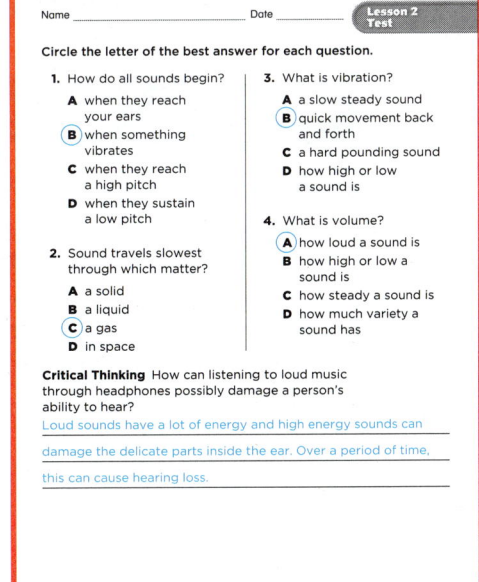

Name _____ Date _____ **Lesson 2 Test**

Circle the letter of the best answer for each question.

1. How do all sounds begin?
   A when they reach your ears
   B when something vibrates
   C when they reach a high pitch
   D when they sustain a low pitch

2. Sound travels slowest through which matter?
   A a solid
   B a liquid
   C a gas
   D in space

3. What is vibration?
   A a slow steady sound
   B quick movement back and forth
   C a hard pounding sound
   D how high or low a sound is

4. What is volume?
   A how loud a sound is
   B how high or low a sound is
   C how steady a sound is
   D how much variety a sound has

Critical Thinking How can listening to loud music through headphones possibly damage a person's ability to hear?

Loud sounds have a lot of energy and high energy sounds can damage the delicate parts inside the ear. Over a period of time, this can cause hearing loss.

Lesson Test, p. 153
Also available as a student workbook

Activity Flipchart
for Science Centers and Workstations

p. 58

ELL Teacher's Guide

p. 140

Transparencies for Visual Literacy

pp. 77–78

School to Home Activities

pp. 101–102

Key Concept Cards

77–78

Vocabulary Cards

182–185

Technology

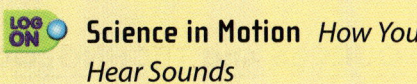

🔴 **Science Activity DVD**

🔴 **Instructional Navigator CD-ROM**

🔴 **Presentation Toolkit CD-ROM**

🔵 **Science in Motion** *How You Hear Sounds*

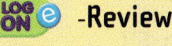🔵 -Review

🔵 **NSDL**

Lesson 2 Sound

Objectives
- Describe how vibrations produce sounds.
- Compare the pitch and volume of a sound.

1 Introduce

▶ **Assess Prior Knowledge**

Have students identify all the sounds they can hear in the classroom. Record responses on the board. Possible responses: teacher talking; students whispering; heater; fan; someone walking in hallway; bird chirping outside. Ask:

- **What enables you to hear all of these sounds?** the ears

- **What do you think causes all of these sounds?** Accept all reasonable answers.

FAST TRACK

Look and Wonder

Invite students to share their responses to the Look and Wonder statement and question.

- **At a parade you hear many different sounds. What makes all of these sounds?** Possible answers: people talking; people singing; musical instruments; horse hooves on the street

Write ideas on the board and note any misconceptions that students may have. Address these misconceptions as you teach the lesson.

RESOURCES and TECHNOLOGY
- **Activity Lab Book,** pp. 216–218
- **Activity Flipchart,** p. 58
- **Science Activity DVD**

Lesson 2

Sound

Look and Wonder

Sounds are all around you. At a parade you hear many different sounds. What makes all of these sounds?

488
ENGAGE

Warm Up

Start with a Demonstration

Ask students who play musical instruments to bring them to class. Have them demonstrate their musical instrument. Then have students describe the sounds made by each instrument and suggest how the instrument made the sound. Record responses on the board. Possible responses: Strings made the sound in a guitar or violin. Wind made the sound in a trumpet.

Explore

Inquiry Activity

How can you make sounds?

Make a Prediction

Look at the paper, ruler, and rubber band. What must you do to make a sound with each object?

Test Your Prediction

⚠️ **Be Careful.** Wear goggles.

1. **Observe** Hold a piece of paper by one corner. Wave it around. What happens?

2. **Observe** Place a ruler on a desk. Extend half of it over the edge of the desk. Hold one end of the ruler down, and tap the other end. What happens?

3. **Observe** Wrap a rubber band around a box. Pluck the rubber band. What happens?

Draw Conclusions

4. What happened when you moved the paper, ruler, and rubber band?

5. **Infer** Can you make a sound with the paper, ruler, or rubber band without making it move? Explain your answer.

6. **Infer** How are sounds made?

Explore More

Experiment Test ways to change the sound you made with each object. Try to make the sounds louder or softer, higher or lower. For example, try pulling the rubber band tighter and then plucking it. Record your results and the steps you follow.

Materials

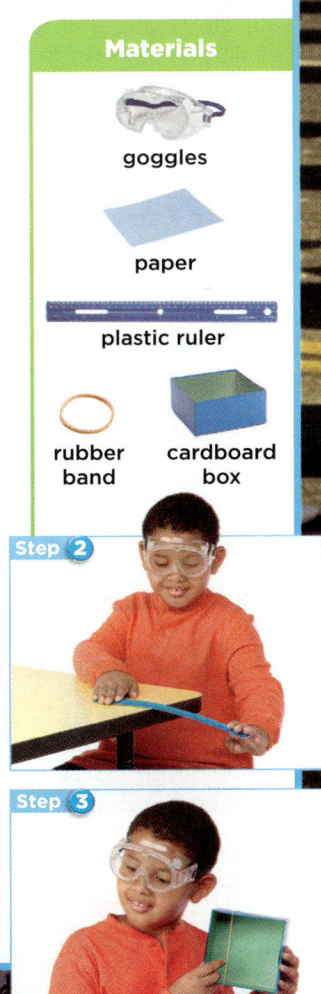

goggles

paper

plastic ruler

rubber band   cardboard box

Step 2

Step 3

489
EXPLORE

Alternative Explore

Can a balloon help you hear?

Materials balloon

Have students work in pairs to determine if sounds passing through a balloon are easier to hear than those not passing through a balloon. Ask students to make a prediction before they begin the activity. **Be Careful!** Instruct students to exercise caution when blowing up their balloons.

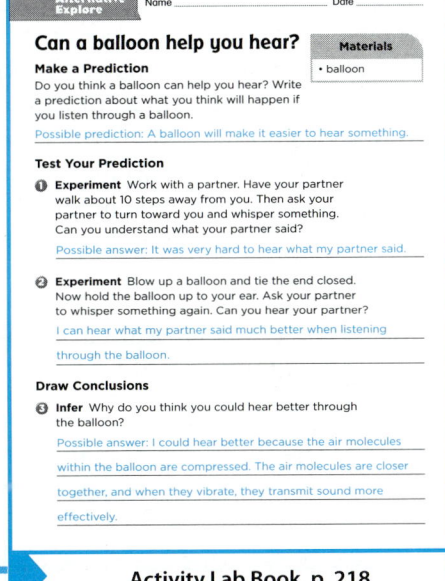

Alternative Explore   Name _____ Date _____

Can a balloon help you hear?

Materials
• balloon

Make a Prediction
Do you think a balloon can help you hear? Write a prediction about what you think will happen if you listen through a balloon.
Possible prediction: A balloon will make it easier to hear something.

Test Your Prediction

1. **Experiment** Work with a partner. Have your partner walk about 10 steps away from you. Then ask your partner to turn toward you and whisper something. Can you understand what your partner said?
Possible answer: It was very hard to hear what my partner said.

2. **Experiment** Blow up a balloon and tie the end closed. Now hold the balloon up to your ear. Ask your partner to whisper something again. Can you hear your partner?
I can hear what my partner said much better when listening through the balloon.

Draw Conclusions

3. **Infer** Why do you think you could hear better through the balloon?
Possible answer: I could hear better because the air molecules within the balloon are compressed. The air molecules are closer together, and when they vibrate, they transmit sound more effectively.

Activity Lab Book, p. 218

Explore

 pairs    30 minutes

Plan Ahead Have all the necessary materials ready before the class assembles.

Purpose In this activity, students demonstrate that movement causes sound.

Structured Inquiry

Make a Prediction Possible prediction: To make a sound, I must make each of the items move.

Test Your Prediction

1. **Observe** Possible answer: The paper wiggles and makes a sound.

2. **Observe** Possible answer: The moving ruler makes a sound.

3. **Observe** Possible answer: The rubber band vibrates and makes a sound.

4. The objects made a sound when they moved.

5. **Infer** No; these objects need to move in order to make a sound.

Guided Inquiry   Explore More

Experiment Waving a short strip of paper will produce a higher sound than a long strip. Tapping a short section of a ruler will produce a higher sound than tapping a long section. Plucking a tighter rubber band will produce a higher sound than a looser one. Waving, tapping, and plucking with more energy will produce louder sounds.

Open Inquiry

Ask students how the length of something changes the sound it makes. Have them think of their own question about how things make sounds. Then ask them to make a plan and carry out an experiment to answer the question. Ask:

Do shorter or longer strings make higher sounds?

2 Teach

Read and Learn

Main Idea Have students read aloud the questions at the beginning of each section of the lesson. Ask them to try to answer each of the questions before reading the lesson.

Vocabulary Have students read aloud the vocabulary words and their definitions. Write the words and their definitions on the board to refer to as you complete the lesson.

Reading Skill Predict
Graphic Organizer 3
Have students fill in a Predict graphic organizer as they read

What I Predict	What Happens

each two pages of the lesson. They can use the Quick Check questions to identify each prediction to make.

What is sound?

▶ Discuss the Main Idea

Explain to students that sounds are caused when something vibrates. Ask:

- **What does the vibration cause?** movement and sounds

FAST TRACK **Develop Vocabulary**

vibrate *Word Origin* Point out to students that the word vibrate comes from the Latin word *vibratus,* which means "to move quickly back and forth."

sound Ask a volunteer to look up two other meanings of the word *sound* in a dictionary. Have students come up with sentences that use the scientific meaning of *sound,* as well as the two other meanings from the volunteer, in three separate sentences.

RESOURCES and TECHNOLOGY

▶ **Reading and Writing,** pp. 233–235

💿 **PuzzleMaker CD-ROM**

💿 **Presentation Toolkit CD-ROM**

📶 **e-Glossary**

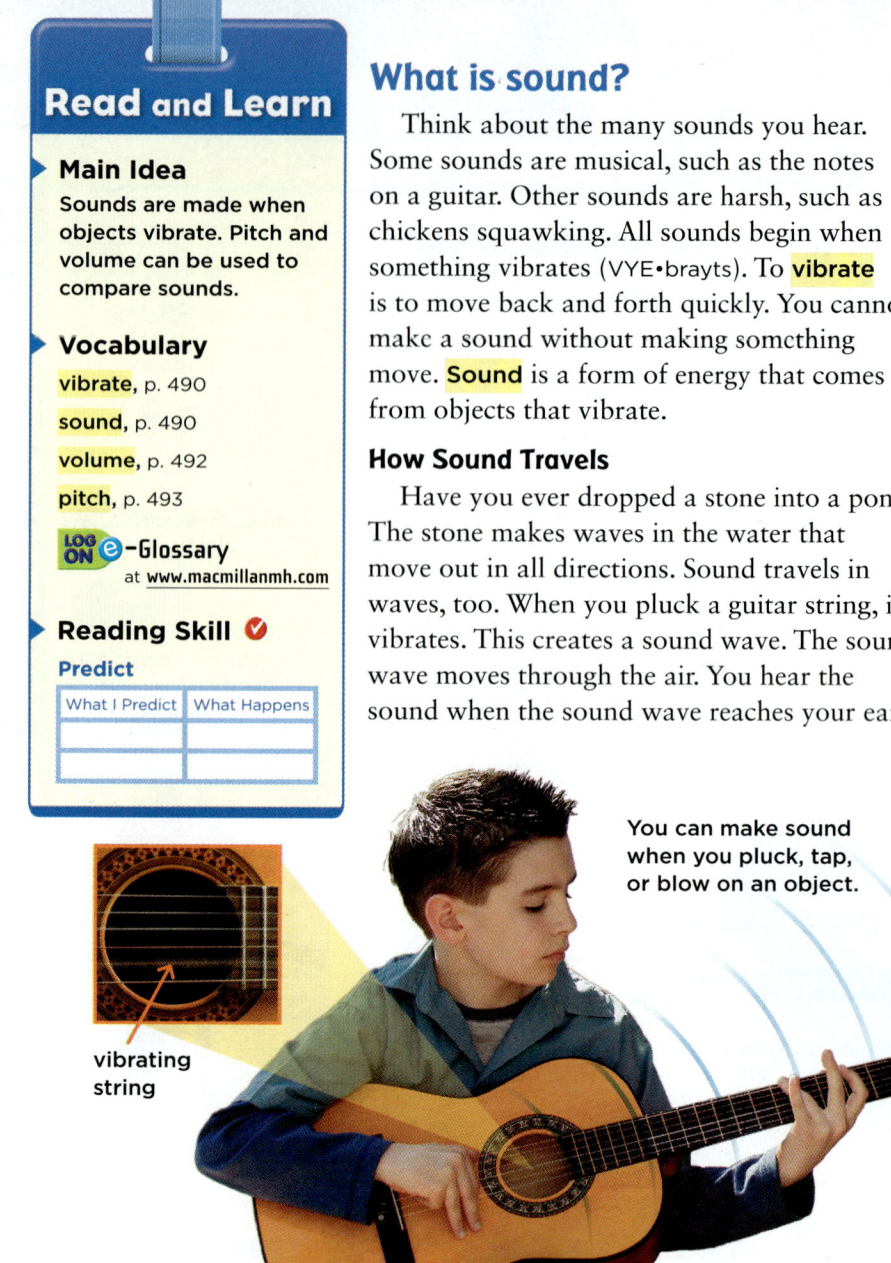

Read and Learn

▶ **Main Idea**
Sounds are made when objects vibrate. Pitch and volume can be used to compare sounds.

▶ **Vocabulary**
vibrate, p. 490
sound, p. 490
volume, p. 492
pitch, p. 493

📶 **e-Glossary**
at www.macmillanmh.com

▶ **Reading Skill** ✓
Predict

What I Predict	What Happens

What is sound?

Think about the many sounds you hear. Some sounds are musical, such as the notes on a guitar. Other sounds are harsh, such as chickens squawking. All sounds begin when something vibrates (VYE•brayts). To **vibrate** is to move back and forth quickly. You cannot make a sound without making something move. **Sound** is a form of energy that comes from objects that vibrate.

How Sound Travels

Have you ever dropped a stone into a pond? The stone makes waves in the water that move out in all directions. Sound travels in waves, too. When you pluck a guitar string, it vibrates. This creates a sound wave. The sound wave moves through the air. You hear the sound when the sound wave reaches your ear.

You can make sound when you pluck, tap, or blow on an object.

vibrating string

490 EXPLAIN

Science Background

The intensity of sounds, which humans perceive as loudness, is measured using the decibel scale. The absence of sound, but just at the point of human hearing, is recorded as 0 dB, or zero decibels. The sound in a library measures around 30 dB. The sound of a human voice during normal conversation is about 60 dB. A jet engine is about 130 dB. This scale can be used to measure extremely low- and high-intensity sounds because each multiple of ten on the scale is 10 times greater than the one before.

See **Science Yellow Pages** in the Teacher Resource section for background information.

📶 **Professional Development** For more Science Background and resources from **NSDL** visit **www.macmillanmh.com/nsdl/**

Orca whales use sounds to communicate.

Sound waves can travel through air, a gas. Sound waves can also travel through liquids and solids. Some sea animals communicate by making sounds underwater. You hear a knock on the door because sound travels through the door. Sound waves travel through matter. Sound cannot travel in space.

Sound does not travel at the same speed through all materials. Sound travels slowest through a gas. It travels faster through a liquid. It travels fastest through a solid.

 **Quick Check**

Predict You hit a drum with a stick. What happens?

Critical Thinking Do you think that you can hear sound in outer space?

Tie a piece of string to two cups. Then talk in one end while a friend listens in the other. Why can you hear your friend?

491
EXPLAIN

▶ **Use the Visuals**

Refer students to the pictures on pages 490–491. Ask:

- **What does sound travel through when you hear the sound from a guitar?** air

- **What does sound travel through when a whale beneath the surface of the ocean hears something?** water

- **What does sound travel through when you talk on a tin can telephone?** Possible answers: a solid; string; cans

✓ **Quick Check Answers**

- **Predict** The drum head vibrates and you hear a sound.

- **Critical Thinking** No; because there is no matter in space. Sounds must travel through a solid, a liquid, or a gas.

Differentiated Instruction

Leveled Questions

EXTRA SUPPORT **What causes sound?** Sounds are caused when something vibrates.

ENRICHMENT **If you were in a swimming pool when someone called you to come out and have lunch, would you be able to hear them call you if your head were under water?** Yes, because sound waves travel through liquids.

Lesson 2   **491**

How are sounds different?

Discuss the Main Idea

FAST TRACK

Write the terms *pitch* and *volume* on the board next to each other. As students answer the following questions, record their responses below each word. Ask:

- **What is the pitch of a sound?** how high or low the sound is

- **What are some sounds with a high pitch? a low pitch?** Accept all reasonable answers. Possible answer: the song of a bird; a bass guitar

- **What is the volume of a sound?** how loud the sound is

- **What are some sounds with high volume?** Accept all reasonable answers. Possible answers: music at a concert; race cars

▶ Develop Vocabulary

volume *Scientific vs. Common Use* Explain to students that the word *volume* also refers to one book in a set of books, such as Volume 1 in a set of encyclopedias that has 20 volumes.

pitch *Scientific vs. Common Use* Point out to students that *pitch* also means when a player on a baseball or softball team throws the ball to the batter. Remind them that in science, *pitch* refers to how high or low a sound is.

RESOURCES and TECHNOLOGY

▶ **Activity Lab Book,** p. 218

Presentation Toolkit CD-ROM

The sounds of jets roaring through the sky are louder than the chirps of a bird.

How are sounds different?

Close your eyes and listen. What sounds do you hear? What makes different sounds different?

Volume

Sounds can have different volumes. Volume describes how loud a sound is. A plane flying overhead is louder than a bird's song. A plane has a greater volume.

Loud sounds are made by objects that vibrate with a lot of energy. The more energy an object vibrates with, the louder the sound it makes. Tap your foot on the floor. Then stomp your foot. You use more energy to stomp your foot than to tap it. Stomping makes a high-energy vibration, so you hear a loud sound. Tapping makes a low-energy vibration, so you hear a soft sound.

492
EXPLAIN

ELL Support

Vocabulary/Use Picture Clues Review lesson vocabulary. Model pronunciation and have students repeat. Resolve any questions about meaning. Direct students' attention to the pictures on pages 490–491. Explain that sound waves can travel through solids, liquids, and gases. Encourage students to use the lesson vocabulary when they speak.

BEGINNING Students point to pictures that illustrate sound waves traveling through solids, liquids, and gases.

INTERMEDIATE Students explain in short sentences that sound waves can travel through solids, liquids, and gases.

ADVANCED Students use complete sentences to explain the different ways sound waves travel through solids, liquids, and gases.

Pitch

Some sounds are high, such as the squeaking of a mouse. Other sounds are low, such as the croaking of a bullfrog. A sound's **pitch** (PICH) is how high or low it is. An object that vibrates quickly has a high pitch. An object that vibrates slowly has a low pitch.

An object's length can affect pitch. Look at the marimba below. When hit, the shorter keys vibrate faster than the longer ones. The shorter a key is, the faster it vibrates and the higher its pitch.

The thickness of an object can also affect pitch. A guitar has both thin and thick strings. Thin strings vibrate faster than thick strings, so they have higher pitches.

✔ Quick Check

Predict How does tightening a rubber band affect its pitch?

Critical Thinking Compare the sound of a bicycle horn to a car horn.

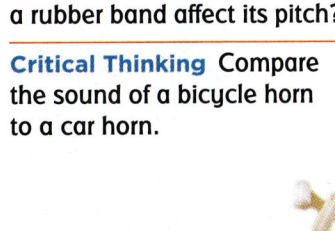

A marimba can make sounds with a high or a low pitch.

493
EXPLAIN

≡ Quick Lab

Changing Sounds

1. **Predict** How can you change the sound a straw makes?

2. Flatten one end of a straw. Then cut across the tip of this end as shown.

3. **Experiment** Cover your teeth with your lips and then blow hard through the cut end of the straw. Describe what you hear. Now blow softer. How does the sound change?

4. **Experiment** Now try using straws of different lengths. Remember to cut the tip before you blow into the straw. Describe what you hear. How does the sound change?

≡ Quick Lab

small groups   15 minutes

Objective Observe how sounds change in pitch and volume.

Materials plastic drinking straws, scissors

1. Accept all reasonable predictions.

2. **Be Careful!** Caution students to be careful using scissors when they cut the straws.

3. Students will have to blow hard into the straw to make sounds. The harder students blow, the louder the sounds they will produce.

4. Changing the length of the straw will make the pitch change. Shorter straws will have a higher pitch than longer straws.

✔ Quick Check Answers

- **Predict** Tightening a rubber band will cause it to become thinner, making its pitch higher.

- **Critical Thinking** The pitch of a bicycle horn is higher than a car horn. The volume of a car horn is louder than a bike horn.

Differentiated Instruction

Leveled Activities

EXTRA SUPPORT Have students use encyclopedias or the Internet, if available, to research the loudest sounds recorded on Earth. Have students record their findings in a poster that lists the sounds in order of increasing loudness.

ENRICHMENT Have students use encyclopedias or the Internet, if it is available, to research the shape and motion of a sound wave as it travels through matter. Encourage students to illustrate a sound wave on a poster and to describe its parts. Also ask students to show how the sound wave changes if the sound becomes louder or its pitch becomes higher.

Lesson 2   **493**

How do you hear sounds?

▶ Discuss the Main Idea

Refer students to the photo on page 494. Ask:

- **How is this person protecting his hearing?**
  He is wearing a device on his head that covers his outer ears.

- **Why is it important for you to protect your hearing?** Possible answer: to protect the ears from loud noises that could damage them

Read a Diagram

Answer Sound waves enter the outer ear. Then they move to the eardrum, the inner ear, and finally to nerves that send messages to the brain.

 **Science in Motion**, *How You Hear Sounds*

▶ Develop Vocabulary

Review lesson vocabulary with a word-study activity. Invite students to write a paragraph about any type of musical concert. Have them use all the vocabulary words in the lesson: *vibrate, sound, volume,* and *pitch.*

✔ Quick Check Answers

- **Predict** Possible answer: Over time, many loud sounds might damage the parts inside the ear and cause hearing loss.

- **Critical Thinking** a high sound

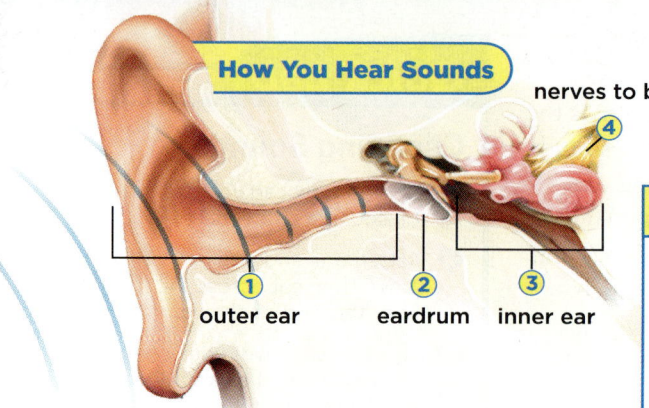

How You Hear Sounds

nerves to brain

① outer ear   ② eardrum   ③ inner ear

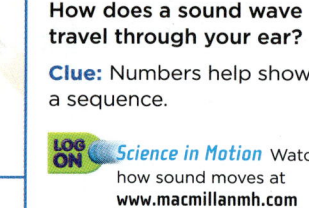

Read a Diagram

How does a sound wave travel through your ear?

Clue: Numbers help show a sequence.

Science in Motion Watch how sound moves at www.macmillanmh.com

How do you hear sounds?

What happens to a sound wave when it reaches your ear? First, the sound wave is collected by the outer ear. Next, the sound wave makes your eardrum vibrate. This causes three tiny bones in your inner ear to vibrate. These vibrations pass through the inner ear to nerves. The nerves send a message to your brain, and you hear a sound.

Protect Your Hearing

It is important to protect your ears. Never put a finger or pencil in your ear. You may hurt the parts inside. Loud sounds can damage your ear as well. Loud sounds have a lot of energy. They can damage the parts inside the ear.

 Quick Check

Predict What might happen to your hearing if you listen to loud music often?

Critical Thinking Which makes your eardrum vibrate faster, a high sound or low sound?

494
EXPLAIN

This construction worker must protect his ears.

Homework Activity

Animal Hearing

Ask students to use encyclopedias, magazines, or the Internet, if available, to research how animals hear. Encourage students to research animals that live in the water, such as whales and sharks, and animals of different sizes that live on land, such as insects, birds, and reptiles. Have students prepare illustrated oral reports to share with the class.

Lesson Review

Visual Summary

 **Sound is produced when** an object vibrates. Sound can travel through solids, liquids, and gases.

 **Sounds can be compared by** using volume and pitch.

 **You hear sounds when** vibrations travel through your ear.

Make a 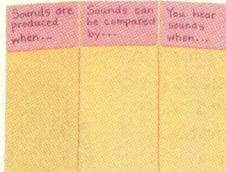 **Study Guide**
Make a Trifold Book. Use it to summarize what you learned about sound.

Think, Talk, and Write

1 Main Idea What do you need to do to make a sound?

2 Vocabulary What is the difference between pitch and volume?

3 Predict How would cymbals sound if you tapped them together lightly? How would they sound if you crashed them together?

What I Predict	What Happens

4 Critical Thinking List five different sounds you hear. How are these sounds the same? How are these sounds different?

5 Test Prep Objects that vibrate quickly make sounds with
A high volume.
B low volume.
C high pitch.
D low pitch.

 Writing Link

Persuasive Writing
It is important to protect yourself from loud sounds. Loud sounds can harm your hearing. Find out how you can protect your hearing. Then write a paragraph telling others what you learned.

Music Link

Make an Instrument
Put on goggles. Stretch rubber bands of different thicknesses around a shoe box. Then use the rubber bands to make sounds. How can you change the pitch? How can you change the volume?

LOG ON e-Review Summaries and quizzes online at www.macmillanmh.com

495
EVALUATE

Formative Assessment

Approaching Have students continue making vocabulary flash cards using index cards. Have them write the vocabulary word on one side of the index card and the definition on the other side. Have students add these flash cards to the cards they made for Lesson 1.

On-Level Have students explain in their own words the difference between the pitch and the volume of a sound.

Challenge Have students research the speed of sound and prepare an oral report to present their findings to the class.

 **Key Concept Cards** For student intervention, see the prescribed routine on **Key Concept Cards 77 and 78.**

---

3 Close

Lesson Review

▶ **Visual Summary**

Have students save their lesson Study Guides. They will use them at the end of the chapter for review.

▶ **Make a FOLDABLES Study Guide**

See pages R27–R28 for instructions.

FAST TRACK **Think, Talk, and Write**

1 Main Idea make an object vibrate

2 Vocabulary Pitch is how high or low a sound is. Volume is how loud or soft a sound is.

3 Predict

What I Predict	What Happens
Lightly = soft	The same
Crash = loud	The same

4 Critical Thinking Possible answer: bird singing, hands clapping, voice shouting, person walking, dog barking; the same: they are all produced by vibrations; different: dog barking, hands clapping, voice shouting are high in volume, the bird singing and person walking are fairly low in volume.

5 Test Prep C

Writing Link The writing should include information about ear plugs and other protective gear and the situations in which ear protection is recommended. They also should indicate how people should avoid loud noises altogether.

Music Link Pitch can be changed by stretching the rubber bands more or using thinner bands; volume can be raised by strumming the rubber bands with more force..

RESOURCES and TECHNOLOGY

▶ Reading and Writing, p. 236

▶ Visual Literacy, p. 78

▶ School to Home Activities, pp. 101–102

LOG ON Science in Motion, *How You Hear Sounds*

LOG ON e-Review Narrated Summary and Quiz

LOG ON Progress Reporter Assessments

Be a Scientist pairs  30 minutes

Skills experiment, interpret data, infer

Objective

- Compare how sound travels through a gas, liquid, and solid.

Materials 3 plastic bags, tuning fork, water, wooden block

Plan Ahead Gather all the materials before the class assembles. Tuning forks might be available from the school's music department or band. Be sure wooden blocks are small enough to fit inside the plastic bags.

EXTEND This activity will show students how sound moves through different types of matter.

Structured Inquiry

How does sound move through different types of matter?

Form a Hypothesis Possible hypothesis: If sound travels through a solid, it will be louder than it would through a liquid or a gas.

Test Your Hypothesis

3 **Be Careful!** To prevent spills, caution students to not overfill the bags with water.

4 **Experiment** Students will observe that the sound of the tuning fork is louder through water than through air.

6 **Experiment** Students will observe that the sound of the tuning fork is louder through the wooden block in the bag than through water or air.

Be a Scientist

Materials

3 plastic bags

tuning fork

water

wooden block

Structured Inquiry

How does sound move through different types of matter?

Form a Hypothesis

You just learned that sound travels through solids, liquids, and gases. How does the state of matter affect how sound travels? Write a hypothesis.

Test Your Hypothesis

1 Fill a plastic bag with air and seal it. Hold the bag against your ear.

2 **Experiment** Tap the tines of the tuning fork against the bottom of your shoe. Then hold the base of the tuning fork against the plastic bag. Listen to the sound it makes.

3 Fill a plastic bag with water. Seal it and hold it against your ear.

4 **Experiment** Tap the tuning fork and hold it against the bag. Record any differences you hear.

5 Place a wooden block in a plastic bag. Squeeze out as much air as you can and seal the bag. Hold the bag against your ear.

6 **Experiment** Tap the tuning fork and hold it against the bag. How is the sound different now? Record your observations.

Step **4**

Integrate Writing

Writing about Sound

Have students think about how they use sound every day and how sound makes their lives more enjoyable.

- **In what ways do you use sound?** Possible answers: to talk with my friends; to listen to music

Have students write a story about a day in their lives that includes details on how they use sound.

Inquiry Investigation

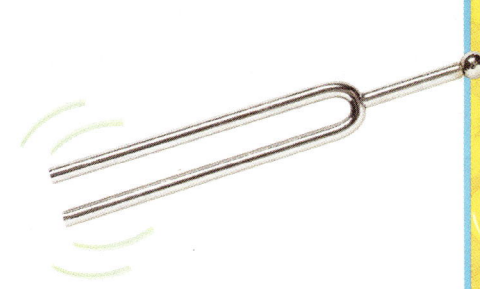

Draw Conclusions

7. How did the tuning fork sound different through the different materials?

8. **Interpret Data** Through which material was the sound loudest?

9. **Infer** Does sound travel best through a solid, a liquid, or a gas?

Guided Inquiry

How does sound move through different solids?

Form a Hypothesis

Sound can be stopped, slowed down, or absorbed by different solids. How does sound travel through different solids?

Test Your Hypothesis

Design an experiment to investigate how sound travels through different solids. Decide on the materials you will need. You may want to try plastic, wooden, and metal objects. Write out the steps you will follow. Record your results and observations.

Draw Conclusions

Did your results support your hypothesis? Why or why not?

Open Inquiry

What other questions do you have about sound? For example, what objects block sound the best? Design an experiment to find out.

Remember to follow the steps of the scientific process.

> Ask a Question
>
> ↓
>
> Form a Hypothesis
>
> ↓
>
> Test Your Hypothesis
>
> ↓
>
> Draw Conclusions

**497
EXTEND**

Draw Conclusions

7. Possible answers: The volume, or loudness was different. The pitch was different.

8. **Interpret Data** The sound was loudest through the wooden block.

9. **Infer** Sound usually travels best through a solid.

Guided Inquiry

How does sound move through different solids?

Form a Hypothesis Possible hypothesis: If sound travels through wood and metal, it will be louder through metal.

Test Your Hypothesis Have students plan to test their hypotheses by gathering information and materials and then recording the steps in, and results of, their experiments.

Draw Conclusions Encourage students to analyze the results of their experiments to see if they support their hypotheses.

Open Inquiry

Help students plan their investigations. Have them begin by deciding what materials and information they will need to answer the questions. Encourage students to share the results of their investigations with the rest of the class.

Be a Scientist

Name _____ Date _____

Structured Inquiry

How does sound move through different types of matter?

Form a Hypothesis

You just learned that sound travels through solids, liquids, and gases. How does the state of matter affect how sound travels? Write a hypothesis.

Possible hypothesis: If sound moves through matter in different states, then it will travel at different speeds and its pitch will change.

Test Your Hypothesis

1. Fill a plastic bag with air and seal it. Hold the bag against your ear.

2. **Experiment** Tap the tines of the tuning fork against the bottom of your shoe. Then hold the base of the tuning fork against the plastic bag. Listen to the sound it makes.

3. Fill a plastic bag with water. Seal it and hold it against your ear.

4. **Experiment** Tap the tuning fork and hold it against the bag. Record any differences you hear.

Possible answer: The pitch changes.

5. Place a wooden block in a plastic bag. Squeeze out as much air as you can and seal the bag. Hold the bag against your ear.

Materials
- 3 plastic bags
- tuning fork
- water
- wooden block

▶ **Activity Lab Book, pp. 220–223**

RESOURCES and TECHNOLOGY

▶ **Activity Lab Book,** pp. 220–223

▶ **Activity Flipchart,** p. 59

💿 **Instructional Navigator CD-ROM**

Plan Your Lesson

Lesson 3 Light

Objective

- Explore how light travels.
- Describe how colors are seen.

Reading Skill Draw Conclusions

Text Clues	Conclusions

Graphic Organizer 13, p. TR15

 Professional Development Look for **NSDL** to find recommended Science Background materials from the National Science Digital Library

FAST TRACK

Lesson Plan When time is short, follow the Fast Track and use the essential resources.

1 Introduce

Look and Wonder, p. 498

Resource Alternative Explore, p. 226

2 Teach

Develop Vocabulary, p. 501

Develop Vocabulary, p. 502

Discuss the Main Idea, p. 504

Resource Visual Literacy, p. 79

3 Close

Think, Talk, and Write, p. 507

Resource Assessment, p. 154

▶ Reading and Writing

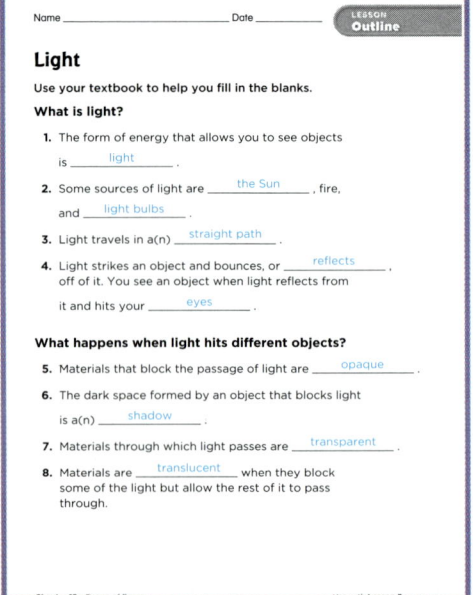

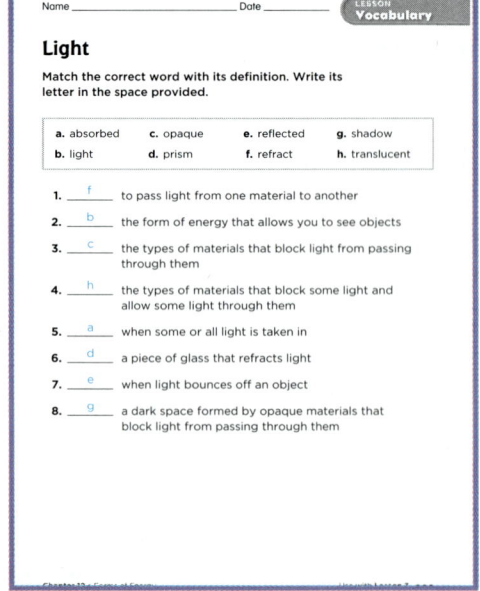

Outline, pp. 237–238
Also available as a student workbook

Vocabulary, p. 239
Also available as a student workbook

▶ Visual Literacy

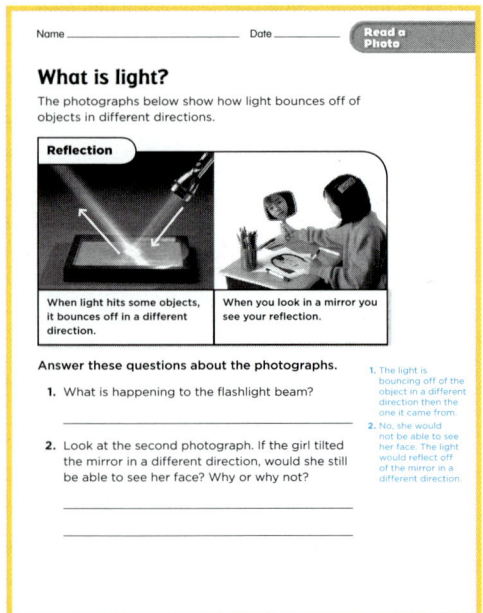

Read a Photo, p. 79
Also available as a transparency

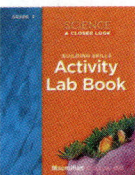

▶ Activity Lab Book

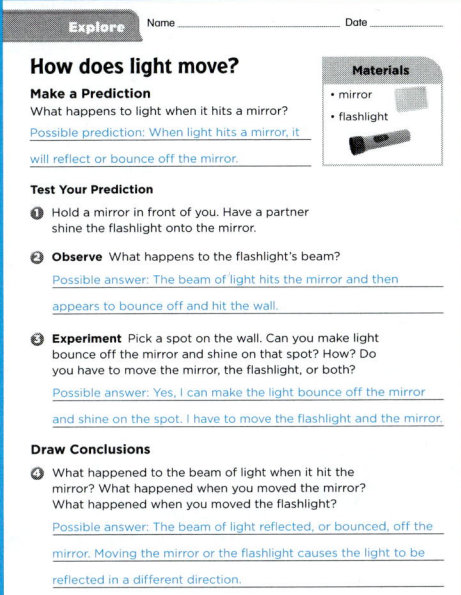

Explore Name _____ Date _____

How does light move?

Make a Prediction
What happens to light when it hits a mirror?

Possible prediction: When light hits a mirror, it
will reflect or bounce off the mirror.

Materials
• mirror
• flashlight

Test Your Prediction

❶ Hold a mirror in front of you. Have a partner
shine the flashlight onto the mirror.

❷ **Observe** What happens to the flashlight's beam?
Possible answer: The beam of light hits the mirror and then
appears to bounce off and hit the wall.

❸ **Experiment** Pick a spot on the wall. Can you make light
bounce off the mirror and shine on that spot? How? Do
you have to move the mirror, the flashlight, or both?
Possible answer: Yes, I can make the light bounce off the mirror
and shine on the spot. I have to move the flashlight and the mirror.

Draw Conclusions

❹ What happened to the beam of light when it hit the
mirror? What happened when you moved the mirror?
What happened when you moved the flashlight?
Possible answer: The beam of light reflected, or bounced, off the
mirror. Moving the mirror or the flashlight causes the light to be
reflected in a different direction.

Explore, pp. 224–225
Also available as a student workbook

Name _____ Date _____ **Quick Lab**

Mixing Colors

❶ **Predict** Look at the photo below and the color version
in your textbook. What happens to the color of the plate
when you spin it?
Possible prediction: If I spin the color wheel, the individual colors
will blend and appear to turn a shade of white.

❷ Divide a white paper plate into eight equal parts. Color
each section of the plate a different color.

❸ **Observe** Carefully push a pencil into the center of the
plate. Hold the plate away from your body. Spin it. What
color do you see when the plate is spinning?
Possible answer: I see a shade of white, which is a mixture of all of
the other colors.

Quick Lab, p. 227
Also available as a student workbook

▶ Assessment

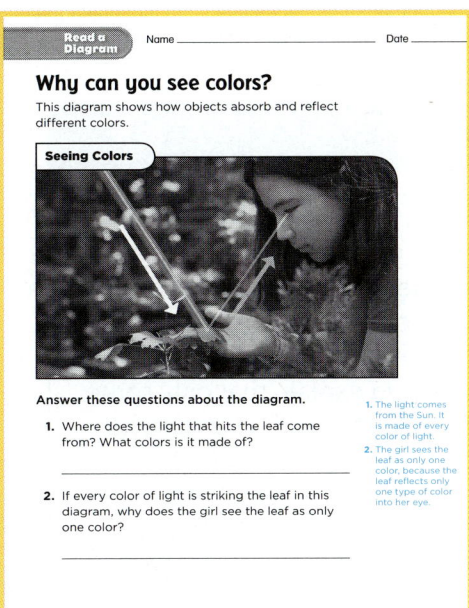

Read a Diagram Name _____ Date _____

Why can you see colors?

This diagram shows how objects absorb and reflect
different colors.

Seeing Colors

Answer these questions about the diagram.

1. Where does the light that hits the leaf come
from? What colors is it made of?

2. If every color of light is striking the leaf in this
diagram, why does the girl see the leaf as only
one color?

1. The light comes
from the Sun. It
is made of every
color of light.
2. The girl sees the
leaf as only one
color, because the
leaf reflects only
one type of color
into her eye.

Read a Diagram, p. 80
Also available as a transparency

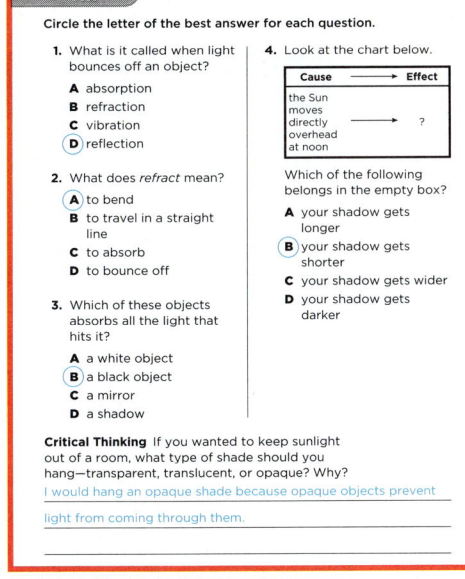

Lesson 3 Test Name _____ Date _____

Circle the letter of the best answer for each question.

1. What is it called when light
bounces off an object?
   A absorption
   B refraction
   C vibration
   D reflection

2. What does *refract* mean?
   A to bend
   B to travel in a straight
   line
   C to absorb
   D to bounce off

3. Which of these objects
absorbs all the light that
hits it?
   A a white object
   B a black object
   C a mirror
   D a shadow

4. Look at the chart below.

Cause	→	Effect
the Sun moves directly overhead at noon	→	?

Which of the following
belongs in the empty box?
   A your shadow gets
   longer
   B your shadow gets
   shorter
   C your shadow gets wider
   D your shadow gets
   darker

Critical Thinking If you wanted to keep sunlight
out of a room, what type of shade should you
hang—transparent, translucent, or opaque? Why?
I would hang an opaque shade because opaque objects prevent
light from coming through them.

Lesson Test, p. 154
Also available as a student workbook

ADDITIONAL RESOURCES

p. 60

p.141

pp. 79–80

pp. 103–104

79–80

186–193

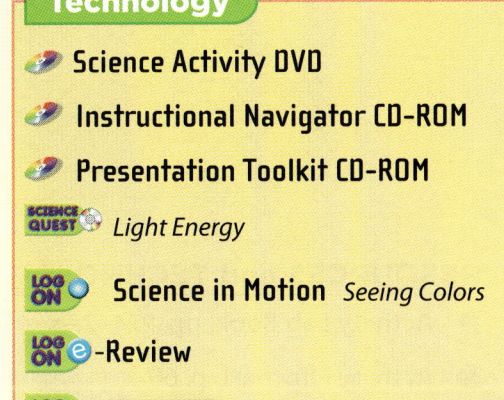

Technology

🔴 **Science Activity DVD**

🔴 **Instructional Navigator CD-ROM**

🔴 **Presentation Toolkit CD-ROM**

SCIENCE QUEST *Light Energy*

LOG ON **Science in Motion** *Seeing Colors*

LOG ON **e-Review**

LOG ON **NSDL**

Lesson 3 Light

Objectives
- Explore how light travels.
- Describe how colors are seen.

1 Introduce

▶ **Assess Prior Knowledge**

Direct students' attention to the sculpture shown on page 498. Ask:

- **What do you think this sculpture is made of?**
  Accept all reasonable answers. Possible answers: mirrors; shiny steel

- **What can you see in the sculpture?** Possible answers: a reflection of the clouds, buildings, and people

- **Why can you see these things in the sculpture?**
  Accept all reasonable answers.

Look and Wonder

Invite students to share their responses to the Look and Wonder statement and question:

- **When you look at a mirror, you can see yourself. Light makes this possible. How does light move?**
  Accept all reasonable responses.

Write ideas on the board and note any misconceptions that students may have. Address these misconceptions as you teach the lesson.

RESOURCES and TECHNOLOGY

▶ **Activity Lab Book,** pp. 224–226

▶ **Activity Flipchart,** p. 60

💿 **Science Activity DVD**

Lesson 3

Light

Millennium Park, Chicago, Illinois

Look and Wonder

When you look at a mirror, you can see yourself. Light makes this possible. How does light move?

498
ENGAGE

Warm Up

Start with a Demonstration

Explain to students that light passes through some materials and is blocked by others. Show students a piece of clear glass or plastic and explain that light passes through this material. Tell students this material is *transparent*. Show students a translucent piece of glass or plastic and explain that only some light passes through this material. This material is *translucent*. Then show students a piece of opaque material, such as a piece of aluminum foil or a block of wood. Explain that this material does not allows light to pass through it. It is *opaque*. Write the three terms on the board. Have students look around the classroom for examples of each type of material. Record each material on the board. Possible responses: transparent—windows, glass jars, aquarium, water; translucent—plastic jars, wax paper; opaque—dark paper, books, desks, chalkboard

Explore

How does light move?

Make a Prediction

What happens to light when it hits a mirror?

Test Your Prediction

1 Hold a mirror in front of you. Have a partner shine the flashlight onto the mirror.

2 **Observe** What happens to the flashlight's beam?

3 **Experiment** Pick a spot on the wall. Can you make light bounce off the mirror and shine on that spot? How? Do you have to move the mirror, the flashlight, or both?

Draw Conclusions

4 What happened to the beam of light when it hit the mirror? What happened when you moved the mirror? What happened when you moved the flashlight?

5 **Communicate** Make a drawing to show how light moves when it strikes the mirror.

Explore More

Experiment Sit next to your partner. Leave a meter of space between you and your partner. Then hold a mirror so that you can see your partner. Can your partner see you in the mirror? Can you see yourself and your partner in the mirror at the same time?

Materials

mirror

flashlight

Step 1

Step 3

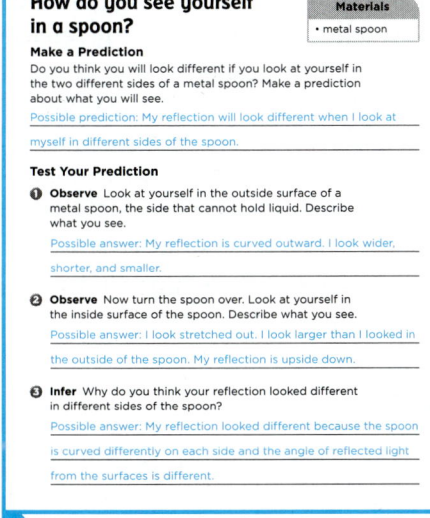

499
EXPLORE

Alternative Explore

How do you see yourself in a spoon?

Materials spoon

Have students look at themselves in the outside, or convex side, of a metal spoon, and then observe themselves in the inside, or concave side, of the same spoon. Have students compare their observations from both sides of the spoon.

Alternative Explore Name _____ Date _____

How do you see yourself in a spoon?

Make a Prediction

Do you think you will look different if you look at yourself in the two different sides of a metal spoon? Make a prediction about what you will see.

Possible prediction: My reflection will look different when I look at myself in different sides of the spoon.

Test Your Prediction

1 **Observe** Look at yourself in the outside surface of a metal spoon, the side that cannot hold liquid. Describe what you see.

Possible answer: My reflection is curved outward. I look wider, shorter, and smaller.

2 **Observe** Now turn the spoon over. Look at yourself in the inside surface of the spoon. Describe what you see.

Possible answer: I look stretched out. I look larger than I looked in the outside of the spoon. My reflection is upside down.

3 **Infer** Why do you think your reflection looked different in different sides of the spoon?

Possible answer: My reflection looked different because the spoon is curved differently on each side and the angle of reflected light from the surfaces is different.

Materials
• metal spoon

Activity Lab Book, p. 226

Explore

 pairs   30 minutes

Plan Ahead Acquire enough mirrors and flashlights before the class assembles.

Purpose In this activity, students demonstrate what happens to light when it hits a mirror.

Structured Inquiry

Make a Prediction Possible prediction: When light hits a mirror, it bounces off in a different direction.

Test Your Prediction

1 **Be Careful!** Caution students not to shine the flashlight or the reflected light directly into another person's eyes.

2 **Observe** The beam of light reflects, or bounces, off the mirror.

3 **Experiment** Try making several target marks on the wall or chalkboard for different student pairs to hit with the light. Both partners will probably need to adjust their positions to get the light to hit the target exactly.

4 It reflected, or bounced, off the mirror. Moving the mirror or the flashlight causes the light to be reflected in a different direction or at a different angle from which it first bounced.

5 **Communicate** Student drawings should show that light bounces off the mirror. Drawings might show that the angle between the incoming beam of light and the mirror equals the angle between the reflected beam of light and the mirror.

Guided Inquiry   Explore More

Experiment Students will observe that they cannot see themselves and their partners at the same time.

Open Inquiry

Ask students what would happen to the reflected light if the mirror were curved. Have them think of their own question about how light is reflected off of a mirror. Then ask them to make a plan and carry out an experiment to answer the question. Ask:

Would the reflected light change direction if the mirror were curved?

2 Teach

Read and Learn

Main Idea Have students preview the lesson by looking at the photos and diagrams. Ask students to write one question they think will be answered in this lesson.

Vocabulary Ask student volunteers to find each of the new vocabulary words in the lesson and then read the term and its definition aloud to the class.

Reading Skill Draw Conclusions
Graphic Organizer 13
Have students fill in a Draw Conclusions graphic organizer as they read each two pages of the lesson. They can use the Quick Check questions to identify each conclusion.

Text Clues	Conclusions

What is light?

▶ **Discuss the Main Idea**

Explain to students that light is another form of energy. Ask:

- **What are some different sources of light?** Possible answers: the Sun; fire; lightning; light bulbs

- **How does light travel?** Students should recognize that light travels in a straight path.

- **How is reflected light similar to a bouncing tennis ball?** Possible answer: Just as light does, when a tennis ball strikes an object, it bounces off the ground in a different direction.

RESOURCES and TECHNOLOGY

📖 **Reading and Writing,** pp. 237–239

💿 **PuzzleMaker CD-ROM**

💿 **Presentation Toolkit CD-ROM**

Light Energy

e-Glossary

Read and Learn

▶ **Main Idea**
Light is a form of energy that allows you to see objects. Light moves in straight paths.

▶ **Vocabulary**

light, p.500

absorb, p.500

reflect, p.501

opaque, p.502

shadow, p.502

transparent, p.503

translucent, p.503

refract, p.503

e-Glossary
at **www.macmillanmh.com**

▶ **Reading Skill** ✓

Draw Conclusions

Text Clues	Conclusions

▶ **Technology** SCIENCE QUEST

Explore light with the Secret Agents.

What is light?

You see and use light every day. **Light** is a form of energy. It allows you to see objects. Light comes from many different sources. The Sun is Earth's main source of light. Fires and light bulbs are some other sources of light.

Light travels away from a source in a straight path. When you turn on a flashlight you can see a straight beam of light. Even light from the Sun travels millions of miles through space in a straight path. Light travels in a straight path until it hits an object.

Absorption

Light can be **absorbed** (ab•ZORBD), or taken in, when it hits an object. Black objects absorb almost all the light that hits them. White objects absorb almost no light.

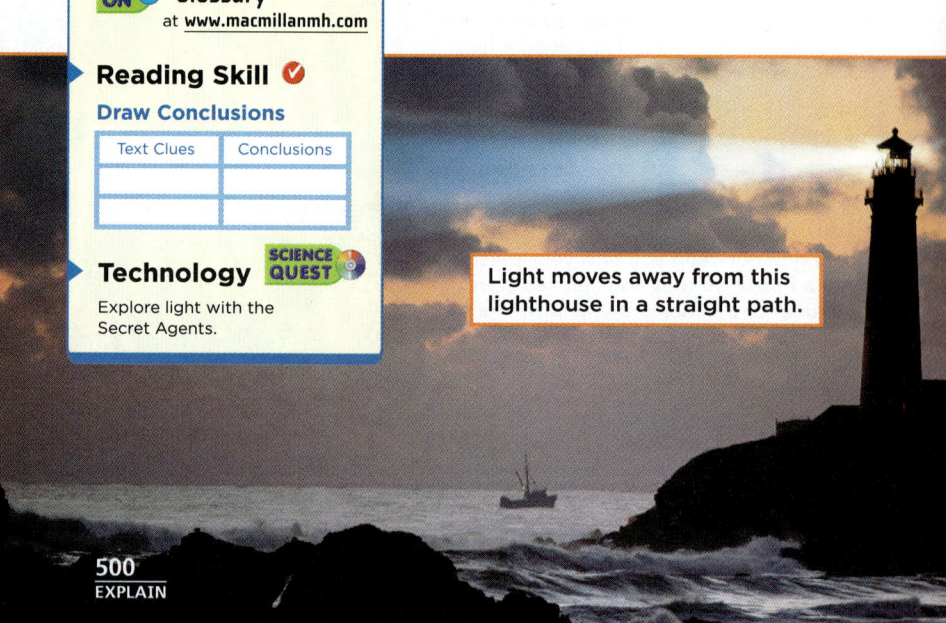

Light moves away from this lighthouse in a straight path.

Science Background

Bioluminescence is the ability of an organism to produce and emit its own light. It is the result of chemical reactions in which some of the energy released is converted to light. The chemical reactions that cause bioluminescence can occur within or outside the organism's cells. Bioluminescent arthropods include fireflies, glow worms, some centipedes, and some millipedes. Some mushrooms are bioluminescent. Many marine organisms, including angler fish, eels, corals, jellyfish, krill, and octopuses, can be bioluminescent.

See **Science Yellow Pages** in the Teacher Resource section for background information.

Professional Development For more Science Background and resources from **NSDL** visit **www.macmillanmh.com/nsdl/**

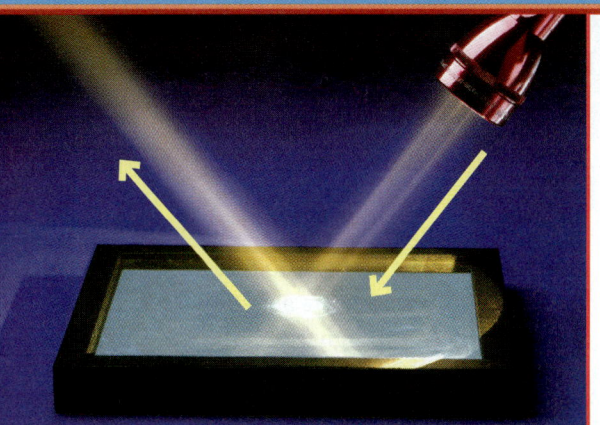

◀ When light hits some objects, it reflects off in a different direction.

Reflection

When light hits some objects, it **reflects** (ri•FLEKTS), or bounces off of them. It changes direction and then continues to move in a straight path.

Light bounces off objects in the same way that a ball bounces. When you hit a ball down, it bounces up. When light hits an object, it bounces off in a different direction.

You see an object when light from the object reaches your eyes. Most objects do not make their own light. You see those objects when light reflects off of them and to your eyes.

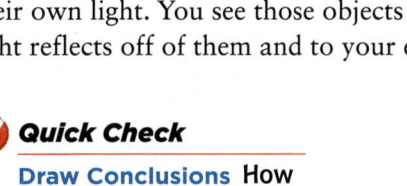

Mirrors are very smooth, shiny surfaces. They reflect almost all the light that hits them. ▼

501
EXPLAIN

✓ Quick Check

Draw Conclusions How can a mirror help you see behind you?

Critical Thinking Is it possible for you to see in the dark? Explain.

Differentiated Instruction

Leveled Activities

EXTRA SUPPORT   Refer students to the photo of the lighthouse on p. 500. Have students use encyclopedias and the Internet, if available, to research how lighthouses work to make such a bright and powerful beam of light. Have students write and illustrate reports to share with the class.

ENRICHMENT   Have students use encyclopedias and the Internet, if it is available, to research the history of mirrors. Encourage students to find out about when mirrors were first used and how they were made, how mirrors were used and made through history, and how mirrors are used and made today. Have students create an illustrated time line of their research.

▶ Explore the Main Idea

ACTIVITY   Use masking tape to tape two mirrors with their reflective sides together so they can be opened and closed like a book. Have students look at their reflections in the mirrors when the mirrors are flat and are partly closed at different angles. Have students share their observations with the class.

FAST TRACK Develop Vocabulary

light Point out to students that the word *light* can be used as a noun when referring to energy that allows you to see, a verb when used to describe the action of igniting a candle or fire, and an adjective when used to describe something that is not heavy or does not contain a lot of fat.

absorb *Word Origin* The word *absorb* comes from the Latin word *absorbere,* which means "to swallow up" or "to suck in." Explain that light that is absorbed appears to be "swallowed up" by an object.

reflect *Word Origin* Point out to students that the word *reflect* comes from two Latin words: *re-,* a prefix meaning "back," and *flectere,* meaning "to bend." Put together, the Latin words mean "to bend back."

▶ Use the Visuals

Refer students to the visuals on pages 500–501. Ask:

■ **What do you notice about the way the beam of light travels from the lighthouse?** The light travels in a straight path.

■ **How does the angle made by the flashlight beam striking the object compare to the angle made by the beam of light reflecting off the object?** The angles are the same.

✓ Quick Check Answers

• **Draw Conclusions** The mirror reflects light from objects behind back toward the eyes.

• **Critical Thinking** No; because we see objects only when light reflects off them and travels to the eyes.

What happens when light hits different objects?

▶ Discuss the Main Idea

Remind students of the demonstration you did at the beginning of the lesson, using opaque, transparent, and translucent materials. Ask:

- **What kinds of materials let light pass through?** transparent and translucent

- **How is a transparent material different from a translucent material?** A transparent material allows all or most light to pass through it. Translucent materials block some light from passing through.

- **What kind of objects will cause shadows to form?** opaque

Develop Vocabulary

opaque *Word Origin* The word *opaque* comes from the Latin word **opacus,** which means "shady" or "dark." Explain that an opaque surface appears dark because no light passes through it.

shadow *Scientific vs. Common Use* The word *shadow* is commonly used to refer to someone or something that follows something else very closely. For example, "the puppy 'shadows' the small boy everywhere he goes."

transparent *Word Origin* The word *transparent* comes from two Latin words: **trans-,** which means "through," and **parere,** which means "come in sight," "appear."

translucent *Word Origin* The word *translucent* comes from the Latin word **translucere,** which means "to shine through."

refract *Word Origin* The word *refract* comes from the Late Latin word **reflexux,** which means "a bending back." Explain that when white light is refracted by a prism, the waves are bent and broken up into the different colors that make up white light.

RESOURCES and TECHNOLOGY
🖊 **Presentation Toolkit CD-ROM**

SCIENCE QUEST *Light Energy*

▲ When the Sun is behind the tree, a shadow forms in front of the tree.

What happens when light hits different objects?

How do you stay dry on a rainy day? You stand under an umbrella. An umbrella blocks raindrops so they do not reach you. Opaque (oh•PAYK) objects act somewhat like an umbrella to light. **Opaque** objects block light from passing through them. A brick wall, a piece of cardboard, and even you are opaque. You cannot see through opaque objects.

Opaque objects can cause shadows to form. A **shadow** is a dark space that forms when light is blocked. You have probably seen your shadow on a sunny day. Your body blocked the sunlight. The shadow that formed had a shape similar to your body.

A shadow's size depends on where a light source is. The closer an object is to a light source, the bigger its shadow. Light coming from above creates a short shadow. As the light source gets lower, the shadow gets longer.

Your shadow follows you everywhere. Your shadow looks like you.

502
EXPLAIN

ELL Support

Vocabulary On the board write *light, reflect, refract, opaque, shadow, transparent,* and *translucent.* Model their pronunciation and have students repeat. Elicit the definitions and record them randomly on the board. Allow volunteers to match the word and its definition. Then have students give examples of the terms or use them in sentences.

BEGINNING Students say two or three of the terms and use simple words or point to pictures that illustrate the term.

INTERMEDIATE Students use each term in short phrases or simple sentences.

ADVANCED Students use each term in complete sentences.

Not all objects are opaque. Light can pass through some objects. **Transparent** (trans•PAYR•uhnt) objects allow light to pass straight through them. Air, glass, and clear plastic are transparent. You can see clearly through these objects because they allow light to pass through them.

Other objects are translucent (trans•LEW•suhnt). **Translucent** objects let light pass through them but scatter the light. You cannot see clearly through translucent objects for this reason. Wax paper and frosted glass are translucent.

Refraction

Light can refract (ri•FRAKT) when it passes from one material to another. To **refract** means to bend. Look at the photograph of the pencil. Is the pencil broken? No, it just looks that way. Light refracts when it passes from the air to the water. Then it bounces off the pencil. The light refracts again when it moves from the water back to the air. The bending light makes the pencil look broken.

 **Quick Check**

Draw Conclusions What three things do you need to make a shadow?

Critical Thinking Why does sunlight pass through a window and not a wall?

Frosted glass is translucent.

Clear glass is transparent.

▲ Bending light rays make this pencil look broken.

503
EXPLAIN

▶ **Explore the Main Idea**

ACTIVITY   Have students collect an assortment of from 15 to 20 different objects from the classroom or from home. Have students work together in pairs in a darkened classroom, shining a flashlight beam at each object to determine if it is transparent, translucent, or opaque. Record student findings in a data table on the board.

✓ **Quick Check Answers**

- **Draw Conclusions** To make a shadow, you need an opaque object, a source of light, and the opaque object in position so that it blocks the light

- **Critical Thinking** A window is made of a material that lets light pass through it. A wall is made of a material that blocks light.

Why can you see colors?

FAST TRACK ## Discuss the Main Idea

To lead a discussion about color, show students a prism with light passing through it to form the colors of the visible spectrum. Ask:

- **What color is the light going into the crystal?**
  Possible answers: white, clear, yellow

- **What color is the light coming out of the crystal?**
  Possible answers: rainbow colors; all the colors; any individual color

- **What do you think happened inside the crystal to change the color of the light?** Accept all reasonable answers.

▶ Develop Vocabulary

After students examine the visuals on pages 504–505, have them do a word-study activity that relates some of the vocabulary words they have learned to why they see color. Have them write a short paragraph describing why they see the color of a classmate's shirt using the vocabulary words *light, reflect,* and *absorb.*

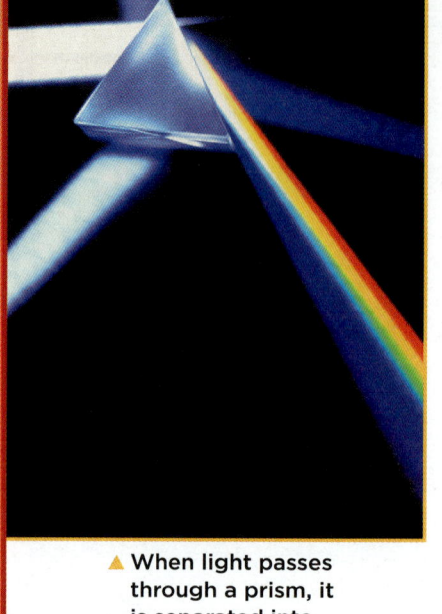

▲ When light passes through a prism, it is separated into different colors.

Why can you see colors?

What color is light from the Sun? You might say yellow or white. In fact it is a mixture of many colors. White light, such as sunlight, is made up of every color of light. To show this you can use a prism (PRIZ•uhm). A *prism* is a piece of glass that refracts light. Prisms separate white light into all of the colors that make it up. They do this by refracting each color of light a different amount.

When white light strikes a colored object, some colors of light are absorbed. Other colors of light are reflected. The reflected light enters your eyes. You see the object as the color of this reflected light.

Water vapor in the sky can act like tiny prisms. When water vapor refracts sunlight, a rainbow forms.

504
EXPLAIN

FACT ▶ White light is made up of every color of light.

RESOURCES and TECHNOLOGY

▷ **Visual Literacy,** p. 80

▷ **Activity Lab Book,** p. 227

🖸 **Presentation Toolkit CD-ROM**

SCIENCE QUEST *Light Energy*

When white light strikes a green leaf, the leaf absorbs all of the colors except for green. Only green light bounces off the leaf. This color is reflected to your eyes. You see the leaf as green. Something different happens when light strikes a red flower. Now the green light is absorbed. All colors are absorbed except for red. Only red light is reflected to your eyes. You see the flower as red. An object that absorbs all light that strikes it looks black. An object that reflects all light that strikes it looks white.

 **Quick Check**

Draw Conclusions What colors make up light from the Sun?

Critical Thinking Why does a banana appear to be yellow?

Seeing Colors

Read a Diagram

Why does the leaf look green?

Clue: Look at the color of light that is reflected.

LOG ON *Science in Motion* Watch colors at www.macmillanmh.com

505
EXPLAIN

≡ Quick Lab

Mixing Colors

1. **Predict** Look at the photo below. What happens to the color of the plate when you spin it?

2. Divide a white paper plate into eight equal parts. Color each section of the plate a different color.

3. **Observe** Carefully push a pencil into the center of the plate. Hold the plate away from your body. Spin it. What color do you see when the plate is spinning?

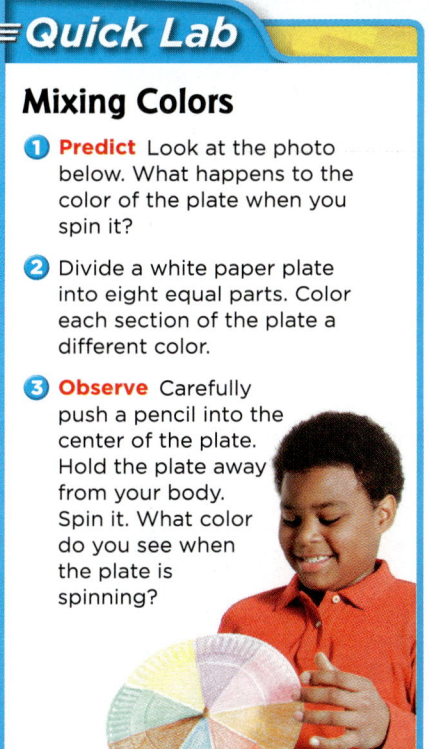

≡ Quick Lab

small groups   15 minutes

Objective Observe how white light is made up of different colors of light.

Materials paper plates, pencil, crayons

1. Accept all reasonable predictions.

2. Show students how to fold the plate in half, in half again, and then in half again to form eight sections. Encourage students to use their creativity when coloring the sections of the plate.

3. **Be Careful!** Remind students to be careful with the pencils so they do not poke their fingers. Students should observe that the colors on the spinning plate merge together and tend to appear white or light-colored.

Read a Diagram

Answer The leaf is green because it reflects green light and absorbs other colors of light.

LOG ON **Science in Motion** *Seeing Colors*

✔ **Quick Check Answers**

- **Draw Conclusions** all the different colors of light
- **Critical Thinking** It reflects yellow light and absorbs the other colors of light.

Differentiated Instruction

Leveled Activities

EXTRA SUPPORT   Have students write a short poem about a rainbow. Encourage them to use their poems to describe what a rainbow looks like or how a rainbow makes them feel.

ENRICHMENT   Have students use the Internet, if available, encyclopedias, other books, or magazines to research the colors of light. Have students create a poster that illustrates primary and secondary colors of light, and how three colors can form white light.

How do you see?

▶ Use the Visuals

Explain to students that they see objects when light that is reflected off the objects enters their eyes. Refer students to the diagram of the eye on page 506. Ask:

- **What is the first thing light passes through in your eye?** the cornea

- **What part of your eye does light pass through to get all the way inside your eye?** the pupil

Discuss the Main Idea

Have students read the paragraph on page 506, then lead a discussion on how the eye perceives objects. Ask:

- **Which parts of the eye does light pass through before it reaches the optic nerve?** cornea, then pupil, then lens

- **Which part of the eye refracts light to a small spot on the back of the eye?** the lens

- **How does the brain receive information to form a picture?** The optic nerve carries the information about the light to the brain.

✓ Quick Check Answers

- **Draw Conclusions** Light is reflected from the page. The light moves through the corneas and enters the pupils of the eyes. Lenses in the eyes then focus the light on the back part of the eyeballs. The brain makes sense of the pictures and words.

- **Critical Thinking** The iris causes the pupil to get smaller in bright light and larger in dim light.

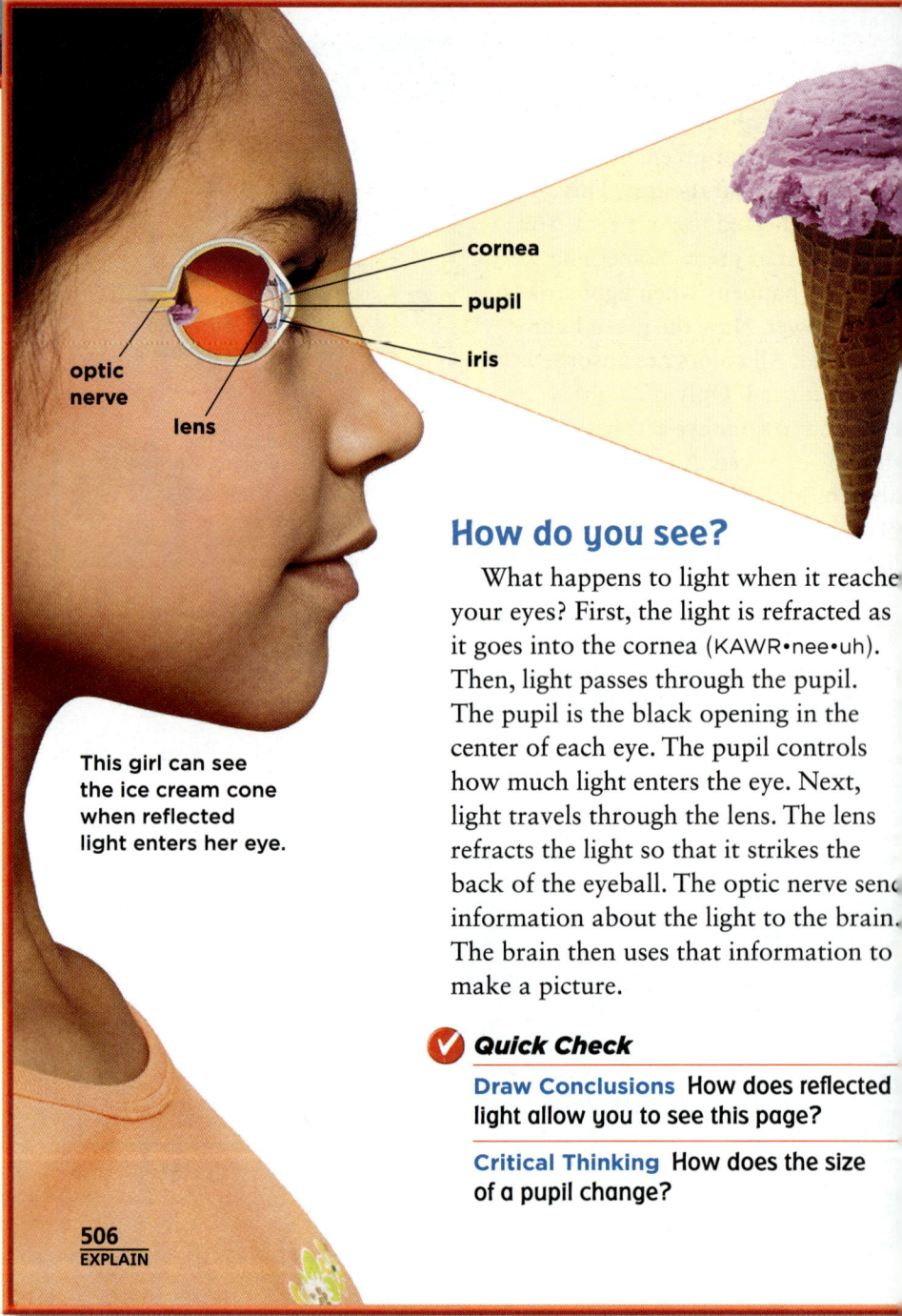

cornea
pupil
iris
optic nerve
lens

This girl can see the ice cream cone when reflected light enters her eye.

How do you see?

What happens to light when it reache[s] your eyes? First, the light is refracted as it goes into the cornea (KAWR•nee•uh). Then, light passes through the pupil. The pupil is the black opening in the center of each eye. The pupil controls how much light enters the eye. Next, light travels through the lens. The lens refracts the light so that it strikes the back of the eyeball. The optic nerve send[s] information about the light to the brain. The brain then uses that information to make a picture.

✓ Quick Check

Draw Conclusions How does reflected light allow you to see this page?

Critical Thinking How does the size of a pupil change?

506
EXPLAIN

Homework Activity

How Do Animals See?

Ask students to use encyclopedias, magazines, or the Internet, if it is available to them, to research how animals see. Encourage students to research animals that live in the water, such as whales and sharks, and different land animals such as birds, insects, spiders, reptiles, and mammals. Encourage them to research the structure of the eye in different animals, and what colors the animals can see. Have students present what they learn in an oral report.

Lesson Review

Visual Summary

Light travels away from a source in a straight path. Objects can reflect or absorb light.

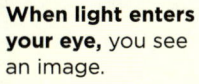

White light is made of many different colors. You see an object as the color of its reflected light.

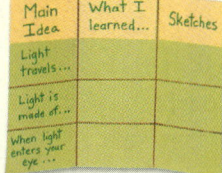

When light enters your eye, you see an image.

Make a  **Study Guide**

Make a Trifold Book. Use it to summarize what you learned about light.

Think, Talk, and Write

1 Main Idea What is light? What are some sources of light?

2 Vocabulary What happens when light is refracted?

3 Draw Conclusions Why does a school bus appear yellow and a fire truck appear red?

Text Clues	Conclusions

4 Critical Thinking How could you make the shadow of a marble look like the shadow of a tennis ball?

5 Test Prep A sheet of aluminum foil is an example of what type of material?
A translucent
B shadow
C transparent
D opaque

 Writing Link

Writing that Informs
Find out how you can protect your skin from absorbing too much sunlight. Find out why wearing white clothes and sunblock can help. Then write about it.

Art Link

Shadow Puppets
Use your hands and a flashlight to make shadows. Try to make different shapes and animals. Move your hand closer and farther away from the light. What happens to the shadow puppet?

LOG ON e-Review Summaries and quizzes online at www.macmillanmh.com

507
EVALUATE

---

3 Close

Lesson Review

▶ **Visual Summary**

Have students save their lesson Study Guides. They will use them at the end of the chapter for review

▶ **Make a** FOLDABLES **Study Guide**

See pages R27–R28 for instructions.

FAST TRACK

Think, Talk, and Write

1 Main Idea Light is a form of energy that allows you to see objects. The Sun, fires, light bulbs are some sources.

2 Vocabulary It bends.

3 Draw Conclusions

Text Clues	Conclusions
School bus reflects yellow light, absorbs other colors.	School bus appears yellow.
Fire truck reflects red light, absorbs other colors.	Fire truck appears red.

4 Critical Thinking Move the marble closer to the light source until the marble's shadow is the same size as the tennis ball's shadow.

5 Test Prep D

Writing Link Accept all reasonable responses based on scientific information.

Art Link The shadow puppet gets larger as the hand moves closer to the light source.

Reading in Science

Objective
- Summarize information in an article.

A Beam of Light

Genre: Nonfiction

Have students read the title and look at the pictures.

- **What can a beam of light help you do?** Answers will vary but should include: see in the dark, read, and so on.

Before Reading

Have students read the captions. Ask:

- **What kind of light will you learn about?**

Call on volunteers to tell what they know about lasers.

Discuss the work of surgeons with students. Ask:

- **How do surgeons perform operations? What tools do they use? Do you think a beam of light can be used to do operations?**

Tell students they will look for the answers to these questions as they read the article.

During Reading

Read the article together. Ask:

- **What is a laser?** a beam of light

- **How do you think the beam of light from a laser is different from the beam of light from a lamp or flashlight?** Possible answers: Light from a laser is more powerful, more concentrated, a thinner beam, more focused

- **How are lasers different from scalpels?** Lasers can cut tissue without causing blood loss.

- **What were lasers first used for?** to remove birthmarks on children's skin

Call on students to describe how lasers help people see better.

Reading in Science

A BEAM OF LIGHT

A laser creates a narrow beam of light.

Surgeons are doctors who perform operations to fix injuries or treat diseases. They can use scalpels—special tools with sharp blades—to cut through skin, muscles, and organs of the human body. Today, surgeons have another tool they can use to do operations. This tool is a beam of light!

This beam of light is called a *laser*. Lasers are very powerful. They can cut though the human body without causing much bleeding.

Lasers were first used to remove birthmarks on children's skin. Today, surgeons also use lasers to treat injuries to the brain, the heart, and many other parts of the body. Lasers are also used to improve people's eyesight.

 Write About It

Summarize Read the article again. List the most important information in a chart. Then use the chart to write a summary of the article.

 **Journal** Write about it online at www.macmillanmh.com

508
EXTEND

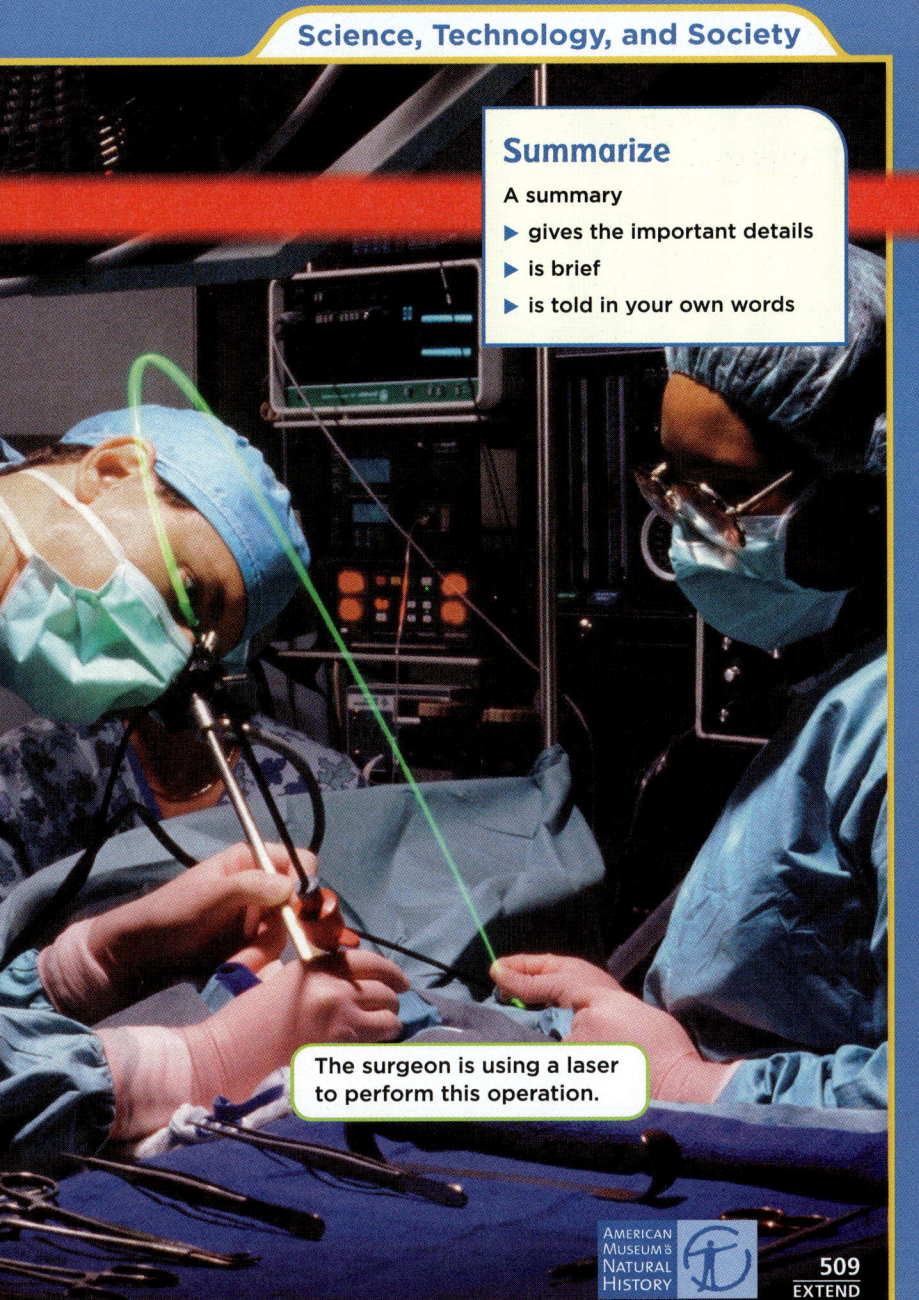

Summarize

A summary
▶ gives the important details
▶ is brief
▶ is told in your own words

The surgeon is using a laser to perform this operation.

AMERICAN MUSEUM & NATURAL HISTORY

509
EXTEND

After Reading

Display Graphic Organizer 5. Discuss with students how facts in the article can be used to make a summary. Have students identify important facts in the article. Write responses in the three smaller boxes of the organizer. Call on students to use those facts to create a summary sentence. Discuss their responses. Choose one summary to complete the organizer.

⊙ Write About It

Charts will vary, but students should note that the laser is a powerful beam of light used to cut through the human body without causing much bleeding. The use of lasers has grown over time, from removing birthmarks to treating injuries of the brain and heart and improving people's eyesight.

Integrate Writing

Multiple-Meaning Words

Write the word *treat* on the board. Ask students if they can think of two meanings for it. "help cure," "a snack" Point out that words spelled the same but having two different meanings are called homographs. Write the words *glasses* and *marks* on the board. Have students work with a partner to determine the multiple meanings of each word. Call on them to write sentences for each meaning of the word. Have partners read them to the class.

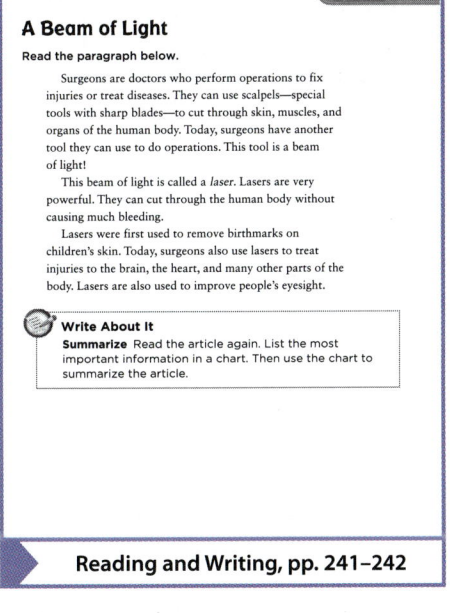

Name _____ Date _____   Reading in Science

A Beam of Light

Read the paragraph below.

Surgeons are doctors who perform operations to fix injuries or treat diseases. They can use scalpels—special tools with sharp blades—to cut through skin, muscles, and organs of the human body. Today, surgeons have another tool they can use to do operations. This tool is a beam of light!

This beam of light is called a *laser*. Lasers are very powerful. They can cut through the human body without causing much bleeding.

Lasers were first used to remove birthmarks on children's skin. Today, surgeons also use lasers to treat injuries to the brain, the heart, and many other parts of the body. Lasers are also used to improve people's eyesight.

⊙ **Write About It**
Summarize Read the article again. List the most important information in a chart. Then use the chart to summarize the article.

▶ Reading and Writing, pp. 241–242

RESOURCES and TECHNOLOGY

▶ **Reading and Writing,** pp. 241–242

▶ **Technology: A Closer Look,** Lesson 4

LOG ON e-**Journal** Online research and writing

p. TR26 **Writing Rubric**

Plan Your Lesson

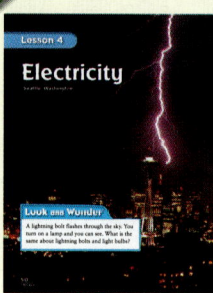

Lesson 4 Electricity

Objective
- Describe electrical charge.
- Identify the parts of a circuit.

Reading Skill Sequence

First

Next

Last

Graphic Organizer 7, p. TR9

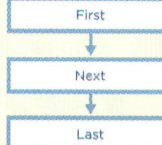 **Professional Development** Look for **NSDL** to find recommended Science Background materials from the National Science Digital Library

FAST TRACK

Lesson Plan When time is short, follow the Fast Track and use the essential resources.

1 Introduce
Look and Wonder, p. 510
Resource Alternative Explore, p. 230

2 Teach
Develop Vocabulary, p. 513
Discuss the Main Idea, p. 514
Resource Visual Literacy, p. 81

3 Close
Think, Talk, and Write, p. 517
Resource Assessment, p. 155

▶ Reading and Writing

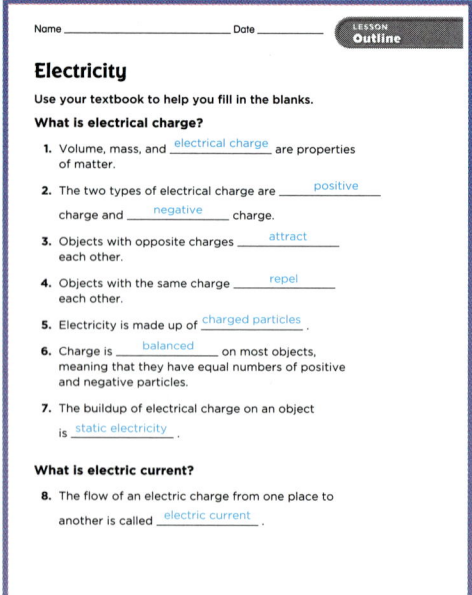

Outline, pp. 243–244
Also available as a student workbook

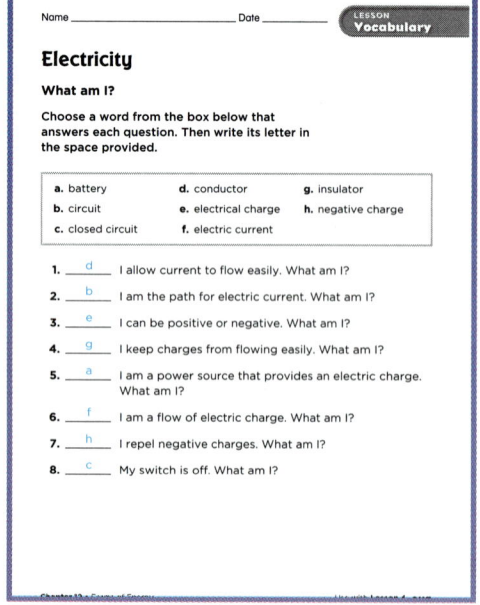

Vocabulary, p. 245
Also available as a student workbook

▶ Visual Literacy

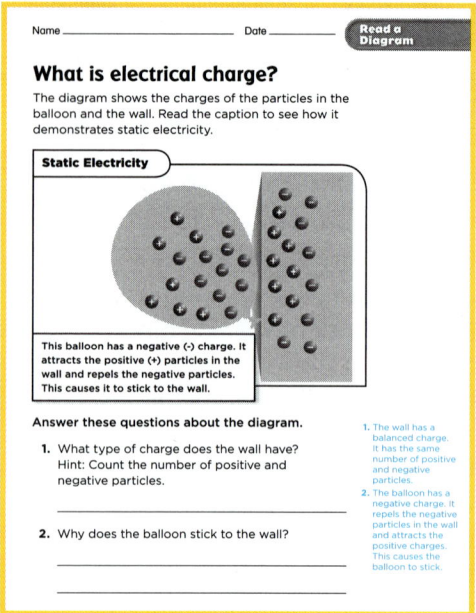

Read a Diagram, p. 81
Also available as a transparency

▶ Activity Lab Book

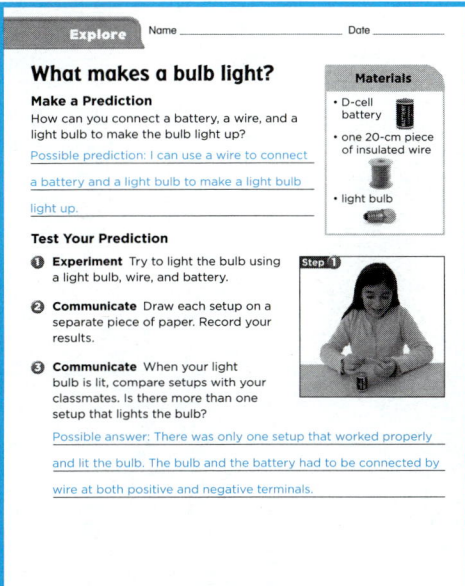

Explore, pp. 228–229
Also available as a student workbook

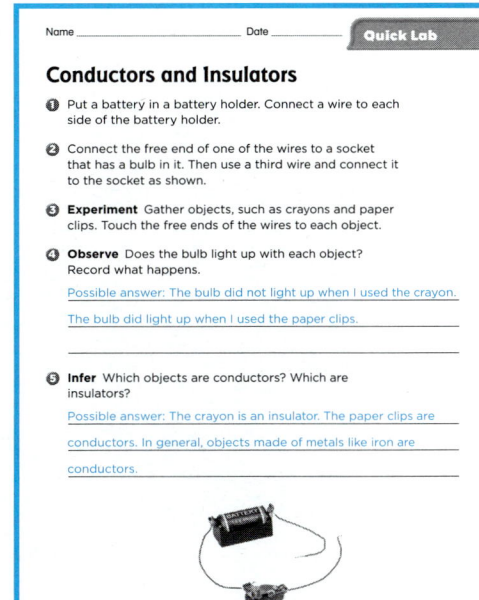

Quick Lab, p. 231
Also available as a student workbook

▶ Assessment

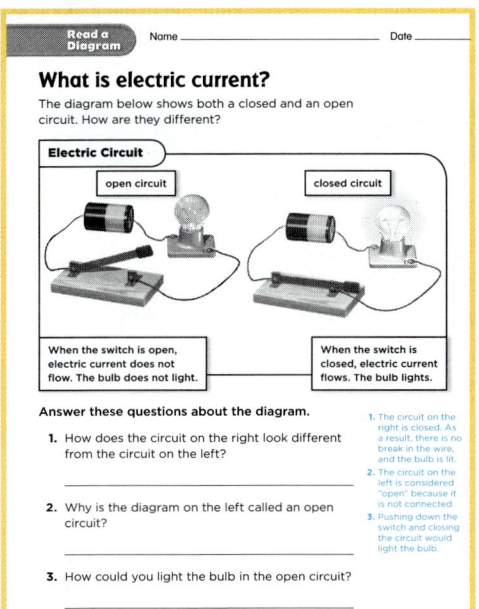

Read a Diagram, p. 82
Also available as a transparency

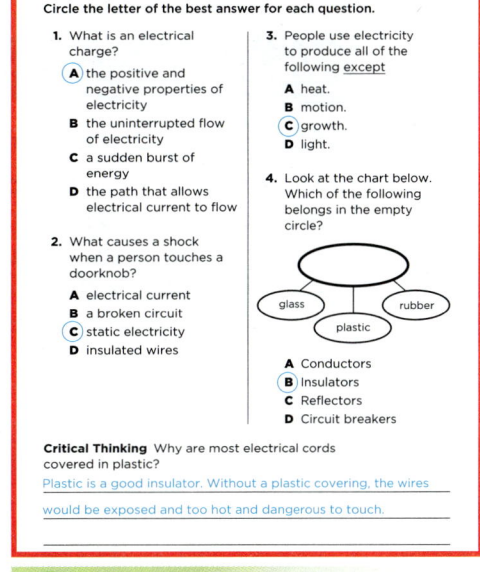

Lesson Test, p. 155
Also available as a student workbook

ADDITIONAL RESOURCES

p. 61

p. 142

pp. 81–82

pp. 105–106

81–82

194–198

Technology

- Science Activity DVD
- Instructional Navigator CD-ROM
- Presentation Toolkit CD-ROM
- 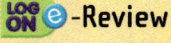 e-Review
-  NSDL

Lesson 4 Electricity

Objectives
- Describe electrical charge.
- Identify the parts of a circuit.

1 Introduce

▶ Assess Prior Knowledge

Ask students what they know about electricity. Ask:

- **What kind of energy is used when you turn on a light?** electricity

- **What kind of energy is used when you turn on the TV?** electricity

- **What other items in your home use electricity?** Possible answers: radio, computer, hair dryer, stove, toaster, fan

Look and Wonder

Invite students to share their responses to the Look and Wonder statement and question.

- **What is the same about lightning bolts and light bulbs?** Possible answers: They are both examples of electricity.

Write ideas on the board and note any misconceptions that students might have. Address these misconceptions as you teach the lesson.

RESOURCES and TECHNOLOGY
- Activity Lab Book, pp. 228–230
- Activity Flipchart, p. 61
- Science Activity DVD

Lesson 4

Electricity
Seattle, Washington

Look and Wonder

A lightning bolt flashes through the sky. You turn on a lamp and you can see. What is the same about lightning bolts and light bulbs?

510
·ENGAGE

Warm Up

Start with a Discussion

Show students photographs of electricity in action, such as dramatic lightning strikes, satellite images of Earth at night showing how cities are illuminated, and members of a family using electricity in their daily lives. Discuss with students how all of the visuals show electricity in action.

Explore
Inquiry Activity

What makes a bulb light?

Make a Prediction
How can you connect a battery, a wire, and a light bulb to make the bulb light up?

Test Your Prediction
❶ **Experiment** Try to light the bulb using a light bulb, wire, and battery.

❷ **Communicate** Draw each setup. Record your results.

❸ **Communicate** When your light bulb is lit, compare setups with your classmates. Is there more than one setup that lights the bulb?

Draw Conclusions
❹ How many setups could you find that made the bulb light?

❺ **Infer** Look at the setups that lit the bulb. What do you think is necessary to make the bulb light up?

Explore More
Experiment How could you light two bulbs using only one battery? Can you think of more than one way? Try it.

Materials

D-cell battery

one 20-cm piece of insulated wire

light bulb

Step ❶

Step ❷

Does not work

511
EXPLORE

Alternative Explore

What does a balloon attract?

Materials balloon, wool cloth, foam peanuts, puffed rice cereal, pieces of paper, salt and pepper

Tell students that objects with opposite electric charge are attracted to each other, and that when they rub a balloon with a wool cloth, the balloon becomes more negatively charged. In this activity, students will observe which materials a charged balloon attracts.

Alternative Explore

Name _____ Date _____

What does a balloon attract?

Make a Prediction
When you rub a balloon with a wool cloth, negative electrical charges are transferred from the cloth to the balloon. What kind of object do you think the balloon will attract?
Possible prediction: The balloon will attract something with positive electric charge.

Materials
• balloon
• wool cloth
• foam peanuts
• puffed rice cereal
• salt and pepper

Test Your Prediction

❶ **Experiment** Blow up a balloon and tie the end closed. Rub the balloon for a few seconds with the wool cloth.

❷ **Experiment** Hold the balloon near some foam peanuts. Are they attracted to the balloon? Then hold the balloon close to other materials. Are they attracted to the balloon? Record your results in the table below.

Materials Attracted to the Balloon	Materials Not Attracted to the Balloon

❸ Was your prediction correct?
Yes. The balloon with the negative charge attracted objects with a positive charge.

▶ **Activity Lab Book, p. 230**

Explore
 small groups   30 minutes

Plan Ahead Acquire and organize all the necessary materials before the class assembles.

Purpose In this activity, students will demonstrate how electric current flows in a circuit.

Structured Inquiry

Make a Prediction Possible prediction: The wire will need to be connected to both sides of the light bulb to make it light up.

Test Your Prediction

❷ **Communicate** Student drawings will vary but should include a wire from the battery to the bulb socket, a bulb screwed into the socket, and a wire leading from the other side of the bulb socket back to the battery.

❸ **Communicate** Students should recognize that a setup similar to that described above is required to make the bulb light up.

❺ **Infer** Students should recognize that a complete circuit is necessary to light the bulb.

Guided Inquiry   Explore More

Experiment Students should recognize that the battery can light more than one bulb as long as the bulbs are arranged in a complete circuit.

Open Inquiry

Ask students what would happen if one of the light bulbs in a circuit was broken. Ask them to think of their own question about how electric current flows. Then have them make a plan and carry out an experiment to answer that question. Ask:

Will other bulbs light if one bulb in the circuit is broken?

Read and Learn

Main Idea Have students take a photo tour of the lesson. Ask them to list three things they think they will learn about.

Vocabulary Write the vocabulary words on the board. Ask students to look for the definitions of these terms in the lesson. Have volunteers read the definitions aloud when they find them.

Reading Skill Sequence

Graphic Organizer 7 Have students fill in a Sequence graphic organizer as they read each two pages of the lesson. They can use the Quick Check questions to identify each sequence.

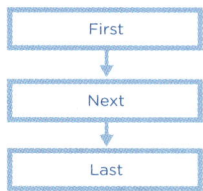

First

↓

Next

↓

Last

What is electrical charge?

▶ **Discuss the Main Idea**

Have students describe what they know about electrical charges. Ask:

- **Have you ever received a shock after you walked across a carpet and then touched an object or person?** Accept all reasonable responses.

- **Have you ever combed your hair and had it stand on end?** Accept all reasonable answers.

- **What do you think causes these events to happen?** Students should recognize that electricity or electrical charge causes these events.

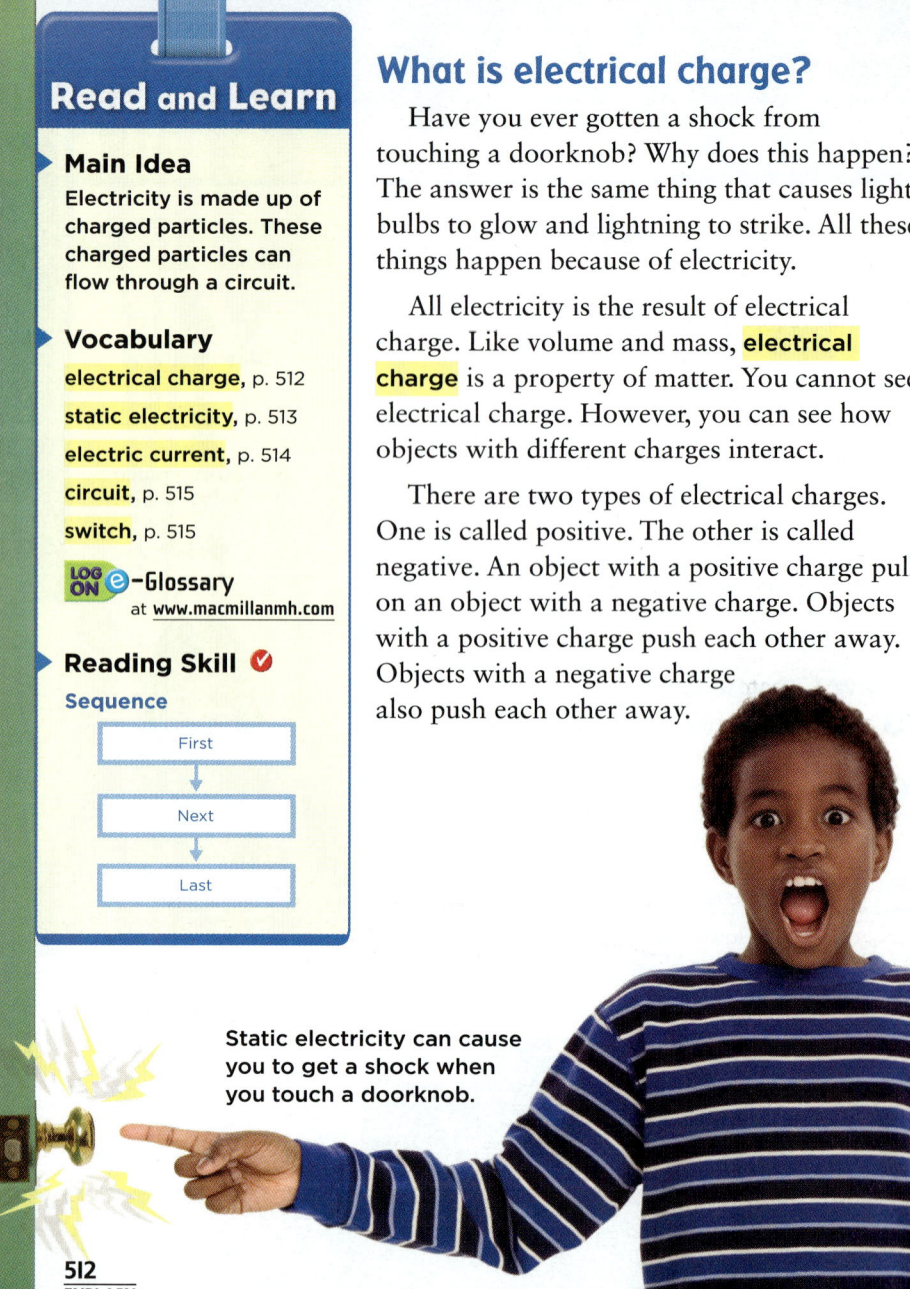

Read and Learn

▶ **Main Idea**
Electricity is made up of charged particles. These charged particles can flow through a circuit.

▶ **Vocabulary**
electrical charge, p. 512
static electricity, p. 513
electric current, p. 514
circuit, p. 515
switch, p. 515

e-Glossary
at www.macmillanmh.com

▶ **Reading Skill** ✓
Sequence

First

↓

Next

↓

Last

What is electrical charge?

Have you ever gotten a shock from touching a doorknob? Why does this happen? The answer is the same thing that causes light bulbs to glow and lightning to strike. All these things happen because of electricity.

All electricity is the result of electrical charge. Like volume and mass, **electrical charge** is a property of matter. You cannot see electrical charge. However, you can see how objects with different charges interact.

There are two types of electrical charges. One is called positive. The other is called negative. An object with a positive charge pulls on an object with a negative charge. Objects with a positive charge push each other away. Objects with a negative charge also push each other away.

Static electricity can cause you to get a shock when you touch a doorknob.

512
EXPLAIN

Science Background

Why Does Lightning Strike? Lightning is a visible bright flash of static electricity that is created from the discharge of a current between areas of opposite charge. Lightning strikes somewhere on Earth approximately 100 times each second. Lightning can be contained within a cloud, can move from cloud to cloud, can move from the cloud to the ground, and can even move from the ground to a cloud.

See **Science Yellow Pages** in the Teacher Resource section for background information.

Professional Development For more Science Background and resources from **NSDL** visit www.macmillanmh.com/nsdl/

Static Electricity

All objects are made up of charged particles. Most objects have the same number of positive particles as negative particles. The charge is balanced. When two objects touch, however, negative particles can move from one object to the other. Negative particles build up on one object. That object now has a negative charge. A buildup of electrical charge is called **static electricity.**

Rub a balloon on a sweater and hold it near a wall. The balloon sticks to the wall! When you rub the balloon, negative particles move from the sweater to the balloon. The balloon gets a negative charge. It repels the negative particles in the wall and attracts the positive particles. This causes it to stick to the wall.

Static electricity is what sometimes causes you to get a shock when you touch a doorknob. When you walk across the floor, negative particles move from the floor to your body. You get a negative charge. When you touch a doorknob, the negative particles move from you to the knob. You feel this as a shock. When static electricity moves from one object to another, it is called a *discharge*.

▲ This balloon has a negative charge. It attracts the positive (+) particles in the wall and repels the negative (-) particles. This causes it to stick to the wall.

Quick Check

Sequence What happens when you rub a balloon with a wool cloth?

Critical Thinking Why do clothes stick together when they come out of the dryer?

513
EXPLAIN

FAST TRACK Develop Vocabulary

electrical charge *Word Origin* Point out to students that the words *electricity* and *electrical* were first used during the 1600s by William Gilbert, scientist to Queen Elizabeth I of England. He based the term on the Greek word *elecktron,* which means "amber." In about 600 B.C., Greek scientist and philosopher Thales described static electricity when he rubbed a piece of amber with a silk cloth.

static electricity *Scientific vs. Common Use* Point out to students that in common usage, the word *static* refers to something that stays the same or shows little movement.

▶ Explore the Main Idea

ACTIVITY Provide balloons to students and have them demonstrate the examples described in the text, such as rubbing a balloon on their clothing and then holding the charged balloon to the wall or their hair.

Quick Check Answers

- **Sequence** Negative particles move from the cloth to the balloon. The balloon gets a negative charge and the wool gets a positive charge.

- **Critical Thinking** The clothes rub together in the dryer. This causes charges to move. Some clothes get a positive charge, and some clothes get a negative charge. Opposite charges attract each other, so the clothes stick together.

Differentiated Instruction

Leveled Activities

EXTRA SUPPORT Ask students to conduct research, using the Internet, if available, or an encyclopedia, to make an illustrated time line of at least 10 events in the history of electricity. Encourage students to share their time lines with the class.

ENRICHMENT Encourage students to conduct research, using encyclopedias or the Internet, into the contribution of several scientists to the study of electricity. Suggest that students research individuals such as Marie Curie, Thomas Alva Edison, Mae Jemison, Benjamin Franklin, and others. Have students prepare short oral reports to share with the class.

What is electric current?

FAST TRACK Discuss the Main Idea

Lead a discussion with students about different forms of energy. Ask:

- **What types of energy have you already learned about?** Possible answers: heat, light, sound

- **How can electricity be used to produce these forms of energy?** Possible answers: Electricity produces heat in a hair dryer. Electricity produces sound in a radio. Electricity produces light in a lamp.

▶ Develop Vocabulary

electric current *Scientific vs. Common Use* Point out to students that the word *current* also refers to the flow of water in a river or the ocean. The word also refers to something that is happening now or has happened recently, such as a *current event*.

circuit *Word Origin* Explain to students that the word *circuit* comes from the Latin word **circuitus,** which means "going around." Ask students to relate this meaning to an *electric circuit*. An electric circuit is made of connected parts that allow the current to flow.

switch *Scientific vs. Common Use* Point out that the word *switch* is also commonly used to indicate a shift or a change. Ask students to relate these meanings of the word to its use in *electric switch*.

RESOURCES and TECHNOLOGY

▷ **Visual Literacy,** pp. 82

💿 **Presentation Toolkit CD-ROM**

What is electric current?

Charged particles can build up on objects. They can also be made to flow. A flow of charged particles is called an **electric current** (KUR•uhnt). You use electric current every day. Electric current provides the energy you need to power lights, radios, computers, hair dryers, and many other products. We use energy from electric current to produce heat, light, sound, and motion.

▲ Electrical energy is changed into heat inside this toaster.

These headphones change electrical energy into sound. ▶

514
EXPLAIN

ELL Support

Use Picture Clues Review the terms *electric current* and *circuit*. Model pronunciation and have students repeat. Point out that *cuit* is pronounced /kit/. Refer students to the photos of open and closed circuits on p. 515. As students look at each picture, have them point to and name the battery, switch, wires, and light bulbs. Then ask students to point to the correct visual when you ask them to identify a closed and an open circuit.

BEGINNING Students can name and point to pictures of a closed circuit and an open circuit and their parts.

INTERMEDIATE Students use short phrases to describe the parts of a closed and open circuit and how they work.

ADVANCED Students use complete sentences to identify the parts of a closed and open circuit and tell how they work.

Circuits

Electric current needs a path, or circuit (SUR•kit), through which to flow. A **circuit** is a path that is made of parts that work together to allow current to flow. Look at the diagram on this page. Wires connect the bulb to a battery. The battery is the circuit's power source.

To keep an electric current moving, a circuit cannot have any breaks. A complete unbroken circuit, like the one shown on top, is called a *closed circuit*. A circuit with breaks or openings is called an *open circuit*.

Switches

You can use a switch (SWICH) to open and close a circuit. A **switch** allows you to control the flow of current. When a switch is in the on position, there is no gap in the path. The circuit is closed and current can flow. Turn the switch off, and there is a gap in the path. The circuit is open, and current does not flow.

✔ Quick Check

Sequence What happens when you close the switch on a circuit?

Critical Thinking Where can you find circuits in your home?

 **FACT** Current flows through the light bulb back to the battery.

Electric Circuit

closed circuit

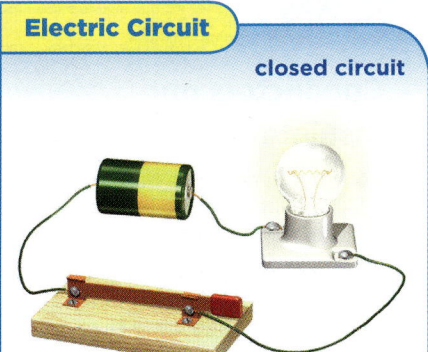

When the switch is closed, electric current flows. The bulb lights.

open circuit

When the switch is open, electric current does not flow. The bulb does not light.

Read a Diagram

Why is the second bulb not lit?

Clue: Compare the paths in each diagram.

515
EXPLAIN

▶ Address Misconceptions

A common misconception is that current stops when it reaches a light bulb or some other part of a circuit.

FACT **Current flows through the light bulb back to the battery.** Remind students that in a closed circuit, the current flows continuously through all the parts of the circuit.

Read a Diagram

Answer The switch is open, so the wires do not make a complete path for the current.

✔ Quick Check Answers

- **Sequence** The circuit is closed and current will flow.
- **Critical Thinking** Possible answers: toys, radios, toasters, lamps, computers

Differentiated Instruction

Leveled Questions

EXTRA SUPPORT How does a switch allow you to control the flow of electricity in a circuit? When the switch is open, the electric current cannot flow. When the switch is closed, the electric current can flow.

ENRICHMENT Imagine that you take out a string of lights and plug them in. You plug the lights in, but they do not work. After unplugging the string of lights, you discover that one of the bulbs is missing. Does the missing bulb explain why the string of lights does not work? Yes; the broken bulb makes the circuit open so the electric current cannot flow.

What are conductors and insulators?

 small groups 15 minutes

Objective Observe that only some materials are good conductors of electricity.

Materials battery, battery holder, wire, light bulb, bulb socket, crayons, paper clips

2 Make sure that students have correctly set up the circuit as shown in the illustration.

4 Students should observe that the light bulb does not light up with every object.

5 Students should infer that metallic objects, such as a copper penny, are good conductors of electricity, and that most nonmetallic objects, such as crayons, are not good conductors.

▶ Develop Vocabulary

Review with students the meaning of a conductor and insulator from Lesson 1. Ask:

- **How is a heat insulator similar to an electric insulator?** A heat insulator does not allow heat to flow through it easily. An electric insulator does not allow electricity to flow through it easily.

- **How are a heat conductor and an electric conductor similar?** A heat conductor allows heat to pass through it easily. An electric conductor allows electricity to pass through easily.

✔ Quick Check Answers

- **Sequence** It is blocked from flowing.

- **Critical Thinking** Copper is a good conductor of electric current, it allows the current to flow easily in a circuit.

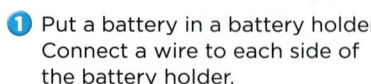

Conductors and Insulators

1 Put a battery in a battery holder. Connect a wire to each side of the battery holder.

2 Connect the free end of one of the wires to a socket that has a bulb in it. Then use a third wire and connect it to the socket as shown.

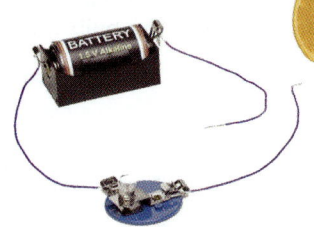

3 **Experiment** Gather objects, such as crayons and paper clips. Touch the free ends of the wires to each object.

4 **Observe** Does the bulb light up with each object? Record what happens.

5 **Infer** Which objects are conductors? Which are insulators?

Copper wires are conductors. The plastic around each wire is an insulator. ▶

What are conductors and insulators?

The electric current in your home flows through wires. These wires are usually made of copper and are wrapped in plastic. Copper is a material that allows current to flow through it very easily. Materials that allow current to flow easily are called conductors (kuhn•DUK•tuhrz). Most metals are conductors.

Plastic is wrapped around the wires in your home because plastic is an insulator (IN•suh•lay•tuhr). An insulator is a material that does not allow current to flow easily through it. The plastic coating on wires does not let current flow through it. This protects you from getting a shock. Glass, plastic, and rubber are good insulators.

✔ Quick Check

Sequence What happens to current when it reaches an insulator?

Critical Thinking Why are the wires in a circuit often made of copper?

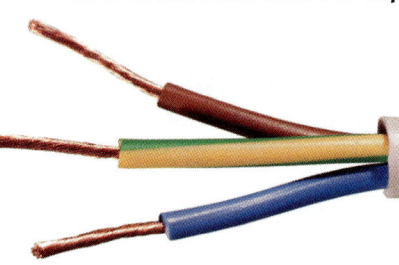

Homework Activity

How Can You Save Electricity?

Have students contact the local electric utility company and find out how to save electrical energy. Have students create posters that illustrate at least three ways that they could save electricity each day in their homes or at school. Encourage students to share their findings with the class. If possible, encourage students to implement their electricity-saving practices in the classroom.

Lesson Review

Visual Summary

 Electricity is made up of charged particles. **Static electricity** is a buildup of charged particles.

 **Electric current** travels in a path called a circuit. A switch can control the flow of current.

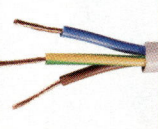 **Conductors** allow electric current to flow through them easily. **Insulators** do not.

Make a FOLDABLES **Study Guide**

Make a Three-Tab Book. Use it to summarize what you learned about electricity.

Static electricity is …

Electric current travels …

Conductors and insulators are …

Think, Talk, and Write

1 **Main Idea** What is electrical charge?

2 **Vocabulary** What is a circuit?

3 **Sequence** How do you get a shock from touching a doorknob?

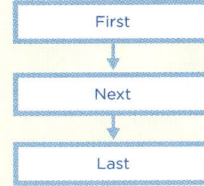

| First |
| Next |
| Last |

4 **Critical Thinking** You turn the switch on a flashlight. The light does not come on. List things that might be wrong with the flashlight.

5 **Test Prep** Which changes electrical energy into motion?
- **A** toaster oven
- **B** paper airplane
- **C** flashlight
- **D** electric train

 **Writing Link**

Expository Writing
Make a list of all the ways your life is made easier by electricity. Describe how you would have to do things differently if there were no electricity.

 Health Link

Make a Poster
Electrical devices can be dangerous if not used properly. Research how to use electricity safely. Make a poster about it.

 e-Review Summaries and quizzes online at www.macmillanmh.com

517
EVALUATE

ENGAGE   EXPLORE   EXPLAIN   **EVALUATE**   EXTEND

Lesson Review

▶ **Visual Summary**

Have students save their lesson Study Guides. They will use them at the end of the chapter for review.

▶ **Make a** FOLDABLES **Study Guide**

See pages R27–R28 for instructions.

FAST TRACK Think, Talk, and Write

1 **Main Idea** An electrical charge is a physical property of a material that gives the material a negative or positive charge. Like charges repel each other, and opposite charges attract each other.

2 **Vocabulary** a path made of parts designed to allow electricity to flow through them

3 **Sequence**

| I run on a carpet, causing negative charges from the carpet to me. |
| I touch the doorknob, the negative charge goes from my hand to the knob. |
| I get a shock. |

4 **Critical Thinking** Possible answers: The circuit is not closed. The battery is dead. The light bulb is burned out.

5 **Test Prep** D

 **Writing Link** Encourage students to be creative in their responses.

 Health Link Posters should include specific examples of how to prevent electrical injuries.

Formative Assessment

Approaching Have students continue making vocabulary flash cards using index cards. Have them write the vocabulary word on one side of the index card and the definition on the other side. Have students add these flash cards to the cards they made for Lessons 1, 2, and 3. Have them quiz each other using the full set of flash cards.

On-Level Ask students to make and label their own drawings of an open and closed electric circuit.

Challenge Have students use reference sources to find out how a battery works. Have students write and illustrate a short report of their research to share with the class.

 **Key Concept Cards** For student intervention, see the prescribed routine on **Key Concept Cards 81 and 82.**

RESOURCES and TECHNOLOGY

▶ **Reading and Writing,** p. 246

▶ **Activity Lab Book,** p. 231

▶ **School to Home Activities,** pp. 105–106

e-Review Narrated Summary and Quiz

Progress Reporter Assessments

Writing in Science

Objective

- Write a persuasive letter to a community leader.

Other Energy Sources

Learn It

Tell students that a good persuasive letter includes facts to convince others to agree with a certain opinion.

Try It

Ask students:

- **What purpose do the questions at the end of the selection serve?** They tie together the information and prompt the reader to make a decision.

Apply It

- Have groups of students look at the editorial page of a local newspaper and analyze a letter to the editor. Ask them to separate fact from argument. Ask:

- **How effective are the different letters in persuading you to adopt their points of view?** Answers will vary. Encourage students to be critical about what they read.

 Write About It

- Help students identify and write a letter to a community leader who is involved in energy issues. Tell students to begin by writing their topic sentence and then listing reasons, facts, and examples to support their opinion and save their best reason for last. Edit the letters for clarity and composition.

RESOURCES and TECHNOLOGY

▶ **Reading and Writing,** pp. 247–248

e-Journal Online research and writing

Writing Rubric P. TR26

Writing in Science

Other Energy Sources

Most of the energy we use to produce electricity comes from burning oil, coal, or natural gas. These energy sources are limited. They cannot be reused or replaced easily. There are other sources of energy that can be replaced in short periods of time. Wind can turn windmills to make energy. Energy from the Sun can be collected by solar panels. Do you think it is important to find other sources of energy? What are some ways you can encourage people to use other sources of energy?

These solar panels use the Sun's energy to produce electricity.

Persuasive Writing

A persuasive letter

▶ clearly states an opinion

▶ supports the opinion with reasons and facts

▶ persuades the reader to agree with that opinion

These windmills use the wind's energy to produce electricity.

518
EXTEND

 **Write About It**
Persuasive Writing
Write a persuasive letter to a community leader. Tell why you think it is important to find other sources of energy. Be sure to follow the form of a formal letter.

e-Journal Write about it online at www.macmillanmh.com

Integrate Writing

Using Solar Energy

- Using reference materials, have students find out more about solar energy and how it is being used. Ask them to form an opinion about whether they would like their area to use solar energy as an energy source.

- Tell students who agree with one another to form groups and produce a persuasive letter to convince local citizens to agree with their opinion. Have groups share letters.

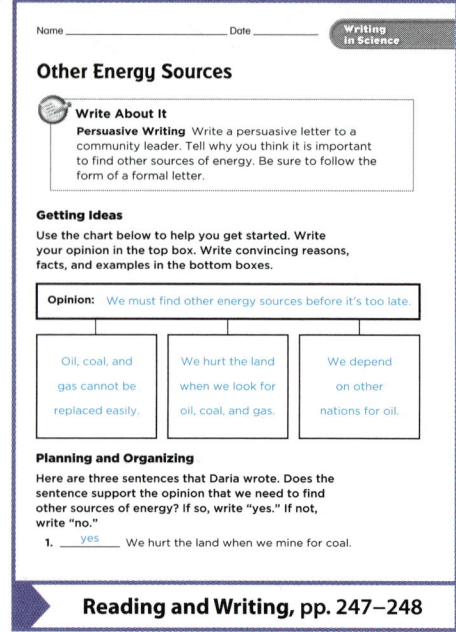

Name _____ Date _____   Writing in Science

Other Energy Sources

Write About It
Persuasive Writing Write a persuasive letter to a community leader. Tell why you think it is important to find other sources of energy. Be sure to follow the form of a formal letter.

Getting Ideas
Use the chart below to help you get started. Write your opinion in the top box. Write convincing reasons, facts, and examples in the bottom boxes.

| Opinion: We must find other energy sources before it's too late. |

| Oil, coal, and gas cannot be replaced easily. | We hurt the land when we look for oil, coal, and gas. | We depend on other nations for oil. |

Planning and Organizing
Here are three sentences that Daria wrote. Does the sentence support the opinion that we need to find other sources of energy? If so, write "yes." If not, write "no."

1. ___yes___ We hurt the land when we mine for coal.

▶ **Reading and Writing,** pp. 247–248

Math in Science

The Cost of Energy

The chart below shows the amount of energy used by several appliances in a year. The energy used is measured in kilowatt-hours (kWh) and costs 10 cents ($0.10) per kWh. Multiply to find out how much it costs to use these appliances for a year.

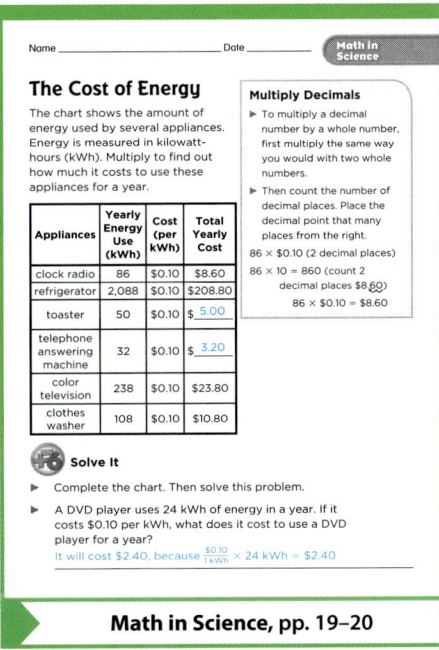

Solve It

Copy the chart and fill in the missing information. Then solve these problems.

▶ A DVD player uses 24 kWh of energy in a year. If it cost $0.10 per kWh, what does it cost to use a DVD player for a year?

▶ If you save $0.50 per week for one year (52 weeks), how much will you have saved in one year?

Multiply Decimals

▶ To multiply a decimal number by a whole number, first multiply the same way you would with two whole numbers.

▶ Then, count the number of decimal places. Place the decimal point that many places from the right.

86 x $0.10 (2 decimal places)

86 x 10 = 860 (count 2 decimal places 8.60)

86 x $0.10 = $8.60

Appliances	Yearly Energy Use (kWh)	Cost (per kWh)	Total Yearly Cost
clock radio	86	$0.10	$8.60
refrigerator	2,088	$0.10	$208.80
toaster	50	$0.10	$ _____
telephone answering machine	32	$0.10	$ _____
color television	238	$0.10	$23.80
clothes washer	108	$0.10	$10.80

519
EXTEND

Integrate Math

The Real Cost

- Have students research the actual cost of electricity in your area.

- Tell them to rewrite the chart using the actual figures.

Name _____ Date _____

Math in Science

The Cost of Energy

The chart shows the amount of energy used by several appliances. Energy is measured in kilowatt-hours (kWh). Multiply to find out how much it costs to use these appliances for a year.

Multiply Decimals

▶ To multiply a decimal number by a whole number, first multiply the same way you would with two whole numbers.

▶ Then count the number of decimal places. Place the decimal point that many places from the right.

86 × $0.10 (2 decimal places)

86 × 10 = 860 (count 2 decimal places $8.60)

86 × $0.10 = $8.60

Appliances	Yearly Energy Use (kWh)	Cost (per kWh)	Total Yearly Cost
clock radio	86	$0.10	$8.60
refrigerator	2,088	$0.10	$208.80
toaster	50	$0.10	$ 5.00
telephone answering machine	32	$0.10	$ 3.20
color television	238	$0.10	$23.80
clothes washer	108	$0.10	$10.80

Solve It

▶ Complete the chart. Then solve this problem.

▶ A DVD player uses 24 kWh of energy in a year. If it costs $0.10 per kWh, what does it cost to use a DVD player for a year?

It will cost $2.40, because $\frac{$0.10}{1 kWh}$ × 24 kWh = $2.40

Math in Science, pp. 19–20

Math in Science

Objective

■ Calculate the cost of several of an item given the price of one.

The Cost of Energy

Learn It

Review multiplying with two decimal places. Tell students that the product will always have the same number of decimal places as both factors together have.

■ **If one factor has two decimal places, and the other factor is a whole number, how many decimal places with the product have?** two

Try It

■ Write several multiplication problems involving decimals on the board.

■ Tell students not to find the products, but simply to tell how many decimal places each product will have.

Apply It

■ Bring in a book, grocery item, or other recently purchased object. Be sure the price is marked somewhere on the item or the price tag is still attached.

■ Have each student roll a number cube, rolling again if it comes up 1. Tell each student to calculate how much it would cost to buy that many of the item you brought in.

Solve It

■ $2.40; $26

RESOURCES and TECHNOLOGY

▶ **Math in Science, pp. 19–20**

CHAPTER 12 Review

▶ Visual Summary

Have students look at the pictures to review the main ideas of the chapter.

Vocabulary

1. shadow
2. circuit
3. light
4. sound
5. reflect
6. heat
7. vibrates
8. insulator
9. electric current
10. temperature

▶ Make a **FOLDABLES** Study Guide

See pages R27–R28 in the back of the Teacher's Edition for more information on Foldables.

RESOURCES and TECHNOLOGY

▶ **Reading and Writing,** pp. 249–250

▶ **Assessment,** pp. 148–151, 156–159

🖱 **Progress Reporter** Assessments

🖱 **PuzzleMaker CD-ROM**

🖱 **Vocabulary Games**

CHAPTER 12 Review

Visual Summary

Lesson 1 Heat affects matter in many ways. Heat always moves from warmer objects to cooler objects.

Lesson 2 Sounds are made when objects vibrate. Volume and pitch can be used to compare sounds.

Lesson 3 Light is a form of energy that allows you to see objects. Light moves in straight paths.

Lesson 4 Electricity is made up of charged particles. These charged particles can flow through a circuit.

Make a **FOLDABLES** Study Guide

Glue your lesson study guides to a piece of paper as shown. Use your study guide to review what you have learned in this chapter.

Vocabulary

Fill each blank with the best term from the list.

circuit, p. 515	**reflect**, p. 501
electric	**shadow**, p. 502
current, p. 514	**sound**, p. 490
heat, p. 480	**temperature**, p. 482
insulator, p. 484	**vibrates**, p. 490
light, p. 500	

1. When light is blocked by an object, a _____ forms.

2. A path that allows electric current to flow is a _____.

3. The form of energy that allows you to see objects is called _____.

4. When a guitar string vibrates, a _____ is made.

5. When light hits an object, it can bounce, or _____, off of the object.

6. Energy that moves from a warm object to a cold object is called _____.

7. When an object moves back and forth very quickly, it _____.

8. A material that heat does not move through easily is an _____.

9. A flow of charged particles is an _____.

10. A thermometer is used to measure _____.

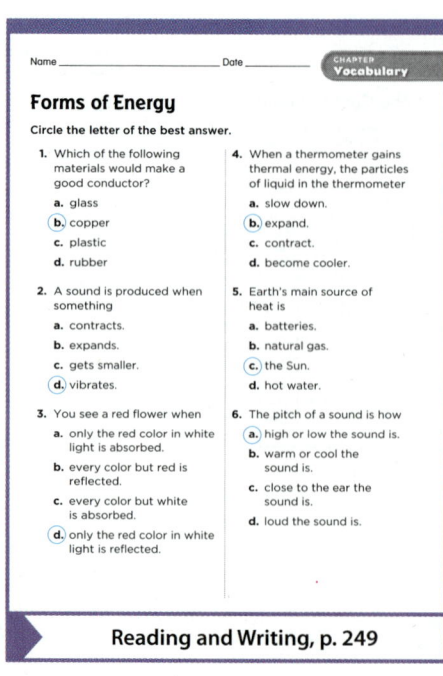

Reading and Writing, p. 249

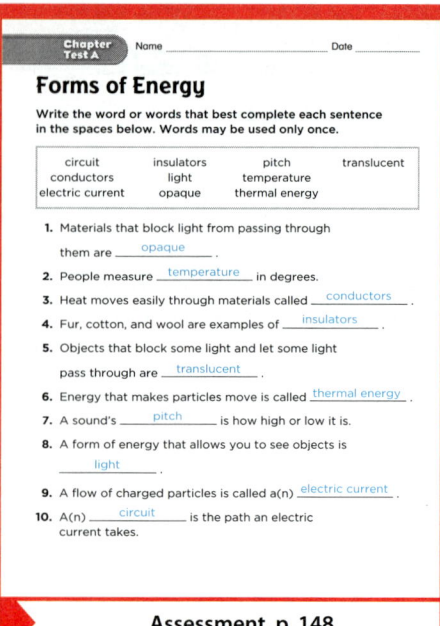

Assessment, p. 148

Answer each of the following in complete sentences.

11. Summarize What happens when an electrical switch is in the off position? What changes when the switch is turned on?

12. Persuasive Writing What is your favorite kind of music? Write a paragraph explaining why you enjoy it. Include volume and pitch in your paragraph.

13. Experiment Cover one thermometer with black paper, and another with white paper. Put them in a warm place for 15 minutes. Then read each thermometer. Which color heats up faster, black or white? Why?

14. Critical Thinking Suppose you see a picture in a book. Then you see the same picture on a computer screen. Where does the light come from to show you each picture?

15. What material is this pot made of? Why?

16. What are the main forms of energy? How are they used?

Producing Energy

Many machines work by changing electricity into other forms of energy. What machines do you use at home or school that use electricity?

▶ List six machines on a chart like the one shown below.

▶ Then list what form of energy they produce. For example, a toaster produces heat. Other forms of energy are sound, light, and motion.

▶ List at least one machine for each of these forms of energy.

Machine	Forms of Energy
Toaster	Heat

1. How are a lamp and the Sun alike?
- **A** Both give off daylight.
- **B** Both produce electricity.
- **C** Both reflect light.
- **D** Both are sources of heat and light.

521

11. Summarize In the off position, the circuit is open because it has a gap in it and no electric current will flow. When the switch is turned on, the circuit is closed and there is no longer a gap in the circuit. Electric current will flow.

12. Persuasive Writing Student answers will vary but should include references to volume and pitch. For example, students might mention that popular music is usually played at a high volume.

13. Experiment The thermometer wrapped in black paper has a higher temperature reading than the thermometer wrapped in white paper. The black paper absorbs heat better than the white paper.

14. Critical Thinking The picture in the book is made visible by light that has been reflected off the page. The computer screen projects light.

15. This pot is made of metal because heat moves quickly through metals.

16. Students should use information from the chapter to answer.

Scoring Rubric

4 Points Student has (1) accurately identified six machines that produce energy; (2) correctly stated what type of energy they produce; (3) listed one machine for heat, one for light, one for motion; (4) identified at least one machine that produces each of these forms of energy.

3 Points Student has correctly carried out three of the four possible tasks.

2 Points Student has correctly carried out two of the four possible tasks.

1 Point Student has correctly carried out one of the four possible tasks.

1. D

Summative Assessment and Intervention

- **Assessment Book** provides a test for Chapter 12.

- **Leveled Readers** may be used to reteach lesson content in an alternative format. The leveled readers deliver chapter content in different readability levels. The back cover of each reader provides comprehension building activities specific to the book content. (see p. 477)

Key Concept Cards 75–82 contain prescribed routines for student intervention.

Careers in Science

Objective
- Apply principles of light and electrical energy to a real-life career.

Lighting Technician

Genre: Nonfiction Draw students' attention to the photos.

- **How do the photos show how fiction and nonfiction can be related?** Setting up lighting is a real-life, or nonfiction, situation. However, the lighting is done for a fictional setting.

Talk About It

- **What part does a lighting technician play in making a movie?** He or she uses light to create a mood for a scene by changing lights or their locations.

Learn About It

- **Why does a lighting technician have to know about electricity?** Possible answer: Electricity supplies energy to the lights.

- **What training do you think a lighting technician must have?** Possible answer: courses in electrical energy and theater, and on-the-job training

✏️ Write About It

Have students research the types of equipment a lighting technician might use. Have them point out how the equipment involves different types of energy, such as electrical energy, light energy, and heat energy. Have them present their findings in a written report.

RESOURCES and TECHNOLOGY

 e-Careers

---

Writing Rubric P. TR26

Careers in Science

Lighting Technician

Have you ever watched a motion picture awards show? If so, you may have heard actors thank members of the film crew. An important part of the film crew is the chief lighting technician.

The chief lighting technician designs the lighting for the scenes of a movie. The lighting must create a mood that matches the action of the scene. The chief lighting technician uses different combinations of lights for different scenes. The technician also changes the location of the light sources to create different moods.

To become a chief lighting technician, you need to know about light and electrical energy. You also should have some experience in drama or filmmaking. Many chief lighting technicians begin their career as members of a lighting crew.

▲ This technician is lighting a set for a motion picture.

▲ A lighting technician knows about light and electrical energy.

> **Here are some other Physical Science careers:**
> - electrician
> - engineer
> - architect
> - car designer

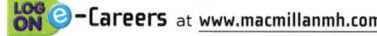
522   **LOG ON** e-Careers at www.macmillanmh.com

Integrate Writing

Classified Ad for a Lighting Technician

Have students work in pairs to write a job description for a lighting technician.

- Have them include details that relate the job to light and electricity.

- If possible, have a stage manager or light specialist from a local high school speak to the students about what is done for lighting for a school performance. Have students use this information in their writing.

- Have students write their job description in the form of a classified ad.

Reference

Science Handbook

Measurements R2

Tools of Science R7

Computers R9

Organizing Data R10

Health Handbook

Human Body Systems R14

Healthy Living R24

FOLDABLES R27

Glossary R29

Index R45

Credits R60

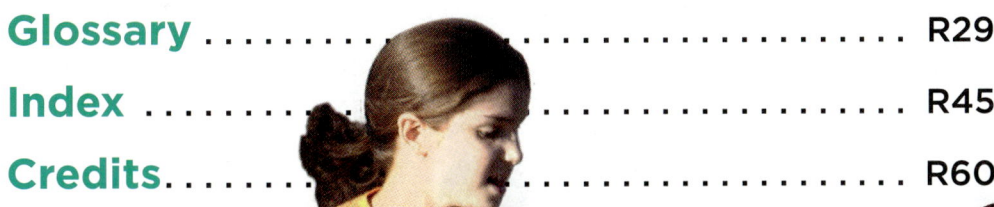

RI

Measurements

Objective

- Review and compare units in the metric and U.S. customary (English) systems.

Units of Measurement

▶ Assess Prior Knowledge

Write the word *ruler* on the board, and then have students add the names of other measuring tools to the list. For each tool, ask:

- **What does the tool measure? What units of measurement are used with this tool?**
  Possible answers: ruler—inch, foot, centimeter; thermometer—degrees

▶ Discuss the Main Idea

Explain that in this lesson students will review some common measuring instruments and two systems of units used to record measurements.

▶ Use the Visuals

Discuss the pictures on pages R2 and R3. For each picture, ask:

- **What is being measured? What measuring tool is being used? What units of measure could be used to record the data?** Possible answers: Temperature: thermometer, degrees Celsius, degrees Fahrenheit; Length and Area: meterstick or ruler, meters, square meters, feet, square feet, inches, square inches, centimeters, square centimeters; Mass: balance, grams; Volume of Fluids: no tool shown, liters, quarts, ounces; Weight/Force: scale, pounds, newtons; Rate: meterstick and stopwatch; meters per second

---

Measurements

Units of Measurement

Temperature

▶ The temperature on this thermometer reads 86 degrees Fahrenheit. That is the same as 30 degrees Celsius.

Length and Area

▶ This student is 3 feet plus 9 inches tall. That is the same as 1 meter plus 14 centimeters.

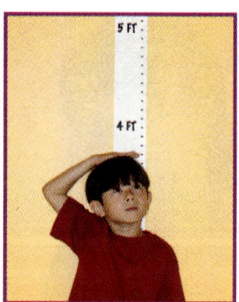

Mass

▶ You can measure the mass of these rocks in grams.

Volume of Fluids

▶ This bottle of water has a volume of 2 liters. That is a little more than 2 quarts.

Weight/Force

▶ This pumpkin weighs about 7 pounds. That means the force of gravity is 31.5 newtons.

Science Background

Systems of Measurement There are two systems of measurement commonly used in the United States—the U.S. customary (English) system and the metric system. Scientists throughout the world use the metric system so that information can be easily shared. The metric system is based on units of length (meter) and mass (kilogram). Temperature is not actually part of the metric system. It is included here for convenience to compare the more-familiar units (U.S. customary system and Fahrenheit scale) and less-familiar units (metric system and Celsius scale). The abbreviation *SI* stands for Système Internationale in French.

Speed

▶ This student can ride her bike 100 meters in 50 seconds. That means her speed is 2 meters per second.

Table of Measures

SI International Units/Metric Units	Customary Units
Temperature Water freezes at 0 degrees Celsius (°C) and boils at 100°C.	**Temperature** Water freezes at 32 degrees Fahrenheit (°F) and boils at 212°F.
Length and Distance 10 millimeters (mm) = 1 centimeter (cm) 100 centimeters = 1 meter (m) 1,000 meters = 1 kilometer (km)	**Length and Distance** 12 inches (in.) = 1 foot (ft) 3 feet = 1 yard (yd) 5,280 feet = 1 mile (mi)
Volume 1 cubic centimeter (cm³) = 1 milliliter (mL) 1,000 milliliters = 1 liter (L)	**Volume of Fluids** 8 fluid ounces (fl oz) = 1 cup (c) 2 cups = 1 pint (pt) 2 pints = 1 quart (qt) 4 quarts = 1 gallon (gal)
Mass 1,000 milligrams (mg) = 1 gram (g) 1,000 grams = 1 kilogram (kg)	**Area** 1 square foot (ft²) = 1 ft x 1 ft 43,560 square feet (ft²) = 1 acre
Area 1 square meter (m²) = 1 m x 1 m 10,000 square meters (m²) = 1 hectare	**Speed** miles per hour (mph)
Speed meters per second (m/s) kilometers per hour (km/h)	**Weight/Force** 16 ounces (oz) = 1 pound (lb) 2,000 pounds = 1 ton (T)
Weight/Force 1 newton (N) = 1 kg x 1m/s²	

R3
SCIENCE HANDBOOK

Integrate Math

Converting Metric Units

Develop A building is 20 meters (m) long. What is its length in centimeters (cm)? 20 m = _____ cm

Think 1 m = 100 cm. Multiply to change from larger to smaller units: 20 × 100 = 2,000, so 20 m = 2,000 cm. The length of the building is 2,000 cm.

Think 1,000 g = 1 kg. To change from smaller to larger units, divide: 5,000 ÷ 1,000 = 5 kilograms

5,000 grams (g) = _____ kilograms (kg)

Practice

1. 3,000 mL = _____ L 3 L

2. 7 kg = _____ g 7,000 g

3. 40 L + _____ mL 40,000 mL

▶ **Use the Visuals**

Have students study the Table of Measures. Ask:

■ **What kinds of information can you find in this table?** the two main systems of measurement; things that are measured; different units of measurement used in each system; examples of measurements; some definitions of units, which can be used in conversions within systems

■ **What are two systems used to report measurements?** the metric system, or SI (International Units) and Customary Units, or English System of Units

■ **Where can you find a Table of Measures?** Possible answers: in the back of a math or science textbook; in a large dictionary; in an encyclopedia

▶ **Discuss the Main Idea**

Write the following words on the board: *distance, mass, temperature, volume.* For each quantity, ask:

■ **What tool could you use to measure this, and what units would you use?** Possible answers:

Distance: tool—ruler, tape measure, meterstick; units—meters, centimeters, inches, feet

Mass: tool—balance; units—grams, kilograms

Temperature: tool—thermometer; units—degrees Celsius, degrees Fahrenheit

Volume: tool—calibrated beaker, graduated cylinder; units—liters, milliliters, cubic centimeters

■ **What metric units can you use to give the speed of a moving object? What customary system units would you use?** Metric: km/h, m/sec, km/min; English: mi/hr, ft/sec, mi/min

Measurements

Objectives

- Use a stopwatch to measure elapsed time.
- Use a centimeter ruler (meterstick) to measure length.

Measure Time

▶ Assess Prior Knowledge

Have students describe different kinds of clocks and watches. Ask:

- **What are some units we use to measure time?**
  days, hours, minutes, seconds

▶ Explore the Main Idea

ACTIVITY Have students complete the activity, which will give them practice measuring time. Ask:

- **What are some advantages in using a stopwatch?**
  Possible answer: With a stopwatch, you don't have to subtract one time from another.

- **Did you get exactly the same time as your classmates?** Answers will vary.

Explain that there is some error in any measurement due to uncertainty and human reaction time.

Measure Length

▶ Assess Prior Knowledge

Hold up a 12-inch ruler. Ask:

- **What units are on this ruler?** inches, half-inches

Explain that 12 inches equals approximately 30 centimeters. Compare the ruler to a meterstick, which is about 100 centimeters long.

▶ Use the Visuals

Refer students to the picture of the caterpillar and the ruler. Ask:

- **What is the name of the animal in the picture? How long is it?** caterpillar; 3 centimeters

---

Measurements

Measure Time

You measure time to find out how long something takes to happen. Stopwatches and clocks are tools you can use to measure time. Seconds, minutes, hours, days, and years are some units of time.

Try it Use a Stopwatch to Measure Time

① Get a cup of water and an antacid tablet from your teacher.

② Tell your partner to place the tablet in the cup of water. Start the stopwatch when the tablet touches the water.

③ Stop the stopwatch when the tablet completely dissolves. Record the time shown on the stopwatch.

0 minutes   **25 seconds**
75 hundredths (0.75) of a second

▲ Push the button on the top right of the stopwatch to start timing. Push the button again to stop timing.

Measure Length

You measure length to find out how long or how far away something is. Rulers, tape measures, and metersticks are some tools you can use to measure length. You can measure length using units called meters. Smaller units are made from parts of meters. Larger units are made of many meters.

Look at the ruler below. Each number represents 1 centimeter (cm). There are 100 centimeters in 1 meter. In between each number are 10 lines. Each line is equal to 1 millimeter (mm). There are 10 millimeters in 1 centimeter.

Try it Find Length with a Ruler

Place a ruler on your desk. Line up a pencil with the "0" mark on the ruler. Record the length of the pencil in centimeters.

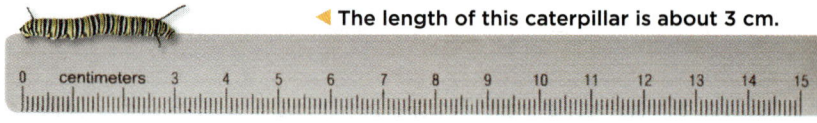

◀ The length of this caterpillar is about 3 cm.

0   centimeters   3   4   5   6   7   8   9   10   11   12   13   14   15

Differentiated Instruction

Leveled Questions

EXTRA SUPPORT   **What kind of clock would you use in a race?** a stopwatch **Why?** so you wouldn't have to subtract to get total times to compare

ENRICHMENT   **When or where could you use a sundial?** outdoors when it is sunny **When is a sundial not useful?** at night or on cloudy days when no shadow is cast on the sundial

Measure Liquid Volume

Volume is the amount of space something takes up. Beakers, measuring cups, and graduated cylinders are tools you can use to measure liquid volume. These containers are marked in units called milliliters (mL).

Try it Measure Liquid Volume

1. Gather a few empty plastic containers of different shapes and sizes.

2. Use a graduated cylinder to find the volume of water each container can hold. To start, fill the graduated cylinder with water, then pour the water into the container. Continue pouring this until the container is full. Keep track of the number of milliliters you add.

10 mL

▲ This graduated cylinder can measure volumes up to 100 mL. Each number on the cylinder represents 10 mL.

Measure Mass

Mass is the amount of matter an object has. You use a balance to measure mass. To find the mass of an object, you compare it with objects whose masses you know. Grams are units people use to measure mass.

Try it Measure the Mass of a Box of Crayons

1. Place a box of crayons on one side of a pan balance.

2. Add gram masses to the other side until the two sides of the balance are level.

3. Add together the numbers on the gram masses. This total equals the mass of the box of crayons.

R5
SCIENCE HANDBOOK

Differentiated Instruction

Leveled Activities

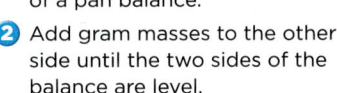 **EXTRA SUPPORT** Have students practice pouring equal volumes of liquid into a graduated cylinder until they become familiar with the different shapes that any volume of liquid can have.

ENRICHMENT Have students pour equal volumes of water into two graduated cylinders. Have them add a small object to one of the cylinders and compare the two volumes. Repeat with two or three other objects.

Objectives

- Use a graduated cylinder to measure volume.
- Use a balance to find the mass of a box of crayons.

Measure Liquid Volume

▶ Assess Prior Knowledge

Hold up a 10 mL graduated cylinder marked in milliliters. Point to the marks having numbers. Ask:

- **What do these marks show?** a number of milliliters

▶ Use the Visuals

Refer students to the picture of the graduated cylinder. Ask:

- **What volume can this graduated cylinder measure?** volumes up to 110 mL

Try It

If milliliters remain in the graduated cylinder after the container is full, subtract the measure of water in the cylinder from the total volume of the cylinder. The result is the amount poured into the container.

Measure Mass

▶ Assess Prior Knowledge

Ask students if they have used a bathroom scale. Ask:

- **What does a bathroom scale tell you?** how much you weigh

Then show them a pan balance. Ask:

- **How is this balance different from a bathroom scale?** It cannot be used to find out how much you weigh.

▶ Explore the Main Idea

ACTIVITY Have students find the mass of a box of crayons in grams. Tell students that *grams* are used for small masses and that *kilograms* are used for more massive things. Explain that the prefix *kilo-* means "thousand." So, 1 kilogram equals 1,000 grams.

Measurements

Objectives
- Use a spring scale to measure force/weight.
- Use a thermometer to measure temperature.

Measure Force/Weight

▶ Assess Prior Knowledge

Have students compare what they know about pan balances and bathroom scales. Ask:

- **How is a pan balance different from a scale?**
  A balance measures mass in grams. A scale gives a number for weight in pounds.

▶ Explore the Main Idea

Explain that weight is a force that is pushing down on a surface. Tell students that a spring scale used in science labs measures force or weight in newtons (N), a metric unit of force. One pound is about 4.5 N. Have students complete the activity described on page R6.

Measure Temperature

▶ Assess Prior Knowledge

Discuss the temperature in the classroom. Ask:

- **What tool measures temperature?** a thermometer

▶ Discuss the Main Idea

Point out that a thermometer can have two scales on it, one for Fahrenheit or one for Celsius. Ask:

- **Suppose you see a temperature written as "35 degrees." Is this a complete measurement? Explain.** No; the measurement must say whether the temperature was measured in Fahrenheit or Celsius degrees.

▶ Use the Visuals

Have students look at the diagram of the thermometer on page R6. Ask:

- **At what temperature does water freeze?** 0°C

- **What is room temperature on this scale?** 25°C

Measurements

Measure Force/Weight

You measure force to find the strength of a push or pull. Force can be measured in units called newtons (N). A spring scale is a tool used to measure force.

Weight is a measure of the force of gravity pulling down on an object. A spring scale measures the pull of gravity. One pound is equal to about 4.5 newtons.

Try it Measure the Weight of an Object
1. Hold a spring scale by the top loop. Put a small object on the bottom hook.
2. Slowly, let go of the object. Wait for the spring to stop moving.
3. Read the number of newtons next to the tab. This is the object's weight.

Measure Temperature

Temperature (TEM•puhr•uh•chuhr) is how hot or cold something is. You use a tool called a thermometer (thuhr•MOM•i•tuhr) to measure temperature. In the United States, temperature is often measured in degrees Fahrenheit (°F). However, you can also measure temperature in degrees Celsius (°C).

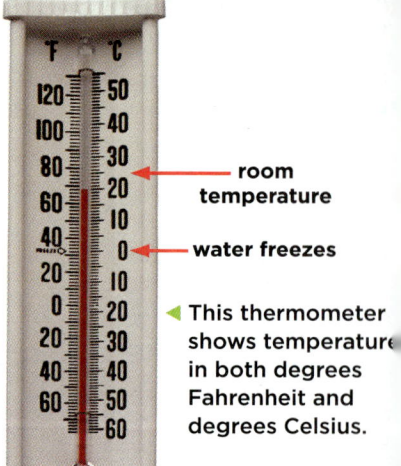

room temperature

water freezes

◀ This thermometer shows temperature in both degrees Fahrenheit and degrees Celsius.

Try it Read a Thermometer
1. Fill a beaker with ice water. Then put a thermometer in the water.
2. Wait several minutes. Read the number next to the top of the red liquid inside the thermometer. This is the temperature.
3. Repeat with warm water.

Science Background

Temperature Scales Two common scales for measuring temperature are the Fahrenheit and Celsius scales. In the Celsius scale, each degree is 1/100 of the difference between the temperature of melting ice and boiling water. In the Fahrenheit scale, each degree is 1/180 of this difference.

Tools of Science

Use a Microscope

A microscope (MYE•kruh•skohp) is a tool that magnifies objects, or makes them look larger. A microscope can make an object look hundreds or thousands of times larger. Look at the photo to learn the different parts of a microscope.

Try it **Examine Salt Grains**

1. Move the mirror so that it reflects light up toward the stage. ⚠ **Be Careful**. Never point the mirror at bright lights or the Sun. This can cause permanent eye damage.

2. Place a few grains of salt on a slide. Put the slide under the stage clips on the stage. Be sure that the salt grains are over the hole in the stage.

3. Look through the eyepiece. Turn the focusing knob slowly until the salt grains come into focus. Draw a picture of what you see.

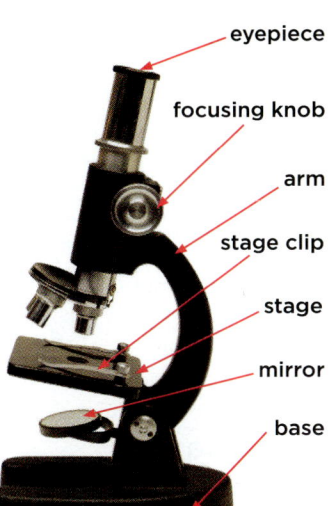

eyepiece

focusing knob

arm

stage clip

stage

mirror

base

Use a Hand Lens

A hand lens is another tool that magnifies objects. It is not as powerful as a microscope. However, a hand lens still allows you to see details of an object that you cannot see with your eyes alone. As you move a hand lens away from an object, you can see more details. If you move a hand lens too far away, the object will look blurry.

Try it **Magnify a Rock**

1. Look at a rock carefully. Draw a picture of it.

2. Hold a hand lens above the rock so that you can see the rock clearly.

3. Fill in any details on your original drawing that you did not see before.

R7
SCIENCE HANDBOOK

Differentiated Instruction

Leveled Activities

EXTRA SUPPORT Have small groups of students use a hand lens to observe three familiar objects. Have each group draw a picture of one of the magnified images. Then have groups exchange pictures and attempt to identify the objects. After they have made their identifications, students can collect and sort pictures that show the same items. They can point out details on various drawings that make the identifications more obvious.

ENRICHMENT Provide each student with a prepared slide of a thin section of onion skin, a lettuce leaf, and/or a leaf of *Elodea*. Have students look at each sample with a hand lens and then with a microscope (under low power). Have them discuss the differences they could see in the detail of the images.

Tools of Science

Objectives
- Practice collecting data using microscopes.
- Practice collecting data using hand lenses.

Use a Microscope

▶ **Use the Visuals**

Ask students to study the diagram of the parts of a microscope on page R7 to find corresponding parts of a real microscope.

▶ **Discuss the Main Idea**

Demonstrate how to hold and carry a microscope. Stress the need for never pointing the mirror directly at the Sun or another bright light. You might need to first teach a small group of students how to use the microscope and then have them work with others. Ask:

- Why are microscopes important in scientific work? Possible answer: Scientists can learn more about objects by finding out what they look like when highly magnified.

Use a Hand Lens

▶ **Assess Prior Knowledge**

Display a hand lens. Have students share past experiences using hand lenses. Ask:

- How is this tool like a pair of eyeglasses? It magnifies objects to help a person see better.

▶ **Explore the Main Idea**

ACTIVITY For the hand-lens activity on page R7, provide each group with hand lenses, rocks, and other objects such as soil and seeds. Have groups read all the directions before they begin working. After students have drawn their pictures of the rock, allow time for them to examine other objects. Ask:

- How do you know how close the hand lens should be to the object? You do not know. You must experiment until a blurry object looks clear.

Tools of Science

Objectives
- Use calculators to analyze collected data.
- Use a camera to record visible changes.

Use a Calculator

▶ Assess Prior Knowledge

Have students share previous experiences using calculators. Ask:

- **Why do you sometimes get the wrong answer when using a calculator?** Possible answers: not entering the correct numbers; choosing the wrong operation

▶ Explore the Main Idea

ACTIVITY Have students read "Use a Calculator" on page R8. Then have students work through the problem with a calculator. First, have them press the On key and make sure the calculator is on.

As they work, explain that to subtract a number they must first enter the larger number, then the minus (–) key, and then the smaller number, as follows: Enter 212, then press the minus (–) key, enter 32, and then press the equals (=) key. The result is 180. Then walk them through the remaining keystrokes for Steps 2 and 3 as follows: $180 \times 5 = 900$; and then $900 \div 9 = 100$.

212°F = 100°C; 100°F = 38°C

Use a Camera

▶ Assess Prior Knowledge

Have students share experiences using cameras. Ask:

- **Can you use pictures of yourself to find out how you have changed from year to year?** Accept all reasonable answers.

▶ Explore the Main Idea

ACTIVITY Have students complete the activity at the bottom of page R8.

Tools of Science

Use a Calculator

Sometimes during an experiment, you have to add, subtract, multiply, or divide numbers. A calculator can help you carry out these operations.

Try it Convert from °F to °C

Water boils at 212°F. Use a calculator to convert 212°F into degrees Celsius.
1. Press the ON key. Then, enter the number 212 by pressing 2 1 2.
2. Subtract 32 by pressing – 3 2.
3. Multiply by 5 by pressing × 5.
4. Finally, divide by 9 by pressing ÷ 9. Press =. This is the temperature in degrees Celsius.

Now, convert 100°F into degrees Celsius.

Use a Camera

During an experiment or nature study, it helps to observe and record changes that happen over time. Sometimes it can be difficult to see these changes if they happen very quickly or very slowly. A camera can help you keep track of visible changes. Studying photos can help you understand what happened over the course of time.

Try it Gather Data From a Photo

The photos below show a panda eight days after birth and then several months later. What differences do you notice? How has the panda changed over those months? Now think of something else that changes over time. With the help of an adult, use a camera to take photos at different times. Compare your photos.

R8
SCIENCE HANDBOOK

Differentiated Instruction

Leveled Activities

EXTRA SUPPORT Have students find numbers to add or subtract by tossing number cubes to form 2- or 3-digit numbers from every 2 or 3 tosses. They will find that they must first enter the larger of two numbers when they form two numbers to subtract one from the other.

ENRICHMENT Have students list examples of when a calculator could be useful in science—for example, adding cups or ounces of fluid when computing the volume of a large container, or finding the difference between today's temperature and yesterday's.

Computers

Use a Computer

A computer has many uses. You can use a computer to get information from compact discs (CDs) and digital videodiscs (DVDs). You can also use a computer to write reports and to show information.

The Internet connects your computer with computers around the world, so you can collect all kinds of information. When using the Internet, visit only Web sites that are safe and reliable. Your teacher can help you find safe and reliable sites to use. Whenever you are online, never give any information about yourself to others.

Try it Use a Computer for a Project

1. Choose an environment to research. Then use the Internet to find out about this environment. Where is the environment located in the world? What is the climate like in the environment? What kinds of plants and animals live there?

2. Use DVDs or other sources from the library to find out more about your chosen environment.

3. Use the computer to write a report about the information you gathered. Then share your report with others.

R9
SCIENCE HANDBOOK

Science Background

Computers Computers are electronic machines that obey commands. A computer is made up of hardware (the machine itself and all of its parts), and software (instructions called programs that tell the computer how to work and what to do). Hardware and software work together to change data into more useful formats. The hardware in a personal computer includes the hard disk, electronic circuits, processor chips, ribbon cables, and the keyboard. Software includes programs for word processing, connecting to the Internet, and playing games.

Computers

Objective

■ Understand that computers can be used to organize information in tables, to write reports, and to collect facts through Internet access.

Use a Computer

Note: It is recommended that students of this age access the Internet with adult supervision.

▶ **Assess Prior Knowledge**

Have students share computer experiences, such as using various application programs. Ask:

■ **How is using a computer like going to the library?** Both are ways to find information on a topic you are learning.

▶ **Discuss the Main Idea**

Explain that in this lesson, students will learn about some ways they can use computers to learn science. Ask:

■ **How is using the Internet different from going to the library?** The computer provides information on many different things in one place.

Explain that using an online encyclopedia or searching on the Internet requires the use of *key words*. Ask:

■ **What key words would you use for a project about the climate in an environment?** Possible answers: *climate, pond, mountains, desert*

■ **How can a computer be used to organize facts or to gather information for a project?** The computer can be used to make tables and graphs; a word processor can be used to write a report; an online encyclopedia or a Web page search can provide facts and pictures.

▶ **Explore the Main Idea**

ACTIVITY Have students complete the activity in the right column. Encourage them to choose a topic from one of the chapters in their science textbook.

Organizing Data

Objective
- Read and make geographical maps and idea maps.

Make Maps

▶ Assess Prior Knowledge

Show students a road map. Ask them to share experiences about using road maps. Ask:

- **How is a road map like the area it represents? How is the map different from the real area?** The map shows the same key features, such as streets and important buildings, that the real area has. The map is smaller, it is two-dimensional, and it does not have as much detail.

✔ Answers to Locate Places
American Falls and Horseshoe Falls

▶ Explore the Main Idea

ACTIVITY Have students work in small groups to complete the activities under "Locate Places" and "Idea Maps" on page R10. Explain that students will use two kinds of maps. The first is like a road map. The second type of map is a way to show how ideas are related.

▶ Use the Visuals

Refer students to the idea map on page R10. Ask:

- **What are two sources of salt water?** oceans, seas

- **Do any sources have both fresh and salt water?** no

Organizing Data

Make Maps

Locate Places

A map is a drawing that shows an area from above. Many maps have numbers and letters along the top and side. The letters and numbers help you find locations. The Buffalo Zoological Garden, for example, is located at D4 below. To find it, place a finger on the letter D along the side of the map and another finger on the number 4 at the top. Then move your fingers straight across and down the map until they meet. Now find B1? What is there?

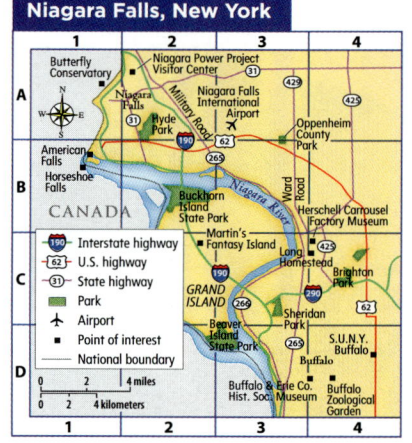

Niagara Falls, New York

Try it Make a Map

Make a map of an area in your community. It might be a park or the area between your home and school. Include numbers and letters along the top and side. Use a compass to find north, and mark north on your map.

R10
SCIENCE HANDBOOK

Idea Maps

The Niagara Falls map shows how places are connected to each other. Idea maps, on the other hand, show how ideas are connected to each other. Idea maps help you organize information about a topic.

Look at the idea map below. It connects ideas about water. This map shows that Earth's water can be fresh water or salt water. The map also shows three sources of fresh water. You can see that there is no connection between "rivers" and "salt water" on the map. This can remind you that salt water does not flow in rivers.

Try it Make an Idea Map

Make an idea map about a topic you are learning in science. Your map can include words, phrases, or even sentences. Arrange your map in a way that makes sense to you and helps you understand the connection between ideas.

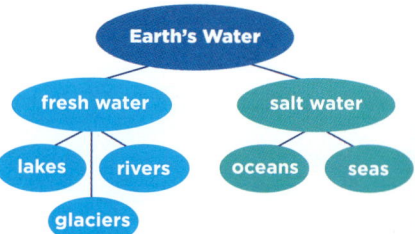

Science Background

Maps Geographical maps are used to show key features of geographical areas. These maps can also present other information, such as elevation above/below sea level. Topographical maps are drawn to show how elevation in an area changes from one location to the next.

Make Charts

Charts are useful for recording information during an experiment and for communicating information. In a chart, only the column or the row has meaning but not both. In this chart, one column lists living things. A second column lists nonliving things.

Living	Nonliving
tree	rock
chipmunk	puddle
bird	cloud

Try it Organize Data in a Chart

Take a survey of your class. Find out each student's favorite kind of pet. Make a chart to show this information. Remember to show your information in columns or in rows.

Make Tables

Tables also help to organize data, or information. Tables have columns that run up and down and rows that run across. Column and row headings tell you what kind of data they hold.

The table below shows the properties of some minerals. Which mineral in the table has a white streak? Which mineral is yellow in color?

Try it Organize Data in a Table

Collect a few minerals from your teacher. Observe the properties of each. Make a table like the one shown. Use the same column headings. Record the properties of each mineral.

Mineral Identification Table				
	Hardness	Luster	Streak	Color	Other
pyrite	6–6.5	metallic	greenish-black	brassy yellow	called "fool's gold"
quartz	7	nonmetallic	none	colorless, white, rose, smoky, purple, brown	
mica	2–2.5	nonmetallic	none	dark brown, black, or silver-white	flakes when peeled
feldspar	6	nonmetallic	none	colorless, beige, pink	
calcite	3	nonmetallic	white	colorless, white	bubbles when acid is placed on it

R11
SCIENCE HANDBOOK

Make Charts

▶ Assess Prior Knowledge

Hold up a chart showing different weather conditions for each day of the week. The chart should have seven columns labeled with the days of the week, and an overall title "Weather Forecast." Use pictures of the Sun, clouds, raindrops, or snowflakes for each day. Ask:

- **On which day was it rainy?** Answers will vary depending on the chart.

▶ Discuss the Main Idea

Explain that charts have only columns or rows.

Make Tables

▶ Assess Prior Knowledge

Write the words *cat, dog, fish,* and *other* in a vertical list on the board. Ask for three volunteers who have at least two pets. Write students' names and how many of each kind of pet they have in a horizontal list. Use the data to make a simple table. Ask:

- **How can you find out how many dogs a particular student has?** Find the number listed under the word *dog* next to the student's name in the table.

Answer Calcite has a white streak; pyrite is yellow.

▶ Discuss the Main Idea

Have students study the "Mineral Identification Table" on page R11. Ask:

- **What color is calcite?** colorless, white
- **Which metal is metallic?** pyrite

Organizing Data

Objective

- Read and make bar graphs, pictographs, and line graphs.

Make Graphs

▶ Assess Prior Knowledge

Take a short poll of the class to find out whether each student would prefer pizza with mushrooms or pizza with onions. Make a bar graph with the data. Ask:

- **How does a bar graph show the results of the survey?** The length of each bar shows the number of students who chose each topping. The longer bar shows that more students chose that topping.

▶ Use the Visuals

Have students read page R12 and study the bar graph. Ask:

- **What are the parts of a bar graph?** the bars, the two axes, the labels on the two axes (month and temperature)

- **What do you do if a bar does not exactly meet one of the lines on a bar graph?** Estimate the number the bar shows.

✔ **Answer to Bar Graphs** 1. 13°C; 2. July; 22°C; 3. January; 6°C

Organizing Data

Make Graphs

Graphs also help organize data. Graphs make it easy to notice trends and patterns. There are many kinds of graphs.

Bar Graphs

A bar graph uses bars to show data. What if you want to find the warmest and coldest months for your city? Every month you find the average temperature in the newspaper. You can organize the temperatures in a bar graph so you can easily compare them.

Month	Temperature (°C)
January	6
February	8
March	10
April	13
May	16
June	19
July	22
August	20
September	19
October	14
November	9
December	7

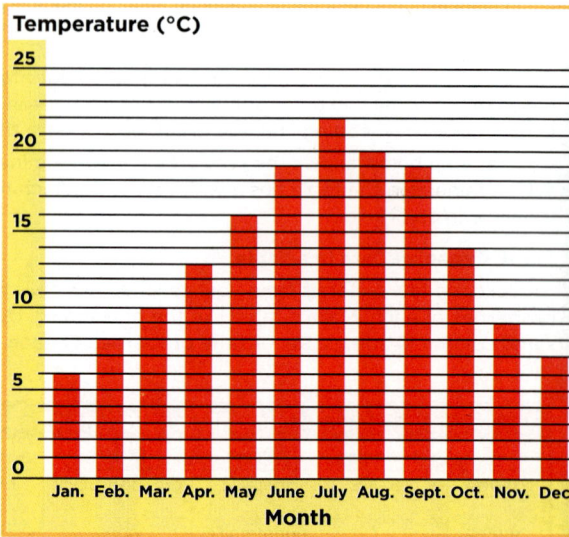

❶ Look at the bar for the month of April. Put your finger at the top of the bar. Move your finger straight over to the left to find the average temperature for that month.

❷ Find the highest bar on the bar graph. This bar represents the month with the highest average temperature. Which month is it? What is the average temperature for this month?

❸ Look at the bars of the graph. What pattern do you notice in the temperatures from January to December?

R12
SCIENCE HANDBOOK

Science Background

When are graphs helpful? Graphs are visual ways of representing quantitative data. Students can use graphs to identify relationships, make comparisons, and predict future occurrences. Often, more than one kind of graph can be used to display the same set of data.

Pictographs

A pictograph uses symbols, or pictures, to show information. What if you collect information about how much water your family uses each day?

Water Used Daily (liters)	
drinking	10
showering	100
bathing	120
brushing teeth	40
washing dishes	80
washing hands	30
washing clothes	160
flushing toilet	50

You can organize this information into a pictograph. In the pictograph below, each bucket means 20 liters of water. A half bucket means half of 20, or 10, liters of water.

1. Which activity uses the most water?
2. Which activity uses the least water?

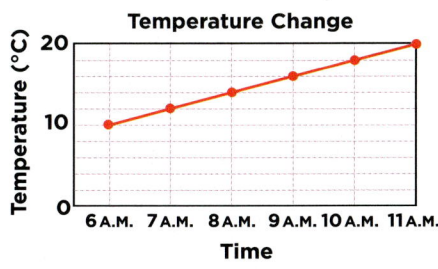

Water Used Daily	
drinking	🪣
showering	🪣🪣🪣🪣🪣
bathing	🪣🪣🪣🪣🪣🪣
brushing teeth	🪣🪣
washing dishes	🪣🪣🪣🪣
washing hands	🪣🪣
washing clothes	🪣🪣🪣🪣🪣🪣🪣🪣
flushing toilet	🪣🪣🪣

🪣 = 20 liters of water

Line Graphs

A line graph can show how information changes over time. What if you measure the temperature outdoors every hour starting at 6 A.M.?

Time	Temperature (°C)
6 A.M.	10
7 A.M.	12
8 A.M.	14
9 A.M.	16
10 A.M.	18
11 A.M.	20

Now organize your data by making a line graph. Follow these steps.

1. Make a scale along the bottom and side of the graph. Label the scales.
2. Draw a point on the graph for each temperature measured each hour.
3. Connect the points.
4. How do the temperatures and times relate to each other?

Temperature Change

(line graph: Temperature (°C) on vertical axis from 0 to 20; Time on horizontal axis 6 A.M. 7 A.M. 8 A.M. 9 A.M. 10 A.M. 11 A.M.; line rises from 10 to 20)

▶ **Discuss the Main Idea**

Explain that students will now read two kinds of graphs: pictographs and line graphs. Have students read the section about pictographs. Ask:

■ **How is a pictograph different from a bar graph?** A pictograph uses symbols; a bar graph uses bars.

■ **Why is the key important on a pictograph?** The key tells what each symbol on the graph stands for.

■ **Why might you choose to make more than one graph for the same set of data?** to find the graph that best shows the data, or to display the data in more than one way

✅ **Answers to Pictographs** 1. washing clothes; 2. drinking

Discuss how reading a pictograph made answering the questions easier than reading the data table.

▶ **Discuss the Main Idea**

Have students read the section about line graphs. Tell students that line graphs often show time on one of the scales. Ask:

■ **How does this line graph show changes?** The line moves up to show that the temperature is changing over a period of 5 hours.

▶ **Discuss the Main Idea**

ACTIVITY Have students complete the activity at the bottom of page R13.

✅ **Answer** The temperature rises by 2 degrees each hour throughout the morning hours.

Differentiated Instruction

Leveled Activities

EXTRA SUPPORT Have students convert the information from the bar graph on page R12 and the information from the pictograph on page R13 to bar graphs.

ENRICHMENT Have students collect graphs of any type from newspapers and magazines. Encourage students to collect and show their graphs on a poster. Have volunteers explain what kind of information is shown in each of the graphs they collected.

Human Body Systems

Objective

- Describe the functions of the skeletal system.

The Skeletal System

▶ Assess Prior Knowledge

Begin a discussion about the systems of the human body. Ask:

- **What systems of the human body can you name?**
  Possible answers: skeletal, muscular, digestive, respiratory, nervous, circulatory

▶ Discuss the Main Idea

Explain to students that the skeletal system is only one of the organ systems of the body. Ask:

- **What is the skeletal system made of?** Most students will answer bones; some may know about cartilage.

- **What are the functions of the skeletal system?** support the body, give the body its shape, protect organs, work with muscles to move the body, store minerals, make blood

- **What works with the bones of your skeleton to make your body move?** muscles

▶ Use the Visuals

Refer students to the diagram and descriptive text on page R14. Ask:

- **What kinds of joints are found in your skull?** immovable joints

- **What kinds of joints are in your knees and elbows?** movable joints

Human Body Systems

The Skeletal System

Feel your elbows, wrists, and fingers. What are those hard parts? Bones! Bones make up the skeletal system. The skeletal system is one of many body systems. A body system is a group of organs that work together to perform a specific job.

The skeletal system is made up of 206 bones. Each bone has a particular job. The long, strong leg bones support the body's weight. The skull protects the brain. The hip bones help you move. Together, bones do important jobs to keep the body active and healthy.

- ▶ Bones support the body and give the body its shape.

- ▶ Bones protect organs in the body.

- ▶ Bones work with muscles to move the body.

- ▶ Bones store minerals and produce blood for the body.

Joints

A joint is a place where two or more bones meet. There are three main types of joints.

Immovable joints form where bones fit together too tightly to move. The 29 bones of your skull meet at immovable joints. Partly movable joints are places where bones can move a little. Ribs are connected to the breastbone with these joints. Movable joints, like the knee, are places where bones can move easily. The knee lets the bones of your leg move.

R14
HEALTH HANDBOOK

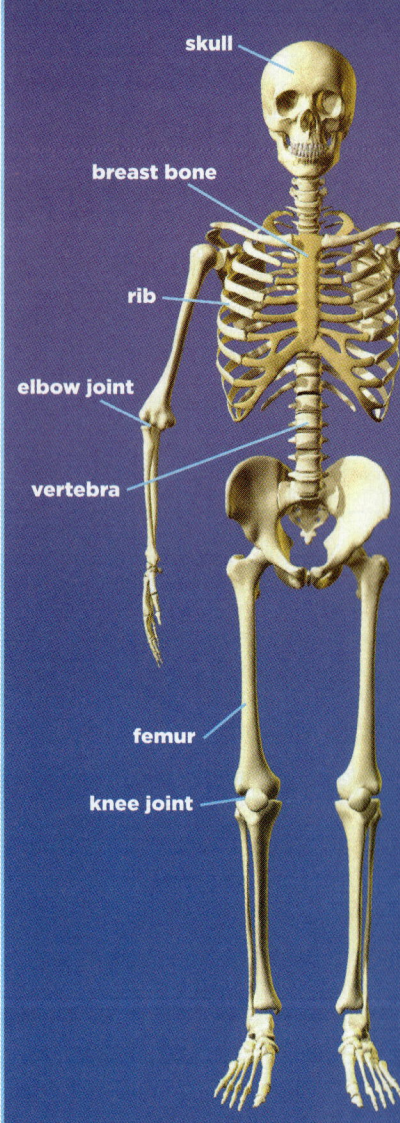

skull

breast bone

rib

elbow joint

vertebra

femur

knee joint

Differentiated Instruction

Leveled Activities

EXTRA SUPPORT Have students write a short paragraph, using their own words, that describes what might happen to their body if it did not have a skeletal system.

ENRICHMENT Have students use encyclopedias or the Internet, if available, to research the different types of joints in the human body, such as ball-and-socket joint, hinge joint, gliding joint, and pivot joint. Encourage students to illustrate where each type of joint is located in the skeleton and how bones move at that joint.

The Muscular System

Together, all the muscles in the body form the muscular system. Muscles allow the body to move. Without muscles, you would not be able to run, smile, breathe, or even blink.

Most muscles are attached to bones and skin. These are called skeletal muscles. To move bones back and forth, skeletal muscles usually work in pairs. Each pulls on a bone in a different direction. When you want to move, your brain sends a message to a pair of skeletal muscles. One muscle contracts, or gets shorter. It pulls on the bone and skin. The other muscle relaxes to let the bone move.

▲ There are 53 muscles in your face. You use 12 of them whenever you smile.

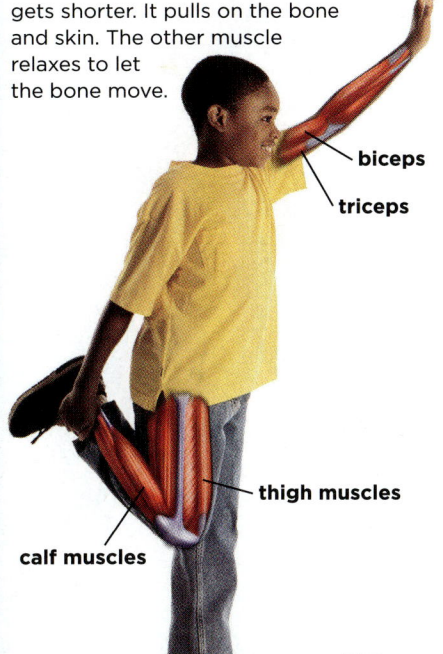

biceps

triceps

◄ To bend his arm, this boy's biceps contract while his triceps relax.

thigh muscles

calf muscles

Some muscles work without you even thinking about it. The heart is made of muscle. It pumps blood throughout the body even while you sleep. Smooth muscle in the lungs helps you breathe. Smooth muscle in the stomach helps you digest food.

R15
HEALTH HANDBOOK

Science Background

The Muscular System The human muscular system has over 640 muscles and makes up about 40 percent of the body's weight. The largest muscle in the body is the gluteus maximus in the buttocks. The smallest muscle is the stapedius, found in the middle ear. The longest muscle in the body is the sartorius in the inner thigh. Tendons attach skeletal muscles to bones.

The Muscular System

▶ Assess Prior Knowledge

Begin a discussion of the muscular system; the discussion will extend as students read the information on page R15. Ask:

■ **What do you already know about the muscular system?** Possible answers: It is made up of muscles. It works with the skeletal system to make the body move.

▶ Discuss the Main Idea

Explain to students that the muscular system is made up of two groups of muscles. Ask:

■ **What do skeletal muscles do?** They work in pairs to cause movement.

■ **Where are smooth muscles found in your body?** lungs, stomach

■ **How are skeletal muscles different from smooth muscles?** You can control the movement of skeletal muscles. You cannot control the movement of smooth muscles.

▶ Use the Visuals

Refer students to the visual of the boy with labeled muscles on page R15. As students look at it, have them feel the muscles and joints in their arms. Ask:

■ **What muscles contract when you curl your arm?** biceps

■ **What happens to the triceps muscles when you curl your arm?** They relax.

■ **What happens to the biceps and triceps muscles when you straighten your arm?** The biceps relax, and the triceps contract.

Human Body Systems

Objective

■ Describe the function of the circulatory system.

The Circulatory System

▶ Assess Prior Knowledge

Begin a discussion about the circulatory system; the discussion will extend as students read the information on page R16. Ask:

■ **What parts of the circulatory system do you already know about?** Possible answers: heart, veins, arteries, capillaries, blood

■ **What jobs are done by the circulatory system?** Possible answers: moving oxygen and carbon dioxide, transporting nutrients

▶ Discuss the Main Idea

Have students look at the inside of their wrists. Ask:

■ **What do you see when you look at your wrist?** Possible answers: blood vessels, veins

■ **Do you think these vessels are arteries or veins? How can you tell?** Possible answer: They are veins because they look blue.

Explain that the blood in the veins is not really blue, but it is a darker red than the blood in the arteries. The veins take on their bluish tinge when viewed through the skin. Also explain that the walls of arteries are much thicker than the walls of veins so the color of arteries is not usually seen through the skin.

▶ Use the Visuals

Refer students to the visual of the circulatory system. Ask:

■ **What color is used to represent blood that is rich in oxygen?** red

■ **What color is used to represent blood that is returning to the heart through veins?** blue

---

Human Body Systems

The Circulatory System

The body's cells need a constant supply of oxygen and nutrients. The circulatory (SUR•kyuh•luh•tawr•ee) system is responsible for sending these things throughout the body. The circulatory system is made up of the heart, blood vessels, and blood.

Blood rich in oxygen travels from the lungs to the heart. The heart is an organ about the size of a fist. It beats about 70 to 90 times each minute, pumping blood through the blood vessels.

Blood vessels are tubes that carry blood. There are two main types of blood vessels. Arteries are blood vessels that carry blood away from the heart. Veins carry blood back to it.

Blood contains plasma, red blood cells, white blood cells, and platelets. Plasma is the liquid part of blood. It carries nutrients and other things the body needs. Red blood cells carry oxygen to all the cells of your body. Red blood cells and plasma also carry wastes, such as carbon dioxide, away from cells. White blood cells work to fight disease. Platelets keep you from bleeding too much when you get a cut.

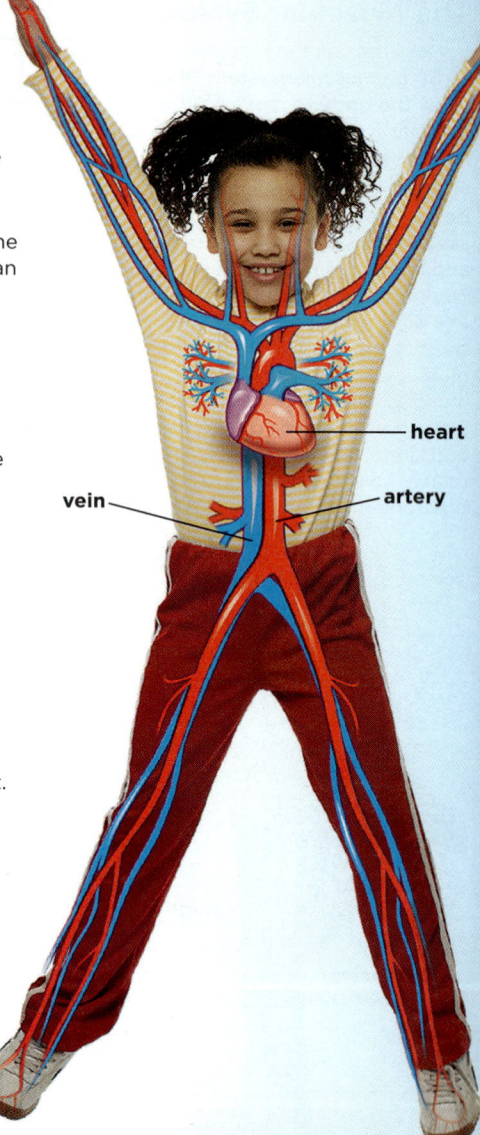

heart

vein

artery

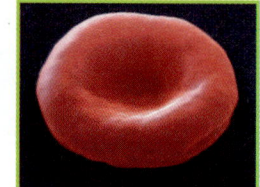

◀ This is how a red blood cell looks through a microscope.

R16
HEALTH HANDBOOK

Science Background

Circulation Blood from the body enters the heart through the vena cava, a large vein. Blood leaves the heart through the aorta, an artery. The right side of the heart receives blood from the body and pumps it to the lungs through the pulmonary artery. The left side of the heart receives blood from the lungs and pumps it to the body through the aorta.

The Respiratory System

The respiratory (RES•puhr•uh•tawr•ee) system helps the body take in oxygen and give off carbon dioxide and other waste gases. All of the cells in your body require oxygen to work properly. You take in oxygen from the air when you breathe.

Every time you inhale, a muscle called the diaphragm (DYE•uh•fram) contracts. This makes room in your lungs for air. Air is taken in through the nose or mouth. This air travels down the throat into the trachea (TRAY•kee•uh).

In the chest, the trachea splits into two bronchial (BRONG•kee•uhl) tubes. Each tube leads to a lung. Inside each lung, the bronchial tube branches off into smaller tubes called bronchioles (BRONG•kee•ohlz). At the end of each bronchiole are millions of tiny air sacs. Here, red blood cells release carbon dioxide, a waste gas, and absorb oxygen. When you breathe out, the diaphragm relaxes. This causes the lungs to deflate and push carbon dioxide out of your body through the nose and mouth.

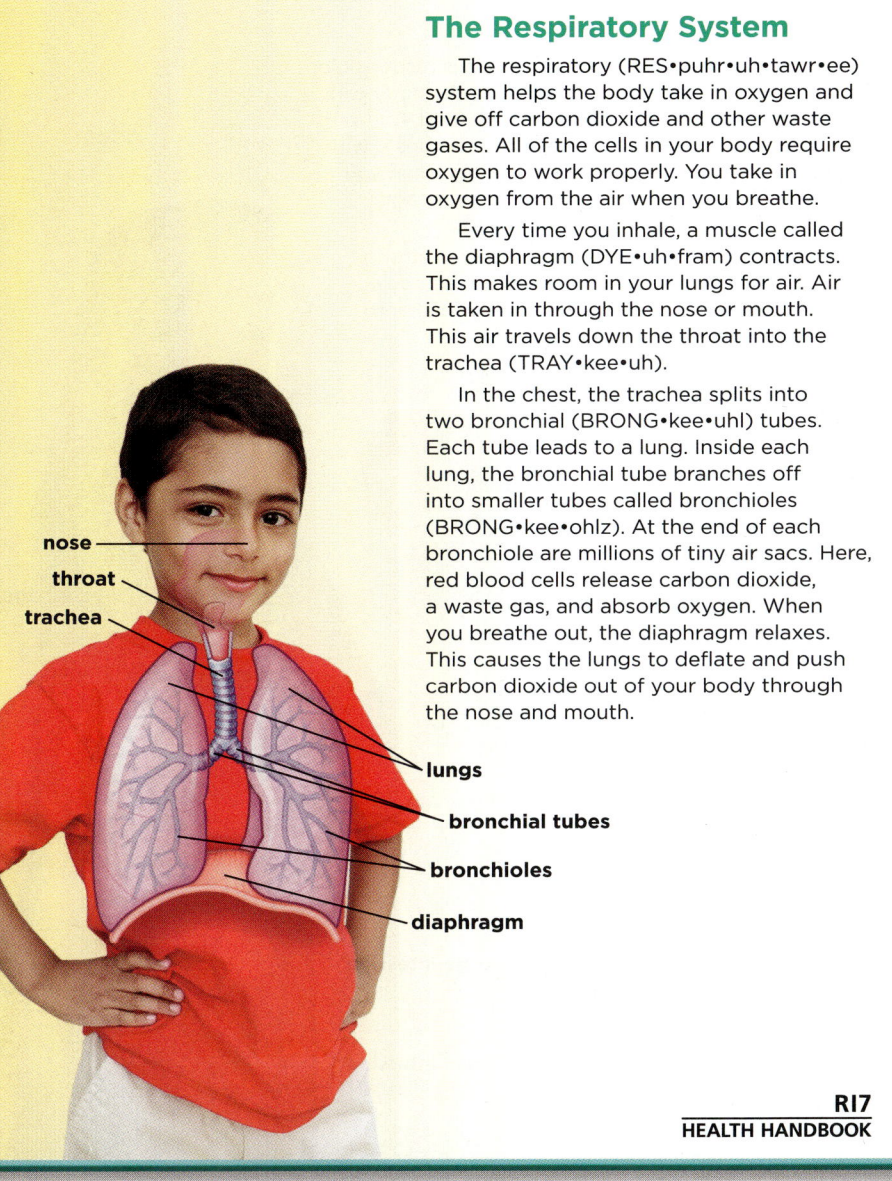

nose
throat
trachea
lungs
bronchial tubes
bronchioles
diaphragm

R17
HEALTH HANDBOOK

Differentiated Instruction

Leveled Activities

EXTRA SUPPORT   Have students draw a diagram of the respiratory system and label the parts identified in the text. Ask students to locate the place where carbon dioxide and oxygen are exchanged in the lungs

ENRICHMENT   Ask students to write a short story about what happens to an oxygen molecule when it goes up a nose. Encourage students to provide details about where the oxygen molecule travels and how its journey might end.

Objective
■ Describe the function of the respiratory system.

The Respiratory System

▶ Assess Prior Knowledge

Begin a discussion about the structure and function of the respiratory system. Ask:

■ **What parts of the respiratory system do you already know?** Possible answers: nose, throat, lungs

■ **What jobs are done by the respiratory system?** Possible answers: taking in oxygen, letting out carbon dioxide, breathing

▶ Discuss the Main Idea

Have students discuss how the exchange of oxygen and carbon dioxide begins. Ask:

■ **Where does the air you inhale go?** from the nose and mouth to the trachea, through bronchial tubes, into the lungs, and then the bronchioles with tiny air sacs

■ **Where in the lungs is oxygen exchanged with carbon dioxide?** in the tiny air sacs

▶ Use the Visuals

Refer students to the diagram of the lungs on page R17. Ask:

■ **Where does inhaled air enter the body?** through the nose or mouth

■ **What structures are between the trachea and the lungs?** the bronchial tubes

■ **Where are tiny air sacs located?** at the ends of the bronchioles in the lungs

Human Body Systems

Objective
- Describe the function of the digestive system.

The Digestive System

▶ Assess Prior Knowledge

Begin a discussion of the digestive system. Ask:

- **What parts of the digestive system do you already know?** Possible answers: mouth, esophagus, stomach, small intestine, large intestine, colon

Draw the parts that students list, to encourage a more complete answer.

▶ Discuss the Main Idea

Explain to students that the function of the digestive system is to break down food into nutrients the body can use. Ask:

- **What is the first step in digestion?** chewing food in the mouth

- **What happens to food in the stomach?** It is mixed with strong acidic juices that break down the food.

- **What is the function of the small intestine?** It takes in, or absorbs, nutrients.

▶ Use the Visuals

Refer students to the diagram of the digestive system on page R18. Ask:

- **In this diagram, which organs make up the digestive system?** esophagus, stomach, liver, large intestine, small intestine

---

Human Body Systems

The Digestive System

The digestive (dye•JES•tiv) system is responsible for breaking down food into nutrients the body can use. Digestion begins when you chew food. Chewing breaks food into smaller pieces and moistens it with saliva. Saliva helps food travel smoothly when you swallow. The food travels down your esophagus (i•SOF•uh•guhs) and into your stomach.

Inside the stomach food is mixed with strong, acidic juices. This causes the food to break down further, making it easier for your body to absorb nutrients from the food.

After passing through the stomach, food moves into the small intestine (in•TES•tin). This is where most nutrients are absorbed. The small intestine is a narrow tube about 6 meters (20 feet) long. It is coiled tightly so it fits inside the body. As food passes through the small intestine, digested nutrients are absorbed into the blood. The blood then carries these nutrients to other parts of the body.

After food has passed through the small intestine, it enters the large intestine. The large intestine removes water from the unused food that is left. Then the unused food is removed from the body as waste.

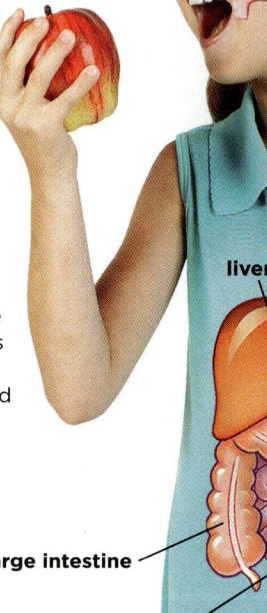

esophagus

liver

stomach

large intestine

small intestine

R18
HEALTH HANDBOOK

The Excretory System

The excretory (EK•skri•tawr•ee) system gets rid of waste products from your cells. Waste products are materials the body does not need, such as extra water and salts. The liver, kidneys, bladder, and skin are some organs of the excretory system.

Liver, Kidneys, and Bladder

The liver filters wastes from the blood. It changes wastes into a chemical call urea and sends the urea to the kidneys. Kidneys turn urea into urine. Urine flows from the kidney to the bladder. It is stored in the bladder until it is pushed out of the body through the urethra.

Skin

The skin takes part in excretion when a person sweats. Sweat glands in the inner layer of skin produce sweat. Sweat is made up of water and minerals that the body does not need. Sweat is excreted onto the outer layer of the skin. Sweating cools the body and helps it maintain an internal temperature of about 98°F (37°C).

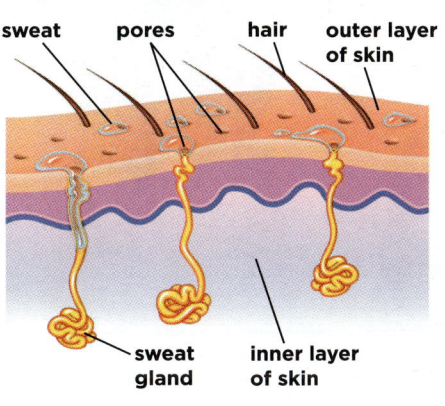

sweat   pores   hair   outer layer of skin

sweat gland   inner layer of skin

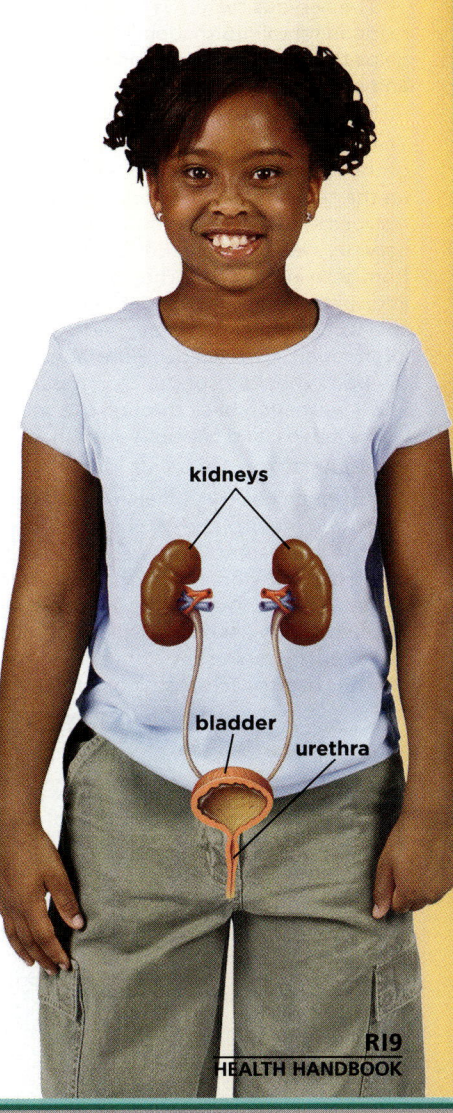

kidneys

bladder

urethra

R19
HEALTH HANDBOOK

Human Body Systems

Objective
- Describe the function of the nervous system.

The Nervous System

▶ Assess Prior Knowledge

Begin a discussion about what students already know about the nervous system. Ask:

- **What do you think is the function of the nervous system?** Possible answers: coordinates the body, allows you to think, controls the senses

- **What parts make up the nervous system?** Possible answers: brain, nerves, spinal cord

▶ Discuss the Main Idea

Explain that the nervous system is made up of two main parts. Ask:

- **What organs make up the central nervous system?** brain, spinal cord

- **What part of the nervous system do nerves belong to?** the peripheral nervous system

- **How is sensory information passed to the brain from cells in the body?** Nerves receive the sensory information and pass it to the brain through the spinal cord.

▶ Use the Visuals

Direct students' attention to the diagram of the brain on page R20. Ask:

- **What is the largest part of the brain? What is its function?** Cerebrum; it stores memories and helps control information from the senses.

- **What part controls heartbeat, breathing, and blood pressure?** the brain stem

Human Body Systems

The Nervous System

The nervous system is responsible for taking in and responding to information. It controls muscles and helps the body balance. It allows a person to think, feel, and even dream.

The nervous system is made up of two main parts. The first part, the central nervous system, is made up of the brain and spinal cord. All other nerves make up the second part, the peripheral (puh•RIF•uhr•uhl) nervous system. Nerves from the peripheral nervous system receive sensory information from cells in the body. They pass this information on to the brain through the spinal cord. When the brain receives this information, it makes decisions about how the body should respond. Then it passes this new information back through the spinal cord to the nerves, and the body responds.

The Brain

The brain has three main parts, the cerebrum (suh•REE•bruhm), the cerebellum (ser•uh•BEL•uhm), and the brain stem. The cerebrum is the largest part of the brain. It stores memories and helps control information received by the senses. The cerebellum helps the body keep its balance and directs the skeletal muscles. The brain stem connects to the spinal cord. It controls heartbeat, breathing, and blood pressure.

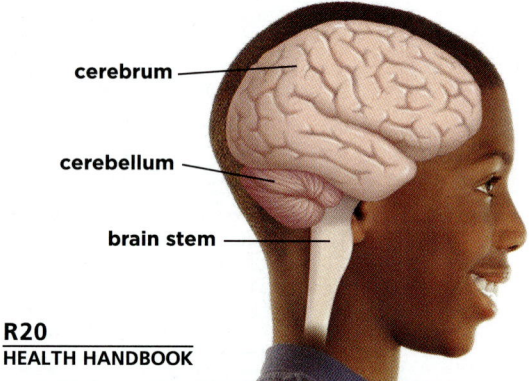

brain

spinal cord

nerves

cerebrum

cerebellum

brain stem

R20
HEALTH HANDBOOK

Differentiated Instruction

Leveled Activities

EXTRA SUPPORT   Have students draw a diagram of the brain, labeling the parts and describing the function of each.

ENRICHMENT   Ask students to write a short story that describes what their life would be like if they did not have one or more of their senses. Have them describe how they would go about their daily activities without the sense or senses.

The Senses

Different nerves in the body take in information from the environment. These nerves are responsible for the body's sense of sight, hearing, smell, taste, and touch.

Sight

Light reflects off an object, such as a leaf, and into the eye. The reflected light passes through the pupil in the iris. Cells in the eye change light into electrical signals. The signals travel through the optic nerve to the brain.

Hearing

Sound waves enter the outer ear. They reach the eardrum and cause it to vibrate. Cells in the ear change the sound waves into electrical signals. The signals travel along the auditory nerve to the brain.

Smell

As a person breathes, chemicals in the air mix with mucus in the upper part of the nose. When they reach certain cells in the nose, those cells send information along the olfactory nerve to the brain.

Taste

On the tongue are more than 10,000 tiny bumps, called taste buds. Each taste bud can sense four main tastes—sweet, sour, salty, and bitter. The taste buds send information along a nerve to the brain.

Touch

Different nerve cells in the skin give the body its sense of touch. They help a person tell hot from cold, wet from dry, and hard from soft. Each cell sends information to the spinal cord. The spinal cord then sends the information to the brain.

R21
HEALTH HANDBOOK

Science Background

The Human Eye The human eyeball is covered by a tough outer layer and is filled with fluid that maintains the eyeball's shape. The eyeballs are protected by bony sockets in the face. The iris controls the amount of light that enters the eye, dilating in dim light and constricting in bright light. The light-sensitive retina contains over 100 million cells, including rods that detect black, white, and low-intensity light, and cones that detect color.

▶ ## Discuss the Main Idea

Before students read the text on page R21, find out what they already know about the senses. Ask:

- **How many senses do you have? What are they?**
  Five senses: sight, hearing, smell, taste, touch

- **What organ of your body allows you to see?**
  the eyes

- **What organ of your body allows you to smell?**
  the nose

▶ ## Use the Visuals

Refer students to the "Sight" photo and text on page R21. Ask:

- **What is the first thing that light passes through when it reaches your eye?** the pupil in the iris

- **What changes light into electrical signals?** cells in the eye

Now refer students to the "Hearing" photo and text on the same page. Ask:

- **What part of the ear funnels sound waves into the eardrum?** outer ear

- **Where are sound waves changed into electrical signals?** in cells in the ear

Refer students to the "Smell" and "Taste" photos and text. Ask:

- **What is sent along the olfactory nerve to the brain?** chemicals in the air mixed with mucus

- **What are the more than 10,000 tiny bumps on the tongue?** taste buds

- **Where do the taste buds send information?** along a nerve to the brain

Lastly, refer students to the "Touch" photo and text. Ask:

- **Does one nerve cell in the skin give the body its sense of touch?** No; different nerve cells in the skin do

- **Do each of the cells send information to the brain?** No; each cell sends information to the spinal cord, which then sends it to the brain.

Human Body Systems

Objective
- Describe the function of the immune system.

Immune System

▶ Assess Prior Knowledge

Find out what students already know about the causes of disease and how the body fights disease. Ask:

- **What causes many kinds of disease?** Possible answers: germs, bacteria, viruses

- **How does the body fight disease?** Possible answers: immune system, fever

▶ Discuss the Main Idea

Explain to students that the immune system is the body system that helps the body to protect itself from germs. Ask:

- **What are the parts of the immune system?** skin, tears, saliva, lymph, lymph vessels, lymph nodes

- **How do you think the circulatory system works with the immune system to prevent and fight germs?** White blood cells, which are part of the circulatory system, help to find and kill germs before you become sick.

Human Body Systems

Immune System

The immune system protects the body from germs. Germs cause disease and infection. Most of the time, the immune system is able to prevent germs from entering the body. Skin, tears, and saliva are parts of the immune system. They work to kill germs and keep them out of the body.

When germs do find a way into your body, white blood cells help find and kill them quickly before you become ill. White blood cells are part of the blood. They travel through blood vessels and lymph (LIMF) vessels. Lymph vessels are similar to blood vessels. However, instead of carrying blood, they carry a fluid called lymph. Many white blood cells are made and live in lymph nodes. Here, they filter out harmful materials from the body.

White blood cells are not always able to kill germs before the germs start to reproduce in your body. When germs reproduce, they cause illness. Even while you feel ill, the immune system works to kill and remove germs until you are well again.

lymph vessels

lymph nodes

◀ This is how a white blood cell looks through a microscope.

R22
HEALTH HANDBOOK

Science Background

The Immune Response When it recognizes or detects a foreign organism such as a bacterium or virus in the body, the immune system mounts an immune response. The immune response happens in two principal events. First, white blood cells called *lymphocytes* attack the foreign substance directly. At the same time, other lymphocytes produce substances called antibodies that help to destroy the invading organisms.

Viruses and Bacteria

One of the main types of germs that makes the body ill are viruses. Illness from a virus like a cold or flu can be a big deal. Yet, viruses themselves are very small. In fact, you need a special microscope, an electron microscope, to look at a virus.

Viruses need to be inside living cells, called hosts, in order to reproduce. As they reproduce, viruses take nutrients and energy from the cell. They can even produce harmful materials that make the body itch or have dangerously high temperatures.

The other main type of germ that can make the body ill is bacteria. Bacteria are tiny, one-celled organisms. They can live on most surfaces and are able to reproduce outside of cells. Some bacteria can have a harmful effect on the body. Other bacteria, however, are good for the body. Some bacteria in your body, for example, help you digest food.

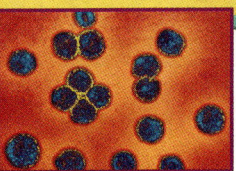

▲ A cold virus as seen through a microscope.

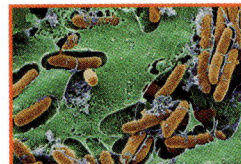

▲ *E. coli* bacteria as seen through a microscope.

You can help your body defend itself against germs. Here's what you can do.

▶ Eat healthful foods. This helps your body get all of the nutrients it needs to stay healthy. A healthy body is better able to fight germs.

▶ Be active. Being active makes your body fit. A fit body is better able to fight germs.

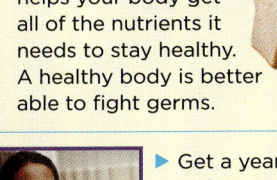

▶ Get a yearly check-up. Make sure you get all of your immunizations. Follow directions when taking medicines given to you by a doctor.

▶ Get plenty of rest. You need about 10 hours of sleep every night. Sleeping helps repair your body. Get extra rest when you are ill.

▶ Do not share cups or utensils with other people. Germs can be on objects you touch. Wash your hands, especially before eating and drinking. By washing your hands, you kill germs and make it harder for harmful things to get into your body.

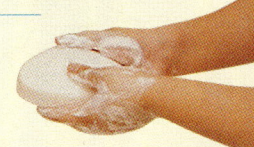

R23
HEALTH HANDBOOK

▶ ## Discuss the Main Idea

Point out to students that viruses and bacteria are both potential causes of disease. Ask:

■ **How are viruses and bacteria similar?** They are both very small and cannot be seen with the naked eye. Both viruses and bacteria can spread when objects are touched.

■ **How are viruses and bacteria different?** Viruses must be inside living cells to reproduce. Bacteria are one-celled organisms that can reproduce outside of cells.

■ **What diseases are caused by viruses?** cold, flu

Explain to students that they can do many things to protect themselves from disease caused by viruses and bacteria. Ask:

■ **Why is it important to eat healthy foods?** Eating healthy foods makes your body healthy. When your body is healthy, it is more able to fight germs.

■ **Why is it important to wash your hands before eating and drinking?** Washing your hands kills germs that might cause disease.

■ **Why is it important to follow directions when taking medicine?** Taking too much medicine or not enough medicine can make you sick.

Differentiated Instruction

Leveled Activities

EXTRA SUPPORT Have students make an illustrated poster that shows the ways in which the body can defend itself against germs.

ENRICHMENT Explain to students that diseases caused by germs such as viruses and bacteria are called infectious diseases. Ask students to use an encyclopedia or the Internet to find out about other kinds of diseases and their causes, such as allergies, nutritional diseases, genetic disorders, and heart disease. Have student volunteers share their findings with the class.

Healthy Living

Objective

- Discuss the importance of nutrients in staying healthy.

Nutrients

▶ Assess Prior Knowledge

Begin a discussion on nutrition to find out what students already know. Ask:

- **What does your body need in order to stay healthy and perform properly?** food, water

- **How does your body get the energy it needs?** from food

▶ Discuss the Main Idea

Explain to students that nutrients are the materials in foods that help the body to grow and stay healthy. Ask:

- **What are the six kinds of nutrients?** carbohydrates, vitamins, minerals, proteins, water, fats

- **Why does the body need carbohydrates to stay healthy?** Carbohydrates provide energy to the body.

▶ Use the Visuals

Direct students' attention to the table on this page. Ask:

- **What vitamins are found in milk?** vitamins A and D

- **What vitamin helps keep eyes, teeth, gums, and skin healthy?** vitamin A

- **Why is it important to eat foods that contain vitamin B?** Vitamin B helps the heart, cells, and muscles function.

Healthy Living

Nutrients

Nutrients are materials in foods that help the body grow, get energy, and stay healthy. By eating a balance of healthful foods, your body gets the nutrients it needs to do all of these things.

There are six kinds of nutrients—carbohydrates, vitamins, minerals, proteins, water, and fats. Each nutrient helps the body in different ways.

Carbohydrates

Carbohydrates are the main source of energy for the body. Starches and sugars are two types of carbohydrates. Starches come from foods like bread, pasta, and cereal. They provide long-lasting energy. Sugars come from fruits and can be used immediately by the body for energy.

carbohydrates

Vitamins

Vitamins help keep the body healthy. They also help to build new cells in the body. The table below shows some vitamins and their sources.

Vitamin	Sources	Benefits
A	milk, fruit, carrots, green vegetables	keeps eyes, teeth, gums, skin, and hair healthy
C	citrus fruits, strawberries, tomatoes	helps heart, cells, and muscles function
D	milk, fish, eggs	helps keep teeth and bones strong

R24
HEALTH HANDBOOK

Differentiated Instruction

Leveled Activities

EXTRA SUPPORT Have students make illustrated charts of the foods that are good sources of vitamins and minerals, as described on pp. R24–R25 of the text. Have students include in their charts drawings of the foods, or pictures cut from magazines.

ENRICHMENT Ask students to use encyclopedias or the Internet, if available, to identify other important vitamins and minerals needed by the body. Have students copy the *Vitamins* and *Minerals* tables from pp. R24– R25. Then ask students to expand the tables with any new information they discover about vitamins and minerals.

Minerals

Minerals help form new bone and blood cells. They also help your muscles and nervous system work properly. Here are some minerals and their sources.

Mineral	Sources	Benefits
calcium	yogurt, milk, cheese, and green vegetables	builds strong teeth and bones
iron	meat, beans, fish, whole grains	helps red blood cells function properly
zinc	meat, fish, eggs	helps your body grow and helps to heal wounds

Fats

Fats help the body use other nutrients and store vitamins. Fats also help the cells of the body to work properly. They even help keep the body warm. Fats can be found in foods such as meats, eggs, milk, butter, and nuts. Oils also contain fats. Though some fats help the body, some fats can cause health problems.

Water

Water is one of the most important nutrients. About $\frac{2}{3}$ of the body is made up of water! Water makes up most of the body's cells. It helps the body remove waste and protects joints. It also prevents the body from getting too hot.

Proteins

Proteins are a part of every living cell. Proteins help bones and muscles grow. They even help the immune system fight diseases. Foods high in protein are milk, eggs, meats, fish, nuts, and cheese.

fats

proteins

R25
HEALTH HANDBOOK

Healthy Living

Objective

- Explain the value of My Pyramid.

Stay Fit

▶ Assess Prior Knowledge

Encourage students to share what they already know about eating healthful foods. Ask:

- **What kinds of foods should you eat to keep your body healthy?** Possible answers: vegetables, fruit, milk, fish, meat

- **What kinds of foods and drinks do you think you should avoid to keep your body healthy?** Possible answers: Foods with too much sugar or solid fats like candy, potato chips, soft drinks, french fries

▶ Use the Visuals

Refer students to the diagram of MyPyramid on page R26. Ask:

- **What kinds of vegetables are especially good for you?** green, yellow, and orange vegetables

- **What kinds of fruit are good for you?** Possible answers: blueberries, strawberries, oranges, cantaloupes, pears, bananas, grapes

Healthy Living

Stay Fit

MyPyramid

You can use MyPyramid as a guide to healthful eating. The pyramid will show you the amounts of foods you should eat from each of the five food groups. A food group is foods with the same kinds of nutrients. To find the correct amounts of foods that are right for you, visit www.MyPyramid.gov

| Grains | Vegetables | Fruits | Milk | & |

Be Drug-Free

Do not use cigarettes, illegal drugs, or alcohol. These things can harm your body. They can keep you from growing properly and becoming fit.

Be Physically Active

You need to be physically active for at least 60 minutes every day. When you are physically active, you become physically fit. When you are physically fit, your heart, lungs, bones, joints, and muscles stay strong. You keep a healthful weight and lower the risk of disease. You do not have to be on a sports team to be physically active. You just need to move your body. Running, biking, and swimming are just some ways to be physically active.

R26
HEALTH HANDBOOK

Differentiated Instruction

Leveled Activities

EXTRA SUPPORT Ask students to convert the MyPyramid into an illustrated table. Encourage them to describe it when they are finished.

ENRICHMENT Have students go to www.MyPyramid.gov and then prepare a checklist of the number of servings of grains, vegetables, and other food types that they find recommended for themselves each day. Then ask students to use the checklist to keep track of how many servings of each food group they eat each day for three days. Encourage students to then analyze their diets to determine if the diets are healthful.

by Dinah Zike

Folding Instructions

The following pages offer step-by-step instructions about how to make Foldables study guides.

Half-Book

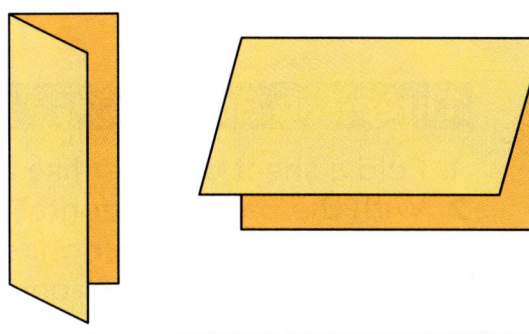

Fold a sheet of paper ($8\frac{1}{2}$″ x 11″) in half.
1. This book can be folded vertically like a hot dog or . . .
2. . . . it can be folded horizontally like a hamburger

Folded Book

1. Make a Half-Book.
2. Fold in half again like a hamburger. This makes a ready-made cover and two small pages inside for recording information.

Pocket Book

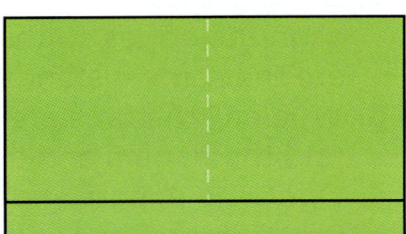

1. Fold a sheet of paper ($8\frac{1}{2}$″ x 11″) in half like a hamburger.
2. Open the folded paper and fold one of the long sides up two inches to form a pocket. Refold along the hamburger fold so that the newly formed pockets are on the inside.
3. Glue the outer edges of the two-inch fold with a small amount of glue.

Shutter Fold

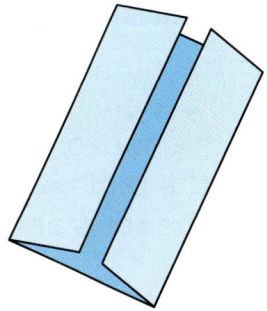

1. Begin as if you were going to make a hamburger, but instead of creasing the paper, pinch it to show the midpoint.
2. Fold the outer edges of the paper to meet at the pinch, or midpoint, forming a Shutter Fold.

R27
FOLDABLES

Trifold Book

1. Fold a sheet of paper ($8\frac{1}{2}''$ x 11") into thirds.
2. Use this book as is or cut into shapes.

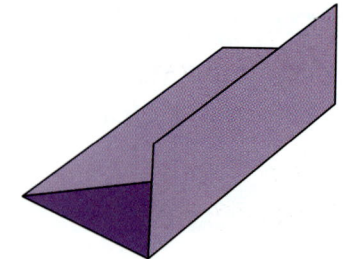

Three-Tab Book

1. Fold a sheet of paper like a hot dog.
2. With the paper horizontal and the fold of the hot dog up, fold the right side toward the center, trying to cover one half of the paper.
3. Fold the left side over the right side to make a book with three folds.
4. Open the folded book. Place one hand between the two thicknesses of paper and cut up the two valleys on one side only. This will create three tabs.

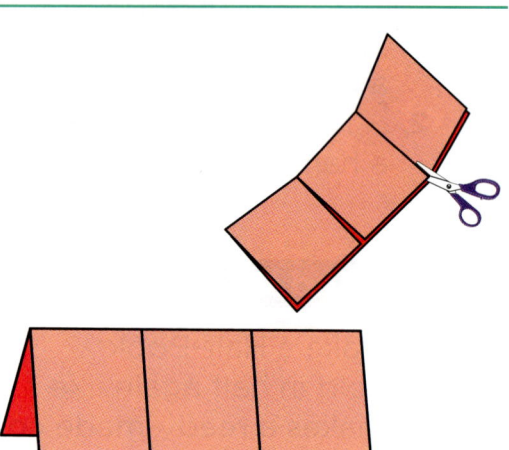

Layered-Look Book

1. Stack two sheets of paper ($8\frac{1}{2}''$ x 11") so that the back sheet is one inch higher than the front sheet.
2. Bring the bottoms of both sheets upward and align the edges so that all of the layers or tabs are the same distance apart.
3. When all the tabs are an equal distance apart, fold the papers and crease well.
4. Open the papers and glue them together along the valley, or inner center fold, or staple them along the mountain.

Folded Table or Chart

1. Fold the number of vertical columns needed to make the table or chart.
2. Fold the horizontal rows needed to make the table or chart.
3. Label the rows and columns.

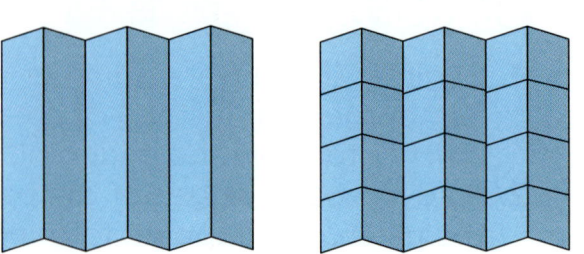

R28
FOLDABLES

Glossary

Use this glossary to learn how to pronounce and understand the meanings of Science Words used in this book. The page number at the end of each definition tells you where to find that word in the book.

absorb (ab·sôrb′) To take in. (p. 500) *A green leaf absorbs all colors of light except for green.*

adaptation (a′dəp·tā′shən) A structure or behavior that helps a living thing survive in its environment. (p. 134) *Sharp spines are one adaptation that helps a cactus survive.*

air pressure (âr presh′ər) The weight of air pressing down on Earth. (p. 283)

amphibian (am·fib′ē·ən) A vertebrate that spends part of its life in water and part of its life on land. (p. 59)

atmosphere (at′məs·fîr′) A blanket of gases and tiny bits of dust that surrounds Earth. (p. 280)

axis (ak′sis) A line through the center of a spinning object. (pp. 305, 319)

bird (bûrd) A vertebrate that has a beak, feathers, wings, and two legs and lays eggs. (p. 58)

blizzard (bliz′ərd) A storm with lots of snow, cold temperatures, and strong winds. (p. 297)

Pronunciation Key

The following symbols are used throughout the Macmillan/McGraw-Hill Science Glossaries.

a	**a**t	e	**e**nd	o	h**o**t	u	**u**p	hw	**wh**ite	ə about
ā	**a**pe	ē	m**e**	ō	**o**ld	ū	**u**se	ng	so**ng**	tak**e**n
ä	f**a**r	i	**i**t	ô	f**o**rk	ü	r**u**le	th	**th**in	penc**i**l
âr	c**a**re	ī	**i**ce	oi	**oi**l	ú	p**u**ll	<u>th</u>	**th**is	lem**o**n
ô	l**a**w	îr	p**ie**rce	ou	**ou**t	ûr	t**u**rn	zh	mea**s**ure	circ**u**s

′ = primary accent; shows which syllable takes the main stress, such as **kil** in **kilogram** (kil′ e gram′).
′ = secondary accent; shows which syllables take lighter stresses, such as **gram** in **kilogram**.

e-Glossary at **www.macmillanmh.com**

boil (boil) To change from a liquid to a gas. (p. 399) *Water will boil if you heat it on the stove.*

camouflage (kam′ə·fläzh′) An adaptation that allows an organism to blend into its surroundings. (p. 134)

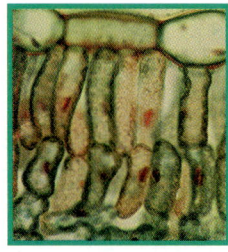

cell (sel) The basic building block that makes up all living things. (p. 26) *You can use a microscope to see that a leaf is made up of many tiny cells.*

chemical change (kəm′i·kəl chānj) A change that causes different kinds of matter to form. (p. 418) *Rust forming on this train is a chemical change.*

circuit (sûr′kit) The path through which electric current flows. (p. 515) *This circuit is made of a battery, wires, a light bulb, and a switch.*

climate (klī′mət) The pattern of weather at a certain place over a long time. (pp. 120, 304) *This desert has a hot, dry climate.*

cloud (cloud) A collection of tiny water drops or ice crystals in the air. (p. 290)

community (kə·mū′ni·tē) All the living things in one place that interact. (p. 166) *All the organisms in this pond make up a community.*

competition (kom′pə·tish′ən) The struggle among organisms for water, food, or other needs. (p. 153) *There is competition for water between these springbok.*

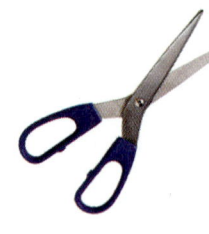

compound machine (kom′pound mə·shēn′) Two or more simple machines put together. (p. 470) *Scissors are a compound machine because they are made of levers and wedges.*

condensation (kon′den·sā′shən) The process through which a gas changes into a liquid. (p. 293)

LOG ON **e-Glossary** at **www.macmillanmh.com**

condense (kən·dens′)
To change from a gas
to a liquid. (p. 400)
*When water vapor in the
air cools and condenses,
it can form dew.*

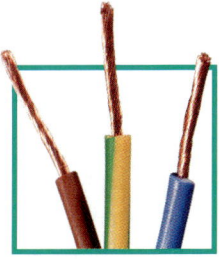

conductor (kən·duk′tər)
A material which heat or
electric current moves
through easily. (pp. 484,
516) *Copper is a good
conductor of heat and
electric current.*

cone (cōn) A plant
structure where seeds
are made in some
nonflowering plants.
(p. 75)

conserve
(kon·sûrv′) To use
resources wisely. (p. 266)
*You conserve gasoline
when you walk, ride a
bicycle, or ride a bus
instead of using a car.*

constellation
(con′stə·lā′shən) A group
of stars that seems to
form a picture. (p. 349)

consumer (kən·sü′mər)
An animal that eats
plants or other animals.
(p. 110) *Eagles eat fish,
snakes, and other small
organisms so they
are consumers.*

continent (kon′tə·nənt)
A great area of land
on Earth. (p. 193) *You live
on the continent of
North America.*

core (kôr) Earth's
deepest and hottest
layer. (p. 198)

crater (krā′tər) A hollow
area, or pit, in the
ground. (p. 332) *There
are many craters on
the Moon.*

crust (krust) Earth's
outermost layer. (p. 198)

D

decomposer
(dē′kəm·pō′zər) An
organism that breaks
down dead plant and
animal material. (p. 111)
*Worms are decomposers
that eat dead leaves that
fall to the ground.*

deposition
(dep′ə·zi′shən) The
dropping off of
weathered rock. (p. 216)

 **e-Glossary** at www.macmillanmh.com

desert (dez′ərt) A sandy or rocky ecosystem with little rainfall and little plant life. (p. 122)

distance (dis′təns) The amount of space between two objects or places. (p. 435) *The distance between these two toys is five centimeters.*

drought (drout) When there is an unusual lack of rain in an area for a long period of time. (p. 162)

earthquake (ûrth′kwāk) A sudden movement in Earth's crust. (p. 204) *An earthquake caused this road to crack.*

ecosystem (ē′kō·sis′təm) The living and nonliving things that share an environment and interact. (p. 108)

egg (eg) An animal structure that protects and feeds some very young animals such as birds. (p. 83)

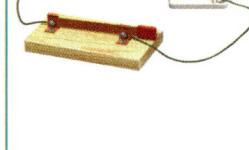

electric current (i·lek′trik kûr′ənt) A flow of charged particles. (p. 514) *When electric current flows from a battery to a light bulb, the bulb glows.*

electrical charge (i·lek′tri·kəl chärj) The property of matter that causes electricity. (p. 512) *Rubbing a balloon on a sweater gives the balloon a negative electrical charge.*

element (el′ə·mənt) A building block of matter. (p. 368) *Gold is an element.*

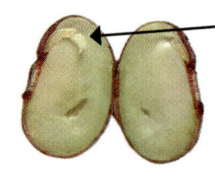

embryo (em′brē·ō) A young organism that is just beginning to grow. (p. 70)

endangered (en·dān′jərd) Having very few of a kind of organism left; close to becoming extinct. (p. 168) *Bengal tigers are endangered animals because there are very few of them left in the world.*

 **e-Glossary** at **www.macmillanmh.com**

energy (en′ər·jē)
The ability to do work.
(p. 456) *Energy from this bowling ball causes the pins to fall over.*

environment
(en·vī′rən·mənt) All the living and nonliving things that surround an organism. (p. 24) *Water, soil, rocks, trees, and zebras are parts of the giraffes' environment.*

erosion (i·rō′zhən)
The movement of weathered rock. (p. 216) *Erosion happens when water in this stream carries rocks away.*

evaporate (i·vap′ə·rāt′)
To slowly change from a liquid to a gas. (p. 399) *These wet clothes will dry when the water in them evaporates.*

evaporation
(i·vap′ə·rā′shən) The process through which a liquid changes into a gas. (p. 292)

exoskeleton
(ek′sō·skel′i·tən) A hard covering, or shell, that holds up and protects an invertebrate's body. (p. 57) *A snail's shell is an exoskeleton.*

extinct (ek·stingkt′)
Died out, leaving no more of that type of organism alive. (p. 174) *Many dinosaurs became extinct millions of years ago.*

fish (fish) A vertebrate that lives in water and breathes oxygen with gills. (p. 59)

flood (flud) Water flowing over land that is usually dry. (pp. 162, 208)

flower (flou′ər) A plant structure where seeds are made. (p. 72)

food chain (füd   chān) A series of organisms that depend on one another for food. (p. 110)

food web (füd   web) Several food chains that are connected. (p. 112)

force (fôrs) A push or a pull. (p. 444)

LOG ON **e-Glossary** at www.macmillanmh.com

forest (fôr′ist) An ecosystem with many trees. (p. 124)

fossil (fos′əl) The trace or remains of something that lived long ago. (pp. 174, 250)

freeze (frēz) To change from a liquid to a solid. (p. 401) *You can freeze juice to make a juice pop.*

friction (frik′shən) A force that occurs when one object rubs against another. (p. 448) *Friction between a break pad and a rim stops a bike.*

fruit (früt) A plant structure that grows around seeds. (p. 73)

fuel (fū′əl) A material that is burned for energy. (p. 252) *Wood and gasoline are examples of fuels.*

gas (gas) Matter that has no definite shape or volume. (p. 387) *These balloons are filled with a gas called helium.*

gills (gilz) A structure some animals use to take in oxygen from water. (p. 47)

glacier (glā′shər) A large sheet of ice that moves slowly over land. (p. 216)

gravity (grav′i·tē) A pulling force between two objects, such as you and Earth. (pp. 378, 447) *Gravity pulls these skydivers toward Earth.*

groundwater (ground′wô′tər) Water that is held in rocks and soil below the ground. (p. 261)

e-Glossary at **www.macmillanmh.com**

H

habitat (hab'i·tat') The home of a living thing. (p. 109) *A coral reef is a habitat for many fish.*

heat (hēt) A form of energy that moves from a hot object to a cold object. (p. 480) *The Sun is Earth's main source of heat.*

heredity (hə·red'i·tē) The passing of traits from parents to offspring. (p. 92)

hibernate (hī'bər·nāt) To rest or sleep through the cold winter. (p. 139)

humus (hū'məs) Decayed plant and animal material in soil. (pp. 120, 240) *Humus makes soil look dark.*

hurricane (hûr'i·kān') A large storm with strong winds and heavy rain. (p. 297)

I

igneous rock (ig'nē·es rok) A rock that forms when melted rock cools and hardens. (p. 231)

inclined plane (in·klīnd' plān) A simple machine with a flat slanted surface that is raised at one end. (p. 468) *A ramp is an inclined plane.*

inherited trait (in·her'i·təd trāt) A characteristic that is passed from parents to offspring. (p. 92) *A flower's color is an inherited trait.*

insulator (in'sə·lā'tər) A material that heat or electric current does not move through easily. (pp. 484, 516) *Plastic is a good insulator of electric current.*

invertebrate (in·vûr'tə·brāt) An animal that does not have a backbone. (p. 55)

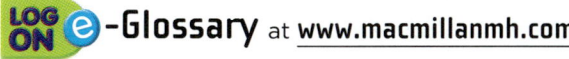

e-Glossary at www.macmillanmh.com

K

kinetic energy (ki·net′ik en′ûr·jē) Energy in the form of motion. (p. 456)

L

landform (land′fôrm′) A feature of land on Earth's surface. (p. 194)

landslide (land′slīd′) The rapid movement of rocks and soil down a hill. (p. 208)

larva (lär′və) The stage in some insects' life cycles which comes after hatching. (p. 83)

lava (lä′və) Melted rock that flows onto land. (p. 206)

leaf (lēf) The plant structure where a plant makes food. (p. 36)

learned trait (lûrnd trāt) Something you are taught or learn with experience. (p. 94) *Riding a bicycle is a learned trait.*

lever (lev′ər) A simple machine that consists of a straight bar that moves on a fixed point, or fulcrum. (p. 466)

life cycle (līf sī′kəl) All the stages in an organism's life. (p. 74)

light (līt) A form of energy that allows you to see objects. (p. 500) *Light travels in a straight path from a lighthouse.*

liquid (lik′wid) Matter that has a definite volume but no definite shape. (p. 386)

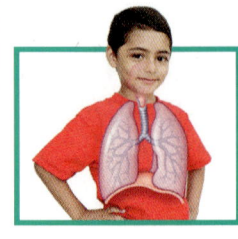

lung (lung) A structure some animals use to take in oxygen from air. (p. 47) *Humans breathe oxygen using lungs.*

LOG ON e-Glossary at www.macmillanmh.com

magma (mag′mə) Melted rock that is below Earth's surface. (p. 206)

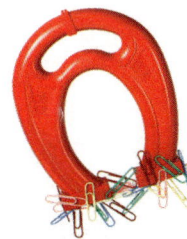

magnet (mag′nit) An object with a magnetic force; magnets can attract or repel certain metals. (p. 446)

mammal (mam′əl) A vertebrate that has hair or fur, is born live, and feeds its young with milk. (p. 60)

mantle (man′təl) The layer of Earth below the crust. (p. 198)

mass (mas) A measure of the amount of matter in an object. (p. 365)

matter (mat′ər) Anything that has volume and mass. (p. 364)

melt (melt) To change from a solid to a liquid. (p. 398) *This snowman will melt on a warm day.*

metamorphic rock (met′ə·môr′fik rok) A kind of rock that has been changed by heating and squeezing. (p. 233)

metamorphosis (met′ə·môr′fə·sis) A series of changes in which an organism's body changes form. (p. 83) *A tadpole becomes a frog through metamorphosis.*

metric system (met′rik sis′təm) A system of measurement. (p. 374) *A centimeter is a unit in the metric system.*

migrate (mī′grāt) To move to another place. (p. 141) *These geese migrate south when the weather gets cold.*

mimicry (mim′i·krē) An adaptation in which one kind of organism looks like another kind in color or shape. (p. 139)

mineral (min′ə rəl) A solid, nonliving substance found in nature. (p. 228)

LOG ON e-Glossary at www.macmillanmh.com

R37
GLOSSARY

mixture (miks′chər) Different kinds of matter mixed together. (p. 410)

motion (mō′shən) A change in the position of an object. (p. 436)

 N

natural resource (nach′ər əl rē′sôrs′) A material on Earth that is necessary or useful to people. (p. 244) *Plants and soil are natural resources.*

nocturnal (nok·tûr′nəl) An adaptation in which an animal is active during the night and asleep during the day. (p. 137)

nonrenewable resource (non′ri·nü′ə·bəl rē′sôrs′) A resource that cannot be replaced or reused easily. (p. 253) *Oil is a nonrenewable resource. Once it is used up, it is gone forever.*

nutrient (nü′trē·ənt) A substance that living things need to grow and stay healthy. (p. 34) *Spinach has many nutrients that people need.*

 O

ocean (ō′shən) A large body of salt water. (pp. 126, 192)

opaque (ō′pāk′) Not allowing light to pass through. (p. 502) *An ice cream cone is opaque. You can't see through it.*

orbit (ôr′bit) The regular path one object travels around another. (p. 320) *Earth's orbit around the Sun is nearly round.*

organism (ōr′gə·niz·əm) A living thing. (p. 22) *Koala bears and eucalyptus trees are organisms.*

 P

pan balance (pan bal′əns) A tool used to measure mass. (p. 376)

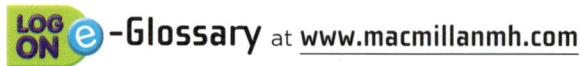
LOG ON e-Glossary at www.macmillanmh.com

phase (fāz) Each shape of the Moon we see. (p. 328)

photosynthesis (fō′tə·sin′thə·sis) The process through which plants make food. (p. 36)

physical change (fiz′i·kəl chānj) A change in the way matter looks. (p. 408) *Shaping clay is a physical change.*

pitch (pitch) How high or low a sound is. (p. 493) *A whistle has a high pitch.*

planet (plan′it) A large body of rock or gas with a nearly round shape that revolves around a star. (p. 338)

pollination (pol′ə·nā′shən) When pollen moves from the male part of a plant to an egg, after which a seed can form. (p. 73)

pollution (pə·lü′shən) What happens when harmful materials get into water, air, or land. (pp. 154, 264)

population (pop′yə·lā′shən) All the members of a single type of organism in an ecosystem. (p. 166)

position (pə·zish′ən) The location of an object. (p. 434)

potential energy (pə·ten′shəl en′ûr·jē) Energy that is stored or waiting to be used. (p. 456) *A sled at the top of a hill has potential energy.*

precipitation (pri sip′i tā′shən) Water that falls to the ground from the atmosphere. (p. 282)

producer (prə·dü′sər) An organism, such as a plant, that makes its own food. (p. 110)

property (prop′ər·tē) Any characteristic of matter that you can observe. (p. 365) *A sweet taste is a property of this pineapple.*

LOG ON e-Glossary at www.macmillanmh.com

pulley (pùl'lē) A simple machine that uses a rope and wheel to move an object. (p. 467)

pupa (pū'pə) The stage of some insects' life cycles before becoming an adult. (p. 83)

R

recycle (rē·sī'kəl) To turn old things into new things. (p. 156) *Plastic can be recycled to make new bottles and other products.*

reduce (ri·düs') To use less of something. (p. 156) *When you fix a leaky faucet, you reduce your use of water.*

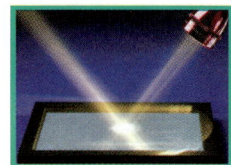

reflect (ri·flekt') To bounce off a surface. (p. 501) *Light reflects off a mirror.*

refract (ri·frakt') To bend. (p. 503) *Light refracts as it moves from the air to the water.*

renewable resource (ri·nü'ə·bəl rē'sôrs') A resource that can be replaced or used again and again. (p. 253) *Wind is a renewable resource.*

reproduce (rē'prə·düs') To make more of one's own kind. (p. 23)

reptile (rep'tīl) A vertebrate that has scaly, waterproof skin, breathes air with lungs, and lays eggs. (p. 58)

resource (rē'sôrs) Substances in the environment that help an organism survive. (p. 152) *Flowers are a resource for butterflies.*

respond (ri·spond') To react to something. (p. 22) *When the weather gets cool in the fall, this tree responds by losing its leaves.*

reuse (rē·ūz') To use something again. (p. 156) *Old bottles were reused to make this building.*

 **e-Glossary** at www.macmillanmh.com

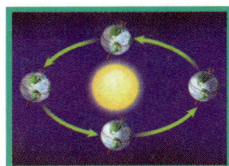 **revolve** (ri·volv′) To move around another object. (p. 320) *Earth revolves around the Sun.*

 **rock** (rok) A nonliving material made of one or more minerals. (p. 230)

 **root** (rüt) A plant structure that takes in water and nutrients and holds a plant in place. (p. 34)

 **rotate** (rō′tāt) To turn or spin. (p. 318) *Earth rotates like a giant top in space.*

 S

 **screw** (skrü) A simple machine made up of an inclined plane wrapped into a spiral. (p. 468)

 **season** (sē′zən) One of four parts of a year that have different weather patterns. (p. 308)

 **sediment** (sed′əmənt) Tiny bits of weathered rock or once-living animals and plants. (p. 232)

 **sedimentary rock** (sed′ə·mən′tə·rē rok) A kind of rock that forms from layers of sediment. (p. 232)

 **seed** (sēd) A structure that can grow into a new plant. (p. 70)

 **shadow** (sha′dō) A dark area that forms when light is blocked. (p. 502)

 **shelter** (shel′tər) A place in which an animal can stay safe. (p. 48) *A nest is a shelter for young birds.*

 **simple machine** (sim′pəl mə·shēn′) A machine with few or no moving parts. (p. 465) *A lever is a simple machine.*

 **soil** (soil) A mixture of minerals, weathered rocks, and decayed plant and animal matter. (pp. 120, 240)

 **LOG ON e-Glossary** at www.macmillanmh.com

solar energy (sol′ər en′ûr·jē) Energy from the Sun. (p. 254)

solar system (sol′ər sis′təm) A system made up of a star and the objects that move around it. (p. 338)

solid (sol′id) Matter that has a definite shape and volume. (p. 384)

solution (sə·lü′shən) A mixture in which one or more kinds of matter are mixed evenly in another kind of matter. (p. 411)

sound (saund) A form of energy that you hear when an object vibrates. (p. 490)

space probe (spās prōb) A machine that leaves Earth and travels through space. (p. 342)

speed (spēd) How fast an object moves over a certain distance. (p. 438)

sphere (sfîr) A body that has the shape of a ball. (p. 305) *A globe is a sphere.*

star (stär) A ball of hot, glowing gases. (p. 322) *Our Sun is a star.*

state of matter (stāt uv mat′ər) A form of matter, such as solid, liquid, and gas. (p. 384)

static electricity (stat′ik i′lek·tris′i·tē) The build up of an electrical charge on a material. (p. 513) *Static electricity can cause you to get a shock from touching a doorknob.*

stem (stem) A plant structure that holds a plant up and helps leaves reach sunlight. (p. 35)

structure (struk′chər) A part of an organism. (p. 33) *Horns are structures that help wild sheep keep safe.*

LOG ON e-Glossary at www.macmillanmh.com

switch (swich) A device that allows you to control the flow of electric current through a circuit. (p. 515)

T

telescope (tel′ə·skōp′) A tool used to make faraway objects appear larger and closer. (p. 342)

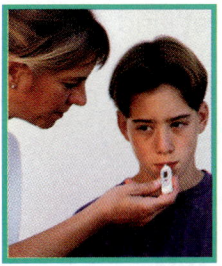

temperature (tem′pər·ə·chər) A measure of how hot or cold something is. (pp. 280, 482) *When you visit the doctor, she measures your body's temperature.*

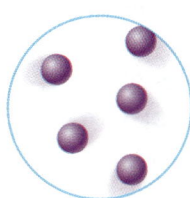

thermal energy (thûr′məl en′ər·jē) The energy of moving particles of matter. (p. 482)

thermometer (thûr′mom′i·tər) A tool that is used to measure temperature. (p. 483)

tornado (tôr·nā′dō) A powerful storm with rotating winds that forms over land. (p. 296)

trait (trāt) A characteristic of a living thing. (p. 92) *Spots are a trait of Dalmatians.*

translucent (trans·lü′sənt) Scattering light that passes through, so that objects on the other side appear blurry. (p. 503) *Frosted glass is translucent.*

transparent (trans·pâr′ənt) Letting all light through, so that objects on the other side can be seen clearly. (p. 503) *Clear glass is transparent.*

V

vertebrate (vûr′tə·brāt′) An animal with a backbone. (p. 54)

vibrate (vī′brāt) To move back and forth quickly. (p. 490) *A guitar string vibrates after you pluck it.*

volcano (vol′kā·nō) A mountain that builds up around an opening in Earth's crust. (p. 206)

volume (vol'ūm)
1. A measure of how much space an object takes up. (p. 365)

2. How loud or soft a sound is. (p. 492) *A whisper has a low volume.*

water cycle (wô'tər sī'kəl) Describes how water moves between Earth's surface and the atmosphere. (p. 294)

water vapor (wô'tər vā'pər) Water in its gas form. (p. 292, 399) *Water vapor is an invisible gas.*

weather (weth'ər) What the air is like at a certain time and place. (p. 280) *The weather shown here is warm and sunny.*

weathering (weth'ər·ing) The breaking down of rocks into smaller pieces. (p. 214) *Weathering caused these interesting rock shapes to form.*

wedge (wej) A simple machine that uses force to split objects apart. (p. 469)

weight (wāt) A measure of the pull of gravity on an object. (pp. 378, 447)

wetland (wet'land) An ecosystem where water covers the soil for most of the year. (p. 128)

wheel and axle (hwēl and ak'səl) A simple machine that consists of a wheel that moves around a post. The post is called an axle. (p. 467)

wind (wind) Moving air. (p. 283)

work (wûrk) What is done when a force changes an object's motion. (p. 454) *You do work when you move a bow to play the violin.*

LOG ON e-Glossary at www.macmillanmh.com

Index

A

Absorption, 500
   of light, 500, 504, 505
Abyssal plain, 197
Acorns, 73
Adaptations, 132–45
   animal, 133*, 134–35, 137, 139,
      142, 165
      being nocturnal, 137
      camouflage, 134, 135, 144*–45*
      hibernation, 139
      migration, 141, 164–65, 274
      mimicry, 139
   climate and, 135
   to cold environment, 132, 133*
   definition of, 134
   in desert, 136–37, 145*
   environmental change and,
      164–65
   in forest, 138–39
   in ocean, 140–41
   plant, 135, 136, 138, 140, 142
   in wetlands, 142
Air, 278
   effect of heat on, 479*
   oxygen in, 25, 260, 388
   as renewable resource, 253, 260
   showing air is around you, 279*
   as solution, 411
   sound wave traveling through,
      490–91,
      496*–97*
   uses of, 260
   wind as moving, 283
Air bladders, 140
Airplane, 441
Air pollution, 264, 268*–69*
   microorganisms that clean up,
      29
   from wildfire smoke, 163
Air pressure, 283
Air temperature, 280–81, 286*–87*
   average, 286*–87*, 311*
   factors affecting, 306–7
Algae, 112
   adaptations of, 140
   in ocean, 126
   as producers, 110
Alligators, 128
Aluminum, 234
Amazon Rain Forest, 124

Amber, fossils preserved in, 248
Amphibians, 59, 83
Analyzing data, 9
Anemometer, 283
Angler fish, 141
Animal rescue workers, 184
Animals, 42–61, 80–87
   adaptations, 133*, 134–35, 137,
      139, 141, 142, 165
      being nocturnal, 137
      camouflage, 134, 135, 144*–45*
      hibernation, 139
      migration, 141, 164–65, 274
      mimicry, 139
   carnivores, 113
   classifying, 50–51*, 52–61, 53*
      invertebrates, 55, 56–57
      vertebrates, 54, 58–60
   compared to plants, 44
   as consumers, 111
   decayed, in soil, 240
   desert, 123
   ecosystems differing by types
      of, 121
   endangered, 168, 171
   extinct, 174–75
   forest, 139
      temperate forests, 125
      tropical rain forest, 124
   fossil fuel from remains of
      ancient, 252
   herbivores, 113, 174
   life cycles of, 80–86
   living in soil, 240
   as living things, 22
   in Madagascar, 2–11
   metamorphosis of some, 83
   movement of, 44, 45
   natural disasters affecting,
      162–63
   needs of, 46–47
      food, 24, 44
      oxygen, 25, 47, 388
      space, 25
      water, 24, 46, 388
   in ocean, 126, 141
   omnivores, 113
   of the past. See Fossils
   pollen carried by, 72
   as renewable resource, 253
   response to changing seasons,
      275
   response to environmental
      changes, 164–65
   seeds carried by, 73
   structures, 44–48, 47*, 54

      to get what they need, 43*,
      46–47
      for staying safe, 48
   traits in common, 44
   weathering caused by, 215
   in wetlands, 128, 142
Animal trainer, 100
Antarctica, 304
   first women to cross, 187
Ants, 240
Aqueducts, 262
Area, units of measurement for, R2,
   R3
Arnesen, Liv, 187
Arteries, R16
Arthropods, 57
Aspirin, 244
Astrophysicist, 352, 370
Atmosphere, 280. See also Air
   water cycle between Earth's
      surface and, 294–95
Attraction of magnets, 367, 446
Auger, 468
Autumn. See Fall season
Average, finding an, 311*
Average air temperature, 286*–87*,
   311*
Aviation meteorologists, 356
Ax, as wedge, 469
Axis, 305, 319
   of Earth, 305, 319
      seasons and tilt of, 320–21
Axle, wheel and, 465, 467, 470

B

Back and forth motion, 436, 437
Backbone, 54, 55*. See also
      Vertebrates
Bacteria, 26, 111, R23
Baking soda-vinegar reaction, 420
Balance, pan, 376, 380, R5
Balanced forces, 445
Bald eagle, 106, 111, 112, 113
Balloons, hot air, 478
Bancroft, Ann, 187
Bar graphs, 89*, 286*–87*, R12
Barometer, 283
Basalt, 231
Bass, largemouth, 111, 112
Bats
   nocturnal, 137
   pollen carried by, 72
Battery, as power source for
      circuit, 515

Note: Page references followed by an asterisk indicate activities.

Bayfield, Wisconsin, seasons in, 308
Beakers, 375
Beaks, bird, 46, 48
Bears, 60
    koala, 42, 170
    panda, 168
Bedrock, 241
Bees, pollen carried by, 72
Beetle, 57
Bengal tigers, 168
Benz, Karl Friedrich, 440
Bicycle, states of matter in, 388
Big Dipper, 349
Binary stars, 352–53
Biomass energy, 257
Birds, 58
    beaks of, 46, 48
    desert, 62*
    eggs laid by, 84, 130–31
    life cycle of, 85*
    as living dinosaurs, 181
    nests of, 48, 84
    of pampas, 130–31
    similarity to ancient reptiles, 178
    as vertebrates, 58
    wings of, 45
Bladder, R19
Blizzards, 297, 298
Blood, R16
Blood cells, R16, R22
Blood vessels, R16
Blubber, 135
Body system, R14
    human. See Human body
        systems
Boiling, 399
Bones
    backbones, 54, 55*. See also
        Vertebrates
    in skeletal system, R14
Brain, R20
    hearing and, 494
    parts of, R20
    seeing and, 506
Brain stem, R20
Brass, as solution, 411
Breathing, structures for, 47, 58, 59,
    60, R17
Breckenridge, Colorado
    climate of, 306
Bronchial tubes, R17
Bronchioles, R17
Buffalo, 121
Bulbs, 76
Butterfly, 16–17, 80, 82

Cactus, 123, 136
Calcite, 242
Calcium, 234
Calculator, using, R8
California, preventing landslides in,
    210
Cameras, using, R8
Camouflage, 134, 135, 144*–45*
Canal, building, 218
Can opener, as compound machine,
    470, 472*
Canyon, 194
    on ocean floor (trench), 196, 197
Carbohydrates, R24
Carbon, 368
Carbon dioxide
    baking soda-vinegar reaction
        and, 420
    photosynthesis and, 36, 37
    plants' need for, 25
Careers in Science
    animal trainer, 100
    environmental chemist, 426
    lighting technician, 522
    mapmaker, 272
    meteorologist, 356
    wildlife manager, 184
Carnivores, 113
Cars, 440
Casts, fossil, 251
Caterpillar, 24, 81*, 82
Catfish, walking, 142
Cattails, 109, 112
Cave, limestone, 200*–201*
Cavern, underground, 185
Cells, 26
    blood, R16, R22
Celsius, degrees, R6
Cenote (water-filled cavern), 185
Centimeter (cm), R4
Central nervous system, R20
Cerebellum, R20
Cerebrum, R20
Chameleon, 58
    giant Madagascan, 4–11
Changes of state, 396–423
    cooling and, 400–401
    heat and, 397*, 398–99, 483
    physical changes, 406–13, 407*,
        422–23*
    in water, 401, 402, 404*–5*
Chapman, Roy, 180
Charts, R11

Cheetah, life cycle of, 86
Chemical changes, 416–21, 419*
    definition of, 418
    how matter is changed in, 417*,
        422*–23*
    signs of, 420, 423*
Chemicals, treating water with, 263
Chemical weathering, 220
Cherry tree, life cycle of, 74
Chicago, climate of, 304
Chlorophyll, 37
Circuits, 515
Circulatory system, R16
Cirrus clouds, 291
Classification
    of animals, 50–51*, 52–61, 53*
        invertebrates, 55, 56–57
        vertebrates, 54, 58–60
    of matter, 367*
    of plants, 38
    of rocks, 230, 231*
    of soils, 243*
    of stars, 348
Classifying, skill at, 12, 50*–51*
Classroom, safety tips in, 14
Claws, 134
Clay soil, 243
Clearing land, environments
    changed by, 154–55
Climate(s), 120, 302–9
    adaptation and, 135
    comparing, 307*
    definition of, 304
    desert, 122
    different, 304
    of different ecosystems, 120
    extinction and change in, 175
    factors affecting, 306–7
    seasons and, 308
    temperate forest, 125
    temperature and precipitation
        patterns, comparing, 303*
    tropical rain forest, 124
Closed circuit, 515
Clouds, 290–93
    cirrus, 291
    cumulus, 291
    formation of, 292–93, 294, 307*
    stratus, 290, 292
Coal, 234
    formation of, 252–53
Coast, 196
Collared lizard, 122
Color(s), 504–5*
    change in, as sign of chemical
        change, 420
    of desert animals, 137, 145*

of leaves, 37, 505
of minerals, 227*, 228
mixing, 505*
seeing, 505
soil, 242
Comets, 344
Communicating, skill at, 12, 116*–17*
Community, 166–67
Competition, 113, 153, 155
Compound machine, 470, 472*
Computers, 6, R9
Conclusions, drawing, 10–11
Condensation, 293, 400, 401*
water cycle and, 294
Conductors, 484, 516*
of electric current, 516
of heat, 367, 484
metals as, 367, 484, 516
Cones, 75
Conifers, life cycle of, 75
Conservation, 266
energy, 257
Constellations, 349*, 350
Consumer, 111
Contact forces, 446
Continental shelf, 196
Continents, 192, 193
ocean floor as continuation of,
196–97
Contour farming, 244
Contraction of matter, heat and,
483
Converting hours to minutes, 325*
Cooking, chemical changes in, 418
Cooling, loss of energy and, 400,
401. See also Heat
Copper, 234
as conductor, 516
Coral reef, 126–27
Core of Earth, 198
Cornea, 506
Cotton, as insulator, 484
Crane fly, 112
Crater, 332
Crayfish, 113
Crescent Moon, 328, 329
Crust of Earth, 198
sudden changes in, 204, 206
Cumulus clouds, 291
Current, electric, 514–16
Curve-billed thrasher, 123
Customary units of measurement,
R3

Dams, 262
Dandelions, 68, 73
Data, 6
analyzing, 9
interpreting, skill at, 13, 286*–87*
organizing, R10–13
charts, R11
graphs, R12–13
maps, R10
tables, R11
testing hypothesis using, 7
Day and night, 318–19
changing length of days, 275
seasons and length of daylight,
321
seeing stars at night, 347*
Dead plant and animal material,
decomposers of, 111, 114*
Death Valley, California, climate of,
307
Decimals, multiplying, 519*
Decomposers, 111, 114*, 152
Decomposition, 74, 82
Definite, 384
Degrees Celsius, R6
Degrees Fahrenheit, R6
Deposition, 216
Descriptive writing, 62*, 236*, 390*
Desert, 122
adaptations for, 136–37, 145*
birds of, 62*
cold, 123
ecosystem, 122–23
Dew, 400
Diamonds, 229, 234
Diaphragm, R17
Digestion, R18
Digestive system, R18
Dinosaur nests, 180
Dinosaurs, 174
changing ideas about, 180–81
fossils of, 174, 176, 180–81, 250
Direction
change in motion and, 445
of force, machines and change
in, 464, 466, 467
Discharge of static electricity, 513
Diseases, environments changed by,
163, 167
Distance, 435
force used on inclined plane and,
468
measuring, 435

pull of magnet and, 450*–51*
speed and, 438
Dolphins, 60, 141
Doppler radar, 300, 301
Dormice, 139
Dotlich, Rebecca Kai, "Jump Rope"
by, 428–29
Dragon trees, 168
Drawing conclusions, 10–11
Drip tips on leaves, 138
Drought, 162–63
Drug-free, being, R26
Dry climate, mountains and, 307

Eagle, bald, 106, 111, 112, 113
Eardrum, 494
Ears
of desert animals, 137
hearing with, 494
Earth
atmosphere of, 280, 294–95
axis of, 305, 319
tilted, seasons and, 320–21
comparing Moon and, 332
distance of Sun from, 322
fossils and clues about history of,
173*, 177
gravity on, 447
layers of, 198
core, 198
crust, 198, 204, 206
mantle, 198, 206
Moon of. See Moon (of Earth)
orbit of, 320–21, 338
as planet, 338, 341
revolving around Sun, 320–21
stars seen during different
seasons and, 350
rotation of, 318–19*
day and night caused by, 319
Moon's position in sky and,
329
as sphere, 305
sudden changes to, 202–9, 203*
earthquakes and, 204–5
floods and, 208
landslides and, 208, 210–11
volcanoes and, 206–7
surface of, 190–97
land features on, 191*, 192, 193,
194–95
landforms, 194–97

ocean floor, 196–97
water features on, 191*, 192, 193, 194–95
weight on Moon vs., 378
Earthquakes, 204–5
Earthworms, 240
Eastern dwarf tree frog, 15
Ecosystems, 108–9, 118–29, 119*
changing, 167*
community in, 166–67
decomposers as important part of, 114
definition of, 108
desert, 122–23
food chain in, 110–11, 112
food web in, 112–13
forest, 124–25
grassland, 116*–17*, 121
how they differ, 120–21
populations in, 166
water, 121
ocean, 118, 121, 126–27
pond, 108–9, 121
wetlands, 128
Eggs
in animal life cycle, 83, 84
bird, 84, 130–31
dinosaur, 180
fish, 84, 85
hatching of, 83, 84
reproduction by laying, 23, 58, 59
reptile, 84
seeds formed by pollen and, 72–73
Electrical charge, 512–13
Electric current, 514–16
in circuit, 515
Electricity, 510–18
conductors of, 516
electrical charge and, 512–13
electric current and, 514–16
energy sources for, 252–53, 254, 256–57, 518*
nonrenewable, 253, 256
renewable, 254, 256–57
heat produced by using, 481
insulators of, 516
making bulb light, 511*
static, 512, 513
Elements, 368
formation in stars, 370
Elephants
migration of, 164–65
similarity to woolly mammoths, 178
trunks of, 46

Elf owls, 62
Embryo, 70
plant, 70, 71
Endangered organisms, 168, 171
Energy, 456–58. See also Heat; Light; Sound; Electricity
biomass, 257
change from one form into another, 458
conserving, 257
cooling and loss of, 400, 401
cost of, 519*
definition of, 456
from electric current, 514
from food, 24, 456, 457*
in food chain, 110–11
from fuels, 252–53, 254, 264, 266
geothermal, 256
growth and, 22
heating and gain of, 398, 399
hydropower, 256
kinetic, 456
potential, 456
sources of, 252–53, 254, 518*
nonrenewable, 256
renewable, 254, 256–57
from the Sun (solar energy), 24, 25, 32, 36, 254, 257
thermal, 482–83
transfer of, 458
using, 457*
Environment, 24. See also Ecosystems
changes in
community and, 166–67
diseases causing, 163, 167
effect on living things, 160–69
how living things cause, 150–59, 151*
natural disasters and, 162–63
organisms endangered by, 168
response of plants and animals to, 164–65
sudden, 174–75
protecting, 156
traits affected by, 94
Environmental chemist, 426
Erosion, 216–17
causes of, 216–17, 218
preventing soil, 240, 241, 244
sedimentary rock formation and, 232
Esophagus, R18
Estimations, making, 221*
Evaporation, 292, 399, 483
separating mixtures by, 412

water cycle and, 294
Excretory system, R19
Exoskeleton, 57
Expansion of matter, heat and, 483
Experimenting, skill at, 13, 486*–87*
Explanatory writing, 472*
Expository writing, 220*
Extinct organisms, 174–75
Eye color, as inherited trait, 92, 93
Eyes, seeing with, 506
colors, 505

F

Fahrenheit, degrees, R6
Fall season, 321
leaves falling off trees in, 23, 138, 150
length of day in, 275
Fantastic leaf-tailed gecko, 10
Farming, contour, 244
Fat, survival in cold environment and body, 133*, 135
Fats, as nutrient, R25
Feldspar, 228
Fence lizard, 96–97
Fern, life cycle of, 76
Fertilizers, pollution from, 264
Fictional narrative, 310*
Field, safety tips in the, 14
Filters, separating mixtures by, 412
Fins, fish, 47
Fires
as source of heat, 481
as source of light, 500
First-quarter Moon, 328, 330
Fish, 52, 59
adaptations to environmental change, 165
eggs laid by, 84, 85
gills for breathing, 47, 59, 141
habitat of, 109
life cycle of, 84–85
in ocean, 118, 126, 141
in pond food web, 111, 112
structures for moving, 47
as vertebrates, 59
Fitness, healthy living and, R26
Flip book of Moon's phases, 331*
Floating
of ice, 402
as property of matter, 366
Floods, 162, 208
environments changed by, 162
in hurricane, 297

plants and, 160, 161*
Flowering plants, 38, 72–73
    life cycle of, 74
Flowers, 33
    female and male parts of, 72
    seeds made by, 72–73
    smell and bright colors of, 72
Fog, 292
Food
    of animals, 24, 44
    animal structures that help them
        get, 46
    chemical changes in, 418, 419
    digestion of, R18
    energy from, 24, 456, 457*
    living things need for, 24
    made by plants, 24, 25, 32, 36,
        110
        in plant roots, 34
        stored in seeds, 70
Food chain, 110–11, 112
    diagram, 116*–17*
Food web, 112–13
"Fool's gold" (pyrite), 226, 228, 229
Forces, 442–51
    balanced, 445
    changes in motion and, 443*,
        445
        work and, 454
    contact, 446
    definition of, 444
    machines and using less force,
        464
    measurement of, R6
        units of, R2, R3
    types of, 446–48
        friction, 448, 458
        gravity, 378, 447
        magnetism, 367, 446, 450*–
            51*
    unbalanced, 445
Forest, 124
    adaptations for, 138–39
        animals, 139
        plants, 138
    cutting down to make wood
        products, 154–55
    ecosystem, 124–25
    kelp or seaweed, 140
    temperate, 125
    tropical rain, 124
Forming a hypothesis, 5
    skill at, 12, 78*–79*
Fossa, 10
Fossil fuels, 252–53, 254
    air pollution from burning, 264
    conserving, 266

as nonrenewable resource, 253
Fossils, 174–78, 249–52
    in amber, 248
        dinosaurs, 174, 176, 180–81, 250
        formation of, 249*, 250–51
        learning about the past from,
            173*, 176–77*
        in sedimentary rock, 232
        similarities between living things
            and, 178
Foxes, 25
Fractions, 237*
Freezing, 401–2, 483
    rock weathered by repeated
        thawing and, 215
    salt water vs. fresh water, 404*–
        5*. See also Ice
Fresh water, 192
    ecosystems, 108–9, 121
    ocean animals and, 119*
    sources of, 261
Friction, 448
    heat from, 458
Frogs, 59, 112
    adaptations of, 134, 165
    eastern dwarf tree, 15
    life cycle of, 82
    Malagasy tree, 11
Frost, Darrel, 96–97
Fruits, 73, 74
Fuels, 252. See also Energy
    fossil, 252–53, 254, 264, 266
Fulcrum, 466, 470
Full Moon, 329, 330, 331
Fur, 48, 60
    as insulator, 484

Gagarin, Yuri, 441
Garbage
    land pollution from, 154
    produced by average American,
        158*–59*
Gas(es), 387
    comparing solids, liquids and,
        387*
    condensation of, 293, 294, 400,
        401*
    formation of, as sign of chemical
        change, 420
    heating liquid to change into,
        399
    living things need for, 24, 25
    natural, 252

particles in, 387
    sound wave traveling through,
        490–91, 496*–97*
    water vapor as, 292
Gas giants (outer planets), 340–41
Gazelle, red, 175
Gecko, fantastic leaf-tailed, 10
Geese, migration in V formation,
    274
Gems, 234
Geothermal energy, 256
Germination, 70–71, 74, 75
Germs, immune system and
    protection from, R22–23
Giant Madagascan chameleon, 4–11
Gibbous Moon, 328, 329
Gills, 47, 59, 141
Glaciers, 192, 216, 261
    erosion caused by, 216, 217
    estimating change in, 221*
Glass, as insulator, 516
Glen Canyon Dam (Arizona), 262
Gneiss, 233
Gold, 228, 368
Goldfish, 25
"Good Ship Popsicle Stick, The,"
    358–59
Graduated cylinder, 375, R5
Grains
    of rock, 230, 231
    of soil, 243
Grams, 376, R5
Grand Canyon, 212
Granger, Walter, 180
Granite, 231, 233
Graphite, 234
Graphs, R12–13
    bar, 89*, R12
    line, R13
    pictographs, R13
Grasshoppers, 110, 113
Grasslands
    dry season in, response of plants
        and animals to, 164
    ecosystem, 116*–17*, 121
    pampas, 130–31
    prairie community in, 166–67
Gravity, 378, 447
    erosion caused by, 216
    observing, 447*
    weight and pull of, 378, 447, R6
    work done by, 454
Great Dark Spot, 341
Great Red Spot, 340
Groundhogs, 48

Groundwater, 261, 262, 264
Growth, energy and, 22
Guitar, pitch of sounds from, 493
Gypsum, 229

Habitat, 109
    environmental change and, 162,
        168
Hail, 282
Hair, 60
Hair color, as inherited trait, 92, 93
Halite, 234
Hand lens, 374, R7
Hardness of mineral, 229
Hatching of egg, 83, 84
Healthy living, R24–26
    nutrients for, R24–25
    to protect against germs, R23
    staying fit, R26
Hearing, 494, R21
Heart
    circulatory system and, R16
    muscles, R15
Heat, 478–87. See also Temperature
    definition of, 480
    electrical energy changed into,
        514
    flow of, 480
        conductors, 367, 484
        controlling, 484
        insulators, 484, 486*–87*
    from friction, 458
    heating objects, 481*
    how heat affects matter, 482–83
        air, 479*
        changing state, 397*, 398–99,
            483
        expanding and contracting,
            483
    as sign of chemical change, 420
    sources of, 480–81
        Sun, 281, 322, 480
        underground, 254
Helium, 387
Hematite, 242
Herbivores, 113, 174
Heredity, 92. See also Traits
Herons, 112
Hibernation, 139
History of Science
    Looking at Dinosaurs, 180–81
    Travel through Time, 440–41
    Turning the Power On, 256–57

Hoodoos, 215
Horsehead Nebula, 370–71
Hosts (of viruses), R23
Hot air balloons, 478
Hours, converting to minutes, 325*
Human body systems, R14–23
    circulatory system, R16
    digestive system, R18
    excretory system, R19
    immune system, R22–23
    muscular system, R15
    nervous system, R20
        senses and, R21
    respiratory system, R17
    skeletal system, R14
Hummingbird, 44
    pollen carried by, 72
Humus, 120, 125, 240, 241, 242
Hurricanes, 297
Hydrogen, 368
Hydropower energy, 256
Hypothesis
    forming, 5, 12, 78*–79*
    testing a, 6–7

Ice, 396
    floating in liquid water, 402
    freezing water into, 261, 401, 402
        salt water vs. fresh water,
            404*–5*
    of glacier, movement of rocks
        by, 217
    heating and melting of, 397*, 398
    measuring mass of ice turning to
        liquid water, 380–81*
Ice Age, 175
Icebergs, 261, 357
Iceland, underground heat used to
    make electricity in, 254
Ice tongue, estimating change in
    size of, 221*
Idea maps, R10
Igneous rocks, 230, 231
Immovable joints, R14
Immune system, R22–23
Imprints, 250, 251*
Inclined planes, 465, 468–69*
    screw, 465, 468
    wedge, 465, 469
Inferring, skill at, 13, 460*–61*
Inherited traits, 92–93*, 97
Inner core, 198
Inner ear, 494

Inner planets, 340–41
Inquiry skills, 12–13
    classifying, 12, 50*–51*
    communicating, 12, 116*–17*
    experimenting, 13, 486*–87*
    forming a hypothesis, 12, 78*–79*
    inferring, 13, 460*–61*
    interpreting data, 13, 286*–87*
    making a model, 12, 200*–201*
    measuring, 13, 380*–81*
    observing, 12, 344*–45*
    predicting, 13, 404*–5*
    using numbers, 12, 158*–59*
    using variables, 13, 246*–47*
Insects
    as decomposers, 114
    as invertebrates, 55
    metamorphosis of, 83
    mimicry used by, 139
Insulator, 484, 486*–87*, 516*
    of electric current, 516
Internet, using the, R9
Interpreting data, skill at, 13,
    286*–87*
Intestine
    large, R18
    small, R18
Invertebrates, 55, 56–57
Iris of eye, 506
Iron, 368
    magnetism of, 367
    rusting of, 419
Island, 194, 197

Jackrabbit, 137
Jefferson Memorial, 236
Jellies, 55, 56
Joints, R14
"Jump Rope" (Dotlich), 428–29
Jupiter, 339, 340

Kangaroos, 48
Kelp, 140
Key, map, 193
Kidneys, R19
Kilogram, 376
Kilowatts/hour (kWh), 519*
Kinetic energy, 456
Knives, as wedges, 469
Koala bears, 42, 170

Kudzu plants, 155

 L

Ladybug, life cycle of, 83
Lakes, 192, 194
Lambs, 88
Land
    changes made by people, 218
    clearing, changing environment
        by, 154-55
    on Earth's surface, 191*, 192, 193,
        194-95
    erosion of, 216-17
    sudden movement and change
        in, 203*, 204-8
    weathering of rocks, 214-15
Landforms, 194-97
    limestone cave, 200*-201*
    on ocean floor, 196-97
    types of, 194-95
Land pollution, 154
Landslides, 208, 210-11
Large intestine, R18
Largemouth bass, 111, 112
Larva, 83
Lasers, 508-9
Last-quarter Moon, 329, 330
Lava, 206, 207, 231, 398
Learned traits, 94
Leaves of plants, 33, 35, 71
    adaptations
        for cold climate, 135
        in desert, 136
    drip tips, 138
    falling off in fall, 23, 138, 150
    functions of, 36-37
    green color, 37, 505
    growth of new plant from, 76
    importance of, 36-37
    shapes and sizes of, 36, 37
Lehman Cave, 200
Lemur, ring-tailed, 11
Length, measuring, 372, 373*, 374,
    R4
    units of measurement, R2, R3
Lens, 342
    of eye, 506
Levers, 463*, 465, 466-67, 473*
    in compound machine, 470
    pulleys, 465, 467
    wheel and axle, 465, 467, 470
Life cycle, 74
    of animals, 80-86
    of plants, 74-76

Life spans, bar graph comparing,
    89*
Lifting objects, simple machines for,
    463*, 466-67
Light, 498-509
    absorption of, 500, 504, 505
    beam of, 500, 508-9
    colors of, 504-5*
    definition of, 500
    how light moves, 499*, 500-501
    lasers, 508-9
    opaque objects and, 502
    reflection of, 499*, 501, 504, 505
    refraction of, 503, 504, 506
    seeing, 505, 506
    shadows and, 316, 317*, 318, 502
    as sign of chemical change, 420
    sources of, 500
        Sun, 322, 500
    translucent objects and, 503
    transparent objects and, 503
    white, 504-5
Light bulb
    making it light up, 511*
    as source of heat, 481
    as source of light, 500
Lighting technician, 522
Lightning, 510
Limestone, 232, 234
Limestone cave, 200*-201*
Lincoln Memorial, 236
Line graphs, R13
Lion, 46
Liquids, 383*, 386
    boiling, 399
    comparing solids, gases, and,
        383*, 387*
    definite volume of, 386
    evaporating, 292, 294, 399, 412,
        483
    freezing to solid, 401, 402,
        404-5*, 483
    melting solids to, 397*, 398, 483
    particles in, 386
    in shape of container, 386
    sound wave travelling through,
        491, 496*-97*
Liquid volume, 386
    measurement of, 375, R5
    units of measurement, R2, R3
Literature
    magazine articles, 102-3, 186-87,
        274-75, 358-59
    poems, 16-17, 428-29
Liters, 375
Liver, R19
Living things, 20-27. See also

    Animals; Plants
    animals as, 22
    cells as building blocks of, 26
    characteristics shared by, 22-23
    in ecosystems, 109
    environmental changes affecting,
        160-69
        community and, 166-67
        diseases and, 163, 167
        endangered organisms, 168,
            171
        natural disasters, 162-63
        responses of organisms to,
            164-65
    how they change environment,
        150-59, 151*
        competition and, 153, 155
        decomposers, 152
        people and, 154-56, 158*-59*
    needs of, 24-25
        in habitat, 109
    nonliving things compared to,
        21*, 23
    of past, 172-81. See also Fossils
    plants as, 22
    in soil, 240
    weathering caused by, 215
Lizards, 48, 96-97, 122
Load, 466
Loam, 243
Loudness, 492
    protecting your hearing, 494
Lungs, 47, 58, 60, R17
Luster of mineral, 229
Lymph, R22
Lymph nodes, R22
Lymph vessels, R22

 M

McDonald, Robert, 358, 359
Machines, 464-72
    compound, 470, 472*
    definition of, 464
    simple, 465-69
        inclined planes, 465, 468-69*
        levers, 463*, 465, 466-67,
            473*
        pulleys, 465, 467
        screw, 465, 468
        wedge, 465, 469, 470
        wheel and axle, 465, 467, 470
    for traveling, history of, 440-41
Madagascar, animals of, 2-11
Magazine articles

"Good Ship Popsicle Stick, The," 358–59
"Once Upon a Woodpecker," 102–3
"One Cool Adventure," 186–87
"What a Difference Day Length Makes," 274–75
Magma, 206, 231
Magnetism, 367, 446, 450*–51*
   magnetic force passing through objects, 451*
   as property of matter, 367
Magnets, 446
   attraction of, 367, 446
   distance and pull of, 450*–51*
   separating mixtures with, 412
Making a model, skill at, 12, 200*–201*
Malagasy web-footed frog, 11
Mammals, 60
   life cycle of, 86
Mammoth, woolly, 176, 178
Manatee, 25
Mangroves, 142
Mantle of Earth, 198, 206
Mapmaker, 272
Maps, R10
   idea, R10
   key of, 193
   land and water features on, 193
   using, 4
   weather, 284
   of your state, 195*
Marble, 234
Marble memorials, 236
Marco, Orsola De, 352*–53*
Mariana Trench in Pacific Ocean, 197
Marimba, pitch of sounds from, 493
Mars, 338, 339, 341
Martens, pine, 125
Mass, 365
   measurement of, 376–77, 377*, R5
      units of, R2, R3
   sinking and floating due to, 366
   volume and, 376–77
   weight compared to, 378
Math in Science
   average, finding an, 311*
   bar graph, making, 89*
   converting hours to minutes, 325*
   estimations, making, 221*
   fractions, 237*
   how to use patterns, 473*
   multiplying decimals, 519*
   order numbers, 63*

   perimeter, finding, 391*
   subtracting multi-digit numbers, 171*
Matter, 362–69
   changes of state. See Changes of state
   classifying, 367*
   definition of, 364
   describing objects, 362, 363*, 390*
   elements as building blocks of, 368
   heat's effect on, 482–83
      air, 479*
      changing state, 397*, 398–99, 483
      expanding and contracting, 483
   mass and, 365
   measuring, 372–81, 380–81*
      length, 372, 373*, 374, R4
      mass, 376–77, 377*, R5
      units of measurement, R2–3
      volume, 375, 377*, R5
      weight, 378, R6
   properties of, 365, 366–67
   sound wave travelling through, 491, 496*–97*
   states of, 384–89
   tiny particles making up, 377, 385, 398
   volume and, 365
   ways of changing, 407*, 417*
Measurement, 372–81, 380–81*
   of distance, 435
   of energy used, 519*
   of force/weight, R6
   of length, 372, 373*, 374, R4
   of mass, 376–77, 377*, R5
   of perimeter, 391*
   of speed, 438*
   table of measures, R3
   of temperature, 280, 281, 482–83, R6
   of time, R4
   units of, R2–3
      standard, 374
   of volume, 375, 377*
      liquid, 375, R5
   of weight, 378, R6
Measuring, skill at, 13, 380*–81*
Measuring cup, 375, 380
Medicines from plants, 244
Meet a Scientist
   Ana Luz Porzecanski, 130–31
   Darrel Frost, 96–97
   Neil deGrasse Tyson, 370–71

   Orsola De Marco, 352*–53*
Megalodon, 176
Melting, 397*, 398, 483
Mercury (planet), 338, 339, 340
Mertz Glacier, ice tongue of, 221*
Mesquite tree, 136
Metals
   brass as solution of, 411
   as conductors, 484, 516
      of heat, 367, 484
   magnetic attraction to, 367, 446, 450*–51*
   from minerals, 234
   ores, mining, 414–15
Metamorphic rocks, 233
Metamorphosis, 83
Meteor, extinction of dinosaurs and, 174
Meteorologist, 356
Meters, 374, R4
Meterstick, 435
Metric system, 374, 435
   measuring length using, 374
   measuring mass using, 376
   measuring volume using, 375
   metric units, R3
Microorganisms, 28
   eating up pollution, 28–29
Microscope, 26, R7
Migration, 141
   environmental change and, 164–65
   of geese, 274
Milk from mammals, 60
Milliliters (mL), R5
Millimeter (mm), R4
Millipede, red, 11
Mimicry, 139
Minerals, 226–35
   as building blocks of rocks, 228, 230
   changes in metamorphic rock, 233
   color and mark, comparing, 227*, 229
   definition of, 228
   fossil formed by, 251
   mining ores for metals, 414–15
   properties of, 228–29
   in soil, 240
   uses of, 234
Minerals (nutrient), R25
Mining ores, 414–15
Mirrors, light reflected by, 498, 499*, 501
Mixtures, 410–12
   separating, 412*

solutions, 411, 412
Model making, skill at, 12, 200*–201*
Mold, as decomposers, 114
Molds and casts, fossil, 251
Mollusks, 57
"Monarch Butterfly" (Singer), 16–17
Monkeys, 124
Moon (of Earth), 326–35
    changing position in night sky,
        329, 335*
    changing shape of, appearance
        of, 326, 327*, 331, 334*–35*
    comparing Earth and, 332
    orbit of Earth, 330, 334*–35*
    phases of, 328–30, 331*
    revolving around Earth, 328, 331
    weight on, 378
Moth, suraka silk, 10
Motion, 436–37
    changes in, 443*, 445
        work and, 454
    definition of, 436
    energy needed for, 456
    speed of, 438
    types of, 436, 437
Mountains, 194
    climates and, 306–7
    landslide and changes in, 208
    volcanic, formation of, 206
Movable joints, R14
Muscles, R15
Muscular system, R15
Mushrooms, as decomposers, 114
MyPyramid, using, R26
Myth, 310*

Narrative. See also Writing in
        Science
    fictional, 310*
    personal, 88*, 324*
Natural disasters, environments
        changed by, 162–63. See also
        Storms, severe
Natural gas, 252
Natural resource, 244. See also Air;
        Soil; Water
    nonrenewable, 253, 256
    renewable, 253, 254, 256–57, 260
Nebula, 370–71
Nectar, 72
Negative charge, 512, 513
Neon, 368
Neptune, 338, 341

Nervous system, R20
    senses and, R21
Nests
    bird, 48, 84
    dinosaur, 180
New Horizon (space probe), 342
New Moon, 328, 330, 331
Newtons (N), R6
Niagara River, 258
Night and day, 319
    seasons and length of daylight,
        321
    seeing stars at night, 347*
Nocturnal animals, 137
Nonflowering plants, 38
Nonliving things. See also Minerals;
        Rocks
    in ecosystems, 109
    living things compared to, 21*, 23
    types of, 23
Nonrenewable resources, 253, 256
    fossil fuels, 252–53, 254, 264,
        266
North America, 193
Northern Hemisphere, 320
    seasons in, 320, 321
Northern saw-whet owl, 101
Number patterns, using, 473*
Numbers. See also Math in Science
    ordering, 63*
    skill at using, 12, 158*–59*
    subtracting large, 171*
Nutrients, 34. See also Food
    for healthy living, R24–25
    plant roots for getting, 34
    plant stem to carry, 35
    in soil, 34, 74, 82, 240
        decomposers and, 111, 114, 152

Observing, skill at, 12, 344*–45*
Ocean(s), 192
    adaptations for, 140–41
        algae, 140
        animals, 141
    climates and, 306
    definition of, 126
    on Earth, names of, 126
    ecosystem, 118, 121, 126–27
Ocean floor, 196–97
    adaptations for living on, 141
Ocean waves, erosion by, 216
Offspring, 93
Oil, 252

Oil spills, pollution from, 264
    microorganisms to clean up, 29
Omnivores, 113
"Once Upon a Woodpecker," 102–3
One-celled organisms, 26, 28–29
    as decomposers, 114
"One Cool Adventure," 186–87
Opaque objects, 502
Open circuit, 515
Optic nerve, 506
Orange clown fish, 52
Orbit, 320
    of Earth around Sun, 320–21, 338
    of Moon around Earth, 330,
        334*–35*
Orca whales, 491
Orchid, 302
Ores, 414
    mining, 414–15
Organisms, 22. See also Animals;
        Living things; Plants
    cells in, 26
    endangered, 168
    extinct, 174–75
    inherited traits of, 92–93*
    life cycle of, 74
    needs of, 24–25
Organizing data, R10–13
    charts, R11
    graphs, R12–13
    maps, R10
    tables, R11
Orion constellation, 350
Ornithologist, 130
Outer core, 198
Outer ear, 494
Outer planets, 340–41
Owen, Richard, 180
Owl pellet, 107*
Owls
    elf, 62
    food needed by, 107*
    northern saw-whet, 101
Oxygen, 368
    in air, 25, 260, 388
    animal structures that help them
        get, 47, 58, 59, 60, R17
    living things need for, 25, 47, 388
    photosynthesis and, 36, 37

Pacific Ocean, 126
    Mariana Trench in, 197
Pampas, 130–31
Panama Canal, 218
Pan balance, 376, 380, R5
Panda bears, 168
Parents, traits passed on to young
        by, 90, 91*, 92–93
Park ranger, 184
Partly movable joints, R14
Patterns, using number, 473*
Peat, 252, 253
Penguins, 58
Peninsula, 194
People
    changes in land made by, 218
    environments changed by,
            154–56
        trash produced, 158*–59*
    organisms endangered as a
            result of, 168
    pollution and, 264
Performance Assessment
    acting out a new term, 425*
    changing Earth, 223*
    clothing, properties of, 393*
    conservation cards, 183*
    conservation poster, 271*
    imaginary organisms, 147*
    life-cycle poster, 99*
    listing living things, 65*
    making book about matter, 393*
    not-so-simple machines, 475*
    producing energy, 521*
    space exploration, 355*
    weather event poster, 313*
Perimeter, finding, 391*
Peripheral nervous system, R20
Personal narrative, 88*, 324*
Persuasive writing, 170*, 518*
Phase(s), 328
    of Moon, 328–30, 331*
Photosynthesis, 36, 37
Phyllite, 233
Physical activity, staying fit with,
        R26
Physical changes, 406–13
    how matter is changed in,
            422*–23*
    mixtures, 410–12
    solutions, 411, 412
    ways of making, 407*, 408
Pictographs, R13

Pineapple, properties of, 365
Pine martens, 125
Pine tree, life cycle of, 75
Pitch of sound, 493
Plain, 194
    abyssal, 197
Planes, inclined, 465, 468–69*
Planets, 336–45
    definition of, 338
    inner, 340–41
    movement through space, 337*
    observing, 342, 344*–45*
    outer, 340–41
    sizing up, 339*
    in solar system, 338–39
Plants, 30–41
    adaptations of, 135, 136, 138, 140,
            142
    animals compared to, 44
    characteristics of, 32–33
    classifying, 38
    decayed, in soil, 240
    in desert, 123, 136
    ecosystems differing by types
            of, 121
    extinct, 175
    food made by, 24, 25, 32, 34, 36,
            70, 110
    in forest, 138
    fossil fuels from remains of
            ancient, 252
    growth of, 70–71
    growth without seeds, 76
    how they are alike, 31*
    life cycle of, 74–76
    as living things, 22
    medicines from, 244
    natural disasters affecting,
            162–63
        floods, 160, 161*
    needs of, 40–41*
        for gases, 25
        for space, 25
        for water, 24
    in ocean, 126
    preventing landslides with, 210
    as producers, 110
    as renewable resource, 253
    response to changing seasons,
            275
    response to environmental
            changes, 164–65
    seeds of, 68, 69*, 70–73
    soils for growing, 240, 242
    structures of, 33, 34–37, 38
        flowers, 33, 72–73

    leaves, 33, 35, 36–37, 71, 76,
            135, 136, 138, 150, 505
    roots, 32, 33, 34, 70, 71, 136,
            142, 155, 156, 215
    stems, 32, 33, 35, 76, 136
    in tropical rain forest, 124
    used for animal shelters, 48
    weathering caused by, 215
    in wetlands, 128, 142
Plasma, R16
Plastic, as insulator, 516
Plateau, 194
    continental shelf as, 196
Platelets, R16
Pluto, 339
Poems
    "Jump Rope" (Dotlich), 428–29
    "Monarch Butterfly" (Singer),
            16–17
Pollen, 72–73, 74
Pollination, 73, 75
Pollution, 154, 264
    air, 29, 163, 264, 268*–69*
    environments changed by, 154
    land, 154
    microorganisms that eat, 28–29
    modeling, 155*
    3 Rs and reducing, 156
    water, 264, 269*
Pond, 192
    ecosystem, 108–9, 121
    food chains, 110–11
    food web, 112–13
Population, 166
Porcupine, 48
Porzecanski, Ana Luz, 130–31
Position, 432–37
    definition of, 434
    describing, 433*, 434–35
    distance and, 435
    motion as change in, 436
    words giving clues about, 434
Positive charge, 512
Potato plants, reproduction of, 76
Potential energy, 456
Prairie community, 166–67
Prairie dogs, 166–67
Precipitation, 282, 294. See also
        Rain; Snow
    from clouds, 290, 291
    patterns of temperature and,
            comparing, 303*
    water cycle and, 294, 295
Predators, 112
    adaptations of, 134
    adjustments to environmental
            change, 165

how prey hide from, 130
Predicting
    skill at, 13, 404*–5*
    tornadoes, 300–301
    weather, 284
Prey, 112, 134
Prism, 504
Producer, 110
Property(ies), 365
    of matter, 365, 366–67
        gases, 387
        liquids, 386
        solids, 384–85
    separating mixtures by, 412
Proteins, R25
Pterodactyls, 176, 178
Pulleys, 465, 467
Pumpkinseed sunfish, 111, 112
Pupa, 83
Pupil of eye, 506
Pushes and pulls, 444. See also
    Forces
Pyrite ("fool's gold"), 226, 228, 229

Quartz, 228, 229

Raccoon, 54
Radar, Doppler, 300, 301
Rain, 282, 288
    in desert, 122
    drought from too little, 162–63
    floods from too much, 162, 208
    how raindrops form, 289*
    in hurricane, 297
    in tropical rain forest, 124
    weathering caused by, 214, 215
Rainbow, 504
Rain forests, tropical, 124, 138
Rain gauge, 282
Ramps, 468
Ranger Rick, articles from
    "Once Upon a Woodpecker,"
        102–3
    "What a Difference Day Length
        Makes," 274–75
Rate of speed, units of
    measurement for, R3
Raxworthy, Chris, 3–11
Razafimahatratra, Paule, 3–11
Recycling, 156

of water, 260
Red blood cells, R16, R22
Red gazelle, 175
Red millipede, 11
Reduce, 156
Reef, coral, 126–27
Reflection, 501
    of light, 499*, 501, 504, 505
    of sunlight off of Moon, 329, 331
    of sunlight off of planets, 339
Refraction, 503
    of light, 503, 504, 506
Regeneration, 96–97
Renewable resources, 253, 260
    for energy, 254, 256–57
Repel, 446
Repelling of magnets, 446
Reproduction, 23
    by amphibians, 59
    by birds, 58, 85
    by fish, 59, 85
    by laying eggs, 23, 58, 59
    by mammals, 60
    plant
        with seeds, 74–75
        without seeds, 76
    by reptiles, 58, 85
Reptiles, 58
    eggs laid by, 84
    life cycle of, 84–85
    species, 63*
    as vertebrates, 58
Reservoirs, 262
Resources, 152
    competition for, 113, 153, 155
    conserving, 266
    natural, 244. See also Air; Soil;
        Water
    nonrenewable, 253, 256
    renewable, 253, 254, 256–57, 260
    using up, 265
Respiratory system, R17
Response of living things, 22, 23
    to environmental change, 164–65
Reuse, 156
Revolving of Earth around Sun,
    320–21
Revolving of Moon around Earth,
    328, 331
Rhinoceros, 172
Ring-tailed lemur, 11
Rio de Janeiro, climate of, 304
Rivers, 192
    erosion by water in, 216
Roadrunners, 62
Rocks, 230–34, 237*
    classifying, 230, 231*

earthquake as sudden movement
    of, 204
erosion of weathered, 216–17
grains of, 230, 231
igneous, 230, 231
landslide of, 208
magma as melted, 206, 231
melting temperature of, 398
metamorphic, 233
minerals as building blocks of,
    228, 230
sedimentary, 232, 253
in soil, 240
texture of, 230
uses of, 234
weathering of, 213*, 214–15, 220
Roots of plants, 32, 33, 34
    adaptations
        for desert, 136
        to wetlands, 142
    functions of, 34
    seed germination and growth of,
        70, 71
    soil and, 155, 156, 240
    weathering caused by, 215
Ross Ice Shelf, 187
Rotation of Earth, 318–19*
    day and night and, 319
Moon's position in sky and, 329
Round and round motion, 436, 437
Rubber, as insulator, 516
Rubies, 234
Ruler, 374, 435, R4
Rust, 419

Saber-toothed cats, 175
Safety tips
    in classroom, 14
    in field, 14
    protecting your hearing, 494
    in severe storms, 298
Saguaro cactus, 136
Saguaro National Park in Arizona,
    122
Saharan cypress trees, 168
Saharan Desert, 168
St. Helena Olive tree, 175
Salamanders, 47
Salt water. See also Ocean(s)
    ecosystems, 121
    freezing, 404–5*
    as solution, 411

Sandstone, 232
Sandy soil, 243
Satellite, 328
    Moon as satellite of Earth, 328
    weather, 284
Saturn, 339, 340
Savanna grasses, 165
Scales
    fish, 59
    reptile, 58
Science, Technology, and Society
    Beam of Light, 508-9
    Eating Away at Pollution, 28-29
    Mining Ores, 414-15
    Slide on the Shore, 210-11
    Tracking Twisters, 300-301
Scientific method, 2-11
    analyzing data, 9
    drawing conclusions, 10-11
    forming a hypothesis, 5
    steps in, 4
    testing a hypothesis, 6-7
Scientists
    scientific method used by, 2-11
    skills used by, 12-13. See also
        Inquiry skills
Scissors, as compound machine,
    470
Screw, 465, 468
Sea ice, 261
Sea otter, 94
Seasons, 308, 324*
    changing length of day and, 275
    Earth's tilted axis and, 320-21
    myth about, 310*
    stars and, 350
    in temperate forests, 125
    temperature and, 308, 321
Sea stars, 56, 126
Seattle, Washington, climate of, 306
Sea turtles, 84
Sea urchins, 56
Seaweed, 140
Sediment, 232
    fossils formed in, 251
Sedimentary rocks, 232
    coal as, 253
Sedona, Arizona, seasons in, 308
Seedling, 71
Seeds, 23, 68, 70-73
    fruit and, 73*
    germination of, 70-71, 74, 75
    growth into plant, 70-71
        needs for, 69*, 70
    how plants make, 72-73
    how they travel, 73
    parts of, 70, 71

in plant life cycle, 74-75
    plants reproducing without
        making, 76
Seeing, 506, R21
    colors, 505
Seesaw, 466
Senses, R21
    animals' use of, 44
    hearing, 494, R21
    sight, 505, 506, R21
Shadows, 316, 502
    changing throughout day, 317*,
        318
Shale, 232, 233
Sheep, 88
Shells
    fossil molds and casts formed
        from, 251
    pupa, 83
    of sea stars and urchins, 56
    of snail, 48
Shelter, 48
Shock, from static electricity, 512,
    513
Sight, 505, 506, R21
Silty soil, 243
Silver, 368
Simple machines, 465-69
    definition of, 465
    inclined planes, 465, 468-69*
        screw, 465, 468
        wedge, 465, 469, 470
    levers, 463*, 465, 466-67, 473*
        pulley, 465, 467
        wheel and axle, 465, 467, 470
    lifting objects with, 463*, 466-67
Singer, Marilyn
    "Monarch Butterfly" by, 16-17
Sinking and floating
    as properties of matter, 366
    separating mixtures by, 412
Skeletal muscles, R15
Skeletal system, R14
Skills. See Inquiry skills
Skin, as part of excretory system,
    R19
Skinks, 23
Skunks, 139
Slate, 233
Sleet, 282
Small intestine, R18
Smell, R21
Snails, 45, 48, 110, 113
    as decomposers, 114
Snake, 45
Snow, 282, 396
    in blizzard, 297

as insulator, 484
Soil, 120, 238-47
    classifying, 243*
    clearing land and loss of, 154-55
    color, 242
    conserving, 266
    decomposition of animals into,
        82
    decomposition of plants into, 74
    of different ecosystems, 120
        desert, 123
        temperate forests, 125
        tropical rain forest, 124
        wetlands, 128
    different types of, 242-43,
        246-47*
    drought and, 162
    erosion of, preventing, 240, 241,
        244
    formation of, 241
    heating, 481*
    humus in, 120, 125, 240, 241, 242
    importance of, 244
    layers of, 241
    living things in, 240
    as natural resource, 244
    nutrients in, 34, 74, 82, 240
        decomposers and, 111, 114*, 152
    roots of plants and, 155, 156, 240
    seed germination in, 70
    texture, 243
    water polluted by, 264
    what makes up soil, 239*, 240
Solar energy, 24, 25, 32, 36, 254,
    257
Solar flares, 322
Solar panels, 518, 518*
Solar system, 338-39, 340. See also
    Planets; Stars; Sun
    observing objects in, 344*-45*
Solids, 383*, 384-85
    comparing gases and, 387*
    comparing liquids and, 383*,
        387*
    definite size and shape of, 384
    freezing liquids to, 401, 402,
        404-5*, 483
    measuring volume of, 375, 377*
    melting of, 397*, 398, 483
    sound wave traveling through,
        491, 497*
Solutions, 411, 412
Sonoran Desert, 122, 304
Sound, 488-97
    changing, 493*
    definition of, 490
    electrical energy changed into,

514
hearing, 494, R21
how it travels, 490–91, 494, 496*–97*
making, 489*
pitch of, 493
speed of, 491
volume of, 492
Sound wave, 490–91, 494
Source, 254, 480
Southern Hemisphere, 320
    seasons in, 320, 321
Space, living things' need for, 25
Space probe, 342
Spaceship, 441
Species, reptile, 63*
Speed, 438
    change in motion and, 445
    forces and, 444
    measuring, 438*
    rate of, R3
    of sound, 491
Sphere, 305
    Earth as, 305
    Moon as, 328, 331
Spiders, 55
Spinal cord, R20
Spines on desert plants, 136
Sponges, 56
Spores, 76
Springbok, 164
Spring scales, 378
Spring season, 321
    length of day in, 275
Squid, 56, 57
Squirrels, 46, 73
Standard unit, 374
Stars, 346–51
    binary, 352–53
    classifying, 348
    constellations of, 349*, 350
    definition of, 348
    elements formed inside, 370
    seasons and, 350
    seeing stars at night, 347*
    sizing up, 348
    Sun as, 322, 348
States of matter, 384–89. See also
    Changes of state; Gas(es);
    Liquids; Solids
    gases, 387
    liquids, 383*, 386
    solids, 383*, 384–85
    uses of, 388
Static electricity, 512, 513
Statue of Liberty, 420
Staying fit, R26

Steam engine, 440
Steel, physical changes in, 409
Stems of plants, 32, 33
    in desert, 136
    functions of, 35
    growth of new plant from, 76
    observing, 35*
Stingray, 59
Stomach, R18
Stony models, fossils as, 250–51
Stopwatch, R4
Storm chasers, 300, 301
Storms, severe, 296–98
    staying safe in, 298
    types of, 296–97
Stoves, as source of heat, 481
Straight line motion, 436, 437
Stratus clouds, 290
    fog as, 292
Streak of mineral, 227*, 229
Streams, 192
    erosion by water in, 216
Structures, 33
    animal, 43*, 44–48, 47*, 54
    plant, 33, 34–37, 38. See also
        Flowers; Leaves of plants;
        Roots of plants; Stems of
        plants
Subsoil, 241
Subtracting large numbers, 171*
Sugars, 36
    elements in, 368
Summer, 308, 320, 321
    stars seen in, 350
Sun, 316–23
    day and night and, 318–19
    distance from Earth, 322
    Earth's orbit around, 320–21, 338
    energy from (solar energy), 24,
        25, 32, 36, 254, 257
        air temperature and, 281
        food chain starting with, 110
    position in sky, 318
    seasons and, 320–21
    as source of heat, 281, 322, 480
    as source of light, 322, 500
    as star, 322, 348
    temperature inside, 322
    water cycle and, 294
Sunlight
    climate and rays striking Earth,
        305
    photosynthesis and, 36, 37
    reflected off the Moon, 329, 331
    reflected off the planets, 339
    shadows and, 316, 317*, 318, 502
Sunrise, 318

Sunset, 318
Suraka silk moth, 10
Surgeons, lasers used by, 508–9
Surveyor, 272
Survival, adaptations for. See
    Adaptations
Sweat, R19
Sweat glands, R19
Switches, 515

T

Tables, R11
Tadpole, 82
Talc, 229
Tape measure, 380
Taproot, 34
Taste, R21
Taste buds, R21
Teeth, 46
    adaptations for eating, 134
    fossil, 176
Telescope, 342
Temperate forest, 125
    animal adaptations in, 139
    plant adaptations in, 138
Temperature, 482
    air, 280–81
        average, 286*–87*, 311*
        factors affecting, 306–7
    boiling, 399
    condensation and cooling, 400
    definition of, 280, 482
    in desert, 122
    freezing, 401
    measurement of, 280, 281,
        482–83, R6
        units of, R2, R3
    melting, 398
    patterns of precipitation and,
        comparing, 303*
    seasons and, 308, 321
    inside the Sun, 322
    water, 127*
    weathering caused by changes
        in, 214
Terraces carved into cliffs,
    preventing landslides with, 210
Testing a hypothesis, 6–7
Test preps, 65, 99, 147, 183, 223, 313,
    355, 393, 425, 475, 521
Texture, 230
    of rocks, 230
    of soil, 243

Thawing and freezing, rock weathered by repeated, 215
Thermal energy, 482–83.
Thermometer, 280, 374, 380, 482, 483, R6
   using, 397*
Thorn-mimic treehopper, 139
Thorns on desert plants, 136
3 Rs, 156, 158*–59*
Thunderstorm, 296, 298
Tigers, Bengal, 168
Time, measurement of, R4
Time for Kids, articles from
   "Good Ship Popsicle Stick, The," 358–59
   "One Cool Adventure," 186–87
Tinamous, 130–31
Tongue, 46
   of frog, 134
   of woodpecker, 103
Tools of Science, R7–8
   for measuring, 374–75, 380*
Topsoil, 241, 242
Tornadoes, 296, 298
   tracking, 300–301
Toucans, 124
Touch, sense of, R21
Trachea, R17
Traits, 90–95, 91*
   changing, 94
   definition of, 92
   inherited, 92–93*, 97
   learned, 94
Translucent objects, 503
Transparent objects, 503
Trash (garbage), 154, 158*–59*
Travel, history of machines for, 440–41
Trees
   as animal shelters, 48
   changes through seasons, 321
   cutting down
      land changed by, 218
      to make wood products, 154–55
   leaves falling off in fall, 23
   planting, to help environment, 156
   reproduction by making seeds, 23
Tree trunks, 35
Trench, 196, 197
Trevithick, Richard, 440
Triceratops, 174
Tropical rain forests, 124
   plant adaptations in, 138
Trunk, elephant, 46

Trunk, tree, 35
Tugboat, work of, 452
Tulips, 92
Tundra, 120
Turtles, 111, 113
   life cycle of, 84
   in ocean, 84, 118
   reproduction in, 23
Tyrannosaurus rex, 176, 180
Tyson, Neil deGrasse, 370–71

Unbalanced forces, 445
Underground heat, 254
Units of measurement, R2–3
   standard, 374
   table of measures, R3
Uranus, 339, 341
Urethra, R19
Ursa Major, 349
U-shaped valley, 217
Using numbers, skill at, 12, 158*–59*
Using variables, skill at, 13, 246*–47*

Valley, 194
   U-shaped, 217
Variables, 5
   skill at using, 13, 246*–47*
Veins, R16
Ventifact, 214
Venus, 336, 338, 339, 340
Vertebrates, 54
   types of, 58–60. See also
      Amphibians; Birds; Fish;
      Mammals; Reptiles
Vibration, 490
   of earthquake, 205
   sound and, 490
      hearing, 494
      pitch and, 493
      volume and, 492
Vinegar-baking soda reaction, 420
Viruses, R23
Vitamins, R24
Volcanoes, 206–7*
   air polluted by, 264
   effects of, 207
   formation of, 206
Volume (measurement), 365, 492
   of gases, 387
   of liquids, 386

   measuring, 375, R5
   units of measurement, R2, R3
mass and, 376–77
sinking and floating due to, 366
of solid, measuring, 375, 377*
Volume of sound, 492

Walruses, 132
Wasps, 44
Waste products, excretory system to get rid of, R19
Water
   animals living in, 58, 59, 60. See also Fish
   animal structures to help get, 46
   changes of state, 401, 402, 404*–5*
   conserving, 266
   on Earth's surface, 191*, 192, 193, 194–95
   elements in, 368
   erosion caused by moving, 216
   evaporation of, 292, 294, 399, 412, 483
   floods, 160, 161*, 162, 208, 297
   fresh, 192
   ecosystems, 108–9, 121
   ocean animals in, 119*
   sources of, 261
   frozen. See Ice
   heating, 399, 481*
   how people get, 262–63
   in living things' bodies, uses of, 24
   making water clean, 259*
   as natural resource, 258
   need for, 24, 46, 388
   as nutrient, R25
   oxygen in, 25
   photosynthesis and, 36, 37
   physical changes in, 408
   plant roots for getting, 34
   plant stem to carry, 35
   precipitation, 282, 290, 291, 294, 295. See also Rain; Snow
   recording water use, 265*
   recycling of, 260
   as renewable resource, 253, 260
      for energy, 254, 256
   sinking and floating in, 366
   soil texture and ability to hold, 243
   storing, 137*

treating, 263
underground, 261, 262, 264
uses of, 260
wasting, 265
weathering caused by, 214, 215
Water cycle, 294–95
Water ecosystems, 121
    ocean, 118, 121, 126–27
    pond, 108–9, 121
Water pollution, 264, 269*
Water temperatures, 127*
Water treatment plant, 263
Water vapor, 292, 399
    condensation of, 293, 294, 400,
        401*
    in sky, rainbow and, 504
Water wheel, making, 460*–61*
Waves, sound, 490–91, 494
Weather, 278–85
    air pressure and, 283
    air temperature and, 280–81
    climate as pattern of, 304
    definition of, 280
    precipitation and, 282, 290, 291,
        294, 295, 303*. See also Rain;
        Snow
    predicting, 284
    severe storms, 296–98
    tools, 282–83, 284
    wind and, 283
Weather balloons, 284
Weathering, 213*, 214–15
    causes of, 214–15
    chemical, 220
    erosion and, 216–17
    sedimentary rock formation and,
        232
    soil formation and, 241
Weather map, 284
Weather vane, 282
Wedge, 465, 469
    in compound machine, 470
Weight, 378, 447
    measurement of, 378, R6
        units of, R2, R3
Wells, 262
Wet climate, mountains and, 307
Wetlands, 128
    adaptations to, 142
    destruction of, 154
    ecosystem, 128
Whales, 25, 60
    orca, 491
"What a Difference Day Length
    Makes," 274–75

Wheel and axle, 465, 467
    in compound machine, 470
Wheelbarrow, 462, 464, 466, 473*

White blood cells, R16
    immune system and, R22
White light, 504–5
Whooping crane, 171
Wildfires, 163
    air polluted by, 264
Wildlife manager, 184
Wind(s), 283
    in blizzard, 297
    as energy source, 254, 256, 518*
    erosion caused by, 216
    force of, 442
    in hurricane, 297
    pollen blown by, 72
    seeds blown by, 73
    tornado, 296
    weathering caused by, 214
Windmills, 518, 518*
Windsock, 283*
Wings, 45
Winter, 308, 320, 321
    hibernation in, 139
    stars seen in, 350
Wires, electric, 516
Wolf, 45
Wood, as conductor, 367
Woodpeckers, 102–3
Wool, as insulator, 484
Woolly mammoth, 176, 178
Work, 452–61
    definition of, 454
    energy as ability to do, 456
    knowing what is work, 453*, 454,
        455
    made easier by machines,
        464–72
Worms, 47
    as decomposers, 111, 114
    how they change environment,
        151*
    as invertebrates, 55, 56
Wright, Wilbur and Orville, 441
Writing in Science
    descriptive writing, 62*, 236*,
        390*
    explanatory writing, 472*
    expository writing, 220*
    fictional narrative, 310*
    personal narrative, 88*, 324*
    persuasive writing, 170*, 518*

Year, 320

Zigzag motion, 436, 437

Credits

Rauschenbach/Zefa/CORBIS. 135: (t) Gregory G. Dimijian/Photo Researchers; (b) Bryan & Cherry Alexander Photography/Alamy. 137: (tl) Dr. Merlin D. Tuttle/Photo Researchers; (b) John Cancalosi/Nature Picture Library. 138: (t) Jacques Jangoux/Photo Researchers; (b) Jonathan Need/GardenWorld Images/Alamy. 139: (t) Bill Beatty/Visuals Unlimited; (c) Paul Sterry/Worldwide Picture Library/Alamy; (b) Digital Zoo/Digital Vision/Getty Images. 140: (t) Ralph A. Clevenger/CORBIS; (b) Marty Snyderman/Stephen Frink Collection/Alamy. 141: (t) Doug Perrine/Nature Picture Library; (b) Peter David/Taxi/Getty Images. 142: (t) M. Timothy O'Keefe/Bruce Coleman Inc.; (b) Blickwinkel/Alamy; 143: (t to b) Bryan & Cherry Alexander Photography/Alamy; Gregory G. Dimijian/Photo Researchers; Doug Perrine/Nature Picture Library. 146: (t to b) Yva Momatiuk/John Eastcott/Minden Pictures; Georgette Douwma/Photographer's Choice/Getty Images; Paul Nicklen/National Geographic/Getty Images. 147: (tl) Annie Griffiths Belt/National Geographic/Getty Images; (tc) Russell Illig/Getty Images; (tr) Image100/PunchStock; (b) Jim Brandenburg/Minden Pictures. 148-149: Jim Brandenburg/Minden Pictures. 149: (t to b) Nigel J. Dennis/Photo Researchers; David Cavagnaro/Visuals Unlimited; T. O'Keefe/PhotoLink/Getty Images; Arco Images/Alamy; Annie Griffiths Belt/CORBIS; Wardene Weisser/Bruce Coleman Inc. 150-151: Claude Ponthieux/Alt-6/Alamy. 154: (t) T. O'Keefe/PhotoLink/Getty Images; (b) Chip Porter/Stone/Getty Images. 155: Robert W. Ginn/PhotoEdit. 156: (tl) Stephen Saks/Lonely Planet Images; (tr) Scott Mitchell/Stone/Getty Images; (b) Photodisc/Getty Images. 157: (t) T. O'Keefe/PhotoLink/Getty Images; (c) Photodisc/Getty Images. 159: Kari Marttila/Alamy. 160-161: Curt Maas/AGStockUSA. 162: Nguyen/Dong/UNEP/Peter Arnold Inc. 163: (t) Bureau of Land Management; (cr) Hal Beral/CORBIS; (br) Mark Turner/PictureQuest. 164: (t) Nigel J. Dennis/Photo Researchers. 164-165: (b) Steve Bloom Images/Alamy. 165: Rod Patterson/ABPL/Animals Animals/Earth Scenes. 168: (t) FAN Travelstock/Alamy; (b) Arco Images/Alamy. 169: (t) Nguyen/Dong/UNEP/Peter Arnold Inc.; (c) Rod Patterson/ABPL/Animals Animals/Earth Scenes. 170: John Cancalosi/Peter Arnold Inc. 171: Lynn M. Stone/Bruce Coleman Inc. 172-173: Annie Griffiths Belt/CORBIS. 174: Russ Merne/Alamy. 175: (t) Topham/The Image Works; (b) Dr. Rebecca Cairns-Wicks/ARKive. 176: (tr) The Natural History Museum, London; (cr) Albert Copley/Visuals Unlimited; (bl) Chris Howes/Wild Places Photography/Alamy. 177: The Natural History Museum, London. 178: (t) Windland Rice/Bruce Coleman Inc.; (b) Wardene Weisser/Bruce Coleman, Inc. 179: (t) Russ Merne/Alamy; (c) The Natural History Museum, London; (b) Wardene Weisser/Bruce Coleman Inc. 180: (bl Owen) The Natural History Museum, London; (bl frame) Image Farm; (c) Bettmann/CORBIS. 180-181: John Sibbick/Natural History Museum Picture Library. 181: Stan Honda/AFP/Getty Images. 182: (t) Claude Ponthieux/Alt-6/Alamy; (c) Curt Maas/AGStockUSA; (b) Annie Griffiths Belt/CORBIS. 183: (t) Photodisc/Getty Images; (b) Chris Howes/Wild Places Photography/Alamy. 184: (t) Penny Tweedie/CORBIS; (b) Paul Nicklen/NGS Images. 185: Pawel Kumelowski/Omni-Photo Communications. 186: yourexpedition.com. 186-187: Higdon Photographic Art Studio/SuperStock. 187: yourexpedition.com. 188-189: Brand X Pictures/PictureQuest. 189: (t to b) Chad Ehlers/Minden Pictures; Lloyd Cluff/CORBIS; Greg Vaughn/Alamy; Yva Momatiuk/John Eastcott/Minden Pictures; Tom & Pat Leeson/Photo Researchers. 190-191: Chad Ehlers/Minden Pictures. 193: Brand X Pictures/PunchStock. 200: (b) Fred Hirschmann/Science Faction/Getty Images. 202-203: Lloyd Cluff/CORBIS. 204: Daniel Sambraus/Photo Researchers. 207: (b) Greg Vaughn/Alamy. 208: (t) Reuters/CORBIS; (b) Josh Ritchie/Getty Images. 209: (t to b) Daniel Sambraus/Photo Researchers; Greg Vaughn/Alamy; Reuters/CORBIS. 210: (t) Craig Lovell/CORBIS; (b) Mark Edwards/Peter Arnold Inc. 210-211: Vince Streano/CORBIS. 212-213: Tom Bean/CORBIS. 214: Scott Darsney/Lonely Planet Images. 215: (t) Yva Momatiuk/John Eastcott/Minden Pictures; (b) Louis Psihoyos/IPN Stock. 216: (t) Tom & Pat Leeson/Photo Researchers; (b) WildCountry/CORBIS. 217: (b) Bernhard Edmaier/Photo Researchers. 218: (l) CORBIS; (r) Danny Lehman/CORBIS. 219: (t to b) Scott Darsney/Lonely Planet Images; Bernhard Edmaier/Photo Researchers; CORBIS. 220: Altrendo Travel/Getty Images. 221: Danita Delimont/Alamy. 222: (t to b) Chad Ehlers/Minden Pictures; Lloyd Cluff/CORBIS; Tom Bean/CORBIS. 223: (l) Tom & Pat Leeson/Photo Researchers. 224-225: Don Smith/Alamy. 225: (t to b) Ian Francis/Alamy; Toyohiro Yamada/Getty Images; B. Runk/S. Schoenberger/Grant Heilman Photography; Owaki-Kulla/CORBIS; T. O'Keefe/PhotoLink/Getty Images. 226-227: GC Minerals/Alamy. 228: (l) B. Runk & S. Shoenberger/Grant Heilman Photography; (c) Roger Weller, Cochise College; (r) Mark A. Schneider/Photo Researchers. 229: (t) Roger Weller, Cochise College; (b) ER Degginger/Photo Researchers. 231: (c) Joyce Photographics/Photo Researchers. 232: (t) Wally Eberhart/Visuals Unlimited; (c) Jonathan Blair/CORBIS. 232-233: William Manning/CORBIS. 233: (t) Jerome Wyckoff/Animals Animals/Earth Scenes; (c) Dr. Parvinder Sethi; (b) Roger Weller, Cochise College. 234: (t) Roger Weller, Cochise College; (c) Richard Cummins/SuperStock; (b) Profimedia.CZ/Alamy. 235: (t to b) Roger Weller, Cochise College; William Manning/CORBIS; Richard Cummins/SuperStock. 236: (l) Lester Lefkowitz/The Image Bank/Getty Images; (r) Richard Nowitz/National Geographic Image Collection. 238-239: Joseph Van Os/Getty Images. 240: Jacana/Photo Researchers; (inset) B. Runk/S. Schoenberger/Grant Heilman Photography. 242: (l) Franck Jeannin/Alamy; (r) Ian Francis/Alamy. 244: (t) Richard Hamilton Smith/CORBIS; (b) David Frazier/The Image Works. 245: (t to b) Jacana/Photo Researchers; Ian Francis/Alamy;

David Frazier/The Image Works. 246-247: Lee White/CORBIS. 247: (b) Brand X Pictures/PunchStock. 248-249: B. Runk/S. Schoenberger/Grant Heilman Photography. 250: (t) Tom Bean/CORBIS; (b) B. Runk/S. Schoenberger/Grant Heilman Photography. 251: (b) B. Runk/S. Schoenberger/Grant Heilman Photography. 252: Andrew J. Martinez/Photo Researchers. 252-253: Owaki-Kulla/CORBIS. 253: Photodisc/Getty Images. 254: (t) Toshiyuki Aizawa/Reuters/CORBIS; (c) Kimimasa Mayama/Reuters/CORBIS; (b) Jeremy Woodhouse/Masterfile. 255: (t to b) Tom Bean/CORBIS; Owaki-Kulla/CORBIS; Jeremy Woodhouse/Masterfile. 256: (l) Wisconsin Historical Society; (r) Klaus Guldbrandsen/Photo Researchers. 256-257: Cristina Pedrazzini/Photo Researchers 257: (l) Tommaso Guicciardini/Photo Researchers; (r) Warren Gretz/NREL/US Department of Energy/Science Photo Library/Photo Researchers. 258-259: Toyohiro Yamada/Getty Images. 260: (t) Digital Vision/Getty Images. 260-261: Hans Strand/CORBIS. 262: Mira/Alamy. 264: Vincent Munier/Nature Picture Library; 264-265: Reuters/CORBIS; 265: C Squared Studios/Getty Images. 266: Jose Luis Pelaez/CORBIS. 267: (t) Digital Vision/Getty Images; (c) Jose Luis Pelaez/CORBIS. 270: (t to b) GC Minerals/Alamy; Joseph Van Os/Getty Images; B. Runk/S. Schoenberger/Grant Heilman Photography; Toyohiro Yamada/Getty Images. 271: (t) Owaki-Kulla/CORBIS; (b) Roger Weller, Cochise College. 272: (t) Tony West/CORBIS; (b) David Hiller/Photodisc/Getty Images. 274-275: nagelestock.com/Alamy. 274-275: Frank Whitney/Photographer's Choice/Getty Images. 275: (t) Alaska Stock/Alamy; (b) Gary W. Carter/CORBIS. 276-277: Tim Fitzharris/Masterfile. 277: (t to b) Erwan Balanca/PictureQuest; Image100/Alamy; David R. Frazier/Photo Researchers; Richard Thom/Visuals Unlimited; Layne Kennedy/CORBIS. 278-279: (t) Bruce Heinemann/Getty Images; (b) Rolf Bruderer/CORBIS. 280-281: First Light/Getty Images. 281: Burke/Triolo/Brand X Pictures/Jupiter Images. 282: (t) Eric Nguyen/Jim Reed Photography/CORBIS; (bl) Tony Freeman/PhotoEdit; (br) Dynamic Graphics Group/Creatas/Alamy. 283: (bl) Matthias Engelien/Alamy; (br) Leonard Lessin/Photo Researchers. 284: (t) David Hay Jones/Science Photo Library/Photo Researchers. 285: (t) Rolf Bruderer/CORBIS; (c) Leonard Lessin/Photo Researchers. 288-289: Erwan Balanca/PictureQuest. 290: (t) John Da Silva/Stone/Getty Images; (b) C Squared Studios/Getty Images. 291: (t) Naoki Okamoto/Getty Images; (b) David R. Frazier/Photo Researchers. 292: (b) Yellow Dog Productions/The Image Bank/Getty Images. 292-293: (t) Darrell Gulin/CORBIS. 293: (b) Michael Newman/PhotoEdit. 296: Jim Zuckerman/CORBIS. 297: (tl) Daniel Aguilar/Reuters/CORBIS; (tr) Royalty-Free/CORBIS; (b) Image100/Alamy. 298-299: Stock Connection/PictureQuest. 299: (t) David R. Frazier/Photo Researchers; (c) Stock Connection/PictureQuest. 300-301: Aaron Horowitz/CORBIS. 301: (t) NOAA; (b) Howard Bluestein/Photo Researchers. 302-303: Neil McAllister/Alamy. 304: (tl) Richard Thom/Visuals Unlimited; (tr) Mark E. Gibson/CORBIS; (bl) Pictor International/ImageState/Alamy; (br) Silvestre Machado/Stone/Getty Images. 306: (l) Kelly-Mooney Photography/CORBIS; (r) Bob Winsett/CORBIS. 307: (l) Blickwinkel/Alamy. 308: (t) Philip & Karen Smith/Lonely Planet Images; (b) Layne Kennedy/CORBIS. 309: (t to b) Silvestre Machado/Stone/Getty Images; Kelly-Mooney Photography/CORBIS; Layne Kennedy/CORBIS. 310: (l) Konrad Wothe/Minden Pictures; (r) Bob Krist/eStock Photo. 312: (t to b) Rolf Bruderer/CORBIS; Erwan Balanca/PictureQuest; Neil McAllister/Alamy. 313: (cr) Stockbyte/PunchStock; (bl) David R. Frazier/Photo Researchers. 314-315: Science Faction. 315: (t to b) Eckhard Slawik/Photo Researchers; StockTrek/Getty Images. 316-317: Altrendo Nature/Getty Images. 321: Herman Eisenbeiss/Photo Researchers. 322: Brand X Pictures/PunchStock. 323: (t) Brand X Pictures/PunchStock. 324: (l) Royalty-Free/CORBIS; (r) D. Hurst/Alamy. 325: Masterfile. 326-327: James Blank/Getty Images. 327: (bkgd) StockTrek/Getty Images. 328: Eckhard Slawik/Photo Researchers. 329: (t) Royalty-Free/CORBIS; (b) Eckhard Slawik/Photo Researchers. 332: (t) Royalty-Free/CORBIS; (b) Brand X Pictures/Punchstock. 333: (tl) Eckhard Slawik/Photo Researchers; (cl) Brand X Pictures/Punchstock. 334: StockTrek/Getty Images. 336-337: Astrofoto/Peter Arnold Inc./Alamy. 340: (tl) USGS/Photo Researchers; (tr) StockTrek/Getty Images; (bl) NASA/Science Source/Photo Researchers; (br) NASA/Photo Researchers. 341: (tl) Stocktrek/Brand X Pictures/Alamy; (tr) StockTrek/Getty Images; (bl) NASA-JPL; (bc) Elvele Images/Alamy; (br) NASA. 342: (t) Charles C. Place/Photographer's Choice/PunchStock; (b) NASA-JPL/Cornell University. 343: (cl) NASA/Photo Researchers; (bl) Charles C. Place/Photographer's Choice/PunchStock. 344 Astrofoto/Peter Arnold Inc./Alamy. 345: (tl) Elvele Images/Alamy; (tc) NASA/Photo Researchers; (tr) StockTrek/Getty Images; (cl) NASA/Science Source/Photo Researchers; (c) Royalty-Free/CORBIS; (cr) Stocktrek/Brand X Pictures/Alamy; (bl) Brand X Pictures/PunchStock; (bc) StockTrek/Getty Images; (br) USGS/Photo Researchers. 346: Daryl Pederson/Alaska Stock. 346-347: (bkgd) Shigemi Numazawa/Atlas Photo Bank/Photo Researchers. 349: (b) Gerard Lodriguss/Photo Researchers. 351: Gerard Lodriguss/Photo Researchers. 352: Denis Finnin/American Museum of Natural History. 352-353: STScI/NASA/Science Source/Photo Researchers. 354 (t to b) Altrendo Nature/Getty Images; James Blank/Getty Images; Astrofoto/Peter Arnold Inc./Alamy; Shigemi Numazawa/Atlas Photo Bank/Photo Researchers. 355: (tr) NASA-JPL/Cornell University; (bl) Eckhard Slawik/Photo Researchers. 356: (t) Joe Raedle/Getty Images; (b) Luiz C. Marigo/Peter Arnold Inc. 357: Alaska Stock. 358-359: Peter Dejong/AP/Wide World Photos. 359: Courtesy Sea Heart Foundation. 360-361: Studio City/eStock Photo/Alamy. 361: (t to b) Paul Barton/CORBIS; Matthias Stolt/FAN Travelstock/

Graphic Organizer 1

Main Idea and Details

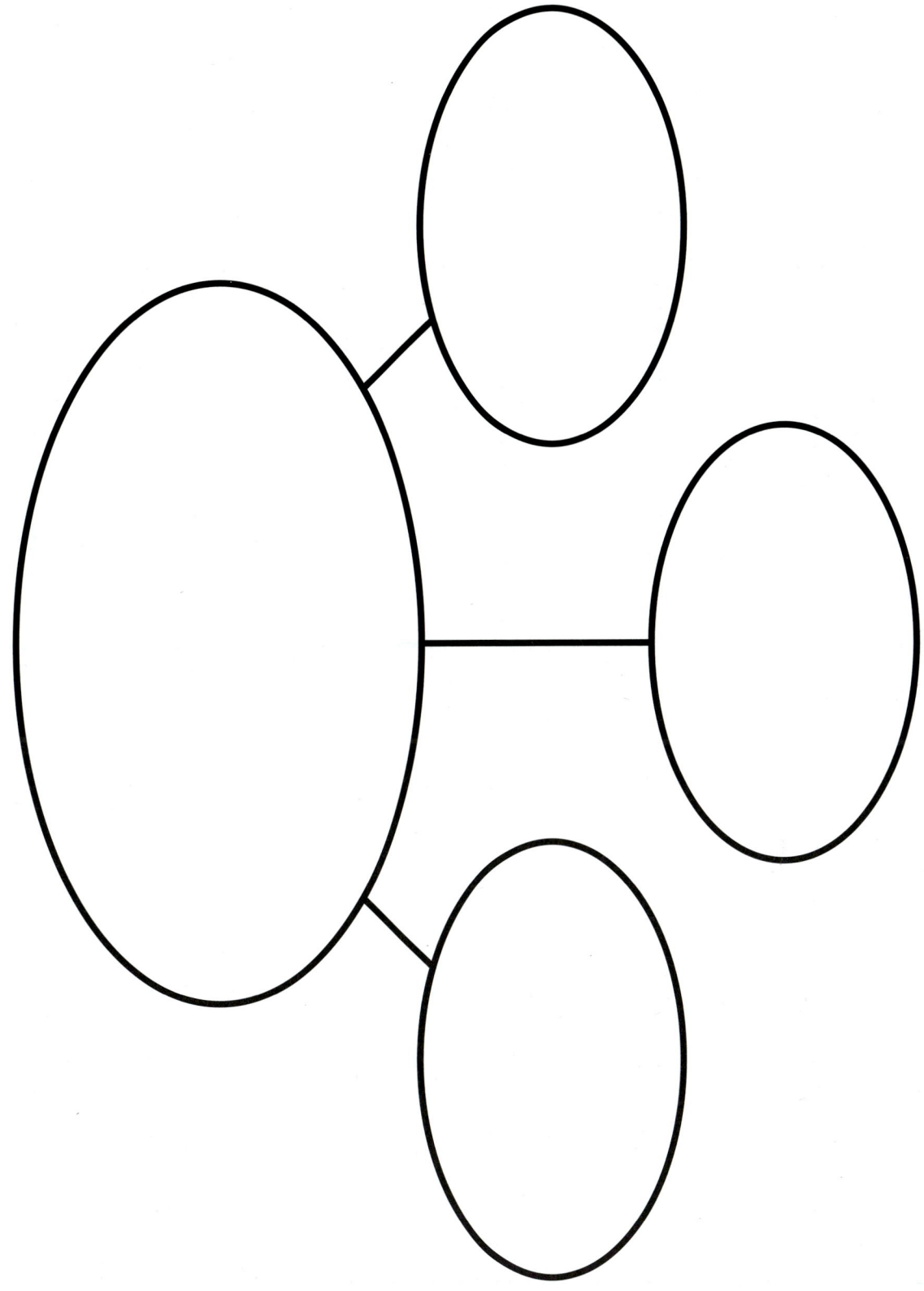

Graphic Organizer 2

Main Idea and Details

Details						

Main Idea	

Graphic Organizer 3

Predict

What Happens		
What I Predict		

Graphic Organizer 4

Predict

My Prediction	What Happens

Graphic Organizer 5

Summarize

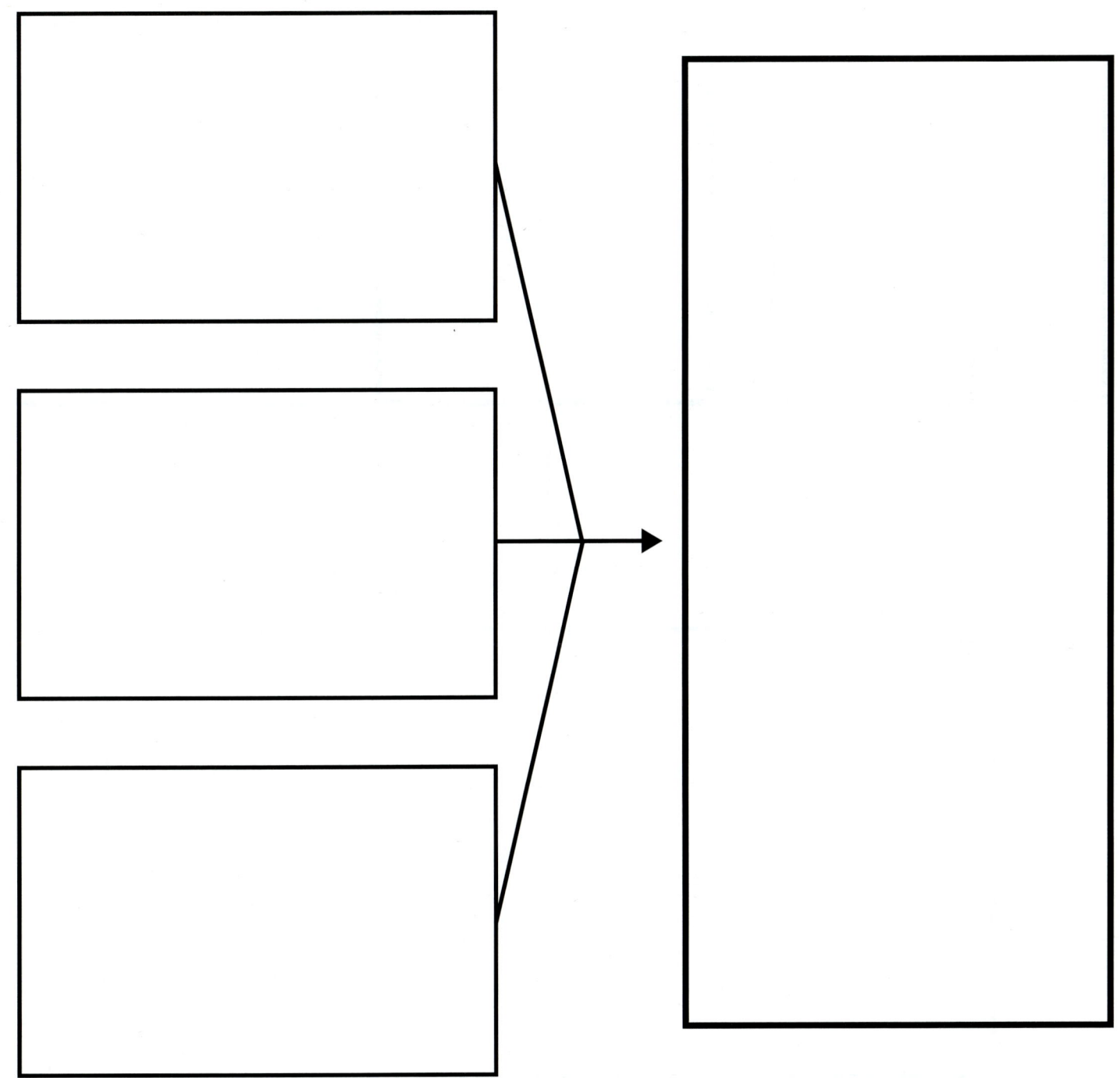

Graphic Organizer 6

Summarize

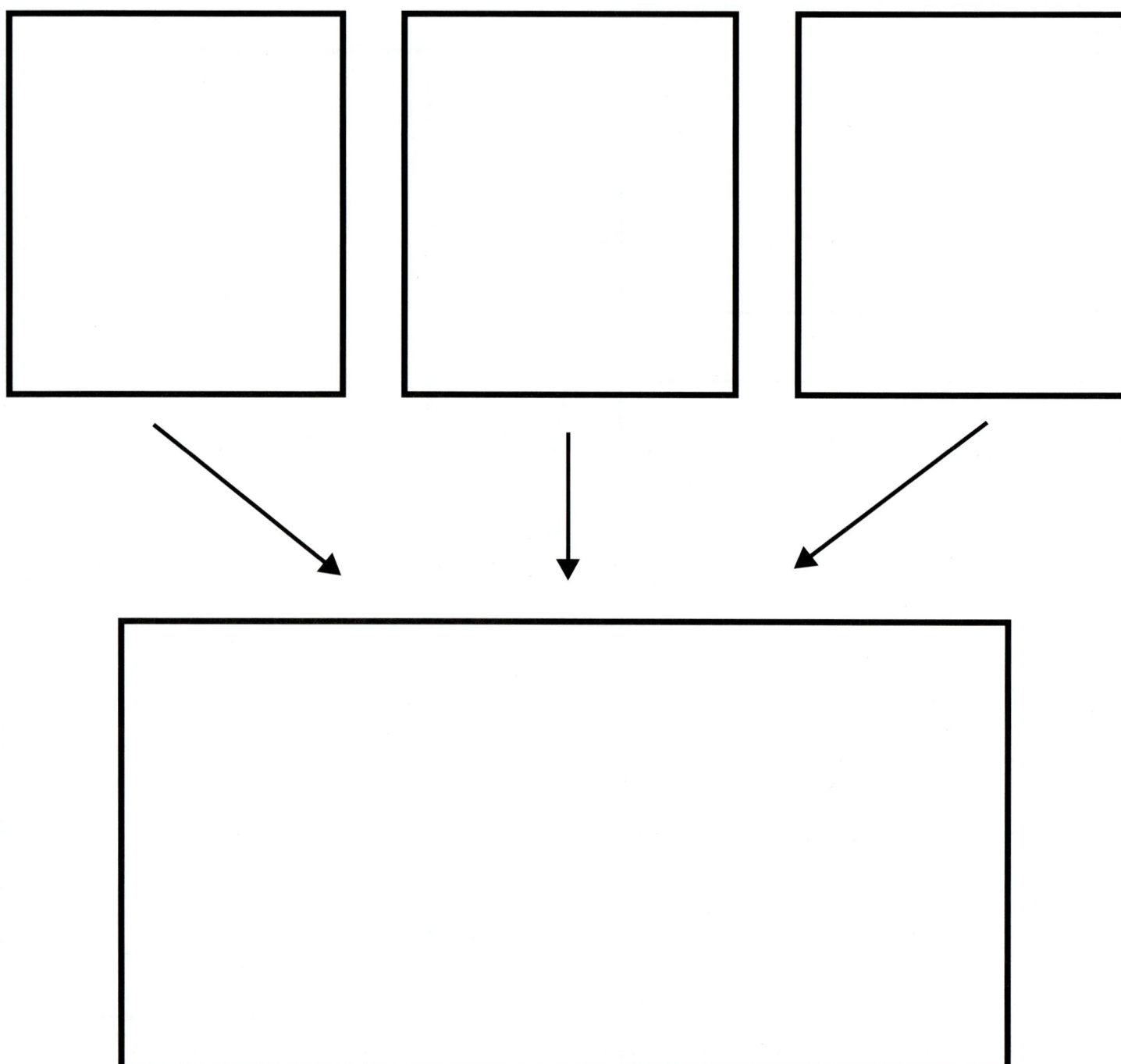

Graphic Organizer 7

Sequence

First

Next

Last

Graphic Organizer 8

Cause and Effect

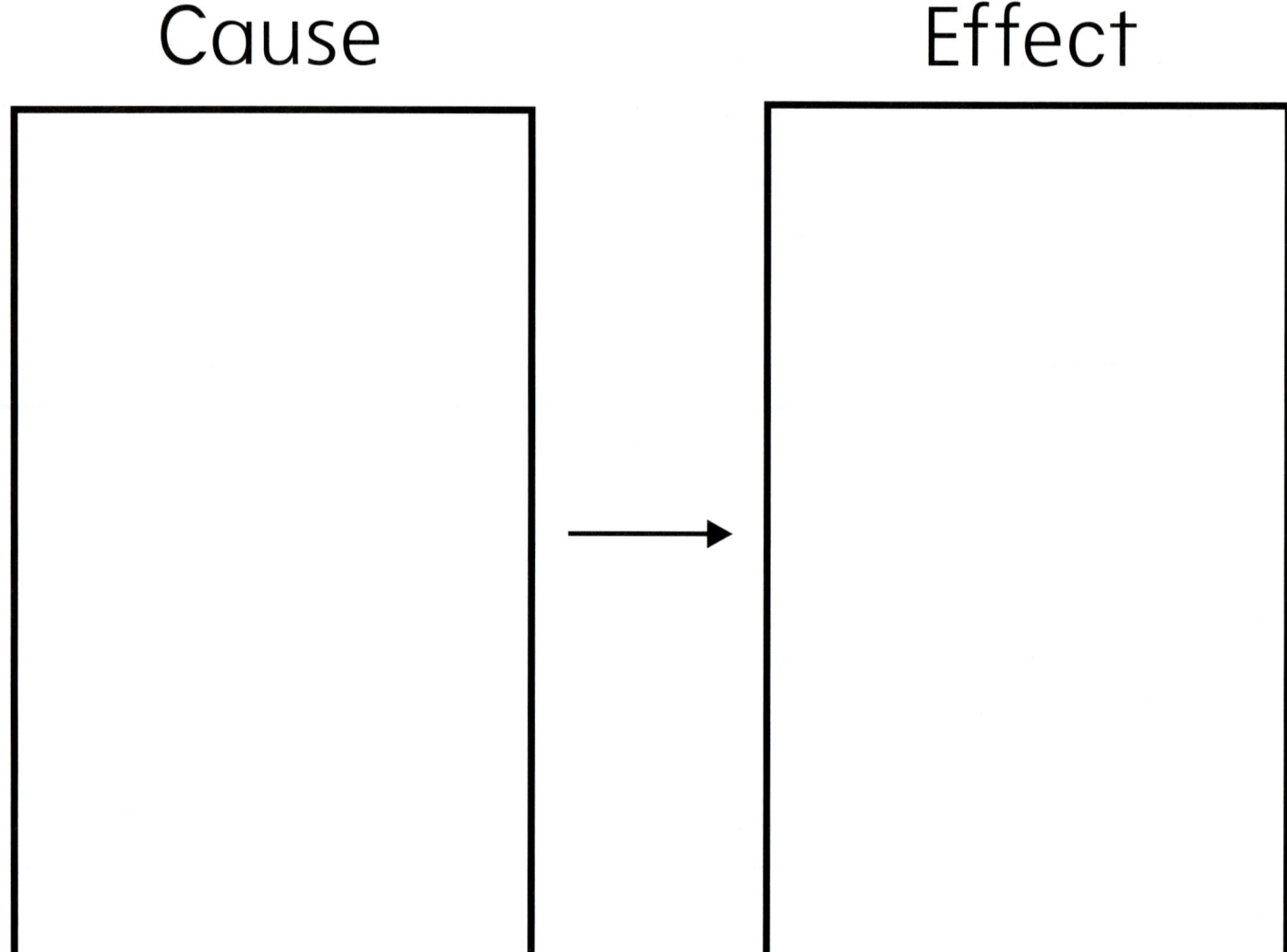

Cause

Effect

Graphic Organizer 9

Cause and Effect

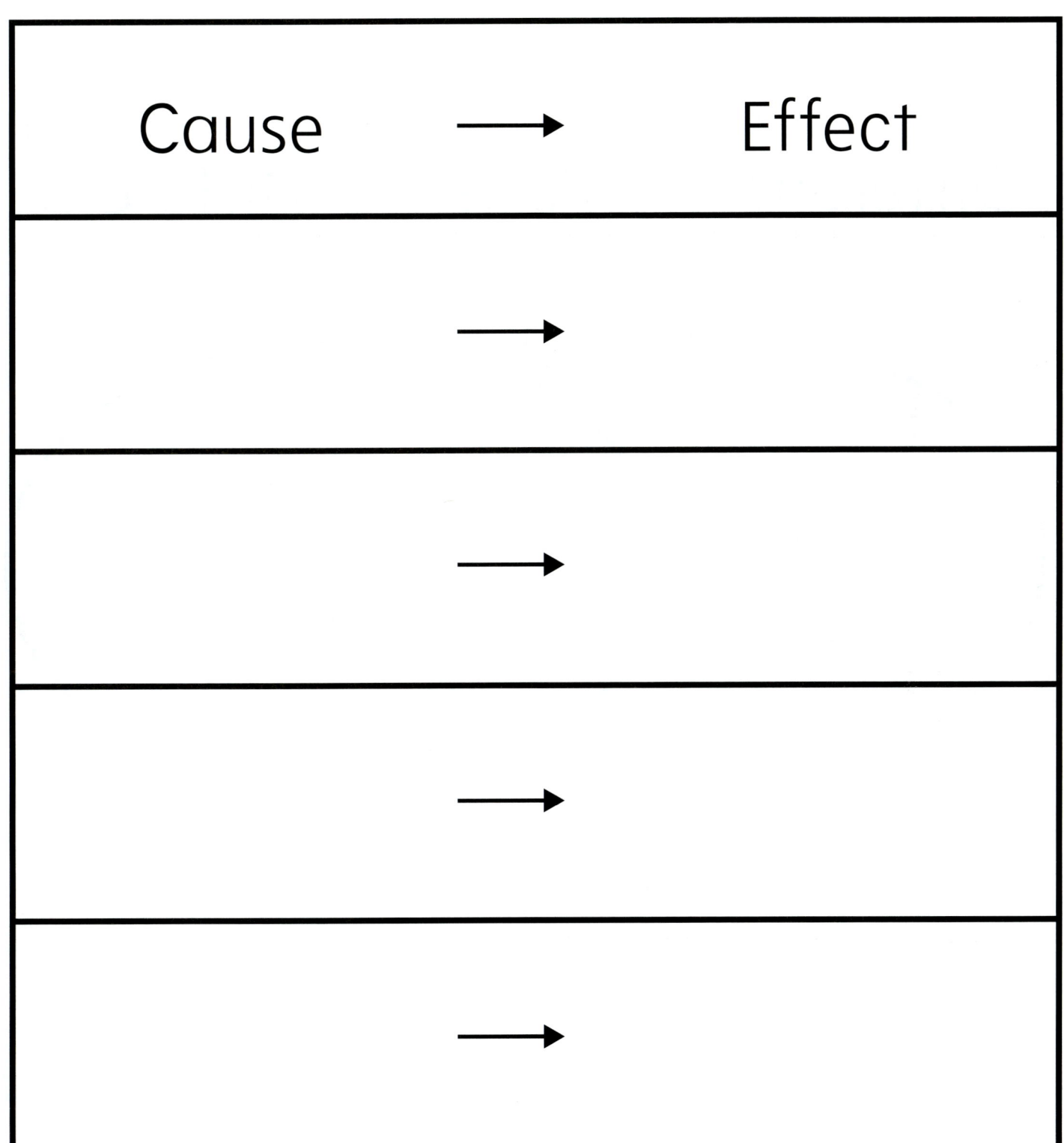

Graphic Organizer 10

Compare and Contrast

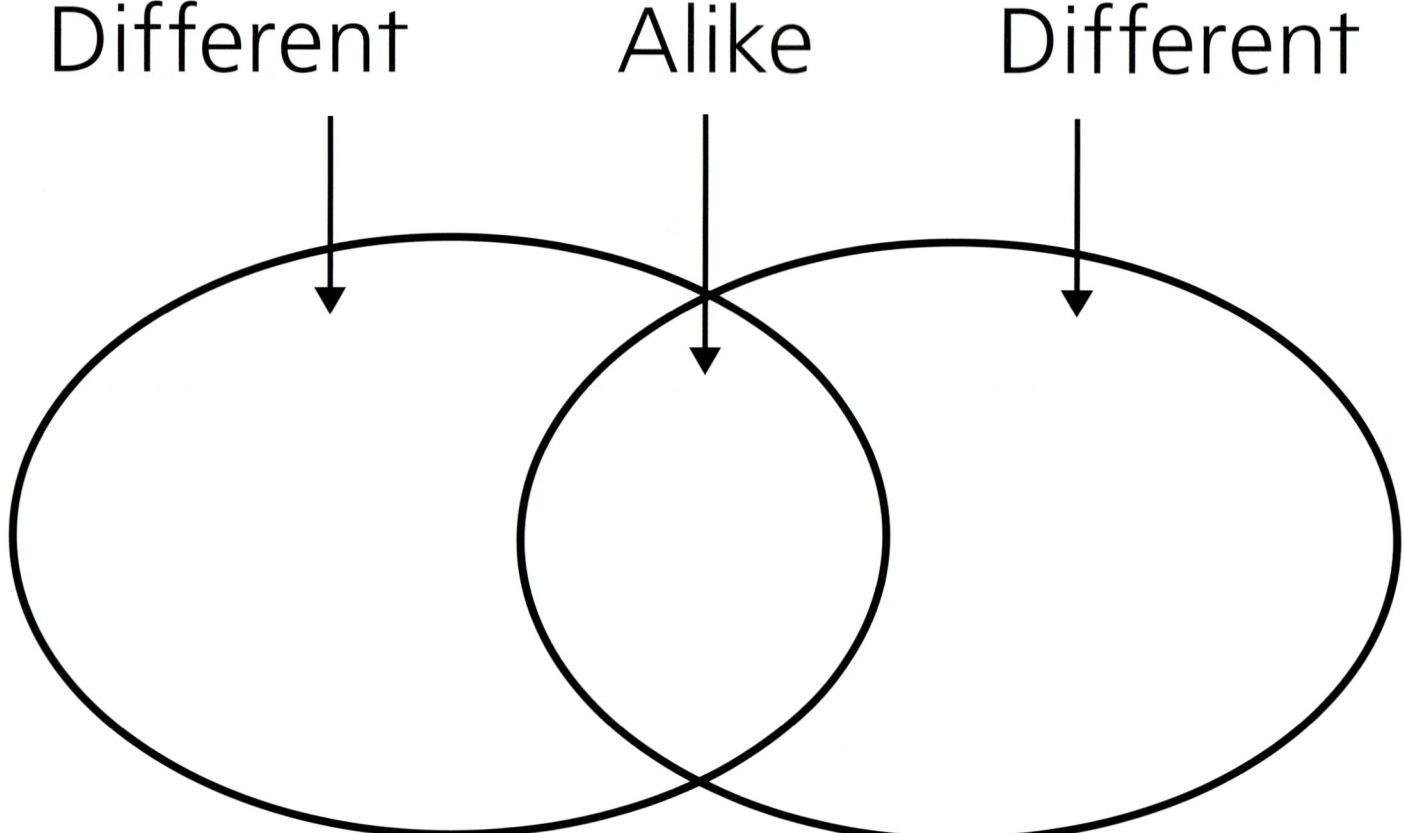

Different   Alike   Different

Name _____ Date _____

Graphic Organizer 11

Classify

© Macmillan/McGraw-Hill

Name ______________________ Date ______________________

Graphic Organizer 12

Problem and Solution

Problem

↓

Steps to Solution

↓

Solution

Name _____   Date _____

Graphic Organizer 13

Draw Conclusions

Conclusions		
Text Clues		

Name _____   Date _____

Graphic Organizer 14

Infer

Clues	What I Know	What I Infer

Graphic Organizer 15

Fact and Opinion

Opinion	
Fact	

Name _____   Date _____

United States Map

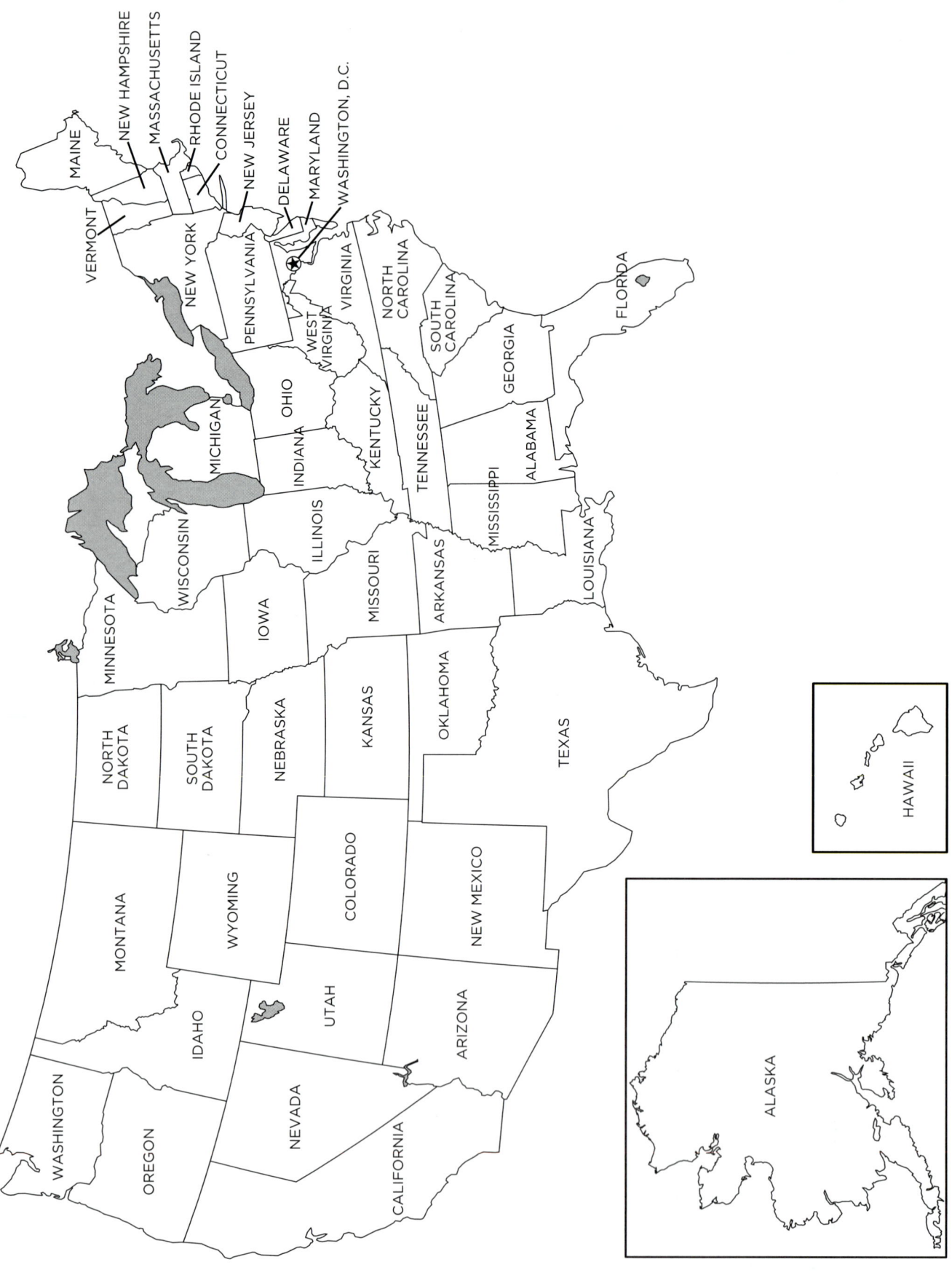

Name _____   Date _____

World Map

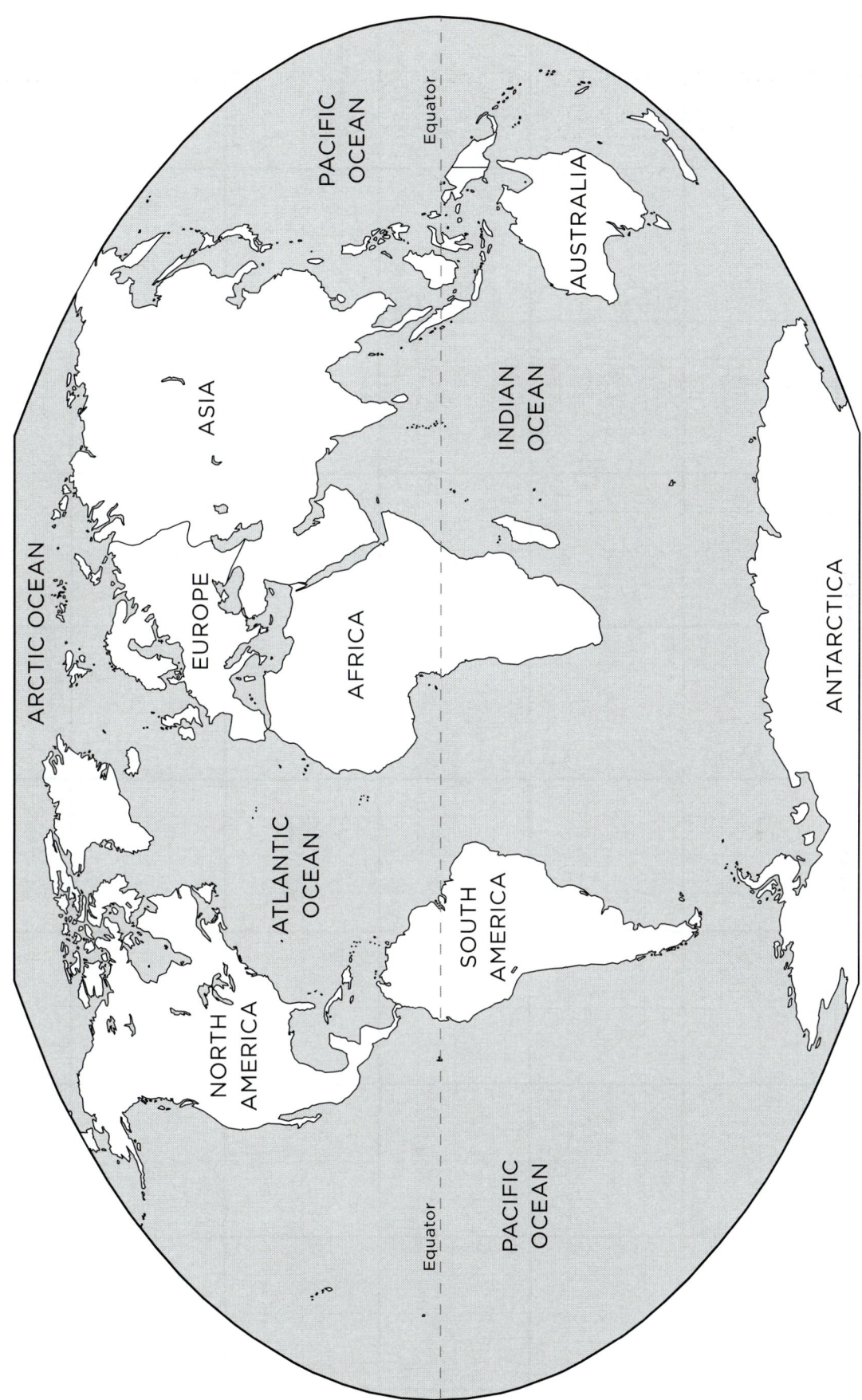

Name _____   Date _____

Graph Paper

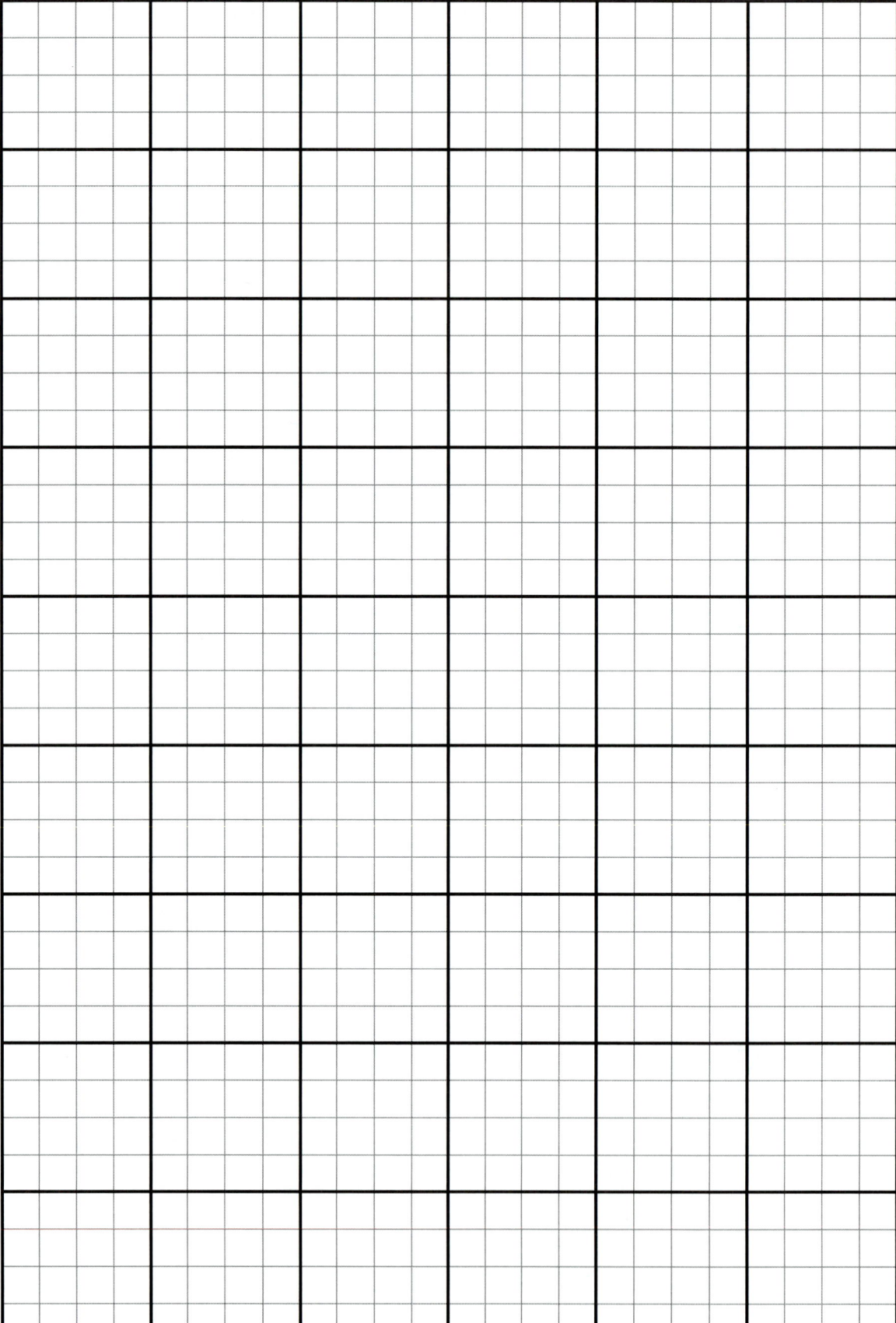

Name _____ Date _____

Calendar

Sunday	Monday	Tuesday	Wednesday	Thursday	Friday	Saturday

Name _____   Date _____

Rulers

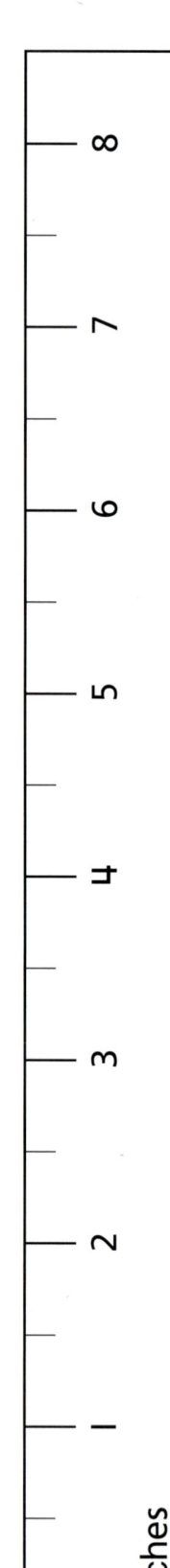

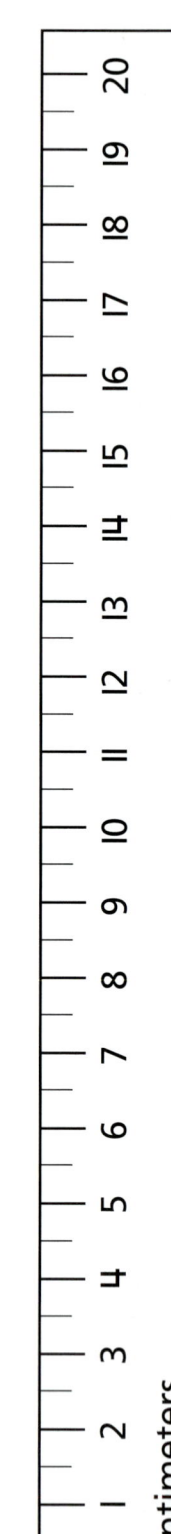

© Macmillan/McGraw-Hill

Periodic Table of the Elements

Key

6	Atomic number
C	Element symbol
Carbon	Element name

Li — Metals
B — Metalloids
C — Nonmetals

	1	2	3	4	5	6	7	8	9	10	11	12	13	14	15	16	17	18
1	1 **H** Hydrogen																	2 **He** Helium
2	3 **Li** Lithium	4 **Be** Beryllium											5 **B** Boron	6 **C** Carbon	7 **N** Nitrogen	8 **O** Oxygen	9 **F** Fluorine	10 **Ne** Neon
3	11 **Na** Sodium	12 **Mg** Magnesium											13 **Al** Aluminum	14 **Si** Silicon	15 **P** Phosphorus	16 **S** Sulfur	17 **Cl** Chlorine	18 **Ar** Argon
4	19 **K** Potassium	20 **Ca** Calcium	21 **Sc** Scandium	22 **Ti** Titanium	23 **V** Vanadium	24 **Cr** Chromium	25 **Mn** Manganese	26 **Fe** Iron	27 **Co** Cobalt	28 **Ni** Nickel	29 **Cu** Copper	30 **Zn** Zinc	31 **Ga** Gallium	32 **Ge** Germanium	33 **As** Arsenic	34 **Se** Selenium	35 **Br** Bromine	36 **Kr** Krypton
5	37 **Rb** Rubidium	38 **Sr** Strontium	39 **Y** Yttrium	40 **Zr** Zirconium	41 **Nb** Niobium	42 **Mo** Molybdenum	43 **Tc** Technetium	44 **Ru** Ruthenium	45 **Rh** Rhodium	46 **Pd** Palladium	47 **Ag** Silver	48 **Cd** Cadmium	49 **In** Indium	50 **Sn** Tin	51 **Sb** Antimony	52 **Te** Tellurium	53 **I** Iodine	54 **Xe** Xenon
6	55 **Cs** Cesium	56 **Ba** Barium	57 **La** Lanthanum	72 **Hf** Hafnium	73 **Ta** Tantalum	74 **W** Tungsten	75 **Re** Rhenium	76 **Os** Osmium	77 **Ir** Iridium	78 **Pt** Platinum	79 **Au** Gold	80 **Hg** Mercury	81 **Tl** Thallium	82 **Pb** Lead	83 **Bi** Bismuth	84 **Po** Polonium	85 **At** Astatine	86 **Rn** Radon
7	87 **Fr** Francium	88 **Ra** Radium	89 **Ac** Actinium	104 **Rf** Rutherfordium	105 **Db** Dubnium	106 **Sg** Seaborgium	107 **Bh** Bohrium	108 **Hs** Hassium	109 **Mt** Meitnerium									

58 **Ce** Cerium	59 **Pr** Praseodymium	60 **Nd** Neodymium	61 **Pm** Promethium	62 **Sm** Samarium	63 **Eu** Europium	64 **Gd** Gadolinium	65 **Tb** Terbium	66 **Dy** Dysprosium	67 **Ho** Holmium	68 **Er** Erbium	69 **Tm** Thulium	70 **Yb** Ytterbium	71 **Lu** Lutetium
90 **Th** Thorium	91 **Pa** Protactinium	92 **U** Uranium	93 **Np** Neptunium	94 **Pu** Plutonium	95 **Am** Americium	96 **Cm** Curium	97 **Bk** Berkelium	98 **Cf** Californium	99 **Es** Einsteinium	100 **Fm** Fermium	101 **Md** Mendelevium	102 **No** Nobelium	103 **Lr** Lawrencium

4-Point Activity Rubrics

Assessing Abilities Necessary to Do Scientific Inquiry

Expresses Natural Curiosity by Manipulating Objects and Ideas

4 Pursues open-ended engagement by manipulating and exploring objects or ideas, tries unusual manipulations, and expresses initial or tentative personal reasoning.

3 Explores and manipulates things or ideas but does not discuss personal reasoning.

2 Relies on others to direct manipulations.

1 Does not engage in manipulations.

Makes and Records Observations, and Notices Both the Expected and Unexpected

4 Describes and accurately records numerous observations using multiple senses.

3 Describes numerous observations, but some observations may be what the student expected to observe.

2 Relies on others to help guide observations.

1 Makes very few observations.

Asking Testable Questions That May Be Explored Through Scientific Investigation

4 Asks testable questions that may be explored scientifically without teacher guidance.

3 Asks testable questions with minimal teacher guidance.

2 Asks testable questions with considerable teacher guidance.

1 Uses teacher-generated questions.

Planning and Conducting an Investigation

4 Investigation has complete logical steps.

3 Investigation has errors in logic.

2 Investigation requires considerable teacher guidance.

1 Uses only teacher-provided investigations.

Using Simple Equipment and Tools to Gather Data and Extend the Senses

4 Consistently chooses appropriate tools and equipment and uses them correctly.

3 Usually chooses appropriate tools and equipment and/or uses them correctly.

2 Sometimes chooses appropriate tools and equipment and/or uses them correctly.

1 Seldom chooses appropriate tools and equipment and/or uses them correctly.

Using Data to Develop a Reasonable Explanation to Answer the Question Being Investigated

4 Consistently records data in a logical manner and develops a reasonable explanation based on collected data and/or information from reliable scientific sources.

3 Usually records data in a logical manner and develops a reasonable explanation based on collected data and/or information from reliable scientific sources.

2 Records of data are incomplete/inaccurate and explanation reflects incomplete/inaccurate data or scientific information.

1 Records of data are missing and explanation, if present, is illogical.

Communicating Procedures, Results, and Explanations of an Investigation

4 Writes precise instructions that others can follow to carry out procedures; makes detailed sketches to aid in explaining procedures or ideas; uses qualitative and quantitative data to describe and compare objects and events.

3 Writes instructions that others can follow in carrying out procedures, but uses mostly qualitative data to describe and compare objects and events.

2 Writes incomplete instructions.

1 Writes incomplete and inaccurate instructions.

Writing Rubrics

Writing Links

Writing Links provide opportunities for teachers to integrate writing into the science curriculum and further prepare students for the writing assessment tests they will be taking. The ✏️ writing logo is used in the Student Edition to identify writing tasks, and can be found in the following places:

- At the end of every lesson there is a writing question in the Think, Talk, and Write lesson review.

- A *Writing in Science* feature is in every chapter. A *Write About It* prompt is found in each feature. Also look for the 🌐 on this page for online writing opportunities for students.

- Look for *Integrate Writing* boxes in the *Be a Scientist*, *Focus on Skills*, and *Writing in Science* features in the Teacher's Edition for more effective ways to include writing throughout each lesson.

Connecting the Rubrics to the Writing Modes

The *Writing Links* four-point rubric features six modes of writing. These writing modes are personal narrative, descriptive writing, writing a story, explanatory (or "how-to") writing, writing that compares, and expository writing. A *7-Trait Writing in Science* rubric is provided to assess each mode of writing.

Each of the six writing modes is designed to build writing skills that are essential to good writing in general and to science in particular, such as developing a clearly organized central (or main) idea with supporting facts and details and using varied sentence structure. These and additional writing skills are stressed on writing assessment tests that your students will be taking and in the *7-Trait Writing in Science* rubrics provided.

Macmillan/McGraw-Hill Writing Modes

▶ **Personal Narrative** mode found in the text will help students craft a detailed true story about a personal experience within the framework of a clearly organized sequence of events. Most writing assessment tests call for the writing of narrative text that is organized in a clear and logical way.

▶ **Descriptive Writing** tasks will help students learn to include vivid, sensory details in their writing, while enabling them to choose colorful or specific vocabulary. These skills will be useful to them in writing observation reports, and in both their narrative and expository writing.

▶ **Writing a Story** found in the text will help students craft fictional narrative—for example, a piece of science fiction, with vivid details and a thoughtfully planned storyline that organizes events from beginning to end. Most writing assessment tests require the writing of a narrative, whether it would be a personal narrative based on a real-life event or a fictional story.

▶ **Explanatory Writing** tasks ask students to explain how to complete a task or a process, such as a science experiment. Students' ability to organize their writing into a step-by-step format is an essential tool for writing in science. Giving clear details and organizing events into a sequence is also required for all good writing.

▶ **Writing That Compares** focuses on skills necessary to write an essay or a report that compares or contrasts two items or findings. This objective mode of writing is often used when writing about Science.

▶ **Expository Writing** focuses on skills necessary to write a summary, an informational or a research report, or an essay. This objective mode of writing is the one used most often when writing about science. It is also consistent with the writing mode tested most frequently on writing assessment tests.

Using the Scoring Rubrics

Use the 4-Point Writing Rubrics to score students' responses to the writing modes and prompts featured in the Writing Links.

4-Point Writing Rubric

To determine the appropriate score:

▶ Find the description of the mode of writing featured in the Writing Link. These six modes are Personal Narrative, Descriptive Writing, Writing a Story, Explanatory Writing, Writing That Compares, and Expository Writing (such as a report).

▶ Then find the description of the writing trait that best expresses the quality of the students' writing in that mode. Assess students' writing as 4 Excellent, 3 Good, 2 Fair, or 1 Unsatisfactory.

▶ Consider how well the response achieves the writer's purpose. Be sure the response addresses the features of **7-Trait Writing in Science:**

- Ideas & Content
- Organization
- Voice
- Word Choice
- Sentence Fluency
- Conventions
- Presentation

▶ Assign a single score of 1–4 based on how well students' writing corresponds to the descriptions that appear in the rubrics.

For remedial purposes:

You can use the 4-Point Writing Rubric to identify specific areas of deficiency (organization, word choice, sentence fluency). Do not, however, assign separate scores for each of the writing traits.

Writing Links: 4-Point Writing Rubric

7-Trait Writing in Science
Personal Narrative

4 Excellent	3 Good	2 Fair	1 Unsatisfactory
Ideas & Content Demonstrates originality in developing ideas or a story drawn from a personal experience.	**Ideas & Content** Develops reasonably clear ideas that develop a true story about the writer.	**Ideas & Content** Displays difficulty in developing content and fails to show a strong sense of purpose.	**Ideas & Content** Makes no attempt to develop ideas or tell about a true event.
Organization Crafts a well-organized personal narrative that flows smoothly and moves the reader through the beginning, middle, and end of the text.	**Organization** Crafts a personal narrative that moves the reader through the text without confusion.	**Organization** Crafts a personal narrative that may exhibit organizational problems, such as lack of follow-through after a good beginning.	**Organization** Shows an extreme lack of organization that interferes with comprehension of the text.
Voice Exhibits a personal voice with a sense of the purpose and audience.	**Voice** Expresses a personal voice and demonstrates an adequate sense of the purpose and audience.	**Voice** Attempts to express a personal voice, but is not fully engaged with the audience.	**Voice** Makes no attempt to express a personal voice or share personal insights.
Word Choice Chooses imaginative words that convey images and feelings in a natural way.	**Word Choice** Makes an effort to choose words that convey images and emotions.	**Word Choice** Chooses words that are often dull and unimaginative.	**Word Choice** Shows an inability to choose words that convey clear pictures or are imaginative.
Sentence Fluency Produces strong, varied, and purposeful sentences that invite expressive oral reading.	**Sentence Fluency** Produces varied sentences that are easy to read aloud with a little practice.	**Sentence Fluency** Produces sentences with some variation but that may lack an easy flow.	**Sentence Fluency** Produces awkward or incomplete sentences that do not invite oral reading.
Conventions Exhibits a good grasp of standard writing conventions, including spelling, capitalization, punctuation, and grammar.	**Conventions** Shows a grasp of most standard writing conventions.	**Conventions** May have trouble with some standard writing conventions, including spelling, capitalization, punctuation, or grammar.	**Conventions** Shows an inability to grasp basic writing conventions severe enough to interfere with readability.
Presentation Uses neat handwriting or fonts to enhance the reader's ability to connect with the message of the text.	**Presentation** Uses readable handwriting or consistent fonts and font sizes that make the text easy to read.	**Presentation** Uses legible handwriting or experimentation with fonts, although the effect is not consistent throughout the text.	**Presentation** Uses inconsistent handwriting or multiple fonts, making the text difficult or impossible to read.

Writing Links: 4-Point Writing Rubric

7-Trait Writing in Science
Descriptive Writing

4 Excellent	3 Good	2 Fair	1 Unsatisfactory
Ideas & Content Shows imagination and originality in developing specific descriptive content that is clear, vivid, and focused.	**Ideas & Content** Develops descriptive content in a general way, using ideas that are reasonably clear and focused.	**Ideas & Content** Has difficulty developing clear, focused ideas and specific descriptive content.	**Ideas & Content** Makes no attempt to present clear ideas and specific descriptive content.
Organization Creates a description that flows smoothly and is well organized in its presentation of details.	**Organization** Organizes a description in a way that groups details, moving the reader through the text without confusion.	**Organization** Creates a description that exhibits organizational problems, such as grouping unlike details together.	**Organization** Demonstrates a lack of organization that interferes with readability and comprehension.
Voice Uses a strong voice that appeals directly to the audience and expresses the writer's personality.	**Voice** Uses a personal voice that connects the writer and audience.	**Voice** Makes an attempt to use an appealing personal voice, but has difficulty maintaining it.	**Voice** Makes no attempt to express a personal voice or appeal to the audience.
Word Choice Chooses vivid, sensory words to create a clear mental picture for the reader.	**Word Choice** Makes an effort to choose words that are clear, vivid, and accurate and that may appeal to the audience's senses.	**Word Choice** Often chooses overused words that fail to capture the audience's imagination.	**Word Choice** Shows an inability to choose words that are correct or appropriate to the description.
Sentence Fluency Crafts varied, well-paced sentences that are a delight to read aloud.	**Sentence Fluency** Crafts sentences that are generally smooth and natural.	**Sentence Fluency** Crafts some sentences that are choppy, rambling, or awkward to read.	**Sentence Fluency** Produces sentences that are incomplete and difficult to read aloud.
Conventions Displays a highly developed grasp of writing conventions that make the description easy to read.	**Conventions** Displays a general understanding of writing conventions and applies them to the description.	**Conventions** Often has trouble with spelling, capitalization, punctuation, and grammar.	**Conventions** Demonstrates an inability to grasp basic writing conventions.
Presentation Consistently uses neat handwriting or appropriate fonts, as well as the correct balance of white space and text, to make the work inviting to the audience.	**Presentation** Uses legible handwriting or consistent fonts, as well as uniform spacing, to invite the reader into the text.	**Presentation** Uses legible handwriting or consistent fonts, as well as uniform spacing, although a different choice of spacing (double spacing) may be preferable.	**Presentation** Uses irregularly formed letters or multiple fonts and font sizes, as well as random spacing, making the text difficult to read and understand.

Writing Links: 4-Point Writing Rubric

7-Trait Writing in Science
Writing a Story

4 Excellent	3 Good	2 Fair	1 Unsatisfactory
Ideas & Content Shows imagination in developing story ideas, structure, and content.	**Ideas & Content** Shows some imagination in developing story ideas, structure, and content.	**Ideas & Content** Adequately develops story ideas, structure, and content.	**Ideas & Content** Makes no effort to develop interesting or imaginative story ideas and content; no story structure is evident.
Organization Exhibits strong organizational skills in creating an interesting beginning, middle, and end of the story.	**Organization** Uses organizational skills to create a beginning, a middle, and an end of the story.	**Organization** Exhibits difficulty in organizing the structure of the story.	**Organization** Shows an inability to create a structure for the story.
Voice Displays a personal voice that echoes the tone of the story and is highly appealing to the audience.	**Voice** Displays a personal voice that is appropriate and appealing to the audience.	**Voice** Displays a personal voice that tries to engage the audience.	**Voice** Makes no attempt to develop a personal voice and shows no awareness of the audience.
Word Choice Chooses words carefully to develop the setting, characters, and sequence of events.	**Word Choice** Chooses colorful, accurate words that are appropriate to developing the story.	**Word Choice** Does not choose colorful or specific vocabulary to develop the story.	**Word Choice** Uses words that are incorrect or that confuse the reader.
Sentence Fluency Crafts interesting and varied sentences that enhance the fluency of the story and invite oral reading.	**Sentence Fluency** Crafts interesting and varied sentences that are easy to read aloud.	**Sentence Fluency** Crafts sentences that can be understood but are sometimes hard to follow or read aloud.	**Sentence Fluency** Writes sentences that are incomplete, confusing, and extremely difficult to read aloud.
Conventions Exhibits firm knowledge of writing conventions, including spelling, capitalization, punctuation, and grammar.	**Conventions** Exhibits a knowledge of standard writing conventions; work needs little editing.	**Conventions** Exhibits a limited grasp of writing conventions; extensive revising and editing are needed.	**Conventions** Exhibits problems with writing conventions that are severe enough to interfere with readability.
Presentation Uses neat handwriting or appropriate fonts and font sizes to enhance comprehension and readability.	**Presentation** Uses readable handwriting or successful experimentation with fonts.	**Presentation** Uses legible handwriting, although discrepancies may exist in letter shape or slant; experimentation with fonts may not be effective.	**Presentation** Creates an unclear or confusing story because of problems relating to handwriting, use of fonts, or spacing.

Writing Links: 4-Point Writing Rubric

7-Trait Writing in Science
Explanatory Writing

4 Excellent	3 Good	2 Fair	1 Unsatisfactory
Ideas & Content Develops a purposeful paper that presents a clear explanation of a task or process.	**Ideas & Content** Develops a paper that presents a reasonably clear explanation of a task or process.	**Ideas & Content** Develops a paper that shows a sense of purpose, but may not explain instructions or a process in a clear way.	**Ideas & Content** Makes no attempt to tell the reader how to do or make something; writing demonstrates no clear purpose.
Organization Organizes the writing in a way that smoothly moves the reader through the text, step by step, while clearly explaining the specific task or process.	**Organization** Presents the steps in a process in a well-planned manner, with clear transitions.	**Organization** Does not present the information clearly; transitions are weak.	**Organization** Shows an inability to organize the writing or provide connected details.
Voice Uses a personal voice that demonstrates a strong involvement with the purpose and audience.	**Voice** Makes an effort to explain ideas in a manner appropriate to the purpose and audience.	**Voice** Uses a voice that does not always involve the purpose of the writing or the audience.	**Voice** Makes no effort to involve self with the purpose and audience.
Word Choice Chooses time-order words, such as *first* and *then*, and spatial words, such as *top* and *bottom*, to present a clear understanding of the steps in the process.	**Word Choice** Chooses functional words that convey the purpose of the paper—to explain a task or process.	**Word Choice** Chooses words that fail to convey a complete understanding of the task or process being explained.	**Word Choice** Shows an inability to choose words that are appropriate to the topic, purpose, and audience.
Sentence Fluency Crafts effective sentences that flow together and support the content and style of the paper; has control of sentence types and lengths.	**Sentence Fluency** Crafts sentences that make sense and flow together; maintains control of simple sentences.	**Sentence Fluency** Crafts sentences that make sense but are short, choppy, or unvaried.	**Sentence Fluency** Uses sentences or sentence fragments that make little sense and are difficult or impossible to follow.
Conventions Implements writing conventions accurately and effectively; work needs little editing.	**Conventions** Uses a variety of writing conventions accurately, but some editing is needed.	**Conventions** Makes frequent mistakes in writing conventions, such as spelling, capitalization, punctuation, and grammar.	**Conventions** Shows an inability to use or grasp writing conventions.
Presentation Uses a pleasing form to present content; successfully aligns text and visuals to support and clarify key information.	**Presentation** Produces easy-to-read text and, for the most part, aligns visuals and content to enable the reader to access the information.	**Presentation** Demonstrates discrepancies in letter shape and slant, as well as spacing; connections between text and visuals are not always clear.	**Presentation** Demonstrates an inability to form regularly shaped and slanted letters, or use consistent spacing; fails to use visuals to support or illustrate key ideas in the text.

Writing Links: 4-Point Writing Rubric

7-Trait Writing in Science
Writing That Compares

4 Excellent	3 Good	2 Fair	1 Unsatisfactory
Ideas & Content Develops ideas and content that make a comparison in an informative, purposeful way.	**Ideas & Content** Develops ideas and content to show similarities and differences effectively.	**Ideas & Content** Develops ideas and content that present a comparison but may not hold the reader's interest.	**Ideas & Content** Makes no attempt to develop a comparison.
Organization Organizes details and information into distinct categories of comparison and contrast.	**Organization** Adequately arranges details and information into categories of comparison and contrast.	**Organization** Arranges some details and information into categories.	**Organization** Shows an inability to organize details or information into categories.
Voice Presents a personal voice that speaks to the audience in an individual and engaging manner.	**Voice** Presents a personal voice that meets the needs of the audience.	**Voice** Lacks an effective personal voice, or presents a voice that is insensitive to the needs of the audience.	**Voice** Makes no attempt to create a personal voice in the writing.
Word Choice Chooses compare-and-contrast words, such as *alike* and *different*, to highlight the points of comparison and contrast.	**Word Choice** Chooses compare-and-contrast words to show similarities and differences between items or ideas.	**Word Choice** Chooses words that attempt to support the comparison and link ideas.	**Word Choice** Makes no effort to use words that compare and contrast.
Sentence Fluency Crafts well-built, interesting sentences that invite expressive oral reading.	**Sentence Fluency** Crafts sentences that may be mechanical but are generally easy to read aloud.	**Sentence Fluency** Crafts short or choppy sentences that may be awkward to read aloud.	**Sentence Fluency** Crafts fragmented, confusing sentences that are difficult to read aloud.
Conventions Exhibits an excellent grasp of writing conventions, including spelling, capitalization, punctuation, grammar, and paragraph indents.	**Conventions** Exhibits an adequate grasp of standard writing conventions.	**Conventions** Exhibits a limited grasp of writing conventions.	**Conventions** Exhibits a severe inability to employ writing conventions.
Presentation Presents text that is pleasing to the eye and easy to read; text enables the reader to access key points of comparison and contrast.	**Presentation** Presents clear text that guides the reader to focus on the points of comparison and contrast.	**Presentation** Produces text that does not demonstrate an effective format for presenting points of comparison and contrast.	**Presentation** Presents text that is difficult or impossible to read and understand.

Writing Links: 4-Point Writing Rubric

7-Trait Writing in Science
Expository Writing

4 Excellent	3 Good	2 Fair	1 Unsatisfactory
Ideas & Content Develops clear content that supports the main idea and is suited to the purpose and audience.	**Ideas & Content** Develops content that is focused on and suited to the purpose and audience.	**Ideas & Content** Develops content that attempts to support the main idea and hold the interest of the audience.	**Ideas & Content** Makes no attempt to develop content that is focused on or suited to the purpose and audience.
Organization Exhibits solid organizational skills, with an effective introduction, body, and conclusion.	**Organization** Exhibits good organizational skills, including an effective introduction and a conclusion that summarizes the information.	**Organization** Exhibits limited organizational skills; does not draw a conclusion based on the facts presented.	**Organization** Exhibits extreme organizational problems that interfere with comprehension and readability.
Voice Expresses a personal voice that is well suited to the topic, purpose, and audience.	**Voice** Expresses a personal voice that is appropriate to the topic, purpose, and audience.	**Voice** Expresses a personal voice that may not fit the topic, purpose, or needs of the audience.	**Voice** Makes no attempt to develop a personal voice.
Word Choice Uses clear, accurate vocabulary that is well suited to the topic, purpose, and audience.	**Word Choice** Uses vocabulary that helps make the topic clear.	**Word Choice** Uses vocabulary that gets the message across in an adequate, yet ordinary way.	**Word Choice** Uses vocabulary that confuses the reader or is inaccurate.
Sentence Fluency Crafts a variety of sentences that enhance the understanding and fluency of the piece.	**Sentence Fluency** Crafts sentences that make sense and are easy to read aloud.	**Sentence Fluency** Crafts sentences that may be awkward at times.	**Sentence Fluency** Writes sentence fragments or sentences that are extremely difficult to read.
Conventions Demonstrates accurate use of standard writing conventions, including spelling, capitalization, punctuation, and grammar.	**Conventions** Demonstrates accurate use of most writing conventions; work needs little editing.	**Conventions** Makes frequent mistakes in spelling, capitalization, punctuation, and grammar; work needs extensive editing.	**Conventions** Makes mistakes in writing conventions that compromise readability and comprehension.
Presentation Presents a pleasing format that integrates text and illustrations, such as charts and maps, to support and enhance key information.	**Presentation** Uses visuals to clarify points in the text, although the visuals do not always support the key information.	**Presentation** Presents a mostly understandable format, but an integration between text and illustrations may be limited.	**Presentation** Presents a confusing format that denies the reader access to the information in the text.

Materials

Materials required to complete activities in the Student Edition are listed below.

Consumable Materials — Based on 6 groups

MATERIALS	QUANTITY NEEDED PER GROUP	KIT QUANTITY	CHAPTER/ LESSON
1 L bottles	3		12/1
Animal photos			1/3
Apple slices			3/1, 6/3
Aquarium gravel	1	1 bag	6/4
Bag, self sealing	2	50	2/1, 6/2
Bag, self sealing	1	50	3/1
Bag, self sealing	3	50	9/3
Baking soda	1	1 box	10/3
Ball, foam 3"	1	6	8/1
Balloons, 9"	2	35	10/3
Batteries, D cell	2	12	7/2, 7/3, 8/4, 12/3, 12/4
Batteries, D cell	1	12	12/4
Brine shrimp eggs	1	1	3/2
Celery			1/2
Chalk	1	1 box	8/1, 8/8, 10/3
Chalk			8/4
Chenille stems, assorted	1	100	1/4
Colored pencils			5/1
Construction paper			3/?, 10/3
Construction paper, black	1	1 pack	8/4
Construction paper, white			8/4
Cotton Swabs	1	72	1/3
Crayons			12/3
Cup, foam	2	25	2/1
Cup, paper	1	25	4/3
Cup, plastic	1	100	7/1, 9/2, 10/1
Cup, plastic	4	100	10/3
Cup, plastic	3	100	12/1
Cup, plastic	2	100	3/?, 10/1
Disposable gloves	1	50 Pair	3/1
Egg, hard-boiled			4/1
Flour	1	1 bag	10/3
Food coloring, red	1	1	1/2
Fruit, 3 types			2/1
Glue			4/2, 4/3, 6/3
Ice			2/1, 3/3, 7/2, 9/2, 10/1, 10/2
Index cards	5	1 pack	4/2
Index cards	8	1 pack	8/2
Leaves			4/1, 6/4
Lettuce			1/3
Live Coupon, Earthworms	1	1	1/4, 4/1

Non-Consumable Materials — Based on 6 groups

MATERIALS	QUANTITY NEEDED PER GROUP	KIT QUANTITY	CHAPTER/ LESSON
1 L bottles			11/2
2 Liter Bottle			10/3
Balance	1	6	9/2
Ball, tennis	1	6	8/3, 11/2
Beaker, 600mL	1	6	9/2, 9/3
Bin, C-thru	1	6	7/1
Blocks	2		11/1
Blocks	4		11/4
Book			11/3
Books	2		11/2, 11/4
Bowl, plastic	1	6	10/2, 10/3
Butterfly pavilion	1	1	2/2
Cardboard	1	6	11/2, 11/4
Cardboard box			12/2
Chair			8/3, 11/3
Classroom objects	3		9/1
Clock			10/1, 12/1
Coin			4/3
Cup, plastic small	2	25	11/4
Droppers	1	6	10/3, 12/1
Flashlight	1	6	7/2, 7/3, 8/1, 8/4, 12/3
Forceps	1	6	3/1
Funnel, plastic	1	6	6/4
Globe			5/1
Golf ball	1	6	9/2
Graduated cylinder, 100mL	1	6	4/2
Hand lens	1	6	1/1, 1/2, 1/3, 1/4, 2/1, 2/2, 3/1, 3/2, 5/3, 6/1, 6/2, 9/1, 10/3
Index cards	1	1 pack	8/3
Jar lid	2	18	3/2
Jar lid	1	18	4/1, 5/3, 7/2
Jar lid	3	18	5/3
Knife (teacher only)			2/1
Magnet, bar	1	6	9/1, 10/2
Meterstick			8/3, 11/1
Microscope	1	6	1/1
Mineral kit	1	1	6/1
Miniature light sockets	2	12	12/4
Mirrors	1	6	12/3
Notebook	6		11/1
Paint brush			4/3
Pan, aluminum	1	6	5/2
Paper clips	20	2 boxes	10/2

TR34

Materials required to complete activities in the Student Edition are listed below.

Consumable Materials

MATERIALS	QUANTITY NEEDED PER GROUP	KIT QUANTITY	CHAPTER/ LESSON
Live Coupon, Lady Bugs	1	1	1/4
Live Coupon, Snails	1	1	1/3
Live Coupon, Snails	1	1	1/4
Long sleeve shirt			7/1
Magazine			1/3
Map			5/1
Marine salt	1	1 bag	3/2
Marker			4/1, 7/3, 10/1, 11/4
Masking tape	1	1 roll	7/8, 8/3, 10/1, 11/1, 11/2
Milk			10/3
Miniature lamps	2	12	12/4
Modeling clay, 4 color	1	3 boxes	1/4, 9/3, 10/4, 11/4
Modeling clay, cream	1	1 box	6/3
Newspaper			5/2
Onion			1/1
Owl Pellets	1	6	3/1
Pan, aluminum	1	6	7/2
Paper			1/3, 5/1, 7/3, 10/2, 12/2, 12/3
Paper clips, box	1	1 box	8/1
Paper plates	1	50	3/1, 6/2, 12/3
Paper towel			1/3, 2/1, 3/3, 6/3, 7/1
Pebbles	1	1 bag	5/3
Pencil			1/3, 8/1, 8/4, 12/3
Penny			10/3
Permanent marker			4/2
Plants			1/2, 4/2
Plastic spoon	1	24	1/2, 2/1, 3/2, 6/2, 6/3, 6/4, 9/3, 10/1, 10/3
Plastic wrap, roll	1	1 roll	7/2, 9/2
Rock, sandstone chips	1	3 bags	5/3
Rubber bands	1	1 bag	7/2, 12/2
Salt	1	1	9/3, 10/1, 10/3
Sand	1	3 bags	5/2, 3, 6/4, 10/2, 12/1
Sand, blue	1	1 bag	4/3
Seed, white bean	1	1 package	3/?
Seeds, black bean	1	1 package	3/?
Seeds, pea	6	1 package	2/1
Shells, assorted	1	30	4/3
Soap, liquid	1	1	9/3
Soil, clay	1	1 bag	6/2
Soil, potting	1	3 bags	4/1, 5/3, 6/2, 6/4, 12/1
Soil, sandy	1	1 bag	6/2
Steel wool pads	1	6	10/3
Stones			4/1

Non-Consumable Materials

MATERIALS	QUANTITY NEEDED PER GROUP	KIT QUANTITY	CHAPTER/ LESSON
Paper clips	1	2 boxes	4/3
Perepared slide, onion	1	6	1/1
Photos, chicken life cycle	10		2/2
Pin			5/2
Plastic bowl	4	24	1/4
Plastic comb	1	6	12/4
Plastic jar	2	18	3/2
Plastic jar	1	18	1/2, 4/1, 5/3, 6/4, 7/2
Plastic jar	3	18	5/3
Plastic tile	1	48	12/1
Prism	1	6	12/3
Rock			7/1, 9/3
Rock, basalt	1	6	6/1
Rock, obsidian	1	6	6/1
Rock, pumice	1	6	6/1
Ruler			1/2, 2/2, 4/2, 9/1, 11/2, 11/4, 12/2
Safety goggles	1		9/3, 10/3
Scissors			1/3, 7/1, 10/2, 11/4, 12/2
Spring Scale, 250g	1	6	11/4
Stapler			8/2
Stopwatch	1	1	3/3, 5/3, 11/1
Strainer	1	6	10/2
Streak plates	1	8	6/1
Tape measure	1	6	8/1, 8/4
Terrarium	1	6	1/3, 4/1
Thermometer	1	18	10/1
Thermometer	3	18	12/1
Thermometer	2	18	3/2
Toy car	1	6	9/2
Toy truck	1	6	11/2
Washers	5	30	11/4
Wind-up toy	1	6	11/1
Wire cutter	1	1	12/4
Wood block	2	12	5/2
Wood block	1	12	9/3

Materials

Materials required to complete activities in the Student Edition are listed below.

Consumable Materials

Based on 6 groups

MATERIALS	QUANTITY NEEDED PER GROUP	KIT QUANTITY	CHAPTER / LESSON
Straws	1	50	12/2
String	1	1 roll	1/1, 7/1, 11/1
Tissue paper			12/4
Toothpaste			5/2
Twigs			5/2
Vegetable shortening			3/3
Vinegar	1	1	4/1, 10/3
Water			1/2, 1/3, 2/1, 3/2, 4/2, 4/3, 5/3, 6/4, 7/1, 7/2, 9/3, 10/1, 12/1
Wax paper, roll	1	1 roll	3/3
Wire, insulated	1	1 roll	7/1, 12/4

Bibliography

Unit A: Living Things

Chapter 1: A Look at Living Things

Bird. by Burne, David. Eyewitness Books, 1988.

Dig and Sow! How Do Plants Grow? Lobb, Janice. Kingfisher, 2000.

From Seed to Pumpkin. Pfeffer, Wendy. HarperCollins, 2004.

Green and Growing: A Book About Plants. Blackaby, Susan. Picture Window Books, 2003.

How Do You Raise a Raisin? Ryan, Pam Muñoz. Charlesbridge, 2002.

Katya's Book of Mushrooms. Arnold, Katya and Sam Swope. Henry Holt, 1997.

Metamorphosis: Changing Bodies. Kalman, Bobbie. Crabtree Publishing Company, 2005.

Plants. Hewitt, Sally. Chrysalis Education, 2003.

Plants with Seeds. Pascoe, Elaine. PowerKids Press, 2003.

Tell Me Tree: All About Trees for Kids. Gibbons, Gail. Little, Brown & Co., 2002.

Tigress. Dowson, Nick. Candlewick Press, 2004.

Chapter 2: Living Things Grow and Change

Bats. Wood, Lily. Scholastic, 2001.

Bluebird's Nest. DePrisco, Dorothea. Piggy Toes Press, 2005.

Growing Frogs. French, Vivian. Candlewick Press, 2000.

Humpback Whale. Douglas, Lloyd. Children's Press, 2005.

Ice Bear: In the Steps of the Polar Bear. Davies, Nicola. Candlewick Press, 2005.

Life Cycle of a Bean. Royston, Angela. Heinemann Library, 1997.

Life Cycles. Ross, Michael Elsohn. The Milbrook Press, 2001.

Near One Cattail: Turtles, Logs, and Leaping Frogs. Fredericks, Anthony D. Dawn Publications, 2005.

Spinning Spiders. Berger, Melvin. HarperTrophy, 2003.

Swing, Slither, or Swim. Stockland, Patricia. Picture Window Books, 2005.

Unit B: Ecosystems

Chapter 3: Living Things in Ecosystems

Animal Survivors of the Wetlands. Somervill, Barbara A. Franklin Watts, 2004.

Arctic Tundra. Forman, Michael H. Children's Press, 1997.

At Home in the Tide Pool. Wright, Alexandra. Charlesbridge Publishing, 1995.

Big Caribou Herd: Life in the Arctic National Wildlife Refuge. Hiscock, Bruce. Boyds Mills Press, 2003.

Claws, Coats, and Camouflage. Goodman, Susan E. Milbrook Press, 2001.

Deserts. Stille, Darlene R. Children's Press, 1999.

Coral Reefs. Earle, Sylvia A. National Geographic, 2003.

Forest Explorer: A Life-Size Field Guide. Bishop, Nic. Scholastic, 2004.

One Night in the Coral Sea. Collard, Sneed B. Charlesbridge Publishing, 2005.

A Walk in the Boreal Forest. Johnson, Rebecca L. Carolrhoda Books, 2001.

Watching Desert Wildlife. Arnosky, Jim. National Geographic, 2002.

Chapter 4: Changes in Ecosystems

African Savanna. Silver, Donald. McGraw-Hill, 1998.

Beavers. Hodge, Deborah. Kids Can Press, 2001.

Birds Build Nests. Winer, Yvonne. Charlesbridge, 2002.

Come to the Ocean's Edge. Pringle, Lawrence. Boyds Mills Press, 2003.

Fossils Tell of Long Ago. Brandenberg, Aliki. HarperCollins Publishers, Inc. 1990.

Hello Fish! Visiting the Coral Reef. Earle, Sylvia A. National Geographic, 2001.

Dory Story. Pallotta, Jerry. Charlesbridge Publishing, 2004.

Nature in the Neighborhood. Morrison, Gordon. Houghton Mifflin, 2004.

One Less Fish. Toft, Kim Michelle. Charlesbridge Publishing, 1998.

One Small Place in a Tree. Brenner, Barbara. HarperCollins, 2004.

What Do You Do With a Tail Like This? Jenkins, Steve. Houghton Mifflin, 2003.

Bibliography

Unit C: Earth and Its Resources

Chapter 5: Earth Changes

Changing Coastlines. Steele, Philip. Franklin Watts, 2003.

Earth. York, Penelope. Dorling Kindersley, 2002.

Earth and Us Continuous: Nature's Past and Future. Lewis, Patrick J. Dawn Publications, 2001.

Earth & You: A Closer View. Lewis, Patrick J. Walker & Co., 2004.

Earthquake! Nicolson, Cynthia Pratt. Kids Can Press, 2002.

Erosion. Winner, Cherie. Carolrhoda Books, 2004.

A Journey Into a Lake. Johnson, Rebecca L. Carolrhoda Books, 2004.

Mountains. Harrison, David L. Boyds Mills Press, 2005.

Our Big Home: An Earth Poem. Glaser, Linda. The Millbrook Press, 2000.

Volcanoes: Nature's Incredible Fireworks. Harrison, David L. Boyds Mills Press, 2002.

A Walk in the Tundra. Johnson, Rebecca L. and Phyllis V. Saroff. Carolrhoda Books, 2000.

Chapter 6: Using Earth's Resources

Air Is All Around. Branley, Franklyn. HarperTrophy, 1986.

Feel the Wind. Dorros, Arthur. HarperTrophy, 1990.

Follow the Water from Brook to Ocean. Dorros, Arthur. HarperTrophy, 1993.

Fossil. Ewart, Claire. Walker Books, 2004.

Fossil. Taylor, Paul D. DK Publishing Inc, 2000.

Fossil Detective: Tyrannosaurus Rex. Schatz, Dennis. Silver Dolphin Books, 2005.

Fossil Fuel Power. Sherman, Josepha. Capstone Press, 2003.

The Great Kapok Tree. Cherry, Lynne. Voyager Books, 2000.

The Magic School Bus Inside the Earth. Cole, Joanna, et al. Scholastic Inc., 1989.

Nature's Green Umbrella. Gibbons, Gail. HarperTrophy, 1997.

Water Dance. Locker, Thomas. Voyager Books, 2002.

Why Should I Save Water? Green, Jen. Barrons Educational Series, 2005.

Unit D: Weather and Space

Chapter 7: Changes in Weather

Four Seasons Make a Year. Rockwell, Anne. Walker & Co., 2004.

Hurricanes. Simon, Seymour. HarperCollins, 2003.

Kids' Book of Clouds and Sky. Staub, Frank. Sterling Publishing, 2003.

The Rainbow and You. Krupp, Edwin C. HarperCollins, 2000.

The Reasons for Seasons. Gibbons, Gail. Holiday House, 1996.

The Seasons. Dolan, Graham. Heinemann Library, 2001.

The Snowflake: A Water Cycle Story. Waldman, Neil. The Millbrook Press, 2003.

Twisters: A Book about Tornadoes. Thomas, Rick. Picture Window Books, 2004.

W is for Wind: A Weather Alphabet. Michaels, Pat. Thomson Gale, 2005.

Wild, Wet, and Windy. Llewellyn, Claire. Candlewick Press, 1997.

Wonderful Weather. Levine, Shar and Leslie Johnstone. Sterling Publishing, 2003.

Chapter 8: Planets, Moons, and Stars

Discover Space. Nicolson, Cynthia Pratt. Kids Can Press, 2005.

Earth Cycles. Ross, Michael Elsohn. Lerner Publishing, 2001.

If You Decide to Go to the Moon. McNulty, Faith. Scholastic, 2005.

The Moon. Simon, Seymour. Simon & Schuster, 2003.

Moon. Tomecek, Steve. National Geographic, 2005.

Next Stop, Neptune: Experiencing the Solar System. Jenkins, Alvin W. Houghton Mifflin, 2004.

Night Wonders. Peddicord, Jane Ann. Charlesbridge, 2005.

On Earth. Karas, G. Brian. G.P. Putnam's Sons, 2005.

The Planets. Gibbons, Gail. Holiday House, 2005.

Sun. Tomecek, Steve. National Geographic, 2003.

Unit E: Matter

Chapter 9: Observing Matter

Experiments with Solids, Liquids, and Gases. Tocci, Salvatore. Children's Press, 2002.

The Facts About Solids, Liquids, and Gases. Hunter, Rebecca. Smart Apple Media, 2004.

Materials, Changes, & Reactions. Oxlade, Chris. Heinemann Library, 2002.

Metals. Oxlade, Chris. Heinemann Library, 2002.

Solids, Liquids, and Gases. Garrett, Ginger. Children's Press, 2004.

States of Matter: A Question and Answer Book. Bayrock, Fiona. Fact Finders, 2006.

Super-Sized Science Projects with Volume. Gardner, Robert. Enslow Publishers, 2003.

Touch It! Materials, Matter and You. Mason, Adrienne. Kids Can Press, 2005.

What Is Matter? Curry, Don L. Children's Press, 2005.

What's the Matter in Mr. Whiskers' Room? Ross, Michael Elsohn. Candlewick, 2004.

What Is the World Made Of? Zoehfeld, Kathleen Weidner. HarperTrophy, 1998.

Chapter 10: Changes in Matter

Chemicals and Reactions. Richards, Jon. Aladdin, 2000.

Forest Fires. Thompson, Luke. Children's Press, 2000.

Freezing and Melting. Nelson, Robin. Lerner, 2003.

I Get Wet. Cobb, Vicki. HarperCollins, 2002.

Let's Try it Out in the Water. Simon, Seymour and Nicole Fauteux. Simon & Schuster, 2001.

The Magic School Bus Gets Baked in a Cake. Cole, Joanna, and Bruce Degan. Scholastic, 1995.

Pop! A Book About Bubbles. Bradley, Kimberly Brubaker. HarperCollins, 2001.

Science School. Manning, Mick and Brita Granstrom. Kingfisher, 1998.

Solid, Liquid, or Gas? Hewitt, Sally. Children's Press, 1998.

Solids, Liquids, and Gases. Garrett, Ginger. Children's Press, 2005.

Unit F: Forces and Energy

Chapter 11: Forces and Motion

Animals in Motion: How Animals Swim, Jump, Slither, and Glide. Hickman, Pamela and Pat Stephens. Kids Can Press, 2000.

Amusement Park Science. Greenberg, Dan. Chelsea Clubhouse, 2003.

Energy and Power. Harlow, Rosie. Kingfisher, 2002.

Energy Makes Things Happen. Bradley, Kimberly Brubaker. HarperTrophy, 2003.

Forces Make Things Move. Bradley, Kimberly Brubaker. HarperTrophy, 2005.

On the Move. Madgwick, Wendy. Raintree Steck-Vaughn, 1999.

Push and Pull. Nelson, Robin. Lerner, 2004.

The Science of a Spring. Stringer, John. Raintree Steck-Vaughn, 2000.

Simple Machines. Mason, Adrienne. Kids Can Press, 2000.

The Wheeling and Whirling Around Book. Hindley, Judith. Candlewick Press, 1994.

Wheels and Axles. Welsbacher, Anne. Bridgestone Books, 2001.

Wind Power. Petersen, Christine. Children's Press, 2004.

Chapter 12: Forms of Energy

All About Light. Trumbauer, Lisa. Children's Press, 2004.

Day Light, Night Light: Where Light Comes From. Branley, Franklyn. HarperCollins, 1998.

Electricity: From Amps to Volts. Cooper, Christopher. Heinemann Library, 2004.

Experiments with Light. Tocci, Salvatore. Children's Press, 2001.

Exploring Light and Color. Gold-Dworkin, Heidi. McGraw-Hill, 1999.

Light. Gish, Melissa. Creative Education, 2005.

Light and Shadow. Livingston, Myra Cohn. Holiday House, 1992.

Magnets: Pulling Together, Pushing Apart. Rosinsky, Natalie. Picture Window Books, 2003.

My Light. Bang, Molly. Blue Sky Press, 2004.

The Science of Noise. Wright, Lynne. Raintree Steck-Vaughn, 2000.

Science Yellow Pages

Life Science • Chapter 1
A Look at Living Things

Lesson 1 **Living Things and Their Needs**

All living things share certain characteristics: organization, growth, reproduction, the need for food, excretion of waste, respiration, and the ability to respond to stimuli. From the simplest microorganisms to the most complex plants and animals, the bodies of all living things consist of cells organized in specific ways. Growth is the increase in size of the total organism, not just of particular parts. Reproduction is the creation of new, similar individuals. All living things require food for energy. Animals ingest food, whereas plants, algae, and some microorganisms use light to produce their own food. Higher animals have specialized organs and systems to handle excretion. Respiration is the exchange of gases with the environment (for plants and animals, the intake of oxygen and release of carbon dioxide). Responses to stimuli generally include movement. Phototropism, a plant's growth toward light, is a form of movement. Plants need sunlight, water, carbon dioxide and oxygen, and nutrients such as nitrogen and phosphorus. Animals need air, water, oxygen, food, and shelter.

The **cell** is the smallest unit of an organism able to carry on the essential life functions. All growth and activities are caused by chemical changes within the cells. Complex organisms are made up of specialized cells. Groups of similar cells form a **tissue**, which is able to carry out one or more kinds of function. Tissues are organized into **organs**, which perform particular functions. Most organs consist of several or more tissues. Organs are organized into **organ systems**. Each organ system consists of organs performing related functions. Medulla cells, for example, form the medulla, the tissue of the inner kidney. The kidney is an organ, and it is part of the body's excretory system.

Lesson 2 **Plants and Their Parts**

Plants are divided into two groups, nonvascular and vascular. The **nonvascular plants** do not have systems to transport water and nutrients. Instead of roots, they have thin growths called rhizoids to help anchor them. Without a system to transport water, they simply soak up water like a sponge. Examples of nonvascular plants are mosses, liverworts, and hornworts. **Vascular plants**, which include trees, most familiar plants, and grasses, have roots, stems, and leaves.

Roots anchor the plant in the soil; absorb water, oxygen, and minerals; and store organic materials. From the roots, the water and minerals enter the **stem**. Running lengthwise through the roots and stem are vascular bundles, which consist of two types of tissues, **xylem** and **phloem**, separated by a layer called the *cambium*. Xylem cells, which are inside the cambium, carry water and mineral nutrients up to the leaves. Phloem cells, which are outside the cambium, carry food made in the leaves to other parts of the plant. Veins carry the water and nutrients throughout the **leaves**. Leaves have small pores called **stomata**. Water evaporates and exits through the stomata in the process called **transpiration**. About 99% of the water that enters a plant through its roots evaporates through the leaves.

Photosynthesis is the process by which plants take energy from sunlight and use it to convert carbon dioxide and water into carbohydrates and oxygen. Small green structures in the leaves called **chloroplasts** capture the sunlight. Chloroplasts get their color from the green pigment chlorophyll, which absorbs red and blue light and reflects green light. The energy stored as carbohydrates feeds the animals, including humans, that eat the plants. It also serves as food for the plant itself. Photosynthesis also supplies all the oxygen required by animals for respiration.

Lesson 3 **Animals and Their Parts**

Each animal species grows to a predetermined size (although some, like the beaver, continue to grow throughout their lifetime) and has a predictable lifespan. Almost all animals can move about on their own. (Some very simple animals, such as sponges, are permanently attached to one place.) Single-celled, animal-like organisms, can exchange gases directly with their environment. Most animals have a respiratory system for the oxygen-carbon dioxide exchange. Insects have air ducts, fish have gills, and all land-dwelling vertebrates have lungs.

An animal's **senses** convey information about the environment that enable the animal to find food, water, and shelter and avoid predators and other dangers. Vision provides the most detailed information. Almost all animals have some form of eyes, although the eyes of some, such as flatworms, detect only light and dark. All but the earliest vertebrates and some insects have some form of ear. The senses of smell and taste provide useful information. For example, the sense of smell can warn an animal that a food is poisonous.

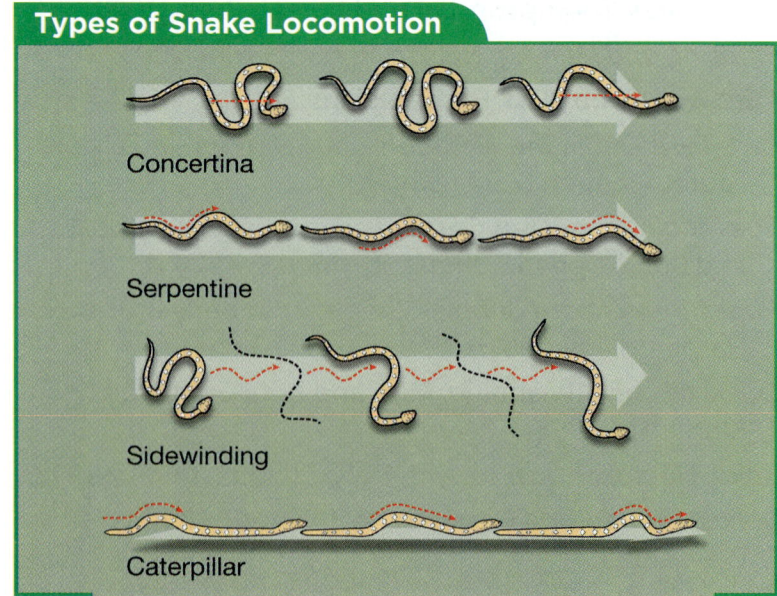

Types of Snake Locomotion

Concertina

Serpentine

Sidewinding

Caterpillar

An animal's means of **locomotion** reflects the medium through which it moves. Animals, such as fish, that move through water are streamlined. Animals, such as flying insects and birds, that move through air have wings. Animals, such as sloths and orangutans, that spend most of their time in trees have long arms that enable them to climb. Most animals that move across land have legs. Some, such as snakes, use their entire body to move. Some animals with legs, such as the giraffe, take longer strides to increase their speed, whereas others, such as the shrew, increase the rate of their stride.

Animals chew their food to break it down, making nutrients available. **Teeth** are the hardest structure in the body. In many vertebrates, teeth are replaced throughout the animal's life. Sharks grow as many as 20,000 teeth in a lifetime. All of a fish's and reptile's teeth are identical. Mammals have different kinds of teeth for specialized tasks: incisors that cut, canines that pierce, and premolars and molars that grind. Herbivores, which eat only plants, have strong incisors that can cut vegetation, small or absent canines, and wide premolars and molars that grind. Carnivores, which eat only meat, have small incisors, strong canines, and sharp molars that can cut through muscle and other tissue. The teeth of omnivores, which eat plants and animals, tend to be less specialized. Most mammals, including humans, have two sets of teeth during their lifetime. Elephants, however, have four sets. Their tusks are very long teeth.

Lesson 4   Classifying Animals

The practice of naming and classifying living organisms is called **taxonomy**. Scientists place all organisms into one of six "kingdoms": plants, animals, fungi, protista, ancient bacteria, and true bacteria. The animal kingdom is divided into two groups: those with backbones, called **vertebrates**, and those without backbones, called **invertebrates**.

Although many familiar animals are vertebrates, 95% of the more than 1.5 million known animal species are invertebrates. Because of their size and mobility, however, vertebrates tend to dominate their environment. **Vertebrates** are divided into **warm-blooded** and **cold-blooded**. Warm-blooded animals maintain a relatively constant body temperature. The body temperature of cold-blooded animals varies according to that of their environment. Only mammals and birds are warm-blooded.

Scientists organize vertebrates into five classes: fish, amphibians, reptiles, birds, and mammals.

Fish are aquatic animals that have fins and internal **gills**. The gills, in slitlike passages between the throat and exterior, enable the fish to breathe in water. Water enters the fish's mouth. As the oxygen-rich water passes over tiny blood-filled filaments in the gills, oxygen crosses from the water into the capillaries of the gills to the body's tissues, and carbon dioxide is released back into the water.

Most fish **eggs** are fertilized after they are deposited in the water. Because most of the eggs get eaten by other animals, only if a fish lays a tremendous number will any survive. In many species, the female produces as many as 5 million eggs during a single spawning period (as long as several months, depending on the species). As few as 10 out of one million may survive.

The **amphibians** include frogs and toads, salamanders and newts, and caecilians, which are limbless animals that resemble large worms. Certain species can regenerate not only amputated tails and limbs, but also parts of the eye, lower jaw, intestine, and heart. A frog develops from a water-dwelling tadpole into a four-legged land-dwelling adult.

Reptiles, which include turtles, crocodiles, lizards, and snakes, are cold-blooded, land-dwelling vertebrates. Although some turtles live in fresh water or in the ocean, all turtles breathe by means of lungs and lay their eggs on land. Some crocodiles and snakes live in the ocean and return to land only to lay eggs. Because they are cold-blooded, they must live in warm climates or bask in the midday Sun to bring up their body temperature. Although most reptiles lay eggs, in some species the eggs are incubated and hatched internally.

Birds are warm-blooded, egg-laying vertebrates. Their bodies are covered with feathers, which protect against cold and water. Although all birds have wings, some birds, such as penguins, do not fly.

Mammals, which include humans, are warm-blooded vertebrates. Mammals have three characteristics not found in other animals: three middle-ear bones, hair, and milk-producing mammary glands. Familiar mammals include cats, dogs, horses, cows, pigs, and rodents. Not all mammals live on land. Aquatic mammals include the sea lion, walrus, whale, porpoise, and dolphin.

The largest group of **invertebrates** is the **Arthropods**. Of all known animal species, about 75% are Arthropods. Arthropods have a segmented body covered by an exoskeleton (external skeleton) and jointed legs. The Arthropods include not only the six-legged insects, but also the eight-legged Arachnids, such as spiders, scorpions, and ticks; Crustaceans, such as shrimp, crabs, and lobsters; millipedes; and centipedes.

Biologists have identified about 800,000 species of insects. There may be as many as 10 million living insect species. Most insects have wings, antennae, and compound eyes (eyes consisting of separate units, each with its own lens).

The second largest group of invertebrates is the **mollusks**. These include the scallop, clam, oyster, mussel, snail, slug, squid, and octopus. All mollusks have a soft body, and many

Fish Gills

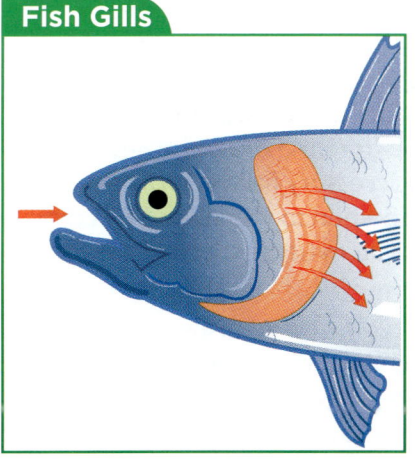

have a hard external shell. The Cephalopods, a class of mollusks that includes the squid, octopus, cuttlefish, and nautilus, can grow very large. Giant squid can reach more than 50 feet in length. Their eyes can be 1 foot in diameter. The octopus is particularly intelligent. It can learn to unscrew a cap to obtain food from inside a jar.

Science Yellow Pages

Life Science • Chapter 2
Living Things Grow and Change

Lesson 1 **Plant Life Cycles**

Plants can reproduce by both asexual and sexual reproduction. Some plants, including ferns, club mosses (which, despite their name, are not really mosses), and horsetails, produce **spores** instead of seeds. Ferns are the most diverse of the spore-producing plants. Unlike a seed, which contains a dormant embryo, or miniature plant, a spore requires an extra step. First it grows into a tiny *plantlet* called a **gametophyte**, which produces both egg and sperm cells. The fertilization of an egg by a sperm produces a sporophyte, which grows into an adult plant.

In some types of asexual reproduction, a small piece of plant tissue, such as a cutting, grows into a complete plant. In other types, a piece of tissue produces embryos, which then grow into adult plants. In asexual reproduction, the new plants are genetically the same as the ones they came from.

Most plants are flowering plants. Botanists divide **flowers** into two kinds: perfect flowers, which contain both female and male parts, and imperfect flowers, which contain either male or female parts. The male part of the flower is the **stamen**, a stalk with an anther on top that produces pollen that contains sperm cells. The female part is the **pistil**, which contains the ovary. Inside the ovary are ovules, which contain egg cells. When an egg cell and sperm cell join, they form an embryo, which is contained within a **seed**. Perfect flowers can produce seeds without any assistance, but imperfect ones need a way to bring sperm and egg together. Some plants depend on the wind to pollinate them, others depend on insects. Those that need to attract insects tend to have more colorful and fragrant flowers.

A **fruit** is a mature ovary that contains seeds. There are four kinds of fruit: simple, aggregate, multiple, and accessory. A simple fruit develops from a single ovary. In the berry type of simple fruit, the entire fruit is soft. Examples are the grape, tomato, banana, and blueberry. In the drupe type, the outer layers are soft or leathery and an inner pit or stone encloses one or more seeds. Examples include the peach, olive, coconut, and peanut. An aggregate fruit, such as the raspberry, consists of a group of small drupes that develop from separate ovaries of the same flower. Multiple fruits, such as the pineapple, develop from the ovaries of several flowers growing together. An accessory fruit, such as the pear, contains other plant parts in addition to the ovary. The edible part of the pear, for example, is stem tissue that surrounds the actual fruit.

The fruit plays an important role in seed dispersal. Plants have a number of ways of dispersing their seeds. Some, like thistles, have seeds that are blown by the wind. Some have seeds with hooks that stick to animals' fur. Still others

Four Kinds of Fruit

Simple

Aggregate

Multiple

Accessory

have fruits that are eaten by animals that either do not eat the indigestible seeds or pass them through their digestive system, dispersing the seeds as they travel.

Some plants, such as the pine tree, form **cones** instead of flowers for reproduction. The same plant usually has male cones for pollen and female cones for ovaries. Most cone-bearing trees are evergreens; as their name indicates, they have green leaves year round.

Plants respond to a number of environmental cues when they begin to **germinate**, or sprout. These include the length of daylight, temperature, moisture, and the presence of animal digestive juices. As the seed develops into a seedling and then a mature plant, the roots grow downward and the stem grows upward. Flowers bloom, and when egg cells inside ovules at the base of the flower are fertilized by sperm, seeds form. The seeds are then dispersed, followed by germination and a new cycle.

Annuals complete the entire cycle in one year, dying after seed formation. They include corn, beans, and petunias. A **biennial** completes its cycle over two years. During the first year it grows roots, a compact stem, and leaves and then survives the winter on stored food. During the second year it grows a vertical stem, flowers, fruits, and seeds and then dies. Biennials include carrots, onions, and raspberries.

Most plants are **perennials**, which can live for many years. Seeds that form the first year do not bloom until the second year. Each year after that, the perennial forms seeds that bloom the following year.

Lesson 2   Animal Life Cycles

Many kinds of insects go through egg, **larva**, and **pupa** stages on the way to adulthood. Many insect larvae are white and soft. An example is the fly larva, the maggot. Caterpillars are the larvae of moths and butterflies. After sufficient growth, larvae enter the inactive pupa stage, which can last for several months. The pupa has a protective case. Some moths enclose the case in a cocoon made from leaves glued together with silk. The butterfly pupa, called a **chrysalis**, is not enclosed in a cocoon. In the pupa the larva, which does not grow larger, is transformed into the adult insect, which then sheds its outer covering.

Although all birds lay **eggs**, they are not the only animals to do so. Almost all fish lay eggs, as do most frogs and toads. Most reptiles also lay eggs (some lizards and snakes give birth to live young). The only egg-laying mammals are the duck-billed platypus and spiny anteater.

Although the shell of a **bird's egg** is fragile, it protects the developing embryo. The shell has tiny pores that allow gases, including the oxygen needed by the embryo, to pass in and out. The yolk supplies nutrients, and the albumen (the white) supplies water and provides shock absorption. Bird eggs must be kept warm, or **incubated**, until hatching. Birds incubate their eggs in nests and hollow trees. One bird, called the *cowbird*, lays its eggs in the nests of smaller species. The nest-builder usually incubates the intruder's eggs and feeds them when they hatch—to the detriment of its own young.

Most **fish** lay eggs that are fertilized after they are deposited in the water. Because many of the eggs are eaten by other animals, only if the fish lays a tremendous number of eggs will any survive. In many species, the female produces as many as 5 million eggs during a single spawning period. As few as 10 out of one million may survive to adulthood.

Like fish eggs, the eggs of **amphibians** such as frogs and toads have no outer protection. They must be laid in the water or they will dry out. A frog spends its early years in the water as a tadpole, breathing, like a fish, through gills. A tadpole has no legs and swims by moving its tail. The *froglet* (the stage between a tadpole and adult) breathes with lungs and develops legs before the tail is absorbed by the body.

Reptilian eggs have a leathery shell that prevents them from drying out. In some species, the eggs are incubated and hatched internally.

Almost all **mammals** give birth to **live young**. The newborn is helpless and depends on its mother or other adults. Because it spends an extended period of time dependent on adults, it has time to learn behaviors that are not purely instinctual. This leads to greater variety in the behavior of mammals than in that of other animals. Play is an important activity for young. Through play, juveniles practice activities they will need as an adult, particularly fighting.

Lesson 3   From Parents to Young

Many traits are inherited. **Genetics** is the study of heredity, and **genes** are the units of inheritance. The nucleus of every living cell contains **chromosomes**, which consist of **DNA** molecules. Segments of the DNA contain genes, which provide our body with the coded information that in humans is unique to each individual—with the exception of identical twins, which result from the splitting of a single egg and therefore have identical genetic information. Every species has a **genome**, which is the total genetic information for members of that species. Each unique individual is a variation on its species' genome. Humans are all part of the same species, yet we all look slightly different.

Offspring receive half of their chromosomes from their mother and half from their father. They display (or express) traits of both parents. Some traits, such as hair color, are determined by single genes that are **dominant** or **recessive**. Brown hair is dominant and blonde hair is recessive. A child who receives one dominant gene and one recessive gene will have brown hair. If both genes are recessive, the child will have blonde hair. Because recessive genes can be passed down unexpressed through generations, sometimes two dark-haired parents will be surprised by a blonde child. Many traits, such as skin color, are determined by more than one gene.

A **population** is a group of organisms that inhabit a specific area. Over many generations, plants and animals develop adaptations that make them better able to survive their environment. In a cold climate, for example, animals with thick fur may be healthier and may have more offspring than animals with sparse fur. If animals with slightly thicker fur are born, they are more likely to live long enough to reproduce. Eventually, more and more of the population will have thick fur, and the animals with sparse fur may die out.

Baby Snakes Hatching

Life Science • Chapter 3
Living Things in Ecosystems

Lesson 1   Food Chains and Food Webs

The energy that living things use comes from the Sun. **Food chains** show how energy passes from one living thing to another. Each ecosystem has its own food chains, starting with **producers**, which are plants or plantlike protists that produce food by photosynthesis. All of the animals in the chain, called **consumers**, depend on the producers. **Herbivores**, which eat producers, are primary consumers. Secondary consumers eat primary consumers, and tertiary consumers eat secondary consumers. Meat-eating animals are **carnivores**. Animals that kill and eat other animals, or **prey**, are **predators**. Animals that eat dead animals are **scavengers**. Each food chain ends with bacteria and fungi, called **decomposers**, which break down and consume the remains of dead animals and plants, absorbing some of the products and returning others to the soil. There, plants reuse them to start new food chains.

Interconnected food chains create **food webs**, which describe the relationships among the populations in a given habitat. A food web is like a map of the energy flow within an ecosystem, while the food chains show individual routes.

Lesson 2   Types of Ecosystems

An **ecosystem** consists of a physical environment and the biological community it supports. It has both **abiotic** (nonliving) and **biotic** (living) components. It can be as small as a rotting log or as large as an ocean. Some primary abiotic factors are sunlight, temperature, water, and soil. A marine ecosystem, for example, contains abiotic factors such as rocks, water temperature, water salinity, and amount of sunlight, and biotic factors such as fish, plankton, and seaweed.

Within each ecosystem are **communities**, which include all the organisms that live together in an area. People, plants, and animals are members of an urban ecosystem. A pond ecosystem might include turtles, fish, raccoons, algae, lilies, worms, and frogs. These organisms are interdependent and make up the food chains and food webs in an ecosystem.

Within a community each type of living thing makes up a **population**, such as all of the moon snails in an intertidal ecosystem. Different ecosystems support different populations. This is why abiotic factors are integral parts of ecosystems. In a polar region, less sunlight and colder temperatures are fine for polar bears but not for the tropical macaw.

Ecosystems exist at many different levels, making it difficult to neatly identify specific types. For example, your backyard could be identified as an ecosystem, as could a temporary vernal pool. However, six major ecosystems include rain forest, deciduous forest, grassland, desert, freshwater, and marine ecosystems.

Rain Forests

Forest ecosystems include the world's rain forests and deciduous forests. The two basic kinds of rain forests are tropical and temperate. Found near the equator, **tropical rain forests** are very warm and humid year-round and have a tremendous diversity of life, including hundreds of species of trees. **Temperate rain forests** are found along coasts in temperate regions. The largest grow along the coast of the Pacific Northwest in the United States and Canada. They have two seasons: a rainy, cooler winter and a short, drier summer. Temperate rain forests have a less complex ecology and fewer species than tropical rain forests. Ten to twenty species of trees, most of them conifers, live in temperate rain forests.

Deciduous Forests

Deciduous forests have four seasons: spring, summer, autumn, and winter. The trees, mainly broadleaf hardwoods, survive freezing winters by dropping their leaves in the fall and producing new ones in the spring. A rich variety of animals live there, including deer, owl, snakes, and bears. Many of these animals hibernate or migrate when temperatures begin to drop.

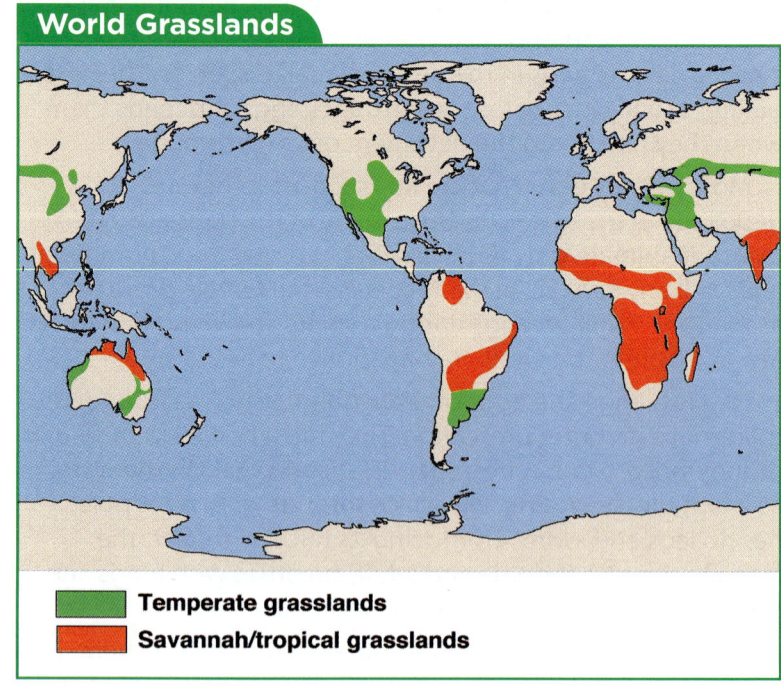

World Grasslands

Temperate grasslands
Savannah/tropical grasslands

Grasslands

Grasslands include prairies, pampas, savannas, and steppes. They are usually found in the interior parts of continents and are characterized by large areas of grasses and nonwoody plants. North American grasslands have hot summers and cold winters, which contributed to the harsh conditions met by early settlers in the area. Tropical grasslands do not have seasonal temperature fluctuations, but they do have wet and dry seasons.

Many animal species live on grasslands. North American prairie states, for example, have hundreds of species of grasshoppers. African savannas are known for their diversity of mammals, including more than 40 species of ungulates, or hoofed animals. Many of these animals are grazers.

Desert Ecosystems

Deserts cover large parts of Earth's surface. They may be sandy or rocky, hot or cold, but all deserts are dry. They receive 25 centimeters (10 inches) or less of rainfall per year.

Plant and animal life are less abundant in deserts than in other habitats. In more moist habitats, most plants are **perennials**, which live for many seasons. In the desert, most plants are **annuals**, which die after one season. Some annuals produce vast numbers of seeds that can survive for years, providing a "seed bank" until conditions turn favorable. Succulents, such as cacti, are able to store large quantities of water. Some desert plants have very long roots that spread out just beneath the surface, where they can catch every last drop of water.

Deserts are cooler at night, a condition that favors **nocturnal** desert animals such as the badger, gopher, and coyote. Others, such as the kit fox, have adaptations that help them moderate body temperature. The extra large ears of the kit fox actually act as "radiators," releasing body heat to the environment.

Kit Fox

Freshwater Ecosystems

Freshwater ecosystems include lakes, ponds, wetlands, rivers, and streams. The term *freshwater* refers to bodies of water that, unlike the ocean, are not salty. All aquatic plant and animal life ultimately depends on photosynthesis. Ponds have submerged plants, such as duckweed, whose leaves float on the surface. Ponds also have emergent plants, such as the broadleaf arrowhead, which rise above the surface. Many populations of animals can be found in freshwater ecosystems, including many species of fish, turtles, frogs, and snakes. Other animals, such as raccoons, ducks, and other waterfowl, live in freshwater ecosystems.

Marine Ecosystems

Oceans cover about 75 percent of Earth's surface. Because the ocean is so deep, most of it is too dark to support rooted plants. But the ocean contains abundant phytoplankton, which are single-celled photosynthetic organisms. These organisms form the base of vast food chains and food webs. Marine animals vary tremendously according to their location. Oceans can be divided into zones depending on depth and distance from shore. The organisms in each zone are uniquely adapted to its conditions.

Lesson 3 Adaptations

Adaptations are characteristics that enable a species to survive and reproduce in its environment. An adaptation may be a physical structure, a behavior, or a special characteristic, such as causing an itch or producing pheromones. But while environments are continually changing, adaptations occur over generations.

Physical structures that can help an organism survive include skin color, fur, beak size and shape, or the length and shape of limbs. The limbs of some animals allow for them to push off on their back legs and gain greater speed. An animal with thick fur is more likely to survive in a cold climate than one with thin fur. Plants also have physical structures that give them a competitive edge in their environments. Waxy cuticles protect leaves from harsh, freezing temperatures. The very small leaves of cacti or conifers minimize water loss.

Instincts are inborn behaviors that increase an organism's ability to survive. Baby birds instinctively open their mouths wide and beg for food and a human baby cries at a high pitch when hungry. Courtship is an instinctive behavior that increases the chances of reproduction. The tropisms of plants, such as growing towards light, are behaviors that help them grow and reproduce.

Other adaptations involve physiological processes. The oils produced by poison ivy or poison oak are defense mechanisms that increase the chances of survival for the plants. Many animals produce toxins, such as venom, that defend them against predators and thus increase their chances to survive, reproduce, and maintain the species.

Life Science • Chapter 4
Changes in Ecosystems

Lesson 1 ### Living Things Change Their Environment

Living things interact with each other and with their environment. These interactions create environments that are in a constant state of flux. Most often these interactions maintain a balance within an ecosystem; however, sometimes the balance shifts or is completely toppled.

Many of the daily activities of living things alter the ecosystem. Some of these changes are visible while others are not quite so. Many organisms make beneficial changes as they carry out their roles in the community. Below ground, decomposers break down wastes and enrich the soil by recycling and returning valuable nutrients. Earthworms and other organisms moving through the soil break it up, providing aeration.

Some species are so beneficial to the ecosystem that they are called **keystone species**. Beavers were called the "sacred centers" of the land by Native Americans who recognized how the animals created rich habitats for other animals. When the beavers build their lodges they create dams across streams, flooding shallow valleys which become wetlands. Wetlands are not only valuable to a wide diversity of animals. They also filter and purify water, soak up potential floodwaters, and prevent erosion.

Beaver Dam and Lodge

Feeding relationships are important in every ecosystem. If the number of predators in an area decreases, then the organisms at lower levels can become overpopulated. If this does not self-correct, then the overpopulated organisms—such as primary consumers—can face starvation as food sources become scarce. On the other hand, an overpopulation of predators can lead to a reduction in the numbers of prey, forcing predators to migrate or die off.

When a non-native, or **invasive**, species is introduced to an ecosystem it often has no natural predators. Sometimes the results can be disastrous. In 1859 twenty-four European rabbits were introduced for hunting purposes in Australia. Within six years the rabbit population grew to 22 million and spread across the continent at a rate of 70 miles (42 km) per year. With no natural predators the rabbit population reached an estimated 750 million by 1930. The rabbits ate everything in sight, devouring vegetation normally eaten by indigenous species and devastating crops. The extinction of 1/8 of the country's mammalian species is attributed to the rabbits. The Australian government has tried numerous ways to control the rabbit population, including the release of Myxomatosis, a disease fatal to European rabbits; yet nothing has succeeded in eradicating them.

A vine covering millions of acres in the southern United States is also the result of the introduction of an invasive species. During the celebration of America's 100th birthday at an exhibit in Philadelphia the Japanese included the large-leafed, aromatic kudzu in their presentation. It caught on in the United States and was used for decorative purposes and erosion control since it grew as much as a foot a day. Resistant to herbicides, and with no natural predators, kudzu grew out of control—competing with native plants.

By far, the species that has created the most changes in the environment is humans. Most of these changes have good intentions but often have resulted in a tremendous price to the environment.

Lesson 2 ### Change Affects Living Things

The **biosphere**, or part of the Earth where life exists, is made up of three main parts—the *lithosphere, atmosphere*, and *hydrosphere*. Each of these parts is a dynamic system, constantly changing as energy flows through them. Some of the changes that occur are natural processes that have been occurring for millions of years. Human activities bring about other changes. Every change affects living things. Since the hydrosphere and atmosphere connect one ecosystem to another, sometimes these effects are very far-reaching.

Some changes occur quickly such as fire or flood. Fires can devastate an area, destroying wildlife habitats and human residences. When this occurs some animals perish, while others migrate in search of vital resources to meet their needs. Other organisms have adaptations that help them survive. Indeed, some even benefit from fire. In forest ecosystems natural fires clear out old growth, reducing competition for resources. As old plant material burns it recycles nutrients and reconditions the forest floor. Some conifers depend on intense heat from fires to burst open their cones and disperse their seeds.

Floods can also quickly wipe out an ecosystem. It will be years before the impact of flooding caused by Hurricane Katrina in New Orleans will be fully understood. People lost homes, valuable habitats were destroyed, and pollution was introduced at catastrophic levels. Yet floods are a natural process that actually helped make areas such as New Orleans rich and fertile grounds. As floodwaters breach the banks of rivers around the world and cover the surrounding land they deposit sediments that were picked up and carried over many miles by the water. As the water recedes the sediments remain and nourish the land with nutrients, making flood plains fertile. The birthplace of agriculture is in the Fertile Crescent between the Tigris and Euphrates Rivers.

Other rapid changes include storms, landslides, earthquakes, and volcanic eruptions. When these changes occur it may take a while for an area to recover. Ecosystems may undergo **ecological succession**, in which, over time, a community is replaced by another. For example, a volcanic eruption can flatten or incinerate everything in the surrounding area. Over time, organisms called *pioneer species* appear and, later, more organisms move in, leading gradually to a **climax community**, a collection of plant and animal populations that make up a new ecosystem.

Other changes on Earth happen much more gradually. Climate changes have occurred over time as evidenced by marine fossils found in sedimentary rock located far from any existing seas. However, human activities alter the natural course of change such as the acceleration of global warming due to burning fossil fuels.

Ecological Succession

Lesson 3 ## Living Things of the Past

Fossils are the preserved remains or traces of organisms from the distant, or geological, past. Most fossils are thousands to billions of years old. Although most are hard animal body parts such as bones, teeth, and shells, some are preserved remains of soft tissue. Rarely, entire animals were preserved, such as the mammoths found in Siberia. Some fossils are tracks or other impressions left in sediment.

Most fossils are found in sedimentary rock. For fossilization to occur an organism must be buried quickly after death. This slows down decay and prevents the remains from being eaten. Burial can happen most quickly in sediments such as mud or sand. As sedimentation continues, layer upon layer of sediments are added. Through compaction and cementation the layers containing the fossils become rock.

Since sedimentary rock forms in layers scientists can use the **law of superposition** to interpret what the fossils tell us about Earth's history. The law states that in a column of layers, the oldest rocks are on the bottom and the younger rocks are on the top. This helps scientists determine the relative ages of fossils.

There are several different types of fossils based on the process by which they formed. In **petrification** mineral-rich water flows into openings in the organism's bones. The minerals eventually replace the organic material, turning the bones to stone and sometimes resulting in exact stone replicas of once-living things.

The most common types of fossils are **molds** and **casts**. If you press a shell into clay and then remove it, what is left is a mold, a space with the same shape as the shell. In nature, the decay of soft tissues removes the object and form a mold. Minerals that fill in the mold can harden into a cast, or three-dimensional replica of the organism.

Imprints are impressions made in soft matter such as mud that hardened and solidified. They can be parts of organisms, such as feathers or leaves, or traces of their activities such as footprints, worm holes, or scratch marks.

Fossils form only when conditions are favorable. Although the chance that any single organism will fossilize is very low, over the billions of years of Earth's history, so many living things have died that there is a rich store of fossils.

Earth Science • Chapter 5
Earth Changes

Lesson 1 **Earth's Features**

The term **topography** refers to all of the characteristics of the land, particularly elevation and shape. Topographic maps, also called topo maps, show land features such as mountains, valleys, and plains, as well as bodies of water, forests, roads, urban areas, and distinctive landmarks. Unlike other types of maps, topographic maps depict relief, or three-dimensionality, through contour lines and, sometimes, cross-hatching. A map on which one inch equals 2,000 feet (a mile is 5,280 feet) can show much more detail than a map on which one inch equals 4 miles.

Topographic Map

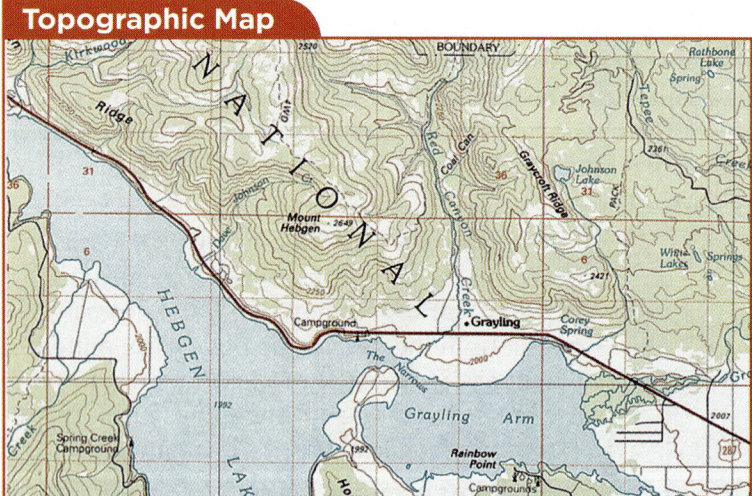

Topographic Map at Three Scales

Three-quarters of Earth's surface is covered by interconnected oceans, called the **world ocean.** Most of the remaining surface is covered by the seven continental landmasses, and a small portion is covered by freshwater lakes. The four main ocean basins are the Pacific, Atlantic, Indian, and Arctic. The first three extend northward from Antarctica to the Arctic Ocean, the area near the North Pole. In 2000, the International Hydrographic Organization recognized a fifth ocean basin, the Southern Ocean, which surrounds Antarctica and extends to 60 degrees latitude. The smaller sections of the oceans that are partially surrounded by land are called seas. The largest seas are the South China Sea, the Caribbean Sea, and the Mediterranean Sea. The average ocean depth is 3 miles, and the deepest point, in the western Pacific, is almost 7 miles. Most of the salt in seawater comes from the land. For millions of years, rainwater and rivers have washed over rocks containing salt, dissolving tiny amounts of salt and carrying the salt to the oceans. When water evaporates from the ocean, the salt remains behind.

Rivers are large freshwater streams that flow from headwater areas to their mouths, where they discharge into an ocean or a major lake. Rivers carry about 68% of Earth's surface runoff to the oceans. **Lakes** are inland bodies of water, and

ponds are small, shallow lakes. Although the majority of lakes are freshwater, there are also salt lakes. Salt lakes tend to be in areas with high evaporation.

Ocean basins formed when tectonic plates slowly split apart, leaving continents separated by a deep ocean floor. Closest to the continent borders are the sloping **continental shelves,** which range from a few miles to hundreds of miles wide. The shelves contain abundant natural resources, including oil and natural gas reservoirs and fishing grounds. At the edge of each shelf is a **shelf break,** and then a steeper **continental slope,** with a gentle **continental rise** at its bottom. The continents rise (not surprisingly) from the continental rise. Together, the shelf, slope, and rise form the **continental margin.** The **ocean basin** begins at the edge of the continental rise. The **ocean floor** is the flat part of the basin. A mountain range in the ocean floor, the **mid-ocean ridge,** rises about one-half mile to two miles above the surrounding plains. Created from rising magma where the plates separated, it extends almost 40,000 miles, through all the major oceans.

Earth's hard outer shell, which averages about 100-km (60 miles) thick, is called the **lithosphere.** The lithosphere consists of the crust, which is from 3 to 42 miles thick, as well as the uppermost part of the mantle. The lithosphere is made up of plates that include the continents and ocean floors. These moving plates rest on top of the **asthenosphere,** the semi-molten portion of the upper mantle. The mantle, which consists mainly of material rich in silicon, iron, and magnesium, is about 1,800 miles deep. The core has a solid center surrounded by an outer liquid core. The liquid core consists of extremely hot, molten metals, mostly iron.

Lesson 2 **Sudden Changes to Earth**

In an **earthquake,** sudden movements within the lithosphere release energy that travels through the surrounding area as **seismic waves.** The largest earthquake in a sequence of earthquakes is called the **main shock.** Smaller shocks called **foreshocks** may occur days to weeks before the main shock. Most main shocks are followed by **aftershocks,** which gradually diminish in size and number; these may continue for days afterwards. Scientists use instruments called **seismographs** to detect seismic disturbances. By measuring the amplitude, or height, of the waves recorded by seismographs, they can determine the magnitude of an earthquake according to the **Richter scale.** The largest recorded earthquake had a magnitude of about 9. The Richter scale is logarithmic, which means that each increase by one unit corresponds to a tenfold increase in the amplitude of the seismic waves. Thus, the waves of a magnitude 7 earthquake have a 10 times larger amplitude than those of a magnitude 6 earthquake, and 100 times larger than those of a magnitude 5 earthquake.

A **volcano** is a steep-sided hill or mountain formed by the accumulation of **magma** (molten lava) that erupts through vents in Earth's crust. Volcanoes cause seismic activity, on the ocean floor as well as on land. When two tectonic plates collide and one goes below the other, the subducted (bottom) plate returns to Earth's mantle, where it is melted down. The melted material mixes with gases and then, because it is less dense and buoyant, it rises as magma. If it doesn't cool down and solidify as crustal rock, it escapes through a volcanic vent as **lava.** Lava flows can travel tens of miles, causing great destruction. Volcanoes can also cause landslides, mudflows, and tsunamis.

Severe thunderstorms sometimes develop into **tornadoes.** Rapidly rising warm air in some thunderstorm cells begins to rotate, producing a tornado funnel. Winds within the funnel can reach 200 miles per hour. The updraft in the center of the funnel can lift cars and even buildings and deposit them hundreds of feet away. **Hurricanes** (also called typhoons) generally form between latitudes of 5 and 20 degrees over warm oceans. These slow-moving, low-pressure systems are energized by rising air carrying large amounts of water vapor that cool and condense higher in the atmosphere, releasing heat energy. Heat released from evaporation causes the swirling winds to move higher and increase in speed. Cooler air rushes in below, causing moist air to rise even higher and to rotate even faster. Soon the hurricane is a giant swirling mass with a calm "eye" at its center. Hurricanes break up soon after reaching land because they can no longer draw energy from the water.

The conditions that contribute to the likelihood of a **landslide,** including deforestation and loss of vegetation from wildfires, can develop over long periods of time. But a landslide, once triggered, can be over within seconds. The most devastating landslides are caused by earthquakes and volcanic eruptions. In a 1998 earthquake in Afghanistan, most of the approximately 4,000 victims were killed in landslides caused by the tremors.

Angle of Repose

Most landslides are triggered by excessive precipitation or melting snow. Water loosens rock, lubricates the surface between the rock or soil and bedrock below, and adds weight to the loose material. Much of the ground in areas prone to landslides consists of fragments of weathered rock that is on its way to becoming soil. This material is less stable than soil, which has already been compacted into horizontal layers. The risk of landslides has risen in recent years, partly due to extensive deforestation and an increase in severe weather caused by global climate changes.

Lesson 3   Weathering and Erosion

Rocks are broken down and changed by the process of **weathering.** The two main types of weathering are physical and chemical. **Physical weathering** breaks down rocks into smaller pieces. The causes of mechanical weathering include the freezing and thawing of water, pressure of plant roots, actions of animals, and wind. Because water expands when frozen, water that freezes in the cracks of rocks can exert enough pressure to split rocks apart. Tree roots can exert enough pressure in cracks and joints to break rocks.

With **chemical weathering,** rainwater, acids, and oxygen interact with minerals in rocks, changing them into other minerals. Rocks have different susceptibilities to chemical weathering. Limestone and marble react chemically with mildly acidic rainwater. The tops of marble tombstones are often heavily weathered. In limestone caves, dripping water carrying dissolved limestone (calcium carbonate) causes the stalactites that hang from the ceiling and the stalagmites that rise from the ground. **Soil,** which consists of loose, disintegrated rock and decomposed plant and animal matter, is caused by weathering.

The term **erosion** refers to the loosening and transporting of soil and rock. The main agents of erosion are moving water, near-shore waves, wind, and gravity. All streams and rivers carry rock fragments and soil acquired from tributaries or rubbed off from their banks. As this material is transported by the water, it grinds against the rock beneath the water, creating even more fragments. Ocean waves near the coast loosen and carry away soil and rock. In desert and beach areas, the wind can remove sand from one area and deposit it in another.

Landforms shaped by rivers and streams, called **fluvial landforms,** are the most common kind of landforms. Two categories of fluvial landforms are erosional and depositional. The main erosional form is the valley. As a stream (a river is a large stream) flows, through erosion, it cuts into the underlying rock. Depositional landforms include floodplains, deltas, and dunes. Sediment that is deposited by a river during periodic floods produces a broad, flat valley floor called a **floodplain.**

Glaciers are large, moving masses of ice that form from compacted, recrystallized snow. Debris carried in the underside of glaciers acts as an abrasive. Through **plucking,** rock pieces are picked up and incorporated into the glacier. Most glacial erosion occurs in valleys, which subsequently deepen and widen. Sediment can be carried many miles before it is deposited.

Exit Glacier in Anchorage, Alaska

Earth Science • Chapter 6
Earth's Resources

Lesson 1 — Minerals and Rocks

Rock is composed of one or more of the **minerals** that make up Earth's crust. A mineral is a naturally occurring material that has a definite chemical composition and, usually, a crystalline structure.

The three major rock classifications are igneous, sedimentary, and metamorphic. **Igneous** rock forms from the cooling of molten material from Earth's interior. **Intrusive** igneous rock is from magma that cools below Earth's surface, often intruding into adjacent rock. Magma that solidifies at the deepest levels is the slowest to cool, taking over tens of thousands of years. It has the coarsest texture (largest grains). Magma that cools closer to the surface has medium-coarse texture (medium-sized grains). Magma that erupts through volcanoes as lava cools the most rapidly, forming **extrusive** rock. It has the most fine-grained texture. The most common mineral in igneous rocks is feldspar, followed by quartz.

Seventy percent of all rocks at Earth's surface are **sedimentary**. Sedimentary rocks form from the weathering of older igneous, metamorphic, or sedimentary rocks and, sometimes, sediment of living organisms. Erosion carries away parts of older rock and transports them elsewhere, where they are deposited in layers, or **stratified**. The weight of the overlying rocks presses the sediment together, causing **compaction**. Some of the water pressed out contains minerals that recrystallize and bind rock particles together. This process is called **cementation**.

Metamorphic rock forms when intense heat and/or pressure within Earth's crust alter the mineral composition and structure of preexisting igneous, sedimentary, or metamorphic rock. When molten rock intrudes into sedimentary rock, the heat causes minerals to change their internal chemistry or grow larger. Heat and pressure from within Earth's crust can also push up and fold sedimentary rock, forming mountain ranges. Common examples of metamorphic rock are marble and slate, which were originally formed as limestone and shale (mud).

Geologists identify minerals by various physical **properties**, including color, luster, hardness, cleavage, and density. The color is as it appears to the unaided eye. Luster, the character of light as reflected from the mineral, is either metallic or nonmetallic. Nonmetallic luster can be further described by such terms as vitreous (glassy). Hardness, a mineral's resistance to scratching, is measured on a 1–10 scale called **Mohs Hardness Scale**. The softest mineral is talc, and the hardest is diamond.

Lesson 2 — Soil

Soil, the uppermost layer of Earth's surface, extends down from about one to six feet. Caused by weathering, it consists of loose, disintegrated rock and decomposed plant and animal matter. The mineral and rock content is grouped according to the size of the grains. In order of decreasing size, these include stones, gravel, sand, and clay.

Topsoil, the uppermost layer of soil, extends down from several inches to several feet. It consists of rock fragments, sand and clay particles, microorganisms, and decayed plant and animal material called **humus**. Humus gives topsoil its dark color. Soil erosion occurs when the topsoil is blown away by wind or washed away by water. **Subsoil**, the layer below the topsoil, usually extends down about one and a half to two feet below the topsoil. It contains more clay particles than the topsoil and very little, if any, humus. The absence of humus makes it lighter in color than topsoil. Subsoil is important for drainage. Below the subsoil is the **parent material**, which consists of the type of rock from which the upper two layers formed.

Lesson 3 — Fossils and Fuels

Fossils are the preserved remains or traces of organisms from the distant, or geological, past. Most fossils are several thousand to several billion years old. Although most fossils are preserved remains of hard animal body parts such as bones, teeth, and shells, some are of soft tissue. In rare cases, entire animals have been preserved. Fossils called **coprolites** are preserved animal feces. Fossils also preserve impressions of plants and animals, as well as tracks left in sediment.

Fossilization begins when an organism is buried rapidly before it can be destroyed. Some plants and animals were buried beneath a body of water after being swept away in a flood. Others were buried in sandstorms or under ash during volcanic eruptions. Still others were enclosed in **amber**,

Mohs Hardness Scale

Rating	Reference Mineral	Reference Objects (approximate value)
1	talc	
2	gypsum	fingernail (2.5)
3	calcite	copper penny (3.5)
4	fluorite	
5	apatite	glass plate (5.5)
6	potassium feldspar	steel file (6.5)
7	quartz	
8	topaz	
9	corundum	
10	diamond	

which is fossilized tree resin that is at least several million years old. Some amber contains a trapped plant or animal that is the sole representative of an extinct species.

The **fossil fuels**—coal, oil, and natural gas—are so called because they were formed by the chemical alteration of ancient plant and animal remains trapped in sedimentary deposits. **Oil** and **natural gas** formed from thick layers of microscopic organisms deposited on the floor of ancient ocean basins. Heat and pressure converted the buried organisms into petroleum, which is separated into different products such as gasoline, heating oil, and asphalt. Other products made from petroleum include bubble gum, crayons, detergents, and plastics. Natural gas, which consists mostly of methane, is found in underground reservoirs, often on top of oil deposits.

Coal, which consists mostly of carbon, comes from decayed forest-type plants that turned to peat. The peat was then compacted beneath layers of sedimentary rock until it turned to coal. Coal, which is removed from the ground by mining, is classified according to the percentage of carbon. The highest grade is 95% carbon. Graphite (pencil lead) is a mixture of carbon and clay. Emissions from the burning of fossil fuels is one cause of air pollution.

Solar Roof Tiles

Renewable energy refers to any energy source that is replenished as quickly as it is used. These include solar, wind, and hydroelectric energy. Energy sources such as wood, which can be replenished within a reasonable period of time, are considered renewable. **Solar energy** is used to generate electricity and to heat homes and provide hot water. The most common way is through the **photovoltaic**, or solar, cell. The cell absorbs sunlight and converts it to electric current. Because the amount of current is small, and the voltage is low, multiple cells are connected in large **solar panels**.

Wind energy is actually converted solar energy. The Sun heats different parts of Earth at different rates. As hot air rises, cool air moves in beneath, and the resulting movement is the wind. The motion of the wind is a form of energy. A windmill or wind turbine is basically the opposite of a fan. A fan uses electricity to move the air. A turbine takes the energy of the moving air and converts it to electricity.

Hydroelectric power takes the energy of flowing water and converts it to electricity. In most hydroelectric plants, a dam on a river stores water in a reservoir. When water is released from the reservoir, it flows through a turbine, which activates a generator that produces electricity. **Nuclear energy** harvests the energy in atomic reactions. It is both renewable, using breeder reactors, and nonrenewable, as it is based on mining uranium.

Hydroelectric Dam

Lesson 4 Air and Water Resources

Most of our water supply comes from rainfall that collects in rivers, lakes, and reservoirs. It is withdrawn through intakes that range from simple pipes to elaborate gated tunnels. Some rainfall penetrates the soil and is pulled down by gravity to a depth where all the pores and openings of the soil and rock are filled with water, which becomes **groundwater**. The upper surface of the groundwater is called the **water table**. One way we withdraw groundwater is from **wells**, vertical holes that are drilled or excavated to below the water table.

Most large water supply systems use conduits called mains, or **aqueducts,** to transport water to treatment plants before it is sent to customers. These plants remove harmful organisms, as well as gases that give the water an unpleasant odor or taste. Gravity allows heavier matter to sink to the bottom. Aeration supplies oxygen to the bacteria that feed on organic pollutants (as organic matter decays, it depletes oxygen). Some treatment plants also filter the water. Finally, the water is disinfected. Most of the drinking water is distributed, and some is stored. Wastewater treatment plants rely on the same basic processes. (Wastewater is liquid waste, including household sewage from toilets, bathtubs, and washers.) Treated wastewater, called **effluent**, is discharged into either a river or the ocean.

Earth Science • Chapter 7
Changes in Weather

Lesson 1 **Weather**

The **atmosphere** is a thin blanket of gases surrounding Earth. Nearly all weather takes place in the lowest level, the **troposphere**, which extends upward about seven miles. The term **weather** refers to the atmospheric conditions at a specific time and place. It is caused by the interaction of several factors including temperature, air pressure, humidity, and winds. Weather describes short-term conditions; **climate** refers to conditions in an area over an extended period of at least several decades. **Meteorology** is the branch of science that deals with the atmosphere. In fact, "*meteor-*" = atmosphere; and "*-ology*" = study of. "The most practical applications of this field are those related to weather prediction.

Weather is driven by energy from the Sun. The amount of solar energy that reaches any given area on Earth's surface depends on the angle at which the Sun's rays reach Earth and for how long a period each day. At the equator the Sun is very high and its light hits this part of Earth at close to a 90° angle. This area receives the most radiant energy and therefore has the highest temperatures. As you move north or south of the equator the Sun's rays are more diffuse. The angle at which the rays hit Earth are less than 90°, making the radiant energy received less direct than at the equator. Less energy results in lower temperatures.

Air pressure is the weight of the air above any point on Earth's surface. Air pressure, also called **barometric pressure**, is measured with a **barometer**. Air pressure is affected by temperature, water vapor, and elevation. Warmer air is less dense than colder air. This is because heat energy causes molecules to move faster and spread farther apart. Molecules that are closer together, as in cold air, are in slower motion and are closer together and, therefore, colder air is denser. Air at higher elevations is also less dense than air at sea level. The air pressure at sea level averages 14.7 pounds (6.7 kilograms) per square inch. At about 5,500 meters (18,000 feet), is air pressure is about half that. Finally, moist air is less dense than dry air.

Humidity is the amount of moisture, or water vapor, in the air. **Relative humidity** is that amount compared to the amount the air can hold. The cliché that "it's not the heat, it's the humidity" is true. High humidity makes warm weather more uncomfortable, because air that contains a lot of water has less room for the evaporation of water given off as perspiration. Humidity is measured with a hygrometer or a sling psychrometer.

Wind is the flow of air relative to Earth. The Sun heats various parts of Earth's surface at different rates, which in turn causes various parts of the atmosphere to heat up differently. Hot air, which has greater energy and movement of its molecules, becomes less dense and rises. Cooler air, which is denser, weighs more than an equal volume of hot air and exerts greater

pressure. As land and water temperatures change, they cause differences in air pressure. The air moves horizontally from areas of higher pressure to areas of lower pressure. This movement causes **convection currents**, which we feel as wind.

Global Wind Patterns

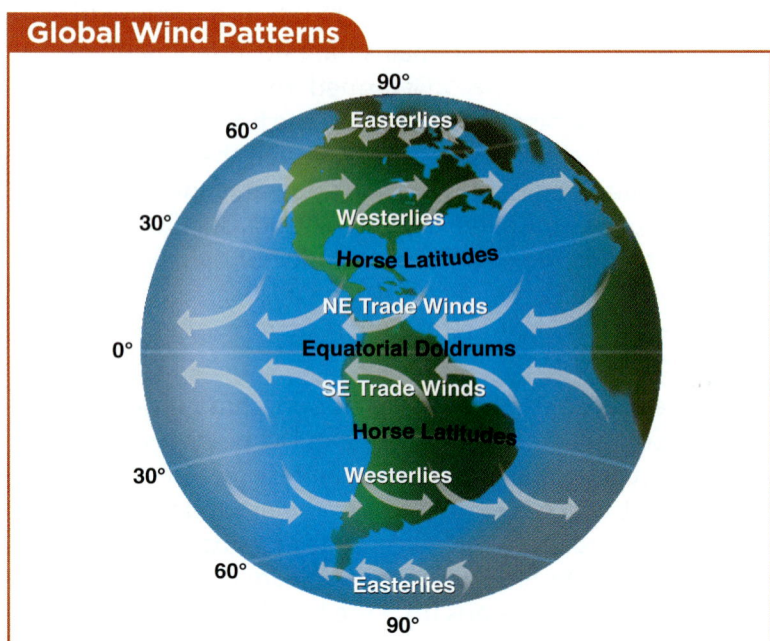

Lesson 2 **The Water Cycle**

The water on Earth is constantly moving through the **hydrologic cycle**, or **water cycle**. This cycle is truly a cycle; it has no beginning or end. The Sun supplies the energy that drives the water cycle. When the Sun heats surface water found on Earth, some of the water evaporates into the air as water **vapor**. Air currents carry the vapor higher up into the atmosphere, where cooler temperatures cause it to condense into tiny water droplets, or to form ice crystals that make up the clouds. Air currents within the clouds cause cloud droplets to collide. As they collide, they grow larger and eventually may fall to Earth as rain. In clouds that contain both water droplets and ice crystals, the ice crystals grow larger and the water droplets become smaller. When the ice crystals are large enough, they fall. If the temperature of the air through which they fall is above freezing, the ice crystals melt and fall as rain. If the air temperature is below freezing, they fall to the ground as snow.

Rain falls into the oceans or onto land, where it soaks into the soil or flows across the ground as **surface runoff**, also called **surface water**. Some of the water that soaks into the ground is taken up by plants or is trapped in pores and cracks of rock as **groundwater**. Some rainwater collects in lakes, and some of it flows into rivers, which then flow into oceans. Over Earth's entire surface, the amount of water that evaporates from the oceans is about the same as the amount that falls back as precipitation. Once evaporated, the average water molecule remains about 10 days in the air.

Plants and animals are an important part of the water cycle. Plants take in water which has soaked into the ground. The water moves upward through the plant's vascular system;

and, eventually, some water is released through pores in the leaves in the process of transpiration. Animals take in water by drinking or through the food they eat. They release water back into the environment by exhaling when they breathe, as well as when they release wastes.

Lesson 3 Climate and Seasons

Weather describes the day-to-day conditions of the atmosphere in a particular location. **Climate** refers to the average conditions for an area over a long period of time. Climate is determined by temperature and precipitation.

Temperatures in a region are influenced by latitude, elevation, and the presence of ocean currents. As you move north or south of the equator the angles at which the Sun's rays hit the Earth's surface get increasingly smaller, making them more diffuse. The radiant energy becomes less intense, which results in lower temperatures. As elevation increases, air molecules are spread farther apart so the air is less dense and, therefore, unable to hold as much heat energy. This means that temperatures at higher elevations are lower.

Ocean currents may be warm water currents or cold water currents, depending on where they originate. Warm water currents warm the air above the surface of the water and cold water currents cool the air, affecting the land areas nearby. Every three to seven years, a surge of warm current off the western coast of South America causes a disruption of normal weather patters called **El Nino**. The far-reaching effects of El Nino include increased rainfall across the southern United States and Peru, and drought in Australia and Indonesia.

Cirrus Clouds

Cirrus clouds are made entirely of ice crystals.

Precipitation in a given area is affected by prevailing winds and mountain ranges. **Prevailing winds** are Global Winds that blow most often across a particular area. Depending on where the winds are blowing from, they carry different amounts of moisture. Mountain ranges can block prevailing winds. As the winds sweep up the mountainside they hit cooler air at higher elevations. Since cooler air cannot hold as much moisture, the

moisture condenses and falls back to Earth as precipitation on the windward side of the mountain. By the time the winds pass over the mountain they have lost most of the moisture they were carrying so there is little precipitation of the other side, or leeward side, of the mountain.

Water Cycle

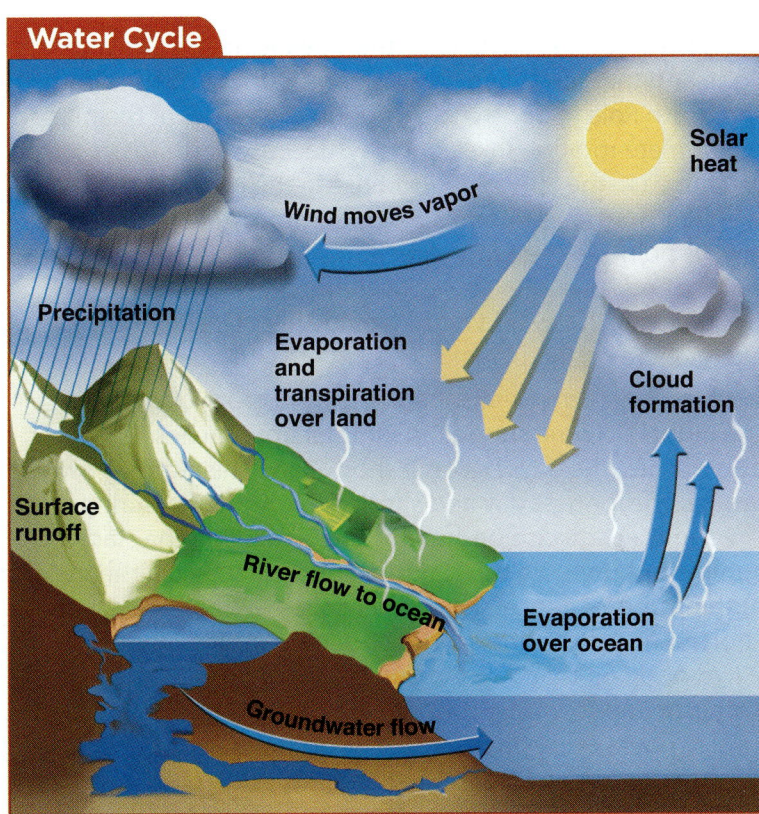

Solar heat

Wind moves vapor

Precipitation

Evaporation and transpiration over land

Cloud formation

Surface runoff

River flow to ocean

Evaporation over ocean

Groundwater flow

Seasons

The amount of seasonal change varies widely according to latitude as well. Near the equator there is very little seasonal change. At high latitudes (near the North and South poles), spring and fall are much shorter than summer and winter. In the temperate zones, between the tropics and the polar regions, there are usually four seasons nearly equal in length.

Seasons occur due to the tilt of Earth's axis relative to the Sun. As Earth orbits the Sun, its **rotation axis** (the line running between the North and South poles) is tilted 23.5 degrees from the vertical. If the axis was not tilted would have the same angle throughout the year with respect to the Sun as well as space, the intensity and duration of sunlight would never change, and there would be no seasons. When the Northern Hemisphere is tilted toward the Sun, the Sun rises higher in the sky, making its rays more direct, and the day is longer. This is spring and summer. When the Northern Hemisphere is tilted away from the Sun, it is fall and winter.

When the Northern Hemisphere is experiencing summer, the Southern Hemisphere is tilted away from the Sun and experiencing winter, and vice versa. Both the angle and direction of the tilt are consistent with respect to a point in space (such as Polaris); it is the orientation of the rotation axis relative to the Sun that changes.

Earth Science • Chapter 8
Planets, Moons, and Stars

Lesson 1 ▸ **The Sun and Earth**

As Earth rotates on its **axis**, the side facing the Sun receives light, and the side facing away is dark. Earth's axis runs from the North Pole to South Pole through the center of the planet. A complete rotation takes one day. Because Earth rotates from west to east, the Sun appears to rise in the east and set in the west. Earth's axis is tilted 23.5 degrees from vertical, and it always points in the same direction. This causes the times of sunrise and sunset to change with the seasons. If Earth were not tilted, both day and night would always be 12 hours long everywhere on the planet.

In the Northern Hemisphere, the days grow longer and the nights become shorter as summer approaches. As winter approaches, the days become shorter and the nights grow longer. The longest day and shortest night occurs during the summer **solstice** on June 21 or 22, marking the start of summer. At noon on this day, the Sun is directly overhead at the Tropic of Cancer. The winter solstice, when the day is shortest and the night is longest, occurs around December 22. At noon on that day, the Sun is directly overhead at the Tropic of Capricorn. **Equinoxes**, when day and night are each twelve hours everywhere on Earth, occur halfway between the solstices. The vernal (spring) equinox is on March 21 or 22, and the autumnal equinox is on September 22 or 23.

The amount of solar energy that reaches Earth's surface depends on the angle at which the Sun's rays reach Earth and for how long each day. The more direct the rays are and the more hours of daylight, the more energy Earth receives. If Earth's axis were not tilted, the intensity and duration of sunlight would never vary and there would be no seasonal changes. When the Northern Hemisphere is tilted toward the Sun, the Sun appears higher in our sky, making its rays more direct. Temperatures get higher, the days are longer, and it is summer. In winter, when the Northern Hemisphere is tilted away from the Sun, the temperatures are lower and the days are shorter. The North and South Poles have six months of daylight followed by six months of darkness. North of the Arctic Circle and south of the Antarctic Circle, there is at least one day a year when the Sun does not rise.

Lesson 2 ▸ **The Moon**

The Moon revolves around Earth and is held in orbit by the mutual gravitational attraction between it and Earth. More specifically, Earth and the Moon revolve around the common center of gravity of the Earth-Moon system, which is located inside Earth. The Moon revolves around Earth because Earth is larger. Because of Earth's rotation, the Moon appears to rise in the east and set in the west.

The Moon orbits Earth every 29.5 days, completing one **lunar cycle**. During each cycle, we see the different **phases** of the Moon. During the new moon, when the Moon is between Earth and the Sun, we do not see the Moon's lighted side at all. For the few days before and after the new moon, we see a crescent lighted on the right. Halfway through the cycle, when the Moon is on the opposite side of Earth from the Sun, we see the Moon's entire lighted side. This is the full moon. When the Moon is halfway between a new moon and a full moon, in either half of the cycle, we see half of its lighted side. Some call this a half-moon. The half-moon between the new moon and full moon is the first quarter, and the half-moon between the full moon and new moon is the last quarter. Between the first quarter and the full moon, and between the full moon and last quarter, we see more than half of the lighted side. The first of these is the **waxing gibbous** moon, and the second is the **waning gibbous** moon. During the first half of the Moon's cycle, we say that it is waxing, and during the second half, we say that it is waning. The same side of the Moon always faces Earth because the Moon rotates once on its axis in each revolution around Earth.

Both Earth and the Moon shine by reflecting light from the Sun, and both cast cone-shaped shadows in the direction away from the Sun. The cone-shaped area that is in total darkness is the umbra. The area surrounding the umbra, which receives partial sunlight, is the penumbra. During a total **lunar eclipse**, the Moon is completely in the umbra of Earth's shadow. During a partial lunar eclipse, the Moon is partly in the penumbra.

Earth's Path

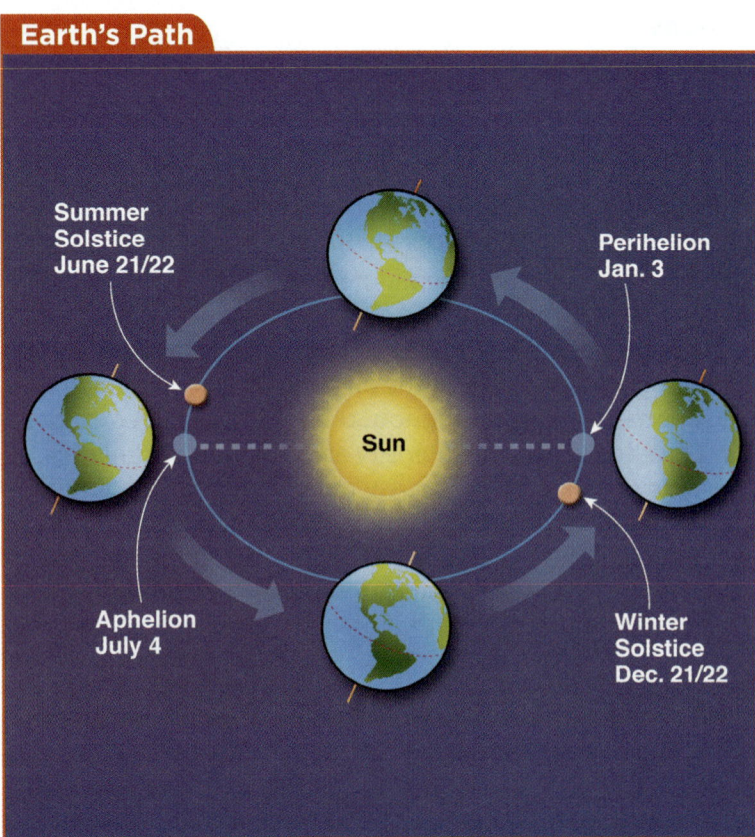

Summer Solstice June 21/22

Perihelion Jan. 3

Sun

Aphelion July 4

Winter Solstice Dec. 21/22

Lesson 3 ## Planets

The Sun is a star and it is at the center of the Solar System. Eight **planets** revolve around the Sun. The mutual gravitational attraction between the Sun and each of the planets keeps the planets in orbit. Planetary orbits are elliptical, and each planet also spins on its axis. The planets do not shine themselves; we see them because they reflect the Sun's light.

The planets are, in order closest to the Sun: Mercury, Venus, Earth, Mars, Jupiter, Saturn, Uranus, and Neptune. Until the summer of 2006, the body called Pluto was believed to be a ninth planet. But astronomers do not now consider Pluto to be a planet—both because of its small icy rock composition and because of its orbit, both of which make it resemble comets in the Kuiper Belt rather than the other planets. In August 2006 the International Astronomical Union redefined Pluto as a "dwarf planet."

Mercury, Venus, Earth, and Mars are called **terrestrial** planets. They have rocky crusts and molten interiors. Venus is about the same size as Earth. Jupiter, Saturn, Uranus, and Neptune are known as **gas giants**. These large, low-density planets, made up mostly of hydrogen, helium, and ice, have numerous moons and spin on their axes more rapidly than the terrestrial planets. Beyond the gas giants is the Kuiper Belt, a region of smaller, icy worlds, which includes Pluto.

Comets are small bodies of ice and dust that travel around the Sun. Their tails can grow up to 100 million miles long. **Asteroids** are tiny rocky bodies that mostly orbit the Sun between the orbits of Mars and Jupiter. **Meteoroids** are particles of dust or rock that enter Earth's atmosphere, appearing as streaks of light (so-called "shooting stars"). As they pass through the atmosphere, where they usually burn up, they are classified as **meteors**. Those that hit the Earth are called **meteorites**.

Lesson 4 ## Stars and Constellations

From Earth we can see thousands of stars with the unaided eye. Stars are celestial bodies that glow because of the nuclear energy produced in their core. They are described by properties such as mass, size, luminosity (brightness), and temperature. Stars exist in galaxies; the Sun is in the Milky Way galaxy. There are billions of galaxies in the Universe, each containing billions of stars.

A star begins its **life cycle** as a cool cloud of gas and dust. As it collapses under its own gravity, it grows hotter and hotter and becomes a **protostar**. If a protostar is large enough, it generates enough heat to begin fusing hydrogen into helium, producing nuclear energy. Then it is considered to be a **main sequence**, or true star, such as our Sun.

The generation of nuclear energy releases heat and light, keeping the main-sequence star from further collapse. When hydrogen begins to run out in the star's core, fusion begins in its outer part, causing its outer shell to expand. When the core no longer produces nuclear energy, it begins to collapse. It produces energy that initiates further expansion of the star's outer shell, and the star becomes a **red giant**.

It will take the Sun 5 billion years to become a red giant. Eventually, all of the nuclear fuel is spent, and gravity causes the star to compress into a **white dwarf**. Astronomers believe that the Sun will end up as a white dwarf.

Earth's Atmosphere

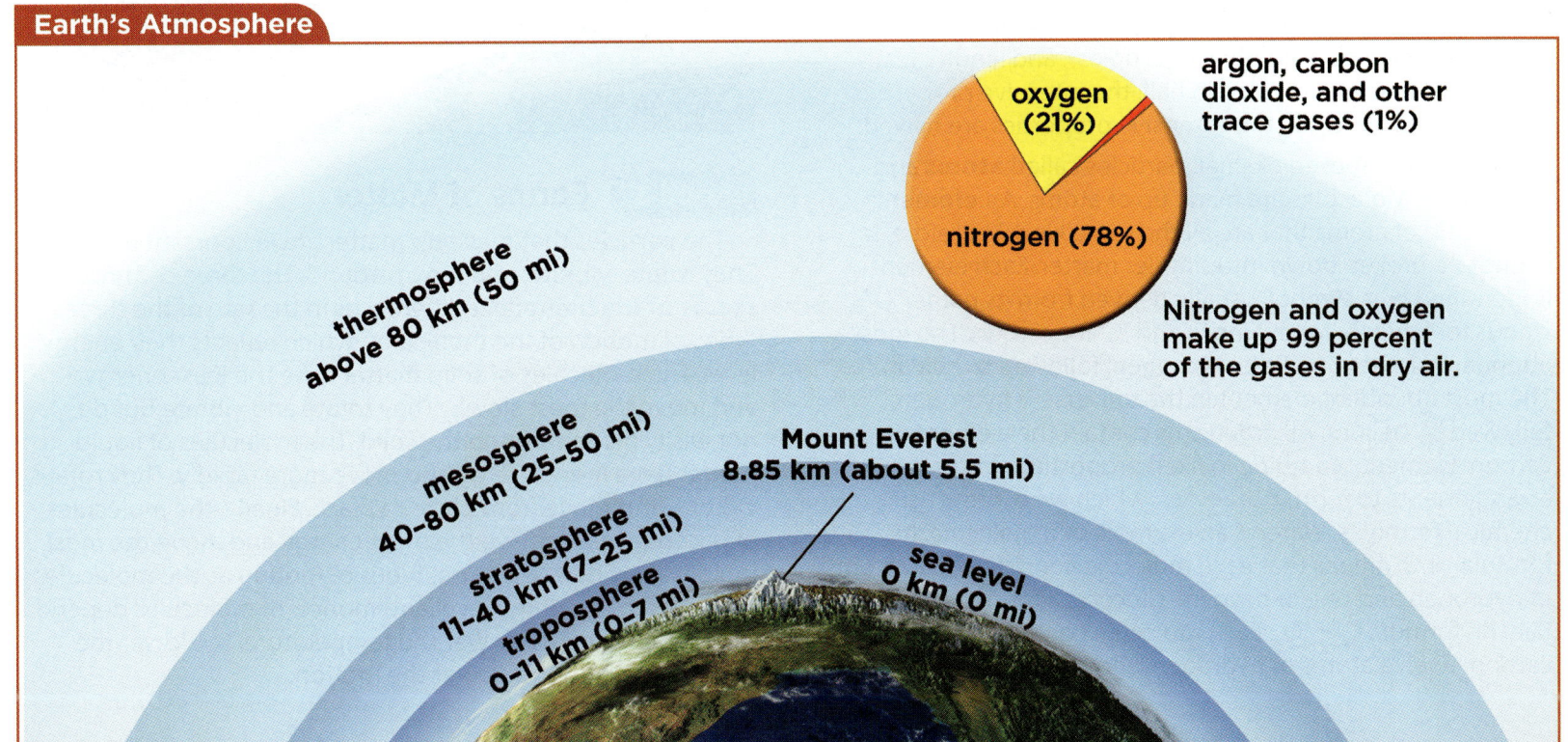

oxygen (21%)

argon, carbon dioxide, and other trace gases (1%)

nitrogen (78%)

Nitrogen and oxygen make up 99 percent of the gases in dry air.

thermosphere above 80 km (50 mi)

mesosphere 40–80 km (25–50 mi)

stratosphere 11–40 km (7–25 mi)

troposphere 0–11 km (0–7 mi)

Mount Everest 8.85 km (about 5.5 mi)

sea level 0 km (0 mi)

Physical Science • Chapter 9
Observing Matter

Lesson 1 — Properties of Matter

Matter occupies space, and it cannot share space with other matter. We describe matter in terms of its **properties**, which we perceive with our senses. These include properties that we can perceive only with senses other than vision. With our eyes, we perceive size, shape, color, and texture. With our ears, we perceive information about such properties as size, texture, and hardness, often by the sound a material makes when coming in contact with another material. A table-tennis ball and a small rubber ball, for example, make different sounds when hitting the floor. This is because the materials they are made of have different properties. Our sense of touch enables us to perceive size, shape, texture, weight, and hardness. With our senses of smell and taste we perceive properties such as bitterness, saltiness, sweetness, and sourness. The sense of smell also enables us to detect the presence of matter, including some dangerous gases, of which we would otherwise be unaware. People who have lost the use of one of their senses use other senses to determine the properties of matter. A person who is sight-impaired, for example, will use the sense of touch to learn an object's size, shape, hardness, and texture.

Physical properties of matter are **intensive** or **extensive**. Intensive properties do not depend on the mass, or amount of matter. They include boiling point, melting/freezing point, hardness, density, and conductivity. Intensive properties that are characteristic of a particular material are useful for identifying unknown matter. Extensive properties vary with the amount of matter. They include mass, weight (measure of the gravitational force exerted by Earth on an object), volume (the amount of space an object occupies), and length. If a quantity of matter is divided in half, the intensive properties remain the same, while the extensive properties are halved.

Matter is composed of small particles called **atoms** and **molecules**. Molecules are made up of atoms. An **element** is made up of atoms that are all the same kind. Therefore, it cannot be broken down into simpler matter. Each element is identified by a symbol, usually derived from its name. "O" stands for oxygen, for example, and "C" for carbon. The most abundant element on Earth is oxygen, followed by silicon. The most abundant element in the universe is hydrogen, followed by helium. All organisms contain the element carbon. Elements combine in fixed proportions and specific arrangements to form **molecules**, which we express using chemical formulas. Water is an example of a molecule. Its formula, H_2O, means that each molecule consists of two atoms of hydrogen and one of oxygen. Glucose, a form of sugar, has the formula $C_6H_{12}O_6$. Each molecule contains six atoms of carbon, twelve atoms of hydrogen, and six atoms of oxygen.

Lesson 2 — Measuring Matter

Two properties of all matter are gravitation and inertia. **Gravity** is the force of attraction that any two objects exert on each other. Because Earth is so massive and so close to us, the only gravitational pull we feel on Earth is that exerted by Earth itself. Just as Earth pulls our bodies toward its center, we pull back on Earth with the same amount of force. Because Earth is so massive, it doesn't budge. **Weight** is a measure of the force of gravity on an object. The farther from Earth you travel, the weaker is the pull of Earth's gravity. Although an object weighs less on a mountaintop than in a valley, the difference is so slight that, again, we don't notice it. **Inertia** is the resistance to change in an object's motion. More specifically, it is the tendency of a body that is not moving to resist moving and of a body that is moving to continue moving at the same speed and in the same direction. Scientifically speaking, any change in a body's state of motion, including causing the body to slow down or stop, is considered a form of acceleration. Inertia is a body's resistance to acceleration. **Mass**, the amount of "stuff" in a body, is a measure of the body's resistance to acceleration. Even in space, where a body is weightless, it will resist acceleration. Therefore, mass is the same everywhere in the universe, whereas an object's weight varies.

Inertia

Lesson 3 — Forms of Matter

The particles that make up matter are in constant motion. They rotate, vibrate, and move around. The three common **states of matter** reflect differences in the rate of the three types of motion of the molecules, which reflects their energy level. The molecules of solid matter have the least energy and move the most slowly. They rotate and vibrate but do not easily move through the solid. The molecules of liquid matter have more energy and move more rapidly. They rotate, vibrate, and move from place to place. Finally, the molecules of gaseous matter have the most energy and move the most rapidly. They have the same types of motion as the molecules of a liquid. The energy level and motion of molecules depend on temperature. The lower the temperature, the lower the energy level and the slower the motion.

A **solid** material has a definite shape and volume (how much space it takes up). External forces may be able to alter its shape, but its new shape will then be another definite shape. A good example of a solid is clay. Whether you stretch it out into a long tube or roll it into a ball, it is still solid. It still has the same mass and volume. One property of matter is **density,** which is expressed in mass per unit volume, such as grams per cubic foot. Forces among the molecules of a solid keep them very close together, in fixed positions. Because they do not have much energy, they do not move about. The closer the molecules are, the denser the material is. Another property of a solid is **hardness**. Hardness is related to the arrangement of the molecules and to how strongly they are held together. The object does not compress or stretch easily. A plate is hard and, as with other solids, its molecules do not move about. A plate might break, but the molecules in the pieces still do not move about. In a soft object, some motion is allowed between molecules. When you compress a soft material, it becomes denser and takes up less space. If you squeeze and roll a slice of soft bread, you end up with a little ball of bread. The weight of the bread is the same, but because the volume is smaller, the density is greater. The pieces of a broken plate, however, have the same weight and volume as the unbroken plate. Therefore, the density is the same as that of the intact plate. Other properties of a solid include elasticity, ductility (ability to bend), and brittleness (how easily it is broken). Some materials such as tar are considered semisolids. They are actually very viscous liquids, but they move so slowly that change is barely visible.

The **liquid** state is partway between solid and gas. Like those of a solid, the molecules of a liquid are close together, but they are not in a fixed position. They are free to move about. A liquid has no definite shape of its own, but takes on the shape of its container. Unlike a gas, however, it maintains a constant volume. In a liquid, the molecules are slightly farther apart than those of a solid. That is why matter in a solid state will sink when placed in the same matter in a liquid state. (Water is the exception. Ice cubes float in liquid because water expands slightly as it freezes. That is, water molecules are slightly farther apart in the solid state than in the liquid state.) The molecules of some liquids may be attracted to those of a solid. They tend to coat the surface of a solid, making it "wet." These liquids are said to be **adhesive**. The molecules of other liquids, however, are more attracted to each other. These liquids are said to be **cohesive**. Water is an example of a cohesive liquid, its tiny drops held together by the cohesive forces between the molecules. Detergents help to make water less cohesive and therefore "wetter." Another cohesive liquid is mercury. Unlike a soft solid, mercury cannot be compressed. A liquid's properties include color, viscosity, and transparency or opaqueness.

A **gas** has neither fixed shape nor volume. If it is enclosed, it will not only take on the shape of its container but also expand to fill the entire container. If it is not enclosed, its molecules will disperse into the surrounding space. There is a lot of space

between the molecules of a gas because the molecules are in rapid motion. Because there are relatively few molecules within a given volume of gas, all gases have low density. A gas's properties include odor. Most, but not all, gases are colorless. Chlorine gas is a greenish-yellow. The element iodine got its name from the Greek word *loeides*, which means "violet," because of its color in gas form.

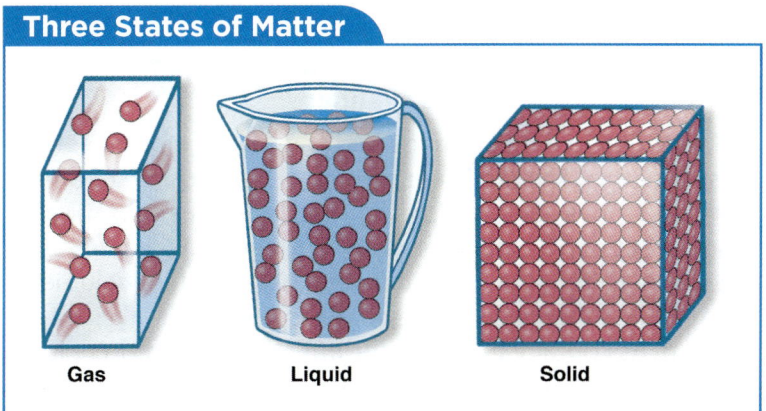

Three States of Matter

Gas Liquid Solid

Plasma is a fourth form of matter that consists of ions and freely moving electrons. Ions are atoms that have acquired an electrical charge through the gain or loss of one or more electrons (which have a negative charge). Plasmas occur naturally when ordinary matter reaches temperatures over 10,000°C (18,032°F). Although plasma is like a very hot gas, it has properties that are distinct from those of solids, liquids, and gases. For this reason, scientists consider plasma to be a fourth state of matter. The matter inside stars and some of the gases between stars, which make up more than 90 percent of the known universe, are plasma. Plasmas are structurally more complex than solids, liquids, and gases. Plasma physics is a field of active research.

Fluorescent light bulbs and neon signs contain plasma. A recent development in plasma technology is the use of plasma in television screens. Conventional televisions use a cathode ray tube, which fires a beam of electrons inside a glass tube. The electrons excite the atoms and molecules in the phosphor coating that covers the wide end of the tube (the screen), causing the phosphor atoms to light up. Different atoms of the phosphor coating react differently to the energy of the electrons, causing the various colors that produce the image. Each tiny dot of color is called a pixel (short for picture element). As the screen size increases, the depth of the television has to be increased proportionately, to give the electron gun sufficient distance to reach the entire screen. The main benefit of a plasma television screen is its flatness. Each pixel on a plasma television screen contains three plasma-containing fluorescent cells, one red, one blue, and one green. A grid of electrodes (conductors) sends current to the cells, allowing the plasma to maintain its charge. A signal causes the pixels to light up in the right combination of colors and intensities to produce the image.

Science Yellow Pages

Physical Science • Chapter 10
Changes in Matter

Lesson 1 Changing States

Matter is anything that takes up space and has mass. Its three common states are solid, liquid, and gas. The state of matter is determined by the speed at which its particles—usually molecules—move. Changes of matter from one state to another are called **phase changes**.

A **solid** has a definite shape and volume. Its molecules are close together. The molecules of a **liquid** are slightly father apart and move somewhat faster than those of a solid. (Water is an exception; it expands slightly as it freezes.) A liquid takes the shape of its container, but maintains a constant volume. A **gas** has neither fixed shape nor fixed volume. If takes the shape of its container and fills the entire container. There is a lot of space between its molecules, which are in rapid motion.

The molecules that make up matter are in constant motion. The energy and motion of molecules depend on temperature. The lower the temperature, the lower the energy and the slower the motion. The three states of matter reflect the energy of the molecules that moves them apart, and the attractive forces that draw them together. The molecules of a solid move the most slowly; they have insufficient energy to overcome the forces that hold them together. The molecules of a liquid move more rapidly than those of a solid; but do not have enough energy to fully overcome the attractive forces. The molecules of a gas move the most rapidly and have enough energy to overcome the attractive forces.

Solid to Liquid

The change in matter from a solid to liquid is called **melting**. As the temperature of a liquid rises, the molecules move faster and faster. Eventually, they have enough energy to overcome the forces that bind them together, and individual molecules begin to escape from the liquid's surface. This process of boiling is called **vaporization**. The term *gas* refers to a material that normally occurs in the gaseous state, and the term *vapor* to the gas form of a material that exists in a liquid or solid state at room temperature.

When a material evaporates, it changes from a liquid to gaseous state at a temperature that may be below its boiling point. **Evaporation** occurs because some of the molecules near a liquid's surface are moving fast enough to escape the forces that bind them to nearby molecules.

When matter changes from a liquid to solid state, we say that it **freezes**. The molecules move more slowly and they move closer together—with the exception of water. Freezing is the opposite of melting.

Condensation, which is the reverse of vaporization, is the change of a material from a gas to a liquid. As temperature decreases, the gas molecules move more and more slowly, until they can no longer resist the forces that attract them to one another. The molecules then cohere into droplets of liquid.

Sublimation is the direct change from a solid state to a gas. The most familiar example is dry ice, which is solid carbon dioxide. The smoke you see is the water fog that condenses when solid carbon dioxide sublimates.

Lesson 2 Physical Changes

Matter can undergo two kinds of changes, physical and chemical. In a **physical change**, the molecules may be rearranged, but the chemical structure (how the atoms are bonded to each other within the molecules) stays the same. Matter that changes from one state to another, as when freezing or melting, undergoes a physical change. A physical change also occurs when two or more substances combine to form a mixture. In a physical change matter changes size, shape, or state without changing identity.

One way that elements can be physically combined is to form a mixture, in which each type keeps its original chemical properties. Ocean water, air, soil, sand, and rocks are all mixtures. In the kitchen, mixtures include salads, soups, casseroles and so on—even the pots and pans are made from special mixtures called *alloys*.

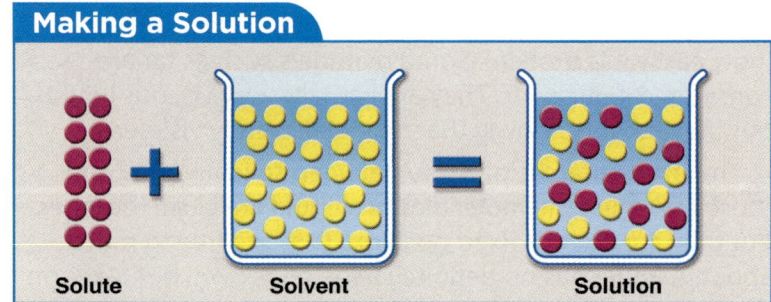

Making a Solution

Solute + Solvent = Solution

A homogeneous mixture is the same throughout. The most common type of homogeneous mixture is called a **solution**. A common example is salt water made from salt and water. The solute (usually a solid) is dissolved in the solvent (usually a liquid). The proportion of solute to solvent in a solution is called the concentration. If there is a relatively large amount of solute, we say that the solution is concentrated. If there is a small amount of solute, we say that it is dilute.

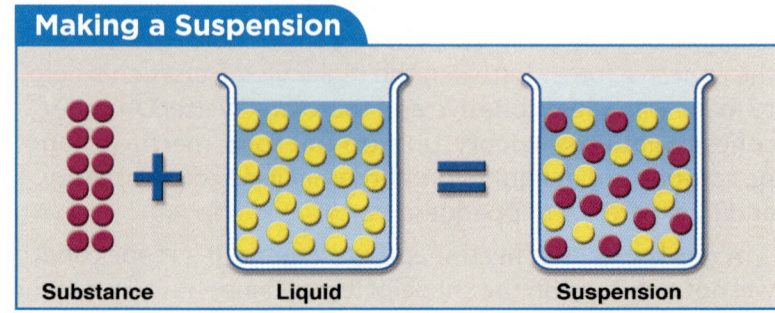

Making a Suspension

Substance + Liquid = Suspension

A heterogeneous mixture is not the same throughout. One example is gravel, which contains rocks of different types and sizes. In the type of heterogeneous mixture called a **suspension**, particles large enough to be seen are mixed into a liquid. An example is muddy water. If left standing long enough, the particles (in this case, of soil) will settle to the bottom.

Mixtures can be formed from endless combinations of solids, liquids, and gases. Sand is an example of a solid in a solid. Sand in water is a solid in a liquid. Smoke consists of solid particles in a gas. The bubbles observed in soda are a gas in liquid mixture. The important thing about all of these mixtures is that they can usually be separated into their original parts and still have their original properties. This is one thing that differentiates mixtures from compounds that are chemically combined.

Sediment Sinking

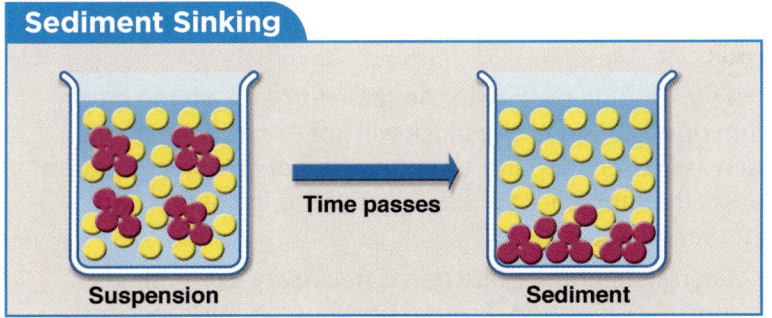

Suspension → Time passes → Sediment

There are several different ways to separate mixtures based on their physical properties. You can simply pluck out different types of matter from a mixture. For example, rocks can be picked out of sand quite easily. Filters can also be used to separate mixtures by size. If sand and rocks are poured through a screen with small holes, the sand will pass through leaving the rocks on top. Some mixtures, such as a salt and water solution, can be separated by evaporation. When the water is allowed to evaporate, the salt remains behind. Magnets can be used to separate magnetic from nonmagnetic materials.

Lesson 3 Chemical Changes

Chemical changes occur all around us in our daily lives. These changes are different from physical changes because original substances lose their identity and new substances are created. For example, if you burn a piece of paper, one product is ashes, or carbon. Other products are gases of carbon and water.

Chemical changes occur as a result of chemical reactions. The original substances are called the reactants and the new substances produced by the reaction are called *products*. During chemical reactions atoms are rearranged in different ways. Sometimes two reactants combine to form a new, more complex substance. As an example, hydrogen and oxygen combine to form water. The reverse can happen as a complex

substance breaks down into simple parts. In other chemical reactions a single uncombined element replaces another or two elements in two different compounds switch places to form new compounds.

Some chemical reactions absorb energy and some release energy. Any type of chemical change that involves combustion, or burning, releases energy. An obvious example of this is burning a campfire, lighting a match, or setting off fireworks.

One of the most important chemical reactions that occur on Earth is photosynthesis. This is the process in which green plants use carbon dioxide and water in the presence of sunlight to produce glucose and oxygen. All of the energy that flows through Earth's ecosystems comes from the process of photosynthesis.

The kitchen is a great place to observe chemical reactions in action. By applying heat energy, a raw egg can be turned into a cooked egg. You know this is a chemical reaction because there is a change in the properties of the egg. It tastes different, smells different, and it cannot be changed back into a raw egg. How can you tell that baking a cake is a chemical reaction? It also tastes and smells different; however, the presence of bubbles in the cake indicates the production of gas. When you light a match the presence of light tells you that a chemical reaction has occurred.

Chemical changes constantly occur in nature. Some weathering of rocks is chemical. When carbon dioxide dissolves in water it forms carbonic acid, which can slowly break down certain rocks such as limestone. Dissolving limestone is how many caverns are formed. Acid rain is the product of chemical reactions between water vapor and certain pollutants that contain sulfur and nitrogen compounds.

Bryce Canyon Limestone Formations

Physical Science • Chapter 11
Forces and Motion

Lesson 1 **Position and Motion**

An object's position is described by comparing it to the position of one or more other objects. Even the vague statement that an object is far away is a qualitative comparison of the object's position relative to a person or object. Many qualitative descriptions of position rely on words such as *above, below, behind, over, between,* and *beyond.*

Another way to express the relationship between two objects is by **direction**. The four **compass points**—north, south, east, and west—can locate objects on a two-dimensional surface. Left and right are other relative directions, but because they depend on the orientation of the speaker they can be misleading. To a person standing in front of a house, a tree may be on the right; to someone behind the house, the tree is on the left.

Directions

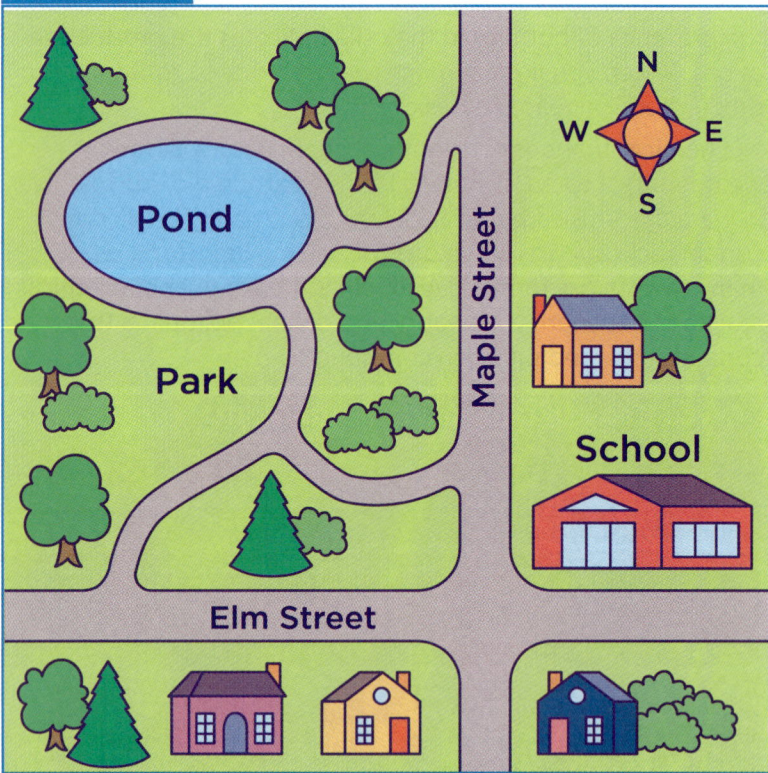

The relationship of objects is also described in terms of **distance**, which is how far away one object is from another. It is expressed in units of measurement such as feet, yards, or metric units. The most complete way to indicate position is through combinations of direction, relationship words, and distance. For example, *"She hung the picture three feet above the couch, six inches to the right of the mirror. The library is two miles east of City Hall."*

Motion

When an object is changing its position it is said to be in motion, or moving. **Speed** is given by the distance something moves divided by the time needed to make the move. Common measurements of speed are miles per hour, kilometers per hour, feet per second, and meters per second. Speed does not express where something is moving, only how fast. **Velocity** is the speed of an object in a specified direction. If a car's velocity is 50 miles per hour northeast, its speed would be 50 miles per hour.

Lesson 2 **Forces**

Objects resist changes to their state of motion. This resistance is **inertia**, which means that an object at rest tends to stay at rest and an object in motion tends to stay in motion at the same speed in the same direction. To set an object in motion or to change the direction of an object already in motion, it is necessary to apply a force, which can be a push or a pull.

If two people push with the same force on a large block from opposite sides, the block will not move. Some forces, such as friction, depend on the motion of an object. Friction resists the movement of an object that is in contact with another object.

A certain amount of friction is necessary to enable us to control our motion. Friction between people's feet and the ground enables them to walk. A person who steps on something slippery loses friction and may fall. Friction is the reason objects in motion, such as a ball rolling on the ground, eventually slow down. Oil is placed on moving parts of machines to reduce the friction between them.

Forces occur in pairs, an action and an equal and opposite reaction, often called the action force and the reaction force. When a person pushes on a ball to throw it, the ball pushes back on the thrower with an equal force. The person isn't pushed back because of friction with the ground.

Lesson 3 **Work and Energy**

Energy is the capacity to do **work**, which is done whenever a force moves matter. It is possible to exert force without doing any work. **Power** is the rate of doing work, or of using energy. Whether you climb stairs slowly or run up the stairs, you make the same change in your energy. But if you do it quickly, you need more power.

Energy can be classified as potential or kinetic. **Potential energy** is the capacity to do work due to position or condition. A child at the top of a slide has potential energy because as she comes down the slide her kinetic energy will increase. This form of potential energy is **gravitational potential energy**. A compressed spring has **elastic potential energy** stored in it because of its squeezed or stretched condition. Chemical, electrical, and nuclear energy are all kinds of potential energy.

Transfer of Kinetic Energy

Kinetic energy is the energy an object has due to its motion. The faster it moves, or the more massive it is, the more kinetic energy an object has. As a child goes down a slide, her potential energy changes to kinetic energy. When a pendulum is at its highest point, all of its energy is potential. As it swings downward, its potential energy changes to kinetic energy. As it swings up the opposite side, the kinetic energy changes back to potential energy. At the highest point, all of its energy is again potential energy.

The law of conservation of energy states that energy cannot be created or destroyed. It can, however, be converted from one form to another or transferred from one object to another.

Lesson 4 Using Simple Machines

Machines are devices that help people do tasks. They can increase the amount of force exerted, increase the speed of an action, or change the direction of a force. Common tools like the hammer are **simple machines**. There are six basic types of simple machines: lever, inclined plane, wedge, screw, wheel and axle, and pulley. More complex machines are made from combinations of these six basic ones.

Types of Simple Machines

When a simple machine multiplies the amount of force used, the force must be applied over a greater distance. That is, the work done by the machine cannot be greater than the work done on the machine. A **lever** has a fulcrum (or pivot), a load, and a place where force is applied. A seesaw is a lever. The farther the child near the ground is from the fulcrum, the

easier it is to raise the other child in the air. A seesaw is a first-class lever. This means the fulcrum is between the two forces, the load is the weight of the child in the air, and the effort force is the weight of the child at the bottom. A wheelbarrow is a second-class lever, with the fulcrum at the wheel and the force applied at the end of the handles. The load is the weight of the material being lifted.

An **inclined plane** moves an object to a higher or lower position without lifting. A wheelchair ramp is an inclined plane. The longer the ramp, the less force is required to move up it. A **wedge** is made of two inclined planes and changes the direction of the force. An axe is a wedge.

A **screw** is a like a nail with an inclined plane going around it. It takes less force to insert a screw than a nail because the force travels over the greater distance of the thread as the screw turns.

The **wheel and axle** of a simple machine move together. When a wheel makes one complete turn, it covers a greater distance than the axle to which it is connected. Therefore, the force placed on the wheel is multiplied on the axle. An example of this is a doorknob. When the knob is turned, it turns the shaft that opens the door. If the knob falls off, it requires more force to turn the shaft with your fingers.

A simple **pulley** consists of a wheel and a rope that goes around it. A fixed pulley changes the direction of the force. The person applying the force on one side of the pulley pulls down on the rope, and the force on the rope on the other side raises the load. Moveable pulleys increase the amount of applied force but require that the distance also be increased. An example is a block and tackle used to hoist a piano to an upper-story window.

Levers

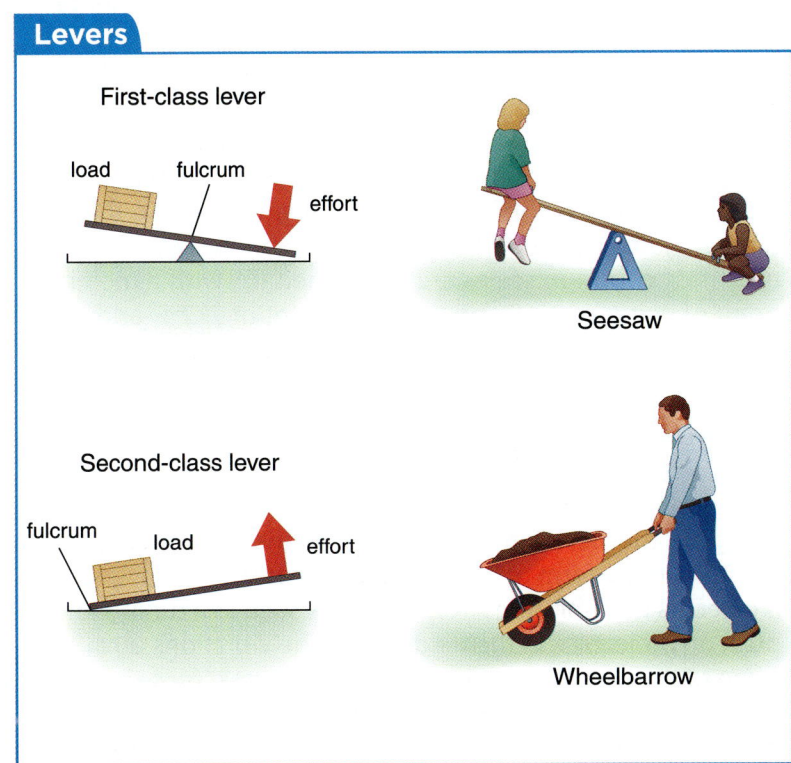

First-class lever

load   fulcrum   effort

Seesaw

Second-class lever

fulcrum   load   effort

Wheelbarrow

Science Yellow Pages

Physical Science • Chapter 12
Forms of Energy

Lesson 1 Heat

As defined by physicists, **energy** is the ability to do work. When you push a book across a table, you expend energy. You move the book over a distance in the direction in which you pushed it. Most often, when energy is used, part of the energy becomes heat. In the previous example, both the book and the table would warm up slightly.

The molecules that make up matter are in constant motion, rotating, vibrating, and colliding with one another with an intensity that depends on the **temperature** of the matter. The lower the temperature is, the less intense the motion. The three common states of matter—solid, liquid, and gas—reflect these differences in temperature. The molecules of a solid, which are closest together, vibrate essentially in place. The molecules of a liquid (except for water) are not as close together; they vibrate more rapidly, have greater energy, and are able to move about. Finally, the molecules of a gas move the most rapidly, the most widely, and with the greatest energy. The vibration and motion of molecules within a substance result in a form of energy that can be transferred between objects.

Heat, or **thermal energy**, is the flow of energy from one substance to another. The energy always flows from a material of higher temperature to a material of lower temperature until, if enough time is allowed, the two will reach the same temperature and no more heat will flow. **Equilibrium** will have been reached.

Temperature is a measure of the average kinetic energy of a material's molecules. It can be measured with a **thermometer**. Most common thermometers are filled with a liquid, such as alcohol or mercury, that quickly expands when heated and contracts when cooled. The **scale** is determined by choosing two reference points, usually the melting of ice and the boiling of water, and dividing the temperatures between them into units called **degrees**. The two most common scales are Fahrenheit (F), with a melting point of water of 32°F and a boiling point of 212°F, and Celsius (C), with a melting point of water of 0°C and a boiling point of 100°C.

Lesson 2 Sound

Sound is a type of energy. When an object vibrates, the energy of the vibrations moves through the surrounding medium in the form of longitudinal **waves**. These waves are created as the vibrations from a source, such as a tuning fork, push on molecules in the medium. As the vibrations push forward they cause the molecules to compress. These regions are called **compressions**. As the vibrations pull back, the pressure is released, causing a region in which the molecules

are spread apart. These regions are called **rarefactions**. A series of compressions and rarefactions create a sound wave. Sound waves can travel through gases, liquids, or solids, such as the Earth. They travel at different speeds in different mediums, slowest through gases, such as air, and fastest through solids.

If the waves reach the ear of the listener, they cause the eardrum, a membrane, to vibrate. The three bones in the ear pass the vibrations to the inner ear. A nerve then sends an impulse to the brain, and the sound is perceived.

The faster the vibration the greater the frequency will be. Frequency is measured in Hertz (Hz), which means "cycles per second." The ear and brain perceive frequency as **pitch.** When a string on a stringed instrument is tightened, it raises the pitch. Thicker strings have a lower pitch, as do longer ones. For example, a cello has a lower pitch than a violin because of the string thickness and length.

Loudness is the way the ear and brain perceive the size, or the distance back and forth, of the vibration. The greater the size of the vibration, the greater the pressure of waves on the eardrum. Loudness is measured in decibels. As sound waves travel, they spread and get smaller and smaller. They can also be absorbed or reflected, just as light and water waves. The closer the source of a sound is, the louder it usually is. Like the ear, a microphone has a membrane that vibrates. The microphone vibrations are changed into an electric signal, which is sent to a processor that increases its size.

Hearing Ranges of Animals in Decibels		
Animals	**Lower Limit**	**Upper Limit**
Elephant	17	10,000
Human	20	20,000
Cat	45	64,000
Horse	55	33,500
Dog	60	45,000
Porpoise	75	150,000
Hedgehog	250	45,000
Rabbit	360	42,000
Mouse	1,000	91,000
Bat	2,000	110,000

Lesson 3 Light

Light is a form of energy made up of waves. Light travels in a straight path until it hits a barrier, when it can be reflected, absorbed, or transmitted. If this barrier is a mirror, the light bounces off, a process called **reflection**.

Light rays reflect from a mirror at the same angle at which they hit it. Since all of the waves coming from an object reflect off a mirror in this way, to the viewer they appear to come from a location behind the mirror. This "virtual" image appears to be the same distance behind the mirror as the object is in front of the mirror. For example, the light rays from a person standing two feet in front of the mirror appear to come from two feet behind it. The image in a plane mirror looks identical to the object viewed, except that right and left appear to be reversed.

When a person looks at an object, some of the light that struck the object is reflected toward their eyes. This reflected light enters the cornea, passes through the pupil, and then through the lens. The cornea and lens focus the light rays, causing them to bend toward each other and converge on the **retina**, a membrane at the back of the eye.

The retina contains two types of light-sensitive cells, **rods** and **cones**. The light that comes through the lens stimulates nerves in the rods and cones, which then transmit the nerve impulse to the brain. The rods, which function in low illumination, respond to light of all wavelengths, and so are not sensitive to color. There are three kinds of cones, each responsive to a range of wavelengths, and so allow us to perceive color.

Lesson 4 Electricity

A flow of electrical charge is called a **current**. An **electrical circuit** is an uninterrupted loop of conducting material that provides a way of transferring energy from a source to a user. A basic circuit consists of four parts: a **source** of electrical energy, conductors, a **load**, and a **switch**. The load changes the electrical energy into another form of energy such as heat, motion, or light. The switch interrupts the current to control the device.

There are two basic types of circuits: series and parallel. A **series circuit** is wired so that the current must go through each component one after another. A **parallel circuit** consists of a group of connected components. The current has a separate path through each component and the current is divided among the components.

All electrical circuits require electric energy to maintain the potential difference that drives the current. Most of the electric energy we use is produced by power plants, where a generator converts mechanical energy into electric energy. Most power plants run on steam. A machine called a **turbine** has blades mounted on a shaft. When steam is forced against the blades the shaft turns. Inside the generator a magnet attached to the shaft rotates inside a set of coils of wire. As the field from the rotating magnet changes, a current is induced in the wire. To produce the steam, the power plants either burn fossil fuels such as coal, oil, and natural gas, or use nuclear fission.

Once generated, the electricity must be transported to consumers. From the power plants it travels along cables to a transformer, which changes the current from low to high voltage. The higher voltage allows the energy to be transmitted over long distances with less loss. The high-voltage electricity travels along transmission lines to a substation. At the substation, transformers change the high-voltage electricity back to low-voltage electricity. Finally distribution lines take the electricity to the homes and businesses where it is used.

Path of Electricity from Power Plant to Residence

Electricity generated at power plant

Transmission line carries electricity long distances

Transformer steps up voltage for transmission

Local transformer steps down voltage

Distribution line carries electricity to house

Transformer steps down voltage before electricity enters house

Scope and Sequence

Inquiry, Technology, History and Nature of Science

Inquiry

Inquiry Skills	K	1	2	3	4	5	6
Observe	I	D	D	D	D	D	D
Infer	I	D	D	D	D	D	D
Predict	I	D	D	D	D	D	D
Communicate	I	D	D	D	D	D	D
Measure	I	D	D	D	D	D	D
Put things in order	I	D	D				
Compare	I	D	D				
Classify	I	D	D	D	D	D	D
Investigate				D	D	D	D
Make models		D	D	D	D	D	D
Draw conclusions	I	D	D				
Use numbers				D	D	D	D
Interpret data				D	D	D	D
Form a hypothesis				D	D	D	D
Experiment				D	D	D	D

Special Abilities for Inquiry	K	1	2	3	4	5	6
Critical thinking	I	I	I	D	D	D	D
Use of mathematics	I	I	I	D	D	D	D
Use of technology	I	I	I	D	D	D	D

Understandings About Inquiry	K	1	2	3	4	5	6
Different questions suggest different kinds of investigations.	I	I	I	D	D	D	D
Current scientific knowledge guides investigations.				D	D	D	D
Investigations may lead to new ideas.				D	D	D	D
Explanations are based on evidence.	I	I	I	D	D	D	D

Science and Technology

Abilities of Technological Design	K	1	2	3	4	5	6
Identify appropriate problems.	I	I	D	D	D	D	D
Design a solution.	I	I	D	D	D	D	D
Implement the solution.	I	I	D	D	D	D	D
Evaluate/communicate the solution.	I	I	D	D	D	D	D

KEY		
I = Introduced	D = Developed	R = Reinforced

Science and Technology *cont.*

Understandings about Science and Technology	K	1	2	3	4	5	6
Contributions made by people in different cultures	I	I	I	D	D	D	D
Tradeoffs: benefits and risks, consequences	I	I	I	D	D	D	D
Reciprocity of science and technology			I	D	D	D	D
Distinguishing natural from human-made objects	I	D	D	R			

Personal and Social Perspectives

Personal Health	K	1	2	3	4	5	6
Practice of healthful behaviors	I	I	I	D	D	D	D
Injury preventions	I	I	I	D	D	D	D

Populations, Resources, Environments	K	1	2	3	4	5	6
Effect of population density on resources and environment	I	I	I	D	D	D	D
Limited supply of many resources	I	I	D	D	D	D	D
Natural and human causes of environmental change	I	I	D	D	D	D	R
Rapid and gradual environmental change	I	I	D	D	R	R	R

Natural Hazards	K	1	2	3	4	5	6
Resulting from Earth processes	I	I	D	D	D	D	D
Induced by human activities			I	D	D	D	D
Challenges to society			I	D	D	D	D
Risks of natural hazards	I	I	D	D	D	D	D

Science, Technology, and Society	K	1	2	3	4	5	6
Effects of science and technology on society	I	I	I	D	D	D	D
Effect of society on science investigation					I	D	D
Many settings for carrying out science and technology					I	D	D
Inability of science to answer all questions					I	I	D

History and Nature of Science

Science as Human Endeavor	K	1	2	3	4	5	6
Carried out by men and women of various backgrounds		I	I	D	D	D	D
The need for different abilities		I	I	I	D	D	D
Varied careers in science	I	I	I	D	D	D	D

Nature of Science	K	1	2	3	4	5	6
Explanations based on observation and experiments	I	I	D	D	D	D	D
Conflicting scientific interpretations					I	D	D
Evaluation of scientific ideas					I	D	D	D

History of Science	K	1	2	3	4	5	6
The change of accepted ideas because of scientific explanations						I	D	D
Contribution of individuals throughout history				I	D	D	D

Scope and Sequence

Life Science

Plants

Needs	K	1	2	3	4	5	6
Identified	I	I	I	I	D	D	R
Related to habitat	I	D	D	D	R	R	R
Related to structures	I	I	I	I	D	D	R

Photosynthesis	K	1	2	3	4	5	6
Raw materials/products		I	I	I	D	D	R
Chemical formula					I	D	D
Producers							

Structures and Functions	K	1	2	3	4	5	6
Roots, stems, leaves	I	I	I	I	D	D	R
Flowering plants	I	I	I	D	D	D	R
▶ Parts of a flower	I	I	I	I	D	D	R
▶ Monocots/dicots					I	D	D
Nonflowering plants				I	D	D	R
▶ Mosses (parts)					D	D	R
▶ Ferns (parts)				I	D	D	R
▶ Gymnosperms (parts)				I	D	D	R

Reproduction (Life Cycles)	K	1	2	3	4	5	6
Flowering plants	I	I	I	I	D	D	R
Gymnosperms				I	I	D	R
Mosses/ferns				I	I	D	R
Bulbs/cuttings/tubers (asexual means)		I		I	D	R	

Regulation and Responses	K	1	2	3	4	5	6
Tropisms				I	I	D	D
Other responses to environment	I	I	I	D	D	R	R

Animals

Needs	K	1	2	3	4	5	6
Identified	I	I	I	I	D	D	R
Related to habitat	I	D	D	D	R	R	R
Related to structures	I	I	I	I	D	D	R

Structures and Functions	K	1	2	3	4	5	6
External characteristics	I	I	I	I	D	D	R
Organ systems					D	D	D

KEY		
I = Introduced	**D** = Developed	**R** = Reinforced

Animals _cont._

Life Cycles	K	1	2	3	4	5	6
Basic examples (e.g., birds, mammals)	I	I	I	I	D	D	R
Complete metamorphosis (insect, amphibian)	I	I	I	D	D	D	R
Incomplete metamorphosis			I	I	D	D	R
Asexual reproduction (regeneration)					I	D	R

Regulation and Behavior	K	1	2	3	4	5	6
Response to environment	I	I	D	D	D	R	R
Internal stimuli				I	D	R	R

Six Kingdom Classification

Binomial Classification	K	1	2	3	4	5	6
6 kingdoms defined					I	D	R
Kingdom to species delineations					I	D	R
Species names					I	D	R

Plant Kingdom	K	1	2	3	4	5	6
Divisions identified					I	D	R
Nonvascular vs. vascular					D	D	R
Seedless vs. seed				I	D	D	R
▶ Nonflowering vs. flowering				I	D	D	R
▶ Monocots/dicots					I	D	R

Animal Kingdom	K	1	2	3	4	5	6
Basic groupings (mammals, insects, reptiles, birds)	I	I	I	I	D	D	R
Invertebrates				I	D	D	R
▶ Basic phyla				I	D	D	R
Vertebrates				I	D	D	R
▶ Classes				I	D	D	R

Microorganisms	K	1	2	3	4	5	6
Bacteria				I	I	D	D
▶ Two kingdoms					I	D	D
Protist Kingdom				I	I	D	D
Fungus Kingdom				I	I	D	D
Reproduction					I	D	D
▶ Asexual vs. sexual						D	D

Scope and Sequence

Cells (Cell Theory of Life)

Cell Structure	K	1	2	3	4	5	6
Cell as unit of life				I	I	D	D
Plant cells vs. animal cells					I	D	R
Multicellular vs. single-celled organism					D	D	R
Levels of organization (cell-tissue-organ-system)					D	D	R

Cell Activities	K	1	2	3	4	5	6
Photosynthesis/respiration		I	I	D	D	D	D
Cell division						D	D
▶ Mitosis/meiosis						I	D
Diffusion/osmosis							D

Heredity and Genetics

Concepts of Heredity	K	1	2	3	4	5	6
Inherited traits	I	I	I	D	D	D	D
▶ In animals	I	I	I	D	D	D	D
▶ In plants	I	I	I	D	D	D	D
Learned behavior/acquired traits	I	I	I	D	D	D	D
Variation	I	I	I	I	D	D	D
Effect of environment	I	I	I	D	D	D	D
Natural selection							D

Mechanisms of Heredity	K	1	2	3	4	5	6
Mendel's patterns						I	D
Chromosomes						I	D
Genes						I	D
DNA							D

Genetics	K	1	2	3	4	5	6
Pedigree						I	D
Sex-linked traits							D
Genetic engineering						I	D

KEY		
I = Introduced	**D** = Developed	**R** = Reinforced

Ecology/Ecosystems

Organization of Ecosystems	K	1	2	3	4	5	6
Habitats	I	I	I	D	D	R	R
Populations and communities			I	D	D	R	R
Niches				I	D	D	R
Land vs. water	I	I	D	D	D	D	R
Biomes					I	D	D
▶ Latitude					I	D	D
▶ Comparisons					I	D	D
▶ Adaptations					D	D	R

Dynamics of Ecosystems	K	1	2	3	4	5	6
Energy transfer		I	I	D	D	D	D
▶ Food chains		I	I	D	D	D	D
▶ Food webs			I	I	D	D	D
▶ Energy pyramids					I	D	D
Cycles of matter				I	I	D	R
▶ Carbon				I	I	D	R
▶ Nitrogen						D	R
Competition				I	D	D	D
Symbiotic relationships					I	D	D

Adaptation and Survival	K	1	2	3	4	5	6
Adaptations	I	I	I	D	D	D	D
▶ Animal behaviors	I	I	I	D	D	D	D
▶ Animal structures	I	I	I	D	D	D	D
▶ Plant responses				I	D	D	R
▶ Plant structures		I	I	D	D	D	R
Changes in ecosystems			I	D	D	D	D
▶ Destructive change			I	D	D	D	D
▶ Effect on populations			I	D	D	D	D
▶ Regrowth/succession						D	D
Extinction			I	D	D	D	D
▶ Endangered species			I	D	D	D	D
▶ Extinct species			I	D	D	D	D
▶ Comparison of present and past species			I	D	R	R	D

Scope and Sequence

Earth Science

Earth Processes and Materials

Earth Structure	K	1	2	3	4	5	6
Landforms	I	I	I	D	D	R	R
▶ Comparisons	I	I	I	D	D	R	R
▶ Classified by Earth processes				I	I	D	D
Features of ocean bottom				I	I	D	D
▶ Spreading zone				I	I	D	D
▶ Subduction						I	D
Crust, mantle, core			I	I	D	D	R
▶ Lithosphere					I	I	D
Hydrosphere	I	I	I	D	D	D	D
▶ Sources of fresh water	I	I	I	D	D	D	D
Glaciers				I	D	D	D

Processes of Change	K	1	2	3	4	5	6
Weathering		I	I	D	D	D	D
▶ Causes		I	I	D	D	D	D
▶ Chemical vs. physical					I	D	D
▶ Soil as product			I	I	I	D	D
Erosion/deposition		I	I	D	D	D	D
▶ Basic agents		I	I	D	D	D	D
▶ River systems					I	D	D
▶ Glacier systems				I	D	R	R
Plate tectonics					I	D	D
▶ Evidence						D	D
▶ Plate model						D	D
▶ Earthquakes				I	I	D	D
▶ Volcanoes				I	I	D	D
▶ Mountain building					I	D	D

Earth History	K	1	2	3	4	5	6
Fossil evidence			I	D	D	D	D
Relative dating					I	D	D
Absolute dating						I	D
Eras of time						I	D
Change of environments over time			I	D	D	D	D

Minerals	K	1	2	3	4	5	6
Examples			I	I	I	D	D
Comparison/observed properties			I	I	I	D	D
Identification via key					I	D	D

KEY		
I = Introduced	D = Developed	R = Reinforced

Earth Processes and Materials *cont.*

Rocks	K	1	2	3	4	5	6
Examples	I	I	I	D	D	R	R
Comparison/observed properties	I	I	I	D	D	R	R
Classification (igneous, sedimentary, metamorphic)				I	D	D	D
Rock cycle					I	D	D

Soil	K	1	2	3	4	5	6
Formation			I	I	D	D	D
Components	I	I	I	D	D	R	R
Horizons				I	D	D	R
Properties	I	I	I	D	D	R	R
▶ Water capacity				I	D	D	R
▶ Permeability					I	D	R
Pollution/conservation				I	D	D	D

Renewable and Nonrenewable	K	1	2	3	4	5	6
Air		I	I	D	D	D	D
▶ Pollution/conservation		I	I	D	D	D	D
▶ Natural recycling				I	D	D	R
Water	I	I	I	D	D	R	R
▶ Pollution/conservation	I	I	I	D	D	D	D
▶ Water cycle		I	I	D	D	R	R
Fossil fuels				I	D	D	R
▶ Formation				I	D	D	D
▶ Alternative energy sources				I	D	D	D

Weather and Climate

The Water Cycle	K	1	2	3	4	5	6
Processes		I	I	D	D	R	R
▶ Evaporation/condensation/precipitation		I	I	D	D	R	R
▶ Runoff/infiltration/transpiration				I	D	D	R
Clouds	I	I	D	D	D	D	R
▶ Formation	I	I	D	D	D	D	R
▶ Classification	I	I	D	D	D	D	R
▶ Shapes	I	I	D	D	R	R	R
▶ Cloud cover (sky conditions)				I	D	D	R	R
Precipitation	I	I	I	I	D	D	D
▶ Forms	I	I	I	I	D	D	D
▶ Formation				I	D	D	D
▶ Relation to clouds			I	I	D	D	R

Scope and Sequence

Weather Variables & Measurement	K	1	2	3	4	5	6
Air temperature	I	I	I	D	D	R	R
▶ Conceptual model			I	I	D	D	R
▶ Thermometer—Celsius/Fahrenheit	I	I	D	D	D	R	R
Air pressure				I	D	D	D
▶ Conceptual model					I	D	D
▶ Barometer				I	D	D	R
Wind	I	I	I	D	D	R	R
▶ Relation to air pressure				I	I	D	D
▶ Direction—vanes	I	I	D	D	D	R	R
▶ Speed—anemometer		I	I	D	D	R	R
▶ Global patterns					I	D	D
▶ Coriolis effect						D	R

Weather Prediction	K	1	2	3	4	5	6
Air masses/fronts					I	D	D
Station models					I	D	R
Weather maps				I	D	D	D
Barometer readings, predictions					I	D	R
Cloud types, trends			I		D	R	R
Severe weather	I	I	D	D	D	D	R
▶ Thunderstorms	I	I	D	D	D	D	D
▶ Hurricanes	I	I	D	D	D	D	D
▶ Tornadoes			I	I	D	D	D
▶ Hailstorms			I	I	D	D	D

Climate							
Components of					I	D	D	D
Comparison/classification					I	D	D	D
Seasons	I	D	D	R	R	R	R
Factors of					I	D	D	D
▶ Latitude					I	D	D	D
▶ Altitude					I	D	D	D
▶ Coastal/inland					I	D	D	D
▶ Mountain ranges					I	D	D	D
▶ Ocean currents					I	D	R
▶ Global winds					I	D	R

KEY		
I = Introduced	**D** = Developed	**R** = Reinforced

Astronomy

Earth-Sun Moon System	K	1	2	3	4	5	6
Day and night	I	I	D	D	D	R	R
▶ Daily observations	I	I	D	D	D	R	R
▶ Rotational model		I	I	D	D	D	R
Year/seasons	I	I	D	D	D	R	R
▶ Orbital model		I	I	D	D	R	R
▶ Solar radiation variations				I	D	R	R
Earth's Moon	I	I	I	D	D	D	R
▶ Basic observations/phases	I	I	I	D	D	D	R
▶ Earth-Sun-Moon model			I	I	D	D	R
▶ Eclipses					I	D	D
▶ Tides						D	D

Solar System	K	1	2	3	4	5	6
Planets		I	I	I	D	D	D
▶ Inner vs. outer					I	D	D
▶ Asteroids					I	D	D
Effects of gravity/inertia					D	D	
▶ Shape of orbital path					I	D	R
▶ Speed of planets						D	R
Comets					I	D	R
▶ Explanation					I	D	R
▶ Historical sightings						D	R
Meteors/meteorites					I	D	D
▶ Explanation					I	D	D
▶ Impact on Earth					I	D	D
Formation of solar system						D	D

Stars/Universe	K	1	2	3	4	5	6
Star properties/comparisons/constellations	I		I	D	D	D	D
▶ Life cycle						D	D
▶ H-R diagram						I	D
Galaxies						I	D
▶ Examples/classification						I	D
▶ Motion/red- vs. blueshift							D
Universe						D	D
▶ Structure						I	D
▶ Change						I	D
▶ Formation						I	D

Physical Science

Matter

Properties of Objects/Materials	K	1	2	3	4	5	6
Observations	I	I	D	D	D	R	R
Classification (physical vs. chemical)					I	D	D
Measurements	I	I	I	D	D	R	R
▶ Length (ruler)	I	I	I	D	D	R	R
▶ Mass (balance)	I	I	I	D	D	R	R
▶ Volume/capacity (containers)	I	I	I	D	D	R	R
▶ Temperature (thermometer)	I	I	I	D	D	R	R

Properties of Matter	K	1	2	3	4	5	6
Density					I	D	R
Conductivity				I	D	D	R
Magnetism	I	D	D	R	R	R	R
▶ Poles	I	D	D	R	R	R	R
▶ Fields			I	I	D	D	R
▶ Interactions	I	D	D	R	R	R	R

States of Matter	K	1	2	3	4	5	6
Descriptions/properties		I	I	D	D	R	R
Particulate models					I	D	D
Melting/freezing point				I	I	D	R
Boiling/condensing point					I	D	R

Classification of Matter	K	1	2	3	4	5	6
Elements				I	I	D	D
▶ Metals				I	I	D	D
▶ Nonmetals					I	D	D
▶ Metalloids						I	D
▶ Periodic table					I	D	R
Mixtures		I	I	I	D	D	R
▶ Solid/liquid/gas combinations		I	I	I	D	D	R
▶ Suspensions			I	I	D	D	R
▶ Solutions					I	D	R
Compounds				I	D	D	R
▶ Compared with mixtures				I	D	D	R
▶ Names for						D	R
▶ Carbon compounds compared							D

KEY		
I = Introduced	**D** = Developed	**R** = Reinforced

Matter *cont.*

Changes of Matter	K	1	2	3	4	5	6
Physical changes	I	I	D	D	D	R	R
▶ Tearing/cutting/molding	I	I	D	D	D	R	R
▶ Change of state	I	I	D	D	D	R	R
▶ Particle/heat model					I	D	R
▶ Temperature/time graph						D	R
▶ Formation of mixtures		I	I	I	D	D	R
▶ Separation of mixtures			I	I	D	D	R
▶ Solubility (means of increasing)						D	D
Chemical Changes			I	I	D	D	D
▶ Chemical reactions					I	D	D
▶ Signs of chemical change					I	D	D
▶ Formation of compounds				I	D	D	D
▶ Acid/base/salts					I	D	D
Nuclear changes							D
▶ Radioactivity							D
▶ Fission vs. fusion							D

Atomic Model of Matter	K	1	2	3	4	5	6
Atoms					I	D	D
▶ Parts of atoms						D	D
▶ Models of atoms of elements						D	D
▶ Model through history						D	D
Molecules						D	D
▶ Bonding within						I	D
▶ Models of compounds						D	D
▶ Models of diatomic elements						D	D

Scope and Sequence

Position, Motion, Forces

Position	K	1	2	3	4	5	6
Position words (*above, below*, etc.)	I	D	D	R			
Relative position, fixed/moving				I	D	D	R

Motion	K	1	2	3	4	5	6
Distance			I	D	R	R	R
Speed	I	I	I	I	D	D	R
▶ Comparing fast and slow	I	I	D	R			
▶ Using mathematical formula						I	D
▶ Compared with velocity					I	D	R
Velocity/acceleration					I	D	R

Forces	K	1	2	3	4	5	6
Pushes and pulls	I	I	D	D	R	R	R
Gravity			I	I	D	D	D
▶ Weight change with distance				I	D	D	D
▶ Universal gravitation						D	D
▶ Weight on other planets					I	D	D
Friction		I	I	D	D	D	R
▶ Reduction of, lubrication			D	D	D	R	R
▶ Source of heat		I	I	D	D	R	R
▶ Kinds (rolling, standing, sliding)						I	D
Magnetism	I	I	D	R	R	R	R
▶ Poles	I	I	D	R	R	R	R
▶ Attraction vs. repulsion	I	I	D	R	R	R	R
▶ Magnetic vs. geographic					I	D	R
▶ Magnetic fields			I	I	D	D	R
▶ Magnetic domains						D	R
Affect on position and motion		I	I	D	D	D	D
▶ Changes in speed and direction (acceleration)		I	I	D	D	D	D
▶ Balanced vs. unbalanced forces				I	D	D	D
▶ Inertia					I	D	D
▶ Circular motion						D	D
▶ Newton's First Law			I	I	I	D	D
Newton's Second Law			I	I	I	D	D
▶ Effect of changing mass on constant force			I	I	I	D	D
▶ Effect of changing force on constant mass			I	I	I	D	D
Newton's Third Law (action-reaction)						D	R
Momentum						I	D
▶ Law of Conservation of						I	D

KEY		
I = Introduced	**D** = Developed	**R** = Reinforced

Work and Energy

Work	K	1	2	3	4	5	6
Examples of				I	D	D	R
Mathematical formula	I	I			I	D	D
Calculations in joules						I	D

Energy	K	1	2	3	4	5	6
Potential				I	D	D	R
▶ Examples of				I	D	D	R
▶ Gravitational potential				I	D	D	R
Kinetic				I	D	D	R
▶ Examples of				I	D	D	R
▶ Relation to temperature					I	D	D
Transformation of potential and kinetic				I	I	D	R
Law of Conservation of Energy					I	D	R

Simple Machines	K	1	2	3	4	5	6
Relation to needs, found in nature		I	I	D	D	R	R
Examples of		I	I	D	D	R	R
▶ Levers		I	I	D	D	D	R
▶ Fulcrum			I	I	D	D	R
▶ Classification			I	I	D	D	R
▶ Pulleys	I	I		I	D	D	R
▶ Fixed vs. movable				I	D	D	R
▶ Wheel-and-axles	I	I	I	I	D	D	R
▶ Inclined planes		I	I	D	D	D	R
▶ Compared with lifting		I	I	D	D	D	R
▶ Effects of texture, angle				D	D	D	R
▶ Wedges		I	I	I	D	D	R
▶ Screws		I	I	I	D	D	R
Mechanical advantage						I	D
Friction/efficiency						I	D
Relation to compound machines				D	D	D	R

Scope & Sequence

Scope and Sequence

Transfer of Energy

Heat

Heat	K	1	2	3	4	5	6
Sources		I	I	D	D	R	R
Transfer				I	D	D	D
▶ Conduction				I	D	D	D
▶ Convection				I	D	D	D
▶ Radiation				I	D	D	D
Related to temperature		I	I	I	D	D	D
Kinetic energy model					I	D	D
Heating of different surfaces, trends					I	D	D

Sound

Sound	K	1	2	3	4	5	6
Sources (vibrations)	I	I	I	D	D	D	R
Transfer of sound			I	D	D	D	R
▶ Through solids, liquids, gases				D	D	D	R
▶ Compression wave model					I	D	D
Pitch variations	I	I	I	D	D	D	R
▶ Frequency					I	D	R
▶ Hertz measurements						I	D
▶ Doppler effect							D
Volume (loudness) variations	I	I	I	D	D	D	R
▶ Amplitude					I	D	D
▶ Decibel measurements						I	D
Absorption (insulation)						D	D
Reflection (echoes)					I	D	R

Light

Light	K	1	2	3	4	5	6
Sources		I	I	D	D	D	R
▶ Natural vs. artificial		I	I	I	D	D	R
Straight-line path		I	I	D	D	D	R
▶ Opaque, transparent, translucent materials				I	D	D	R
▶ Shadows	I	I	I	D	D	D	R
Reflection		I	I	D	D	D	D
▶ Law of reflection					I	D	D
▶ Mirrors			I	I	D	D	D
▶ Convex vs. concave					I	D	D
Refraction			I	I	D	D	D
Lenses			I	I	D	D	D
▶ Convex vs. concave					I	D	D

KEY		
I = Introduced	D = Developed	R = Reinforced

Light *cont.*	K	1	2	3	4	5	6
Prism			I	I	D	D	D
▶ Visible spectrum (colors)				I	I	D	D
▶ Mixing colors						D	R
▶ Mixing pigments						D	R
Wave model						D	D
▶ Parts (crest, trough)						D	D
▶ Photons						D	D
Electromagnetic spectrum						I	D
▶ Descriptions, uses						I	D
▶ Wavelengths/frequencies compared						I	D

Electricity	K	1	2	3	4	5	6
Static			I		D	D	R
▶ Attraction vs. repulsion			I		D	D	R
▶ Induced charge					I	D	D
▶ Discharge					I	D	D
▶ Insulators/conductors				I	D	D	R
Current/circuits			I	I	D	D	R
▶ Closed vs. open				I	D	D	R
▶ Parts of		I	I	D	D	D	R
▶ Batteries/wet-dry cells		I	I	I	D	D	R
▶ Transformation of energy in		I	I	I	D	R	R
▶ Series vs. parallel					D	D	D
Electromagnetism					D	D	D
▶ Making electromagnets					D	D	D
▶ Increasing strength of					D	D	D
▶ Common uses				I	D	D	D
▶ Motors				I	D	D	R
▶ Generating current					D	D	R
▶ AC vs. DC					D	D	R
▶ Power transmission					D	D	R
Household uses	I	I	I	D	D	R	R
▶ Safety					D	D	D

Correlations to National Science Education Standards

Grades K-4

The Science content standards are presented in the *National Science Education Standards*. These standards offer a structured guide to what scientifically literate students should know, understand, and be able to do at different grade levels. The standards are clustered for Grades K–4, 5–8, and 9–12.

The following table shows where the content standards are met in Science: *A Closer Look*, Grades K–4. Grade K page numbers refer to the Teacher's Edition. Grades 1–2 page numbers refer to the Student Edition or, when preceded by AF, to the Activity Flipchart. Grades 3–4 page numbers refer to the Student Edition.

Science As Inquiry

Abilities Necessary to Do Scientific Inquiry

Ask a question about objects, organisms, and events in the environment. This aspect of the standard emphasizes students asking questions that they can answer with scientific knowledge, combined with their own observations. Students should answer their questions by seeking information from reliable sources of scientific information and from their own observations and investigations.	**Grade K:** 12, 52, 65, 82, 89, 96, 97, 102, 116, 127, 130, 131, 136, 144, 150, 158, 161, 164, 202, 218, 237, 254, 255, 257, 260, 266, 269, 272, 273 **Grade 1:** 3, 11, 23, 29, 37, 53, 59, 67, 87, 95, 103, 109, 127, 133, 141, 163, 171, 179, 195, 201, 209, 229, 235, 241, 249, 265, 271, 279, 299, 307, 315, 329, 333, 341, 361, 367, 373, 381, 397, 403, 411, 419, AF4, AF11, AF16, AF27, AF30, AF34, AF45, AF50, AF54 **Grade 2:** 3, 11, 23, 29, 39, 55, 61, 69, 89, 95, 103, 121, 129, 135, 155, 163, 171, 187, 195, 201, 223, 231, 237, 253, 259, 267, 275, 295, 301, 309, 325, 331, 339, 361, 367, 377, 385, 399, 405, 415, 421, AF4, AF10, AF16, AF19, AF22, AF26, AF28, AF34, AF38, AF41, AF46, AF50, AF54, AF56, AF59 **Grade 3:** 3, 21, 31, 43, 53, 69, 81, 91, 107, 119, 133, 151, 161, 173, 191, 203, 213, 227, 239, 249, 259, 279, 289, 303, 317, 327, 337, 347, 363, 373, 383, 397, 407, 417, 433, 443, 453, 463, 479, 489, 499, 511 **Grade 4:** 3, 21, 33, 45, 59, 77, 89, 99, 109, 129, 137, 149, 165, 175, 183, 203, 213, 225, 237, 251, 263, 273, 285, 295, 313, 323, 335, 345, 359, 369, 379, 393, 411, 421, 431, 445, 457, 467, 483, 493, 503, 513, 529, 539, 551, 563, 575
Plan and conduct a simple investigation. In the earliest years, investigations are largely based on systematic observations. As students develop, they may design and conduct simple experiments to answer questions. The idea of a fair test is possible for many students to consider by fourth grade.	**Grade K:** 30, 38, 44, 52, 58, 68, 76, 82, 88, 96, 102, 110, 116, 130, 136, 144, 150, 158, 164, 176, 182, 188, 196, 202, 216, 222, 228, 236, 246, 254, 260, 266, 272 **Grade 1:** 59, 99, 195, 201, 209, 271, 279, 329, 331, 333, 335, 341, 361, 367, 370, 373, 381, 397, 403, 419, AF4, AF11, AF27, AF30, AF34, AF50, AF54 **Grade 2:** 39, 73, 132, 137, 163, 325, 339, 344, 380, 388, AF22, AF28, AF34, AF46, AF50, AF54, AF56 **Grade 3:** 40–41, 144–145, 268–269, 334–335, 422–423, 450–451, 496–497 **Grade 4:** 56–57, 106–107, 244–245, 270–271, 282–283, 352–353, 400–401, 472–473, 522–523, 560–561, 572–573

Abilities Necessary To Do Scientific Inquiry continued

Employ simple equipment and tools to gather data and extend the senses. In early years, students develop simple skills, such as how to observe, measure, cut, connect, switch, turn on and off, pour, hold, tie, and hook. Beginning with simple instruments, students can use rulers to measure the length, height, and depth of objects and materials; thermometers to measure temperature; watches to measure time; beam balances and spring scales to measure weight and force; magnifiers to observe objects and organisms; and microscopes to observe the finer details of plants, animals, rocks, and other materials. Children also develop skills in the use of computers and calculators for conducting investigations.	**Grade K:** 52, 58, 68, 76, 82, 102, 110, 117, 127, 130, 131, 133, 136, 139, 144, 150, 151, 153, 164, 165, 168F, 176, 177, 197, 199, 202, 203, 222, 223, 229, 231, 243, 246, 255, 261, 267, 268, 269, 272, 273, TR7, TR8, TR9, TR10, TR11 **Grade 1:** 29, 53, 87, 95, 127, 171, 229, 235, 249, 271, 307, 311, 315, 329, 341, 343, 376, 381, 384, 397, 403, 411, 419, R3, R4, R5, R6, AF4, AF27, AF34, AF45, AF50, AF54 **Grade 2:** 3, 29, 39, 69, 89, 95, 103, 129, 135, 171, 187, 195, 223, 231, 253, 259, 267, 275, 301, 309, 325, 339, 367, 377, 385, 399, 405, 415, 421, R3, R4, R5, R6, AF4, AF10, AF16, AF22, AF28, AF34, AF46, AF50, AF54, AF56, AF59 **Grade 3:** 26, 31, 40, 41, 53, 69, 81, 107, 119, 133, 144, 145, 161, 173, 200–201, 213, 231, 239, 243, 259, 268, 269, 307, 317, 334–335, 339, 347, 349, 363, 373, 380, 381, 383, 397, 407, 417, 422, 423, 438, 450, 460–461, 481, 486–487, 489, 496–497, 499, 511, 516, R2, R4, R5, R6, R7, R8, R9 **Grade 4:** 21, 30, 33, 39, 45, 56, 93, 99, 106–107, 109, 129, 134–135, 137, 143, 149, 172–173, 175, 237, 241, 251, 263, 267, 270–271, 285, 289, 295, 332–333, 352–353, 359, 363, 369, 379, 393, 400, 411, 421, 425, 431, 445, 461, 464–465, 467, 483, 490–491, 493, 503, 513, 522–523, 536–537, 539, 551, 560–561, 575, 580, R4, R5, R6, R7, R8, R9
Use data to construct a reasonable explanation. This aspect of the standard emphasizes the students' thinking as they use data to formulate explanations. Even at the earliest grade levels, students should learn what constitutes evidence and judge the merits or strength of the data and information that will be used to make explanations. After students propose an explanation, they will appeal to the knowledge and evidence they obtained to support their explanations. Students should check their explanations against scientific knowledge, experiences, and observations of others.	**Grade K:** 7, 30, 58, 76, 79, 99, 113, 158, 176, 203, 228, 243, 246, 260 **Grade 1:** 11, 141, 397, 403, 411, 419, AF4, AF8, AF27, AF30, AF34, AF38, AF40, AF45, AF50, AF54 **Grade 2:** 11, 129, 171, 201, 237, 259, 267, 275, 309, 325, 361, AF4, AF10, AF22, AF26, AF28, AF34, AF38, AF41, AF46, AF50, AF54, AF56, AF59 **Grade 3:** 21, 31, 41, 43, 53, 69, 81, 91, 107, 119, 133, 145, 151, 161, 173, 191, 203, 213, 227, 239, 249, 259, 269, 279, 289, 303, 317, 327, 335, 337, 347, 363, 373, 383, 397, 407, 417, 422, 423, 433, 443, 450, 453, 463, 479, 489, 497, 499, 511 **Grade 4:** 21, 33, 45, 57, 59, 77, 89, 99, 107, 109, 129, 137, 149, 165, 175, 183, 203, 213, 225, 237, 245, 251, 263, 271, 273, 283, 285, 295, 313, 323, 335, 345, 352, 359, 369, 379, 393, 401, 411, 421, 431, 445, 457, 467, 473, 483, 493, 503, 513, 523, 529, 539, 551, 561, 563, 573, 575

Correlations to National Standards

Abilities Necessary To Do Scientific Inquiry continued

Communicate investigations and explanations. Students should begin developing the abilities to communicate, critique, and analyze their work and the work of other students. This communication might be spoken or drawn as well as written.

Grade K: 26, 30, 32, 38, 40, 44, 46, 52, 54, 58, 64, 68, 70, 76, 78, 82, 84, 88, 90, 96, 98, 102, 104, 110, 112, 116, 126, 130, 132, 136, 138, 144, 146, 150, 152, 158, 160, 164, 170, 176, 178, 182, 184, 188, 190, 196, 198, 202, 212, 216, 218, 222, 224, 228, 230, 236, 242, 246, 248, 254, 256, 260, 262, 266, 268, 272

Grade 1: 3, 29, 32, 37, 56, 59, 67, 95, 103, 127, 133, 141, 163, 195, 235, 271, 279, 299, 315, 329, AF4, AF11, AF16, AF27, AF30, AF34, AF50, AF54, AF58

Grade 2: 11, 29, 55, 59, 63, 89, 93, 95, 97, 106, 163, 187, 195, 201, 261, 331, 361, AF4, AF10, AF16, AF22, AF28, AF34, AF38, AF45, AF46, AF50, AF54, AF56, AF59

Grade 3: 26, 35, 43, 47, 53, 85, 91, 93, 114, 116–117, 144, 151, 173, 203, 207, 239, 251, 265, 268, 327, 334, 349, 363, 367, 373, 387, 397, 423, 433, 438, 453, 463, 479, 499, 511

Grade 4: 12, 21, 45, 57, 77, 129, 151, 203, 225, 251, 260, 261, 273, 277, 295, 299, 323, 339, 349, 352, 359, 467, 472, 523, 539, 561, 563, 573, 575

Understanding About Scientific Inquiry

Scientific investigations involve asking and answering a question and comparing the answer with what scientists already know about the world.

Grade K: 30, 38, 44, 52, 58, 68, 76, 82, 88, 96, 102, 110, 116, 130, 136, 144, 150, 158, 164, 176, 182, 188, 196, 202, 216, 222, 228, 236, 246, 254, 260, 266, 272

Grade 1: 3, 11, 23, 29, 37, 53, 59, 67, 87, 95, 103, 109, 127, 133, 141, 163, 171, 179, 195, 201, 209, 229, 235, 241, 249, 265, 271, 279, 299, 307, 315, 329, 333, 341, 361, 367, 373, 381, 397, 403, 411, 419, AF4, AF30, AF34, AF50, AF54, AF55

Grade 2: 3, 11, 23, 29, 39, 55, 61, 69, 89, 95, 103, 121, 129, 135, 155, 163, 171, 187, 195, 201, 223, 231, 237, 253, 259, 267, 275, 295, 301, 309, 325, 331, 339, 361, 367, 377, 385, 399, 405, 415, 421, AF4, AF10, AF26, AF28, AF34, AF46, AF50, AF54, AF56, AF59

Grade 3: 3, 21, 31, 40–41, 43, 53, 69, 81, 91, 107, 119, 133, 144–145, 151, 161, 173, 191, 203, 213, 227, 239, 249, 259, 268–269, 279, 289, 303, 317, 327, 334–335, 337, 347, 363, 373, 383, 397, 407, 417, 422–423, 433, 443, 450–451, 453, 463, 479, 489, 493, 496–497, 499, 511

Grade 4: 3, 21, 33, 45, 56–57, 59, 77, 89, 99, 106–107, 109, 129, 137, 149, 165, 175, 183, 203, 213, 225, 237, 244–245, 251, 263, 270–271, 273, 282–283, 285, 295, 313, 323, 335, 345, 352–353, 359, 369, 379, 393, 400–401, 411, 421, 431, 445, 457, 467, 472–473, 483, 493, 503, 513, 522–523, 529, 539, 551, 560–561, 563, 572–573, 575

Understanding About Scientific Inquiry continued

Scientists use different kinds of investigations depending on the questions they are trying to answer. Types of investigations include describing objects, events, and organisms; classifying them; and doing a fair test (experimenting).

Grade K: 30, 38, 44, 52, 58, 68, 76, 82, 88, 96, 102, 110, 116, 130, 136, 144, 150, 158, 164, 176, 182, 188, 196, 202, 216, 222, 228, 236, 246, 254, 260, 266, 272

Grade 1: 3, 11, 23, 25, 29, 32, 37, 40, 53, 56, 59, 63, 67, 70, 87, 91, 95, 99, 103, 107, 109, 113, 127, 130, 133, 135, 141, 144, 163, 166, 171, 173, 179, 183, 195, 198, 201, 204, 209, 211, 229, 232, 235, 238, 241, 243, 249, 251, 265, 267, 271, 273, 279, 281, 299, 301, 307, 311, 315, 319, 329, 331, 333, 335, 341, 343, 361, 364, 367, 370, 373, 376, 381, 384, 397, 399, 403, 405, 411, 414, 419, 421, AF8, AF27, AF30, AF34, AF38, AF50, AF54

Grade 2: 3, 11, 23, 26, 29, 32, 39, 42, 55, 59, 61, 63, 69, 73, 89, 93, 95, 97, 103, 106, 121, 124, 129, 132, 135, 137, 155, 161, 163, 166, 171, 173, 187, 191, 195, 199, 201, 207, 223, 226, 231, 234, 237, 240, 253, 256, 259, 261, 267, 272, 275, 278, 295, 298, 301, 304, 309, 313, 325, 328, 331, 335, 339, 344, 361, 364, 367, 371, 377, 380, 385, 388, 399, 403, 405, 407, 415, 418, 421, 424, AF4, AF8, AF22, AF26, AF28, AF34, AF38, AF50, AF54, AF56, AF59

Grade 3: 3, 21, 26, 31, 35, 43, 47, 53, 55, 69, 73, 81, 85, 91, 93, 107, 114, 119, 127, 133, 137, 151, 155, 161, 167, 173, 177, 191, 195, 203, 207, 213, 217, 227, 231, 239, 243, 249, 251, 259, 265, 279, 283, 289, 293, 303, 307, 317, 319, 327, 331, 337, 339, 347, 349, 363, 367, 373, 377, 383, 387, 397, 401, 407, 412, 417, 419, 433, 438, 443, 447, 453, 457, 463, 469, 479, 481, 489, 493, 499, 505, 511, 516

Grade 4: 3, 21, 27, 33, 39, 45, 53, 59, 64, 77, 80, 89, 93, 99, 102, 109, 116, 129, 131, 137, 143, 149, 151, 165, 169, 175, 177, 183, 186, 203, 207, 213, 219, 225, 231, 237, 241, 251, 255, 263, 267, 273, 277, 285, 289, 295, 299, 313, 317, 323, 328, 335, 339, 345, 349, 359, 363, 369, 373, 379, 384, 393, 397, 411, 415, 421, 425, 431, 435, 445, 449, 457, 461, 467, 470, 483, 487, 493, 498, 503, 507, 513, 515, 529, 533, 539, 544, 551, 557, 563, 569, 575, 580

Correlations to National Standards

Understanding About Scientific Inquiry continued

Simple instruments, such as magnifiers, thermometers, and rulers, provide more information than scientists obtain using only their senses.	**Grade K:** 52, 58, 68, 76, 82, 102, 110, 117, 127, 130, 131, 133, 136, 139, 144, 150, 151, 153, 164, 165, 168F, 176, 177, 197, 199, 202, 203, 222, 223, 229, 231, 243, 246, 255, 261, 267, 268, 269, 272, 273, TR7, TR8, TR9, TR10, TR11 **Grade 1:** 29, 53, 87, 95, 127, 171, 229, 235, 249, 271, 307, 311, 315, 329, 341, 343, 376, 381, 384, 397, 403, 411, 419, R3, R4, R5, R6, AF4, AF34 **Grade 2:** 3, 29, 39, 69, 89, 95, 103, 129, 135, 171, 187, 195, 223, 231, 253, 259, 267, 275, 301, 309, 325, 339, 367, 377, 385, 399, 405, 415, 421, R3, R4, R5, R6, AF28, AF34, AF46, AF50, AF54, AF56, AF59 **Grade 3:** 26, 31, 40, 41, 53, 69, 81, 107, 119, 133, 144, 145, 161, 173, 200–201, 213, 231, 239, 243, 259, 268, 269, 307, 317, 334–335, 339, 347, 349, 363, 373, 380, 381, 383, 397, 407, 417, 422, 423, 438, 450, 460–461, 481, 486–487, 489, 496–497, 499, 511, 516, R2, R4, R5, R6, R7, R8, R9 **Grade 4:** 21, 30, 33, 39, 45, 56, 93, 99, 106–107, 109, 129, 134–135, 137, 143, 149, 172–173, 175, 237, 241, 251, 263, 267, 270–271, 285, 289, 295, 332–333, 352–353, 359, 363, 369, 379, 393, 400, 411, 421, 425, 431, 445, 461, 464–465, 467, 483, 490–491, 493, 503, 513, 522–523, 536–537, 539, 551, 560–561, 575, 580, R4, R5, R6, R7, R8, R9
Scientists develop explanations using observations (evidence) and what they already know about the world (scientific knowledge). Good explanations are based on evidence from investigations.	**Grade K:** 7, 30, 58, 76, 79, 99, 113, 158, 176, 203, 228, 243, 246, 260 **Grade 1:** 11, 141, 397, 403, 411, 419, AF4, AF11, AF16, AF30, AF50, AF54 **Grade 2:** 11, 129, 171, 201, 237, 259, 267, 275, 309, 325, 361, AF4, AF10, AF22, AF26, AF28, AF34, AF38, AF46, AF50, AF54, AF56, AF59 **Grade 3:** 21, 31, 41, 43, 53, 69, 81, 91, 107, 119, 133, 145, 151, 161, 173, 191, 203, 213, 227, 239, 249, 259, 269, 279, 289, 303, 317, 327, 335, 337, 347, 363, 373, 383, 397, 407, 417, 422, 423, 433, 443, 450, 453, 463, 479, 489, 497, 499, 511 **Grade 4:** 21, 33, 45, 57, 59, 77, 89, 99, 107, 109, 129, 137, 149, 165, 175, 183, 203, 213, 225, 237, 245, 251, 263, 271, 273, 283, 285, 295, 313, 323, 335, 345, 352, 359, 369, 379, 393, 401, 411, 421, 431, 445, 457, 467, 473, 483, 493, 503, 513, 523, 529, 539, 551, 561, 563, 573, 575

Understanding About Scientific Inquiry continued

Scientists make the results of their investigations public; they describe the investigations in ways that enable others to repeat the investigations.	**Grade K:** 6, 7, 12, 13, 19, 26, 27, 30, 31, 32, 33, 38, 39, 41, 44, 45, 46, 47, 52, 53, 54, 55, 58, 59, 64, 65, 68, 69, 70, 71, 76, 77, 79, 83, 88, 90, 91, 97, 98, 99, 102, 103, 104, 105, 110, 111, 113, 116, 117, 126, 127, 132, 133, 136, 137, 138, 139, 146, 147, 150, 151, 152, 153, 158, 159, 160, 161, 164, 165, 170, 171, 177, 178, 179, 182, 183, 184, 185, 188, 189, 190, 191, 196, 197, 198, 199, 203, 212, 213, 217, 219, 222, 223, 224, 225, 229, 231, 236, 237, 242, 243, 246, 247, 248, 255, 256, 257, 261, 262, 263, 266, 267, 269, 273 **Grade 1:** 3, 29, 32, 37, 56, 59, 67, 95, 103, 127, 133, 141, 163, 195, 235, 271, 279, 299, 315, 329, AF11, AF16, AF30, AF34, AF50 **Grade 2:** 11, 29, 55, 59, 63, 89, 93, 95, 97, 106, 163, 187, 195, 201, 261, 331, 361, AF10, AF54 **Grade 3:** 7, 26, 35, 43, 47, 53, 85, 91, 93, 114, 116–117, 144, 151, 173, 203, 207, 239, 251, 265, 268, 327, 334, 349, 363, 367, 373, 387, 397, 423, 433, 438, 453, 463, 479, 499, 511 **Grade 4:** 12, 21, 45, 57, 77, 129, 151, 203, 225, 251, 260, 261, 273, 277, 295, 299, 323, 339, 349, 352, 359, 467, 472, 523, 539, 561, 563, 573, 575
Scientists review and ask questions about the results of other scientists' work.	**Grade 2:** 280–281, AF16 **Grade 3:** 10–11, 180–181 **Grade 4:** 106–107, 387

Physical Science

Properties of Objects and Materials

Objects have many observable properties, including size, weight, shape, color, temperature, and the ability to react with other substances. Those properties can be measured using tools, such as rulers, balances, and thermometers.	**Grade K:** 210E, 211, 212, 213, 214, 215, 218, 223, 224, 226, 227, 228, 230, 231, 232, 233, 234, 235, 236, 237, 238, 239, 240J, 249, 254, 255, 260, 268, 269, 270, 271, 273, 274, 275, 276 **Grade 1:** 298, 299, 300, 301, 302, 303, 304, 305, 306, 307, 308–311, 314, 315, 316–319, 320–323, 324, 325, 328, 329, 330–331, 332–337, 338, 339, 341, 342–345, 346, 347, 348–351, 352, 353, AF45, AF54 **Grade 2:** 295, 296, 297, 298, 299, 300, 301, 302, 303, 304, 305, 306–307, 308, 309, 310, 311, 312, 313, 314, 315, 316–319, 320, 321, 324, 325, 326–329, 331, 332–335, 336–337, 348–351, 352, 353, AF45 **Grade 3:** 358–359, 362, 363, 364, 365, 366, 367, 368, 369, 372, 373, 374, 375, 376, 377, 378, 379, 380–381, 383, 384–389, 390, 391, 392, 393, 406, 407, 408–413, 414–415, 416, 417, 418–421, 422–423, 424, 425, 479, 482, 483, 485 **Grade 4:** 411, 412, 413, 414, 415, 416, 417, 418, 419, 421, 422, 423, 424, 425, 426, 427, 428–429, 431, 432, 433, 434, 435, 436, 437, 440, 441, 445, 446, 447, 448, 449, 450, 451, 452, 453, 454–455, 456, 463, 467, 468, 469, 470, 471, 472–473, 474, 475

Correlations to National Standards

Properties of Objects and Materials continued

Objects are made of one or more materials, such as paper, wood, and metal. Objects can be described by the properties of the materials from which they are made, and those properties can be used to separate or sort a group of objects or materials.	**Grade K:** 210E, 210J, 210, 212, 214, 215, 216, 217, 218, 219, 220, 221, 222, 225, 226, 229, 239, 271 **Grade 1:** 312–313, 325, 346–347 **Grade 2:** 306–307, 336–337, 339, 340, 341, 342, 343, 344, 345, 346, 347, 348–351, 352, 353 **Grade 3:** 366, 367, 368, 369, 370–371, 409, 410, 411, 412, 413, 414–415, 424, 425 **Grade 4:** 54, 406–407, 444, 456, 457, 458, 459, 460, 461, 462, 463, 474, 475
Materials can exist in different states—solid, liquid, and gas. Some common materials, such as water, can be changed from one state to another by heating or cooling.	**Grade K:** 230, 231, 232, 233, 237, 238, 239 **Grade 1:** 294, 295, 306, 307, 308, 309, 310, 311, 314, 315, 316, 317, 318, 319, 320–323, 324, 325, 341, 342, 343, 344, 345, 352, 353, AF50, AF54 **Grade 2:** 299, 300–305, 308, 309, 310–313, 314, 315, 316–319, 320, 321, 327, 330, 331, 332–335, 336–337, 348–351, 352, 353, 400, AF41, AF46 **Grade 3:** 380–381, 382, 383, 384, 385, 386, 387, 388, 389, 392, 393, 396, 397, 398, 399, 400, 401, 402, 403, 404–405, 408, 424, 425, 483, 485 **Grade 4:** 323, 324, 325, 331, 410, 414, 415, 417, 446, 447, 448, 449, 453, 534, 535

Position and Motion of Objects

The position of an object can be described by locating it relative to another object or the background.	**Grade 1:** 361, 362, 363, 365, 392, 393 **Grade 2:** 360, 361, 362, 363, 365, 394, 395 **Grade 3:** 432, 433, 434, 435, 439 **Grade 4:** 484, 485, 503
An object's motion can be described by tracing and measuring its position over time.	**Grade K:** 240F, 242, 243, 250, 251, 252, 253, 255 **Grade 1:** 356–357, 364, 365, 392 **Grade 2:** 363, 364, 365, 367, 372, 373, 383, 390–393, 394, 395 **Grade 3:** 429, 436, 437, 438, 439, 443, 449 **Grade 4:** 482, 483, 484, 485, 486, 487, 488, 489, 490–491, 496, 497, 498, 499, 503, 524, 525
The position and motion of objects can be changed by pushing or pulling. The size of the change is related to the strength of the push or pull.	**Grade K:** 240, 241, 244, 245, 246, 247, 248, 249, 252, 253, 255, 274, 275, 277 **Grade 1:** 366, 367, 368, 369, 370, 371, 373, 388–391, 392, 393 **Grade 2:** 367, 368, 369, 370, 371, 372, 373, 390–393, 394, 395 **Grade 3:** 442, 443, 444, 445, 446, 447, 448, 449, 450–451 **Grade 4:** 486, 487, 488, 489, 490–491, 492, 493, 494, 495, 496, 497, 498, 499, 500, 501, 503, 504, 505, 521, 524, 525

Position and Motion of Objects continued

Sound is produced by vibrating objects. The pitch of the sound can be varied by changing the rate of vibration.	**Grade K:** 264, 265, 266, 267, 274, 275 **Grade 1:** 402, 403, 404, 405, 406, 407, 428, 429 **Grade 2:** 356–357, 405, 406, 407, 409, 410, 411, 412, 429, 432 **Grade 3:** 489, 490, 491, 492, 493, 494, 495, 496–497, 520, 521 **Grade 4:** 538, 539, 540, 541, 542, 543, 544, 545, 546, 547, 548, 549, 588, 589

Light, Heat, Electricity, and Magnetism

Light travels in a straight line until it strikes an object. Light can be reflected by a mirror, refracted by a lens, or absorbed by the object.	**Grade K:** 198, 199, 200, 201, 202, 203 **Grade 1:** 411, 412, 413, 414, 415, 416, 417, 426, 429 **Grade 2:** 414, 416, 417, 419, 430, 432 **Grade 3:** 498, 499, 500, 501, 502, 503, 504, 505, 506, 507, 508–509, 520, 521 **Grade 4:** 506, 509, 550, 551, 552, 553, 554, 556, 557, 558, 559, 560–561, 588, 589
Heat can be produced in many ways, such as burning, rubbing, or mixing one substance with another. Heat can move from one object to another by conduction.	**Grade 1:** 396, 397, 400, 401, 428, 429, AF47, AF58 **Grade 2:** 399, 400, 401, 403, 428, 433 **Grade 3:** 478, 479, 480, 481, 482, 483, 484, 485, 486–487, 520, 521 **Grade 4:** 507, 509, 530, 531, 532, 533, 534, 535, 536–537, 588, 589
Electricity in circuits can produce light, heat, sound, and magnetic effects. Electrical circuits require a complete loop through which an electrical current can pass.	**Grade 1:** 418, 419, 420, 421, 422, 423, 427, 428, 429 **Grade 2:** 420, 421, 422, 423, 425, 426–427, 431, 432, 433 **Grade 3:** 511, 514, 515, 516, 517, 518, 520, 521 **Grade 4:** 508, 509, 562, 566, 567, 568, 569, 570, 571, 580, 581, 582, 583, 584, 585, 586–587, 588
Magnets attract and repel each other and certain kinds of other materials.	**Grade K:** 240J, 268, 269, 270, 271, 272, 273, 274, 275 **Grade 1:** 380, 381, 382, 383, 384, 385, 386, 387, 392 **Grade 2:** 384, 385, 386, 387, 388, 389, 393, AF54 **Grade 3:** 367, 446, 450–451 **Grade 4:** 574, 575, 576, 577, 578, 579, 580, 581, 582, 583, 585, 588, 589

Life Science

The Characteristics of Organisms

Organisms have basic needs. For example, animals need air, water, and food; plants require air, water, nutrients, and light. Organisms can survive only in environments in which their needs can be met. The world has many different environments, and distinct environments support the life of different types of organisms.	**Grade K:** 26, 32, 33, 34, 35, 37, 38, 39, 43, 44, 45, 60, 61, 67, 70, 71, 72, 73, 74, 75, 77, 82, 83, 92, 93, 98, 120, TR3, TR4, TR5, TR6 **Grade 1:** 24, 26, 27, 30, 42–43, 49, 50, 56, 58, 59, 67, 95, 96, 97, 99, 100, 104, 118, 135, 202, AF11 **Grade 2:** 23, 24, 25, 26, 27, 47, 51, 56, 59, 77, 89, 90, 93, 94 **Grade 3:** 20, 24, 25, 27, 32, 33, 36, 37, 39, 40–41, 43, 46, 47, 49, 64, 65, 69, 70, 71, 78–79, 82, 119, 120, 121, 122–128 **Grade 4:** 22, 46, 48, 49, 50, 51, 55, 59, 64, 72

The Characteristics of Organisms continued

Each plant or animal has different structures that serve different functions in growth, survival, and reproduction. For example, humans have distinct body structures for walking, holding, seeing, and talking.	**Grade K:** 5, 10–11, 26, 27, 28, 29, 30, 31, 42, 43, 60, 61, 78, 80, 81, 86, 87, 88, 90, 91, 92, 93, 94, 95, 97, 99, 100, 101, 103, 105, TR9 **Grade 1:** 19, 28, 29, 31, 32, 33, 35, 49, 52, 54, 55, 56, 57, 59, 60, 62, 63, 64–65, 68, 69, 70, 71, 78, 79, 82–83, 89, 90, 92, 93, 95, 98, 99, 102, 103, 105, 106, 107, 114, 115, 123, 129, 130, 131, 133, 136, 137, 138, 139, 154, AF13, AF16, AF52 **Grade 2:** 11, 12, 13, 26, 27, 30, 31, 32, 33, 34, 35, 36, 39, 42, 46–49, 50, 57, 58, 59, 69, 70, 71, 72, 73, 74, 77, 78, 80, 89, 93, 123, 125, 129, 130, 131, 132, 133, 137, 139, R8, R9, R10, R11, AF4 **Grade 3:** 28–29, 31, 32, 33, 34, 35, 36, 37, 38, 39, 42, 43, 44, 45, 46, 47, 48, 49, 50–51, 53, 54, 55, 56, 57, 58, 59, 60, 61, 62, 64, 65, 70, 72, 73, 75, 76, 77, 102–103, 132, 133, 134, 135, 136, 137, 138, 139, 140, 141, 142, 143, 144–145, 146, 147, 449, 494, 506, R14, R15, R16, R17, R18, R19, R20, R21, R22 **Grade 4:** 17, 27, 45, 46, 47, 49, 50, 51, 52, 53, 55, 56–57, 60, 61, 62, 63, 70, 71, 72, 82, 83, 84, 85, 88, 89, 90, 91, 92, 93, 95, 98, 100, 101, 102, 103, 104, 105, 106–107, 110, 120, 121, 124–125, 164, 165, 166, 167, 168, 169, 171, 172–173, 174, 178, 179, 194, 195, 528, 529, R14, R15, R16, R17, R18, R19, R20, R21, R22
The behavior of individual organisms is influenced by internal cues (such as hunger) and by external cues (such as a change in the environment). Humans and other organisms have senses that help them detect internal and external cues.	**Grade 1:** 17, 19 **Grade 2:** 42, 58, 72, 73, R9, AF4 **Grade 3:** 24, 25, 44, 49, R21 **Grade 4:** 99, 100, 101, 105, 120, 478–479

Life Cycles of Organisms

Plants and animals have life cycles that include being born, developing into adults, reproducing, and eventually dying. The details of this life cycle are different for different organisms.	**Grade K:** 40, 41, 42, 43, 44, 47, 104, 105, 106, 107, 108, 109, 110, 111, 119 **Grade 1:** 54, 57, 60, 61, 62, 63, 78, 79, 91, 110, 111, 112, 113, 114, 115, 116, 117, 123 **Grade 2:** 8, 9, 25, 30, 31, 34, 35, 48, 49, 50, 62, 63, 64, 65, 67, 78, 80, 81, AF10, AF13 **Grade 3:** 17, 23, 27, 64, 68, 70, 71, 72, 73, 74, 75, 76, 77, 80, 81, 82, 83, 84, 85, 86, 87, 88, 98, 99 **Grade 4:** 47, 52, 53, 58, 59, 62, 63, 64, 65, 68, 69, 72, 94, 108, 109, 110, 111, 112, 113, 114, 115, 117, 118–119, 120, 158
Plants and animals closely resemble their parents.	**Grade K:** 104, 106, 107, 108, 109, 115 **Grade 1:** 108, 109, 110, 111, 112, 113, 123 **Grade 2:** 24, 35, 40, 41, 43, 47, 61 **Grade 3:** 22, 84, 85, 86, 88, 90, 91, 92, 93 **Grade 4:** 66, 69, 112

Life Cycles of Organisms continued

Many characteristics of an organism are inherited from the parents of the organism, but other characteristics result from an individual's interactions with the environment. Inherited characteristics include the color of flowers and the number of limbs of an animal. Other features, such as the ability to ride a bicycle, are learned through interactions with the environment and cannot be passed on to the next generation.	**Grade K:** 109 **Grade 1:** 113 **Grade 2:** 42, 43, 47, 75, 78 **Grade 3:** 22, 23, 86, 91, 92, 93, 94, 95, 96–97, 98, 99 **Grade 4:** 66, 67, 69, 114, 116, 117, 478–479

Organisms and Their Environments

All animals depend on plants. Some animals eat plants for food. Other animals eat animals that eat the plants.	**Grade K:** 56, 57, 58, 59 **Grade 1:** 12, 13, 14, 104, 105, 142, 144, 145, 154, 155, 198 **Grade 2:** 27, 92, 94, 95, 96, 97, 98, 99, 100, 101, 116, 117, 136, 138 **Grade 3:** 106, 107, 108, 109, 110, 111, 112, 113, 114, 115, 116–117, 146, 147 **Grade 4:** 44, 54, 124–125, 131, 148, 149, 150, 151, 152, 153, 154, 155, 156, 157, 158, 159, 160, 161, 164, 198–199
An organism's patterns of behavior are related to the nature of that organism's environment, including the kinds and numbers of other organisms present, the availability of food and resources, and the physical characteristics of the environment. When the environment changes, some plants and animals survive and reproduce, and others die or move to new locations.	**Grade K:** 89, 102, 103 **Grade 1:** 11, 12, 13, 14, 15, 66, 67, 68, 69, 70, 71, 97, 126, 127, 128, 130, 131, 132, 133, 134, 135, 136, 137, 138, 141, 144, 145, 150–153, 154, 155, 241, AF19 **Grade 2:** 42, 43, 90, 91, 92, 104, 105, 106, 108, 109, 110–111, 112–115, 116, 117, 123, 130, 131, 132, 133, 135, 136, 137, 138, 139, 142–145, 147, 150–151 **Grade 3:** 107, 108, 109, 115, 123, 125, 126, 128, 136, 137, 138, 139, 141, 142, 143, 146, 147, 152, 153, 164, 165, 166, 167, 168, 169, 170, 171, 174, 175, 179, 182, 183, 274–275 **Grade 4:** 23, 96, 98, 99, 118–119, 121, 124–125, 141, 142, 143, 144, 155, 158, 168, 170, 171, 175, 176, 177, 179, 180, 181, 185, 188, 189, 194, 195, 238
All organisms cause changes in the environment where they live. Some of these changes are detrimental to the organism or other organisms, whereas others are beneficial.	**Grade K:** 124–125, 154, 155, 160, 161, 162, 163, 164, 165 **Grade 1:** 143, 146, 147, 195, AF30 **Grade 2:** 105, 106, 107, 109, 126–127 **Grade 3:** 28–29, 114, 115, 128, 150, 151, 152, 153, 154, 155, 156, 157, 182, 183 **Grade 4:** 42–43, 124–125, 185, 191
Humans depend on their natural and constructed environments. Humans change environments in ways that can be either beneficial or detrimental for themselves and other organisms.	**Grade K:** 124–125, 150, 152, 153, 154, 155, 156, 157, 158, 159, 160, 161, 162, 163, 164, 165 **Grade 1:** 64–65, 146, 147, 195, 204, 205, 246–247 **Grade 2:** 44, 105, 106, 107, 109, 126–127 **Grade 3:** 28–29, 154, 155, 156, 157, 158–159, 170–171, 182, 183, 184, 210–211, 218, 219 **Grade 4:** 54, 96, 118–119, 146–147, 159, 186, 187, 190, 191, 232, 233

Correlations to National Standards

Earth and Space Science

Properties of Earth Materials

Earth materials are solid rocks and soils, water, and the gases of the atmosphere. The varied materials have different physical and chemical properties, which make them useful in different ways, for example, as building materials, as sources of fuel, or for growing the plants we use as food. Earth materials provide many of the resources that humans use.	**Grade K:** 124E, 126, 127, 128, 129, 130, 131, 132, 133, 134, 135, 136, 137, 146, 147, 148, 149, 151, 166, 167 **Grade 1:** 64–65, 158–159, 162, 163, 164, 165, 166, 167, 168, 169, 170, 171, 172, 173, 174, 175, 176–177, 179, 185, 186–189, 190, 191, 196, 197, 198, 199, 200, 201, 202, 203, 204, 205, 206–207, 312–313, AF23 **Grade 2:** 155, 162, 163, 164, 165, 166, 167, 168, 169, 171, 172, 173, 178–181, 182, 183, 186, 187, 188, 189, 190, 191, 192, 193, 194, 195, 196, 197, 198, 199, 200, 201, 202, 203, 210–213, 214, 215, 244–247, 401 **Grade 3:** 190, 191, 192, 193, 194, 195, 196, 197, 199, 200–201, 213, 214, 215, 216, 219, 220, 221, 222, 223, 226, 227, 228, 229, 230, 231, 232, 233, 234, 235, 236, 238, 239, 240, 241, 242, 243, 244, 245, 251, 258, 259, 260, 261, 262, 263, 267, 268–269, 270, 271, 278, 279, 280, 281 **Grade 4:** 250, 251, 252, 253, 254, 255, 256, 257, 258, 259, 260–261, 262, 263, 264, 265, 266, 267, 268, 269, 270–271, 284, 285, 286, 287, 288, 289, 290, 291, 304, 305, 314, 315, 319, 428–429, 438–439
Soils have properties of color and texture, capacity to retain water, and ability to support the growth of many kinds of plants, including those in our food supply.	**Grade K:** 126, 127, 128, 129, 130, 131 **Grade 1:** 174, 175, 190, 198, 199, AF27 **Grade 2:** 194, 195, 196, 197, 198, 199, 203, 210, 211, 214, 215, AF28 **Grade 3:** 238, 239, 240, 241, 242, 243, 244, 245, 246, 247 **Grade 4:** 262, 263, 264, 265, 266, 267, 268, 269, 270–271, 304, 305
Fossils provide evidence about the plants and animals that lived long ago and the nature of the environment at that time.	**Grade 1:** 148–149 **Grade 2:** 108, 109, 110, 111, AF16 **Grade 3:** 172, 173, 174, 175, 176, 177, 178, 179, 180–181, 182, 183, 248, 249, 250, 251, 255, 270 **Grade 4:** 189, 255, 272, 273, 274, 275, 276, 277, 281, 282–283, 304

Objects in the Sky

The sun, moon, stars, clouds, birds, and airplanes all have properties, locations, and movements that can be observed and described.	**Grade K:** 190, 191, 192, 193, 194, 195, 196, 197, 203, 205, 256, 257, 258, 259, 261 **Grade 1:** 235, 237, 238, 239, 262, 264, 265, 266, 267, 268, 269, 270, 272, 273, 276, 278, 279, 280, 281, 286–289, 290, 291, AF38, AF40 **Grade 2:** 232, 233, 234, 236, 237, 238, 239, 248, 249, 253, 254–257, 262, 263, 265, 266, 267, 268, 269, 270, 271, 272, 273, 280–281, 286, 287, 374–375, AF38 **Grade 3:** 273, 290, 291, 292, 293, 316, 317, 318, 319–321, 322, 323, 326, 327, 328, 329, 330, 331, 332, 333, 334–335, 346, 347, 348, 349, 350, 351, 352–353, 354, 355, 370–371 **Grade 4:** 326, 328, 329, 331, 343, 358, 361, 363, 364, 365, 368, 369, 370, 371, 372, 373, 374, 375, 390–391, 392, 393, 394, 395, 396, 397, 398, 399, 400–401, 402
The sun provides the light and heat necessary to maintain the temperature of the earth.	**Grade K:** 198, 199, 200, 201, 202, 203 **Grade 1:** 243, 245, 267, 268, 269, 271, 273, 286–289, 290, 291, AF40, AF58 **Grade 2:** 254–257, 262, 263, 265, 273, 400, 416, AF59 **Grade 3:** 294, 305, 316, 319, 320, 321, 322, 323, 480 **Grade 4:** 48, 50, 360, 366, 367, 398, 506

Changes in the Earth and Sky

The surface of the earth changes. Some changes are due to slow processes, such as erosion and weathering, and some changes are due to rapid processes, such as landslides, volcanic eruptions, and earthquakes.	**Grade K:** 138, 140, 141, 142, 143, 144, 148, 149 **Grade 1:** 179, 180, 181, 182, 183, 184, 191 **Grade 2:** 170, 171, 172, 173, 174, 175, 176–177, 178–181, 182, 183 **Grade 3:** 202, 203, 204, 205, 206, 207, 208, 209, 210–211, 212, 213, 214, 215, 216, 217, 219, 220, 221, 222, 223 **Grade 4:** 3–11, 184, 197, 198–199, 202, 204, 206, 210–211, 213, 214, 215, 216, 217, 218, 220, 221, 224, 225, 226, 227, 228, 229, 230, 231, 233, 234, 235, 236, 237, 238, 242, 243, 244–245, 246, 247

Correlations to National Standards

Changes in the Earth and Sky continued

Weather changes from day to day and over the seasons. Weather can be described by measurable quantities, such as temperature, wind direction and speed, and precipitation.	**Grade K:** 4–5, 6, 7, 168E, 168F, 168J, 168, 169, 170, 171, 172, 173, 174, 175, 176, 177, 178, 179, 180, 181, 182, 183, 184, 185, 186, 187, 188, 189, 204, 205, 206, 207 **Grade 1:** 228, 229, 230, 231, 232, 233, 235, 236, 237, 238, 239, 240, 241, 242, 243, 244, 246–247, 248, 249, 250, 251, 252, 253, 254, 255, 256–259, 260, 261, 275, AF32, AF34 **Grade 2:** 176–177, 220, 222, 223, 224, 225, 226, 227, 228, 229, 236, 237, 240, 241, 242–243, 248, 249, 258, 259, 260, 261, 262, 263, 264, 265, AF32 **Grade 3:** 274–275, 278, 280, 281, 282, 283, 284, 285, 286–287, 288, 289, 290, 291, 292, 293, 294, 295, 296, 297, 298, 299, 300–301, 303, 304, 305, 306, 307, 308, 309, 312, 313, 321, 324 **Grade 4:** 240, 241, 243, 312, 313, 316, 317, 318, 319, 320, 321, 325, 327, 328, 329, 330, 331, 334, 335, 336, 337, 338, 339, 340, 341, 342–343, 345, 346, 350, 351, 352–353, 354, 355
Objects in the sky have patterns of movement. The sun, for example, appears to move across the sky in the same way every day, but its path changes slowly over the seasons. The moon moves across the sky on a daily basis much like the sun. The observable shape of the moon changes from day to day in a cycle that lasts about a month.	**Grade K:** 190, 192, 193, 194, 195, 196, 197, 199, 200, 201, 203, 205 **Grade 1:** 272, 273, 274, 275, 276, 279, 280, 281, 282, 283, 284–285, 286–289, 290, 291, AF38 **Grade 2:** 253, 254, 255, 256, 257, 262, 263, 268, 269, 270, 271, 273, 275, 276–279, 280–281, 282–285, 286, 287, AF34, AF38 **Grade 3:** 316, 317, 318, 319, 320, 321, 323, 324, 327, 328, 329, 330, 331, 332, 333, 334–335, 350, 354, 355 **Grade 4:** 358, 359, 360, 361, 362, 363, 364, 365, 372, 373, 374, 375, 376–377, 402, 403

Science and Technology

Abilities of Technological Design

Identify a simple problem. In problem identification, children should develop the ability to explain a problem in their own words and identify a specific task and solution related to the problem.	**Grade 1:** 229, 249, 271, 403, 421 **Grade 2:** 377, 380, 405, 421 **Grade 3:** 144–145, 289, 443, 460–461, 463, 511 **Grade 4:** 106–107, 210–211, 335, 483, 503 See also *Technology: A Closer Look*, Grades K–2 and 3–4.
Propose a solution. Students should make proposals to build something or get something to work better; they should be able to describe and communicate their ideas. Students should recognize that designing a solution might have constraints, such as cost, materials, time, space, or safety.	**Grade K:** 247 **Grade 1:** 271, 373, 376, 405 **Grade 2:** 234, 367, 380, 405, 411 **Grade 3:** 203, 289, 443, 450–451, 463, 475, 486–487, 489, 511 **Grade 4:** 89, 102, 335, 483, 493, 503, 525 See also *Technology: A Closer Look*, Grades K–2 and 3–4.

Abilities of Technological Design continued

Implementing proposed solutions. Children should develop abilities to work individually and collaboratively and to use suitable tools, techniques, and quantitative measurements when appropriate. Students should demonstrate the ability to balance simple constraints in problem solving.	**Grade K:** 246 **Grade 1:** 204, 249, 373, 376, 403, 405 **Grade 2:** 171, 267, 380, 421 **Grade 3:** 55, 293, 463, 469, 486–487 **Grade 4:** 102, 210–211, 483, 503, 522–523 See also *Technology: A Closer Look*, Grades K–2 and 3–4.
Evaluate a product or design. Students should evaluate their own results or solutions to problems, as well as those of other children, by considering how well a product or design met the challenge to solve a problem. When possible, students should use measurements and include constraints and other criteria in their evaluations. They should modify designs based on the results of evaluations.	**Grade 1:** 229, 249, 373, 379, 403 **Grade 2:** 129, 367, 411 **Grade 3:** 203, 283, 443, 463, 469, 486–487, 495, 511 **Grade 4:** 89, 106–107, 210–211, 483, 493, 503, 522–523 See also *Technology: A Closer Look*, Grades K–2 and 3–4.
Communicate a problem, design, and solution. Student abilities should include oral, written, and pictorial communication of the design process and product. The communication might be show and tell, group discussions, short written reports, or pictures, depending on the students' abilities and the design project.	**Grade 1:** 229, 249, 373, 379, 403 **Grade K:** 150 **Grade 1:** 95, 271 **Grade 2:** 95, 421 **Grade 3:** 289, 443, 460–461, 463 **Grade 4:** 106–107, 490–491, 522–523 See also *Technology: A Closer Look*, Grades K–2 and 3–4.

Understanding About Science and Technology

People have always had questions about their world. Science is one way of answering questions and explaining the natural world.	**Grade K:** 4–5, 6, 7, 10–11, 12–13, 16–17, 18–19 **Grade 1:** 206–207, 284–285, 292 **Grade 2:** 110–111, 242–243, 276, 277, 280–281 **Grade 3:** 4, 130–131, 352–353, 370–371 **Grade 4:** 3–11, 38, 276–277, 282–283, 381, 382, 406–407, 438–439, 552
People have always had problems and invented tools and techniques (ways of doing something) to solve problems. Trying to determine the effects of solutions helps people avoid some new problems.	**Grade K:** 150 **Grade 1:** 374–377, 378–379, 398, 399, 408–409, 414, 415, 421, 422 **Grade 2:** 176–177, 189, 242–243, 280–281, 354, 426–427 **Grade 3:** 28–29, 210–211, 254, 262, 263, 300–301, 342, 440–441, 471 **Grade 4:** 43, 218, 219, 279, 280, 288, 289, 302–303, 382, 383, 389, 390–391, 397, 477, 510–511, 586–587 See also *Technology: A Closer Look*, Grades K–2 and 3–4.
Scientists and engineers often work in teams with different individuals doing different things that contribute to the results. This understanding focuses primarily on teams working together and secondarily, on the combination of scientist and engineer teams.	**Grade 2:** 82, 110–111, 288 **Grade 3:** 3–11, 180–181, 187 **Grade 4:** 3–11, 119 See also *Technology: A Closer Look*, Grades K–2 and 3–4.

Correlations to National Standards

Understanding About Science and Technology continued

Women and men of all ages, backgrounds, and groups engage in a variety of scientific and technological work.	**Grade K:** 116, 252 **Grade 1:** 80, 116–117, 148–149, 156, 176–177, 206–207, 222, 284–285, 292, 430 **Grade 2:** 66–67, 82, 110–111, 126–127, 148, 216, 288, 354, 374–375, 434 **Grade 3:** 96–97, 100, 130–131, 184, 272, 352–353, 356, 370–371, 426, 522 **Grade 4:** 118–119, 122, 196, 222–223, 306, 404, 438–439, 476, 590 See also *Technology: A Closer Look*, Grades K–2 and 3–4.
Tools help scientists make better observations, measurements, and equipment for investigations. They help scientists see, measure, and do things that they could not otherwise see, measure, and do.	**Grade K:** TR7, TR8 **Grade 1:** 266, 284–285, 292, R2, R3, R4, R5, R6, R7 **Grade 2:** 224, 225, 226, 227, 242–243, 374–375, R3, R4, R5, R6 **Grade 3:** 6, 7, 26, 209, 281, 282, 283, 284, 300–301, 342, 344–345, 356, 379, R2, R4, R5, R6, R7, R8, R9 **Grade 4:** 28, 218, 219, 276, 318, 319, 342–343, 381, 382, 383, 390–391, 392, 397, 531, R2, R3, R4, R5, R6, R7, R8, R9 See also *Technology: A Closer Look*, Grades K–2 and 3–4.

Abilities to Distinguish Between Natural Objects and Objects Made by Humans

Some objects occur in nature; others have been designed and made by people to solve human problems and enhance the quality of life.	**Grade K:** 276–277 **Grade 1:** 195, 312–313, 325, 414 **Grade 2:** 188, 189, 190, 202, 203, 208–209, 216, 306–307 **Grade 3:** 210–211, 252, 253, 262, 263, 414–415 **Grade 4:** 54, 258, 406–407, 459, 463, 474, 476, 506 See also *Technology: A Closer Look*, Grades K–2 and 3–4.
Objects can be categorized into two groups, natural and designed.	**Grade K:** 276–277 **Grade 1:** 312–313, 414 **Grade 2:** 190 **Grade 3:** 409, 414–415 **Grade 4:** 54, 463, 476 See also *Technology: A Closer Look*, Grades K–2 and 3–4.

Science in Personal and Social Perspectives

Personal Health

Safety and security are basic needs of humans. Safety involves freedom from danger, risk, or injury. Security involves feelings of confidence and lack of anxiety and fear. Student understandings include following safety rules for home and school, preventing abuse and neglect, avoiding injury, knowing whom to ask for help, and when and how to say no.	**Grade K:** TR13, TR14 **Grade 1:** 16, 103, 127, 163, 403, R16, R17 **Grade 2:** R16, R17, R18 **Grade 3:** 14, 49, 299, 323, 517 **Grade 4:** 239, 340, 341, 398, 570, 571

Personal Health continued

Individuals have some responsibility for their own health. Students should engage in personal care—dental hygiene, cleanliness, and exercise—that will maintain and improve health. Understandings include how communicable diseases, such as colds, are transmitted and some of the body's defense mechanisms that prevent or overcome illness.	**Grade K:** TR9, TR10 **Grade 1:** R14 **Grade 2:** R9, R14, R15 **Grade 3:** 323, R23, R26 **Grade 4:** 319, 398, R23, R26
Nutrition is essential to health. Students should understand how the body uses food and how various foods contribute to health. Recommendations for good nutrition include eating a variety of foods, eating less sugar, and eating less fat.	**Grade K:** TR11 **Grade 1:** R13 **Grade 2:** R12, R13 **Grade 3:** 27, R23, R24, R25, R26 **Grade 4:** R23, R24, R25, R26
Different substances can damage the body and how it functions. Such substances include tobacco, alcohol, over-the-counter medicines, and illicit drugs. Students should understand that some substances, such as prescription drugs, can be beneficial, but that any substance can be harmful if used inappropriately.	**Grade 1:** R14, R15 **Grade 2:** R15 **Grade 3:** R23, R26 **Grade 4:** R26

Characteristics and Changes in Populations

Human populations include groups of individuals living in a particular location. One important characteristic of a human population is the population density—the number of individuals of a particular population that live in a given amount of space.	**Grade 4:** 145, 186
The size of a human population can increase or decrease. Populations will increase unless other factors such as disease or famine decrease the population.	This standard is covered in the Macmillan/McGraw–Hill *Health & Wellness* program.

Types of Resources

Resources are things that we get from the living and nonliving environment to meet the needs and wants of a population.	**Grade K:** 146, 152, 153, 154, 155, 156, 157, 158, 159, 160, 161, 162, 163, 164, 165 **Grade 1:** 176–177, 195, 196, 197, 198, 199, 200, 201, 202, 203, 216–219, 220, 221, 222 **Grade 2:** 162, 163, 164, 165, 167, 168, 184, 188, 189, 190, 191, 200, 201, 202, 203, 204, 206, 207, 210–213, 214, 215 **Grade 3:** 152, 153, 244, 245, 252, 253, 254, 255, 256–257, 258, 259, 260, 261, 262, 263, 264, 265, 267, 268–269, 270, 271, 518 **Grade 4:** 248, 258, 259, 278, 279, 280, 281, 290, 291, 292, 293, 304, 305 See also *Technology: A Closer Look*, Grades K–2 and 3–4.
Some resources are basic materials, such as air, water, and soil; some are produced from basic resources, such as food, fuel, and building materials; and some resources are nonmaterial, such as quiet places, beauty, security, and safety.	**Grade K:** 146, 152, 153, 154, 155, 156, 157, 158, 159, 160, 161, 162, 163, 164, 165, 276, 277 **Grade 1:** 64–65, 194, 195, 196, 197, 198, 199, 200, 201, 202, 203, 216–219, 220, 221, 222, 312–313 **Grade 2:** 162, 163, 164, 165, 168, 188, 189, 190, 191, 194–199, 200, 201, 202, 203, 204, 208–209, 210–213, 214, 401, 426 **Grade 3:** 26, 27, 244, 252, 253, 254, 255, 256–257, 259, 260, 261, 262, 263, 267, 270, 271, 518 **Grade 4:** 54, 278, 279, 280, 281, 290, 291, 294, 296, 304, 305 See also *Technology: A Closer Look*, Grades K–2 and 3–4.

Correlations to National Standards

Types of Resources continued

The supply of many resources is limited. If used, resources can be extended through recycling and decreased use.	**Grade K:** 154, 155, 158, 159, 160, 161, 162, 163, 164, 165 **Grade 1:** 192, 197, 201, 203, 205, 209, 210, 211, 212, 213, 214, 215, 216–219, 220, 221 **Grade 2:** 200, 201, 203, 204, 205, 206, 207, 210–213, 215 **Grade 3:** 156, 157, 158–159, 183, 253, 256–257, 264, 265, 266, 267, 270, 271, 415, 426, 518, 519 **Grade 4:** 187, 292, 293, 298, 299, 300, 301, 302–303, 304, 305, 416 See also *Technology: A Closer Look*, Grades K–2 and 3–4.

Changes in Environments

Environments are the space, conditions, and factors that affect an individual's and a population's ability to survive and their quality of life.	**Grade K:** 87, 89, 124–125 **Grade 1:** 11, 66, 68, 69, 70, 71, 126, 128, 129, 130, 131, 132, 133, 134, 135, 136, 137, 141, 150–153 **Grade 2:** 90–93, 120, 121, 122–125, 128, 130–133, 134, 135, 136–139, 140, 141, 142–145, 146 **Grade 3:** 24, 25, 28–29, 108, 109, 118, 120, 121, 122, 123, 124, 125, 126, 128, 129, 146, 147, 184 **Grade 4:** 128, 129, 130, 131, 132, 133, 144, 145, 146–147, 159, 176, 177, 179, 180, 181, 192–193
Changes in environments can be natural or influenced by humans. Some changes are good, some are bad, and some are neither good nor bad. Pollution is a change in the environment that can influence the health, survival, or activities of organisms, including humans.	**Grade K:** 138, 140, 141, 142, 143, 144, 148, 149, 154, 155, 162, 163, 164, 165 **Grade 1:** 131, 146, 147, 155, 184, 204, 205, 206–207, 209, 220 **Grade 2:** 44, 102, 104, 105, 106, 107, 108, 109, 112–115, 116, 117, 126–127, 176–177, 204, 205, 207, 215 **Grade 3:** 28–29, 142, 150, 151, 152, 153, 154, 155, 156, 157, 162, 163, 168, 169, 170, 174, 175, 182, 183, 218, 264, 265, 266, 267, 426 **Grade 4:** 42–43, 96, 118–119, 124–125, 146–147, 159, 183, 184, 185, 186, 187, 188, 189, 190, 191, 192–193, 194, 195, 232, 233, 244–245, 294, 295, 296, 297, 298, 301, 302–303
Some environmental changes occur slowly, and others occur rapidly. Students should understand the different consequences of changing environments in small increments over long periods as compared with changing environments in large increments over short periods.	**Grade K:** 138, 140, 141, 142, 143, 144, 148, 149, 154, 155, 162, 163, 164, 165 **Grade 1:** 131, 146, 147, 155, 182 **Grade 2:** 104, 105, 108, 109, 112–115, 176–177, 178–181 **Grade 3:** 28–29, 114, 150, 151, 152, 153, 154, 155, 162, 163, 168, 169, 170, 174, 175, 179, 182, 183, 202–223 **Grade 4:** 42–43, 118–119, 124–125, 146–147, 190, 191, 192–193, 232, 233, 244–245

Science and Technology in Local Challenges

People continue inventing new ways of doing things, solving problems, and getting work done. New ideas and inventions often affect other people; sometimes the effects are good and sometimes they are bad. It is helpful to try to determine in advance how ideas and inventions will affect other people.	**Grade K:** 244, 245 **Grade 1:** 284–285, 346–347, 378–379, 398, 399, 408–409, 414–415, 422 **Grade 2:** 126–127, 148, 176–177, 189, 242–243, 374–375, 426–427 **Grade 3:** 210–211, 254, 256–257, 300–301, 342, 440–441, 508–509, 518 **Grade 4:** 218, 219, 279, 280, 288, 289, 290, 342–343, 375, 381, 382, 383, 389, 390–391, 397, 501, 510–511, 586–587 See also *Technology: A Closer Look*, Grades K–2 and 3–4.
Science and technology have greatly improved food quality and quantity, transportation, health, sanitation, and communication. These benefits of science and technology are not available to all of the people in the world.	**Grade K:** 150, 244, 245 **Grade 1:** 42–43, 64–65, 222, 312–313, 346–347, 378–379, 408–409, 422 **Grade 2:** 242–243, 280–281, 306–307, 336–337, 354, 426–427 **Grade 3:** 28–29, 300–301, 440–441, 508–509 **Grade 4:** 43, 54, 218, 219, 280, 342–343, 489, 500, 501, 506, 507, 508, 510–511, 513, 584, 585, 586–587 See also *Technology: A Closer Look*, Grades K–2 and 3–4.

History and Nature of Science

Science as a Human Endeavor

Science and technology have been practiced by people for a long time.	**Grade 1:** 284–285, 346–347, 378–379, 398, 399, 408–409, 414–415, 422 **Grade 2:** 126–127, 148, 176–177, 189, 242–243, 374–375, 426–427 **Grade 3:** 210–211, 254, 256–257, 300–301, 342, 440–441, 508–509, 518 **Grade 4:** 218, 219, 279, 280, 288, 289, 290, 342–343, 375, 381, 382, 383, 389, 390–391, 397, 501, 510–511, 586–587 See also *Technology: A Closer Look*, Grades K–2 and 3–4.
Men and women have made a variety of contributions throughout the history of science and technology.	**Grade K:** 252 **Grade 1:** 116–117, 148–149, 176–177, 206–207, 284–285, 378–379 **Grade 2:** 66–67, 110–111, 126–127, 280–281, 374–375 **Grade 3:** 96–97, 130–131, 180–181, 256–257, 333, 352–353, 355, 370–371, 440–441 **Grade 4:** 48, 118–119, 222–223, 375, 381, 382, 390–391, 434, 436, 438–439, 552, 586–587

Correlations to National Standards

Science as a Human Endeavor *continued*	
Although men and women using scientific inquiry have learned much about the objects, events, and phenomena in nature, much more remains to be understood. Science will never be finished.	**Grade K:** 4–5, 6, 7, 10–11, 12–13, 16–17, 18–19 **Grade 1:** 148–149, 284–285 **Grade 2:** 4–9, 44–45, 279, 374–375 **Grade 3:** 4, 11, 96–97, 172, 177, 180–181, 342, 352–353, 355, 370–371 **Grade 4:** 10–11, 34–35, 40, 222–223, 279, 383, 389
Many people choose science as a career and devote their entire lives to studying it. Many people derive great pleasure from doing science.	**Grade K:** 116 **Grade 1:** 80, 116–117, 148–149, 156, 176–177, 206–207, 222, 284–285, 292, 430 **Grade 2:** 66–67, 82, 110–111, 126–127, 148, 216, 288, 354, 374–375, 434 **Grade 3:** 96–97, 100, 130–131, 184, 272, 352–353, 356, 370–371, 426, 522 **Grade 4:** 118–119, 122, 196, 222–223, 306, 404, 438–439, 476, 590

Teacher's Edition Credits

Cover Photos

Gary Bell/oceanwideimages.com

Illustrations

TR40: Joe LeMonnier; TR41: Accurate Art, Inc.; TR42: Virge Kask; TR44: Accurate Art, Inc.; TR46: Paul Mirocha; TR47: Greg Harris; TR50: MMHE; TR52: Barb Cousins; TR53: (r) Barb Cousins; TR54: Barb Cousins; TR55: Sebastian Quigley; TR56: Stephen Durke; TR57: Barb Cousins; TR58: (l to r) Barb Cousins; TR59: (l, t to b) Barb Cousins; TR60: Jean Wisenbaugh; TR61: (l) Barb Cousins; TR61: (r) Kenneth Batelman; TR62: Accurate Art, Inc.; TR63: Kenneth Batelman

Photography

All photographs are by Macmillan/McGraw-Hill (MMH) and Ken Karp for MMH, Janette Beckman for MMH, and Jacques Cornell, John Flournoy for MMH, except as noted below:

Tii: (tl) Erik Stenbakken; (tc) Scott Stewart; (tr) Scott Schauer; (cl) Dan Donovan; (c) Ken Cavanagh for MMH; (cr) Nancy Brown; (bl) Deborah Attoinese/Overall Productions; (bc) Joshua P. Roberts; (bkgd) ©2006 Creatas/PunchStock. Tvii: Stockdisc/Punchstock. TR1: (bc) Ingram Publishing/Alamy. TR2: (bc) ©Creatas/PunchStock. TR43: (br) Zigmund Leszczynski/Animals Animals—Earth Scenes. TR45: (bl) Digital Vision/PunchStock. TR48: (cl) USGS. TR49: (br) Dr. Parvinder Sethi. TR50: (1) Dr. Parvinder Sethi; (2) Visuals Unlimited; (3) Albert J. Copley/Visuals Unlimited; (4) Mark A. Schneider/Visuals Unlimited; (5) Harry Taylor © Dorling Kindersley; (6) Mark Schneider/Visuals Unlimited/Getty Images; (7) Colin Keates © Dorling Kindersley, Courtesy of the Natural History Museum, London; (8) Marli Miller/Visuals Unlimited; (9) Charles D. Winters/Photo Researchers, Inc.; (10) PHOTOTAKE Inc./Alamy. TR51: (l) David R. Frazier/Photo Researchers, Inc.; (r) Russell Illig/Getty Images. TR53: (l) NOAA (National Oceanic and Atmospheric Administration). TR59: (br) Brand X Pictures/PunchStock

Teacher's Notes

Teacher's Notes

Teacher's Notes

Teacher's Notes

Teacher's Notes

Teacher's Notes

Teacher's Notes

Teacher's Notes

Teacher's Notes

Teacher's Notes

Teacher's Notes